Lecture Notes in Mathematics

Edited by A. Dold and B. Eckmann

1180

R. Carmona
H. Kesten
J.B. Walsh

École d'Été de Probabilités de Saint Flour XIV – 1984

Édite par P.L. Hennequin

Springer-Verlag
Berlin Heidelberg New York Tokyo

Auteurs

René Carmona
University of California, Irvine
Department of Mathematics
Irvine, California 92717, USA

Harry Kesten
Cornell University
Department of Mathematics
White Hall
Ithaca, New York 14853, USA

John B. Walsh
The University of British Columbia
Department of Mathematics
121 – 1984 Mathematics Road
Vancouver, B.C., Canada, V6T 1Y4

Editeur

P.L. Hennequin
Université de Clermont II, Complexe Scientifique des Cézeaux
Département de Mathématiques Appliquées
B.P. 45, 63170 Aubière, France

Mathematics Subject Classification (1980): 60-02, 35P05, 35R60, 47F05, 60K35, 60H15, 81C05

ISBN 3-540-16441-3 Springer-Verlag Berlin Heidelberg New York Tokyo
ISBN 0-387-16441-3 Springer-Verlag New York Heidelberg Berlin Tokyo

Printing and binding: Beltz Offsetdruck, Hemsbach/Bergstr.
2146/3140-543210

INTRODUCTION

La Quatorzième Ecole d'Eté de Calcul des Probabilités de Saint-Flour s'est tenue du 19 Août au 5 Septembre 1984 et a rassemblé, outre les conférenciers, une cinquantaine de participants dans les locaux historiques du Foyer des Planchettes.

L'Ecole d'Eté de Saint-Flour a été durement touchée pendant l'année scolaire 1984-85 puisque nous avons perdu brusquement, à quelques mois d'intervalle Gérard COLLOMB, Maître de Conférences à l'Université Paul Sabatier à Toulouse qui avait assisté à l'Ecole en 1977 et en 1979, et Antoine EHRHARD, Docteur ès-sciences de l'Université de Strasbourg et Assistant-chercheur à l'Université de Clermont-Ferrand I , qui avait participé à l'Ecole en 1980, 1981 et 1983.

Les trois conférenciers, Messieurs CARMONA, KESTEN et WALSH nous ont remis rapidement la rédaction définitive de leurs cours, ce qui nous permet de publier ce volume un an après la tenue de l'école.

En outre, plusieurs exposés ont été faits par les participants durant leur séjour à Saint-Flour :

J. AZEMA "Une jolie martingale pour les problèmes d'examen"

A. BADRIKIAN "Eléments de calcul différentiel et intégral par rapport à une mesure gaussienne dans un Hilbert"

A. BENASSI et J.P.FOUQUE "Processus d'exclusion simple asymétriques. La loi des grands nombres. Fluctuations"

M. CHIAROLLA "Martingale methods for stochastic multicompartmental systems"

R. COHEN "Théorème de séparation pour le contrôle mixte, impulsionnel et continu, avec observation partielle"

F. COMETS "Un modèle de champ moyen : nucléation et bifurcations"

N. EL-KAROUI "Une classe de processus de branchement à valeurs-mesures"

M. EMERY "Une définition faible de l'espace BMO"

J.P. FOUQUE	"Processus à valeurs distributions. Régularité et convergence en loi"
L. GALLARDO	"Lois stables sur les groupes d'Heisenberg et sur les groupes Diamant"
Y. GUIVARC'H	"Frontière de Martin pour le cylindre et le laplacien hyperbolique $\Delta = (1-\lvert z\rvert^2)^2 \frac{\partial^2}{\partial z\, \partial \bar{z}} + \frac{\partial^2}{\partial t^2}$ "
R. HÖPFNER	"Population-size dependent Galton-Watson processes"
I. ISCOE	"Large deviations for measures-valued critical branching Brownian Motion"
F. LEDRAPPIER	"Entropie, exposants et dimension"
J. LEON	"Une mesure de la déviation quadratique d'estimateurs non paramétriques". Recherche faite en collaboration avec P. Doukhan et F. Portal
B. MAISONNEUVE	"Processus de Markov et mesures excessives à naissance et mort aléatoires"
D. NUALART - M. SANZ	"Calcul de Malliavin pour les processus à deux indices"
B. PRUM	"Croquis impressionniste : transition de phase"
G. ROYER	"Fonctions log-concaves et distance de Fortet-Mourier"
M. ROSSIGNOL	"Processus de saut avec intéraction selon les plus proches particules"
D. ROUX	"Semi-stabilité"
A. SZNITMAN	"Un résultat de propagation du chaos pour l'équation de Burger"

Certains de ces exposés seront publiés dans les Annales de l'Université de Clermont-Ferrand.

LISTE DES AUDITEURS

Mr. AZEMA J.	Université de Paris V
Mr. BADRIKIAN A.	Université de Clermont-Ferrand II
Mr. BENASSI A.	Université de Paris VI
Mr. BERNARD P.	Université de Clermont-Ferrand II
Mme BERTEIN F.	Université de Paris VI
Mme CHALEYAT-MAUREL M.	Université de Paris VI
Mle CHIAROLLA M.	Université de Bari (Italie)
Mr. COHEN R.	Université de Paris VI
Mme COCOZZA C.	Université de Paris VI
Mr. COMETS F.	Université de Paris XI
Mr. DALANG R.	Ecole Polytechnique Fédérale de Lausanne (Suisse)
Mr. DERRIENNIC Y.	Faculté des Sciences de Brest
Mme EL KAROUI N.	Ecole Normale Supérieure de Fontenay-aux-Roses
Mr. EMERY M.	Université de Strasbourg I
Mle FARRE CERVELLO M.	Université de Barcelone (Espagne)
Mr. FOUQUE J.P.	Université de Paris VI
Mr. FOURT G.	Université de Clermont-Ferrand II
Mr. GALLARDO L.	Université de Nancy II
Mr. GUIVARC'H Y.	Université de Rennes
Mr. GOLDBERG J.	I.N.S.A. de Villeurbanne
Mr. HACHEM B.	Université de Brest
Mr. HENNEQUIN P.L.	Université de Clermont-Ferrand II
Mr. HÖPFNER R.	Université de Freiburg (R.F.A.)
Mr. ISCOE I.	Université d'Ottawa (Canada)
Mle LAWNICZAK A.	Université de Toronto (Canada)
Mr. LEDRAPPIER F.	Université de Paris VI
Mr. LEON J.	Université de Paris XI
Mr. MAISONNEUVE B.	Université de Grenoble II
Mr. NUALARD D.	Université de Barcelone (Espagne)
Mr. PRUM B.	Université de Paris XI
Mr. REMILLARD B.	Université Laval (Canada)
Mr. ROUSSEAU-EGELE J.	Université de Rennes I
Mr. ROUSSIGNOL M.	Université de Paris VI
Mr. ROUX D.	Université de Clermont-Ferrand II

Mr. ROYER G.	Université de Clermont-Ferrand II
Mr. RUSSO F.	Ecole Polytechnique Fédérale de Lausanne (Suisse)
Mme SANZ M.	Université de Barcelone (Espagne)
Mr. SZNITMAN A.	Université de Paris VI
Mle TIBI D.	Université de Paris VI
Mr. TOUATI A.	Ecole Normale Supérieure de Bizerte (Tunisie)
Mr. URBINA Wilfredo	Université de Caracas (Vénézuela)
Mr. VAN DER WEIDE H.	Université de Delft (Pays-Bas)

En outre, ont rendu visite à l'Ecole

Mme BADRIKIAN J.	Université de Clermont-Ferrand II
Mr. BOUGEROL P.	Université de Paris VII
Mr. CREPEL P.	C.N.R.S. à Rennes
Mr. McGILL	Université de Rennes

H. KESTEN : "ASPECTS OF FIRST PASSAGE PERCOLATION"

TABLE DES MATIERES

J.B. WALSH : "AN INTRODUCTION TO STOCHASTIC PARTIAL DIFFERENTIAL EQUATIONS"

RANDOM SCHRÖDINGER OPERATORS

René CARMONA

PREFACE

This set of notes is an outgrowth of a series of lectures given during the fourteenth Saint Flour probability summer school.

Obviously the study of random Schrödinger operators comes from physical considerations, and we try to explain in the INTRODUCTION why one is interested in Schrödinger operators, why one would like to define them in order to be able to prove that hey are self adjoint, why one would like to understand their spectral characteristics and finally, why one would like to do the same thing when these operators depend on a random parameter. This introduction should be regarded as a motivation: why are we doing all this? Nevertheless, no knowledge of physics or even of what is commonly called mathematical physics is assumed.

All the students were probabilists and the course has been designed accordingly: we tried to recall all the definitions and results from functional analysis which were needed. This analysis background is given in chapters I and III. Since the material may not be familiar to many probabilists we decided to give precise references, and very often several references for the same result. This has nothing to do with any desire to pad the bibliography: since one seldom likes to have to go and look for books and papers in a different field, I thought that giving several options could be helpful.

Most of the material presented in this course is taken from the published litterature and the huge quantity of preprints which are released on the subject. Nevertheless the reader will find some new results in chapter II and some essentially known results and proofs which never appeared in print in chapters III and V. As a rule I tried to present in a coherent way material which I gathered and reorganized. The last three chapters may suffer from the fact that the state of the art is not completely satisfactory. Indeed, rather than trying to present abstract theorems with a wide range of applicability, I tried to treat classes of examples on which the nature of the phenomena and the techniques involved in their discoveries could be enhanced. I also believe that this way should be more stimulating for further research.

Part of this work has been prepared while I was teaching a graduate course in U.C.Irvine and I would like to thank the students for their patience and their numerous comments. Also I am very much indebted to the organizor P.L.Hennequin and to the scientific committee of the "Ecole d'Eté de Probabilités de Saint Flour" for giving me the opportunity to lecture on a subject that I like and in a place I love. Moreover, I wish to acknowledge the assistance of Peter Hislop, Abel Klein, Bernard Souillard and especially of Barry Simon, in proofreading the first draft of this set of notes and suggesting many improvements. Finally, I would like to thank the N.S.F. for partial financial support during the preparation of the manuscript and René CARMONA for his fine job at typing the final version of the manuscript.

0. INTRODUCTION

This course is devoted to the study of the spectral properties of random Schrödinger operators. This subject is very challenging because very little is known except in the one dimensional case and in the lattice case. The challenge comes from the physics of disordered systems. Even though the pioneering work of Ph.Anderson [Anderson (1958)] and the wave of publications it initiated are fundamental in the formulation of the mathematical problems to tackle, we will not discuss this important aspect of the problem in detail. Rather, we refer the interestd reader to the review [Kunz-Souillard (in prep)]. We merely present some particular examples to illustrate the possible needs to

i) define Schrödinger operators as self adjoint operators

ii) investigate their spectral properties

iii) tackle these very problems when the potential functions are random.

Our discussion is loose and very naive from the point of view of physics. We are just looking for some incentive to attack these difficult mathematical problems.

Let us consider the example of an atom with N electrons. We first assume that the nucleus is infinitely heavy so that it will not move and stays at the origin of our space $\mathbf{R}^3$. If our system could be studied in the realm of classical mechanics, we would consider the following Hamilton function for the energy of the system:

$$h = \sum (2m_j)^{-1}|p_j|^2 \quad - \sum Ne^2\, |x_j|^{-1} + \quad \sum e^2\, |x_j - x_k|^{-1}$$

where m_j denotes the mass of the j-th electron, $\mathbf{p}_j$ its momentum (i.e. mass times velocity) and e its charge. We know that we have to go to a quantum theory of our system if we want to predict realistically any of its properties. The quantization procedure tells us that the positions $\mathbf{x}_j$ become operators of multiplications by the functions x_j on the appropriate Hilbert space to be chosen later, and the momenta $\mathbf{p}_j$ become i times the operators of partial derivations with respect to the variables $\mathbf{x}_j$ (here and in the following, i denotes the complex number $\sqrt{-1}$ and the constant h is set equal to 1). Consequently the Hamilton function for the energy becomes the formal partial differential operator

$$H = H_0 + q(x)$$

where:

$$H_0 = -\sum (2m_j)^{-1}\, \Delta_j$$

with Δ_i denoting the Laplacian operator in the variable $\mathbf{x}_i$, and where $q(\mathbf{x})$ denotes the operator of multiplication by the function :

$$q(x) = -\sum Ne^2|x_j|^{-1} + \sum e^2|x_j - x_k|^{-1}$$

where $x = (x_1, x_2, \ldots\ldots, x_N)$ is now an element of $\mathbf{R}^{3N}$. As usual we will make the convention $m_j = 1$ so that our operator H_0 representing the quantum mechanical kinetic energy will simply be minus one half the usual Laplacian operator in $\mathbf{R}^{3N}$. The Hilbert space on which these operators are acting will naturally chosen to be $H = L^2(\mathbf{R}^{3N}, dx)$. This is in full agreement with the fact that a quantum mechanical state is no longer an element of $\mathbf{R}^{3N}$ but rather an element of the Hilbert space **H**. More precisely it is a unit vector, say f, in this Hilbert space, hence a funcion on $\mathbf{R}^{3N}$, the integral of the square of the modulus of which being equal to one. The square of this modulus is interpreted as a density of probability: if A is any Borel subset of $\mathbf{R}^{3N}$ the quantity $\int_A |f(x)|^2 dx$ represents the probability that the system described by f is in A.

To describe the time evolution of our system we need to define its dynamics. It is given by the famous Schrödinger equation:

$$i\,\partial f / \partial t = Hf \tag{0.1}$$

Hence, if the state of the system is f_0 at time $t = 0$, the system at a later time t should be given by the solution of (0.1) satisfying the initial condition $f(0,.) = f_0$.

A time honored method to study equations of this form is to look first for solutions of a particular form, the so called stationary solutions. This leads to an equivalent form which is now time independent:

$$Hf = Ef \tag{E.V.E.}$$

where both the (possibly generalized) **eigenfunction** f and the **eigenvalue** E are to be found. We will call this stationary form of the Schrödinger equation the **EigenValue Equation** (E.V.E. for short) and, as pointed out by Schrödinger himself in his original work, this form of the problem is fundamental. Nevertheless, we will restrict ourselves to the equation (0.1) to introduce the mathematical problems we want to investigate.

Let us assume for the sake of simplicity that some existence and uniqueness theorem holds for our Schrödinger equation (0.1). Consequently, for each real numbers s and t there exists an operator U(s,t) on **H** such that $f_t = U(s,t)f_s$ is the solution at time t if f_s is the solution at time s. $\{U(s,t);\ s,t\in\mathbf{R}\}$ is usually called the propagator of the equation (0.1). Our existence and uniqueness assumption forces the propagator to be a family of bounded (one-to-one onto) operators on H which satisfy:

$$U(s,t)\,U(t,u) = U(s,u) \tag{0.2}$$

for all real numbers s,t and u and obviously $U(s,s)=I$ for each real number s. Moreover, saying that a state at a given time will remain a state at all later times simply means that these operators are norm preserving, hence unitary. Finally, if we assume that this propagator is strongly continuous (in fact, weakly measurable is enough) then Stone's theorem tells us that

$$U(s,t) = e^{-i(t-s)\mathbf{H}} \tag{0.3}$$

for some self adjoint operator $\mathbf{H}$. Now putting together (0.1) and (0.3) shows that this last operator $\mathbf{H}$ has to be equal to our Schrödinger operator H, or at least to a self adjoint extension of it whenever we do not have such a strong uniqueness result. Consequently our first task should be to

> **define Schrödinger operators as self adjoint operators, or at least make sure that they have self adjoint extensions, and if more than one such extension exist, try to find and control those which are relevent to the physics of the problem.**

Once this is done we can solve the Schrödinger equation (0.1) for each initial state f_0 in a very weak sense if f_0 is arbitrary and in an L^2-sense if f_0 is in the domain of the self adjoint operator H, and obtain a state f_t at every later time t. We are now interested in this time evolution of the states and especially in the asymptotic behavior of these states as t tends to infinity. For example, we would like to find out that:

$$\lim_{T\to\infty} (2T)^{-1} \int_{[-T,+T]} \int_{|x|\le r} |f_t(x)|^2 dx\, dt = 0 \tag{0.4}$$

for each real $r>0$, or that:

$$\lim_{r\to\infty} \sup_{-\infty<t<+\infty} \int_{|x|\ge r} |f_t(x)|^2 dx = 0. \tag{0.5}$$

Relation (0.4) says that for any given ball, the probability that the state is in this ball for large time is, at least in a Cesaro average sense, going to zero. We say that such an initial state f_0 is an **unbound state**.

Relation (0.5) tells us that up to errors as small as we want, we can find a large enough ball in which the system stays at all times. Such an initial state f_0 is naturally called a **bound state.**

Wouldn't it be nice if, given the Schrödinger operator we could divide the whole Hilbert space H into bound states and unbound ones? It turns out that this is possible, at least theoretically. Indeed, according to a result of Ruelle (see the section on the so-called RAGE Theorem in [Reed-Simon(1979)]), the scattering theoty takes place in the absolutely continuous subspace of H in the spectral decomposition of H relative to the operator H, and the bound states are nothing but the unit vectors of the pure point subspace of H in this very same decomposition. Consequently, knowing that the operator H is self adjoint is not enough. We need to

> **study its spectral characteristics and the decomposition of the Hilbert space according to the various components of the spectrum in the so-called Lebesgue decomposition of the operator.**

The functional analysis we just outlined is described in Chapters I and III.

Let us consider now another example. Let us assume that a particle x_j is sitting at each site of the lattice $\mathbf{Z}^d$, and that each of them creates a potential v_j. If we throw a particle whose position will be denoted x in the system, it will interact with a total potential of the form

$$q(x) = \sum g_j v_j(x-j)$$

where the "coupling constants " g_j are real numbers. If we assume that the particles in the lattice are identical, as is the case in the study of crystals which are perfectly periodic structures, then all the functions v_j are equal and so are all the constants g_j, and the potential function q ends up being periodic. The spectral properties of Schrödinger operators with periodic potentials are well known (see for example Section 16 in Chapter XIII of [Reed-Simon(1978)]) and the spectrum is, as a set, a union of intervals $[a_n,b_n]$ with $b_n \leq a_{n+1}$. The intervals $[a_n,b_n]$ are called **bands** of allowed energies while the intervals (b_n,a_{n+1}) are called **gaps** of forbidden energies. Moreover the spectrum is **purely absolutely continuous** so that the absolute continuous part of the Hilbert space is actually equal to the whole Hilbert space. For applications of the spectral theory of periodic Schrödinger operators to solid states physics and the properties of conductors and insulators, the interested reader may want to look at [Reed-Simon(1978)], Chapter XIII.

We finally come to the discussion of some of the ideas of Ph.Anderson. Let us imagine that the coupling constants are now random and form a sequence of independent identically distributed (i.i.d. for short) random variables, or let us imagine that these constants are equal but that the particles x_j are now randomly placed (according to a Poisson point process in $\mathbf{R}^d$ for example) instead of sitting at the very sites of the lattice $\mathbf{Z}^d$. In both cases the potential function is now random, whether it is of the form

$$q(x) = \sum g_j v(x-j)$$

or

$$q(x) = \sum g v_j(x-x_j),$$

it is a stationary random process, and our Schrödinger operator is now a random self adjoint operator whose spectral properties are drastically different from the ones of the deterministic case. Indeed, according to the predictions of the physicists, one expects **dense pure point spectrum with exponentially decaying eigenfunctions** in one and two dimensions (so, no scattering states, only bound states), and, in higher dimensions, **dense pure point in some regions of the spectrum and purely absolutely continuous spectrum in the rest of the spectrum**, the transitions between these two regimes being se two regions being called **mobility edges**.

We are now ready to explain precisely the realm of these lecture notes.

Study the self adjointness and the spectral characteristics of Schrödinger operators with random functions as potentials and try to account rigorously for the physicists' predictions.

This is the program of the course. After the *soft analysis of general ergodic self adjoint operators in* Chapter II, we tackle the *meaurability* and the *self adjointness* problems of Schrödinger operators with random potentials in Chapter IV. Then we consider the so-called *"integrated density of states"* in Chapter V, and we concentrate on the one dimensional case in the last two chapters. First we try to understand the *absolutely continuous spectrum* and then the *pure point* one.

Many results appear here for the first time in print, and this could well be one of the attractions of this set of notes. Either because the proof was simple enough and I did not feel like writing a special paper for it (this is the case for example of the determinism of the essential support of the absolutely continuous spectrum proved in Proposition II.17) or because the tools were essentially known to many experts but never appeared in print (this is the case for example of what we call the Molcanov formula and its application to the analyticity of the density of states).

Also many open problems are discussed. I did not gather them in a special section. They simply appear in the text in the very sections dealing with the areas in which they arise.

I. SPECTRAL THEORY OF SELF ADJOINT OPERATORS.

The present chapter does not contain any probability theory. We state the notations and the definitions and most of the results from functional analysis which we will need in this course. Because of lack of time and space, we refrain from explaining the motivations behind the numerous definitions to be introduced and we postpone the examples to Chapter III below. Rather than a serious introduction to the spectral theory of self adjoint operators, this chapter should be understood as a glossary and a summary of the terms and results to be used in the sequel.

Our goal is to present the functional calculus for self adjoint operators on an infinite dimensional (complex) Hilbert space. This is done in Sections I.2 and I.5 below where we state the spectral theorem and where we discuss some questions of integration with respect to projection valued measures, of multiplicity and of canonical forms. We use the classic approach based on the representation theorems recalled in Section I.3. Unfortunately the operators we want to study will not always be given as self adjoint operators but merely as formally symmetric partial differential operators. The usefulness of the spectral theory of self adjoint operators will depend on our being able to construct self adjoint extensions for these symmetric operators. We will tackle this problem in Section I.6.

I.1. Domains, adjoints, resolvents and spectra:

An **operator** H on a Hilbert space **H** is a linear map from a vector subspace $\mathbf{D}(H)$ of **H** into **H**. $\mathbf{D}(H)$ is called the **domain** of H.

> *All the abstract Hilbert spaces* **H** *to be considered below will be implicitly assumed separable and unless stated otherwise, all the operators will be assumed to be densely defined.*

Consequently, to be given an operator is to be given a dense vector subspace and an action on the elements of this vector space. For example $H=\Delta$ does not define an operator on $\mathbf{H}=L^2(\mathbf{R}^d,dx)$ until one chooses a domain. See Section III.1.

The **graph** of the operator H is the subset $\{(f,Hf);\ f\in\mathbf{D}(H)\}$ of $\mathbf{H}\times\mathbf{H}$. It will be denoted $gr(H)$. $\mathbf{H}\times\mathbf{H}$ is a Hilbert space for the inner product:

$$\langle(f_1,g_1),(f_2,g_2)\rangle = \langle f_1,f_2\rangle + \langle g_1,g_2\rangle$$

and the operator H is said to be **closed** if $gr(H)$ is a closed subset of $\mathbf{H}\times\mathbf{H}$. Another operator, say H_1, is said to be an **extension** of H if $gr(H) \subset gr(H_1)$, that is if $\mathfrak{D}(H) \subset \mathfrak{D}(H_1)$ and if $Hf = H_1 f$ whenever $f \in \mathfrak{D}(H)$. We will use the notation $H \subset H_1$. H is said **closable** if it has at least one closed extension. When this is the case, the smallest one of these extensions is denoted by $\overline{H}$ and is called **the closure** of H. A given operator H does not always have closed extensions. Indeed, the closure $\overline{gr(H)}$ of the graph of H need not be the graph of an operator. Nevertheless we will see later that symmetric operators always have such closed extensions. Also it is easy to see that the graph of the closure of H is always the closure of the graph of H, i.e. $gr(\overline{H}) = \overline{gr(H)}$, whenever H is closable.

We now come to the important concept of adjoint of an operator H. This is the operator H^* defined by:

$$\mathfrak{D}(H^*) = \{ f \in \mathbf{H};\ \exists g \in \mathbf{H}, \forall h \in \mathfrak{D}(H),\ \langle Hh, f\rangle = \langle h, g\rangle \}$$

and (1.1)

$$H^* f = g$$

if f and g are as in the definition of $\mathfrak{D}(H^*)$. Note that H^* is well defined on $\mathfrak{D}(H^*)$ because of the implicit denseness of $\mathfrak{D}(H)$. In fact, it is easy to see that H^* **is always closed** and that H is **closable if and only if $\mathfrak{D}(H^*)$ is dense** in which case $\overline{H} = H^{**}$ and $(\overline{H})^* = H^*$.

Example I.1:

Let μ be a nonnegative σ-finite measure on a separable measurable space $(\Omega, \mathcal{B})$ and let $\mathbf{H} = L^2(\Omega, \mathcal{B}, d\mu)$. For each measurable function $\mathfrak{f}: \Omega \dashrightarrow \mathbf{C}$ we define the multiplication operator by $\mathfrak{f}$, which we will denote $M_{\mathfrak{f}}$, by:

$$\mathfrak{D}(M_{\mathfrak{f}}) = \{ f \in \mathbf{H};\ \mathfrak{f} f \in \mathbf{H} \}$$

and

$$M_{\mathfrak{f}} f = \mathfrak{f} f \quad , \quad f \in \mathfrak{D}(M_{\mathfrak{f}})$$

One easily checks that $\mathfrak{D}(M_{\mathfrak{f}})$ is dense, that $M_{\mathfrak{f}}$ is closed and that $M_{\mathfrak{f}}^* = M_{\overline{\mathfrak{f}}}$. When Ω is a subset of $\mathbf{C}$ and when $\mathcal{B}$ is the trace σ-field, the multiplication operator by the identity function will be simply denoted by M.

If H is an operator on $\mathbf{H}$ and if $z \in \mathbf{C}$, the operator zH is defined by $\mathfrak{D}(zH) = \mathfrak{D}(H)$ and $[zH]f = z(Hf)$ whenever f is in $\mathfrak{D}(H)$. One easily checks that $[zH]^* = \bar{z}H^*$.

Now, if H and K are operators on $\mathbf{H}$, the operator H+K is defined by $\mathfrak{D}(H+K) = \mathfrak{D}(H) \cap \mathfrak{D}(K)$ and $[H+K]f = Hf + Kf$ whenever f is in $\mathfrak{D}(H) \cap \mathfrak{D}(K)$. Note that we may have $\mathfrak{D}(H) \cap \mathfrak{D}(K) = \{0\}$ even though $\mathfrak{D}(H)$ and $\mathfrak{D}(K)$ are dense. Hence **the sum of densely defined operators is not necessarily densely defined**. Nevertheless we always have $[H+K]^* = H^* + K^*$ if either H or K is bounded. Also, the operator HK is defined by $[HK]f = H(Kf)$ whenever f is in $\mathfrak{D}(HK) = \{f \in \mathfrak{D}(K);\ Kf \in \mathfrak{D}(H)\}$, and one can check that one always has $K^*H^* \subset (HK)^*$ while the equality $K^*H^* = (HK)^*$ holds whenever H is bounded.

We will use the notation $\mathbf{I}$ for the identity operator of $\mathbf{H}$ and $\mathbf{L}(\mathbf{H})$ for the Banach space (for the uniform norm) of bounded operators on $\mathbf{H}$.

If the operator H is one-to-one, the operator H^{-1} is defined by $\mathfrak{D}(H^{-1})=\{Hf;\ f\in\mathfrak{D}(H)\}$ and $H^{-1}g=f$ if $g=Hf$. Note that H^{-1} is closed if H is so (but of course, $\mathfrak{D}(H^{-1})$ need not to be dense).

Let us now assume that the operator H is closed. The set of complex numbers z for which the operator $z\mathbf{I}-H$ is one-to-one from $\mathfrak{D}(H)$ onto $\mathbf{H}$ (in which case $[z\mathbf{I}-H]^{-1}$ is bounded by the closed graph theorem) is denoted by $\rho(H)$ and is called **the resolvent set** of H. The bounded operator $[z\mathbf{I}-H]^{-1}$ is often denoted by $R(z,H)$ and is called **the resolvent operator** of H at z. **The resolvent set is an open** subset of $\mathbf{C}$ and **the function** $z \dashrightarrow R(z,H)$ **is a strongly analytic** function on $\rho(H)$ with values in $\mathbf{L}(\mathbf{H})$. Moreover, for all z and z' in $\rho(H)$, the operators $R(z,H)$ and $R(z',H)$ commute and satisfy the so-called **first resolvent identity** :

$$R(z,H)-R(z',H) = (z'-z)R(z',H)R(z,H) \tag{1.2}$$

The complement $\mathbf{C}\setminus\rho(H)$ of the resolvent set is called the **spectrum of** H and will be denoted $\Sigma(H)$. The operator H is said to be **symmetric** if $H\subset H^*$, i.e. $\mathfrak{D}(H)\subset\mathfrak{D}(H^*)$ and $Hf=H^*f$ whenever f is in $\mathfrak{D}(H)$, or equivalently if

$$\langle Hf,g\rangle = \langle f,Hg\rangle$$

for all f and g in $\mathfrak{D}(H)$. It is said to be **self adjoint** if $H^*=H$, that is if H is symmetric and if $\mathfrak{D}(H)=\mathfrak{D}(H^*)$. A symmetric operator is always closable since H H^* and H^* is closed. Moreover, since H^* is a closed extension of H, the smallest one of them, namely H^{**}, has to be smaller than H^*. Hence, we have:

$$H\subset H^{**}\subset H^*$$

if H is symmetric and:

$$H = H^{**}\subset H^*$$

if H is closed and symmetric and:

$$H = H^{**} = H^*$$

if H is self adjoint.

A symmetric operator is said to be **essentially self adjoint** if its closure is self adjoint. Hence, **it has a unique self adjoint extension** . In fact, the converse statement is also true. Moreover, H is essentially self adjoint if and only if H^* is symmetric in which case one has $\bar{H}=H^*$.

If H is symmetric, if $z\in\mathbf{C}$ and if $f\in\mathfrak{D}(H)$, then:

$$\|\,[z\mathbf{I}-H]f\,\| \geq |\operatorname{Im} z|\ \|f\| \tag{1.3}$$

because:

$$\|\,[z\mathbf{I}-H]f\,\|^2 = \langle [z\mathbf{I}-H]f\,,[z\mathbf{I}-H]f\rangle$$

$$= ||[H-x\mathbf{I}]f||^2 + y^2\, ||f||^2$$

if $z=x+iy$ with x and y real,

$$\geq y^2\, ||f||^2.$$

It follows that, if H is closed, the range of $zI-H$ is also closed if z is not real.

We close this section by introducing the concept of **convergence** which we will use in the sequel. Let $\{H_k; k\geq 1\}$ be a sequence of self adjoint operators on $\mathbf{H}$. We would like to define the convergence of H_k to some operator H on $\mathbf{H}$ and to be able to conclude properties of H from the corresponding properties of the approximants H_k. The first idea which comes to our mind is to define this convergence by the convergence in $\mathbf{H}$ of the $H_k f$'s to Hf. Though natural, this idea cannot be implemented because the set of f's for which $H_k f$ and Hf make sense may be very small. Indeed, even though the domains $\mathfrak{D}(H_k)$ are dense, their intersection may well reduce to $\{0\}$. This last point is the main reason why the use of the convergence of unbounded operators requires a modicum of care. Since it is obviously easier to deal with bounded operators we choose the following definition:

the sequence $\{H_k; k\geq 1\}$ *of self adjoint operators is said to converge to the self adjoint operator* H, *in the sense of* ***strong resolvent convergence***, *if, for each* z *in* $\mathbf{C}\setminus\mathbf{R}$, $R(z,H_k)$ *converges strongly to* $R(z,H)$, *that is if we have, for all* f *in* $\mathbf{H}$:

$$\lim_{k\to\infty} R(z,H_k)f = R(z,H)f \tag{1.4}$$

Our first try to define the concept of convergence was too naive. Nevertheless it makes perfectly good sense in some cases in which it turns out to be equivalent to the more sophisticated definition which we chose. This is particularly the case when all the domains have in common a set of elements f for which the convergence of $H_k f$ to Hf occurs, this set being rich enough to determine H. This justifies the following definition:

if H *is a closed operator and if* D *is a subset of its domain, we say that* D *is a* ***core*** *for* H *if the closure of the restriction of* H *to* D, *which we will denote* H_D, *is equal to* H *itself.*

Hence, the complete knowledge of an operator is given by knowledge of it on a core. Now it is easy to show that:

if H *and* H_k *are self adjoint operators on* $\mathbf{H}$ and if D *is a core for* H *which is contained in all the* $\mathfrak{D}(H_k)$ *and for which we have:*

$$\lim_{k\to\infty} H_k f = Hf \tag{1.5}$$

for all f *in* D, *then* H_k *converges to* H *in the strong resolvent sense.*

I.2. Resolutions of the identity:

Throughout this section **H** will be a fixed (separable) complex Hilbert space and $(\Omega, \mathcal{B})$ a fixed measurable space and we will say projection in **H** for orthogonal (i.e. self adjoint) projection in **H**.

Definition I.2:
A function E *on* $\mathcal{B}$ *with values in the space of projections in* **H** *is called a resolution of the identity (resp. subresolution of the identity) of* **H** *on* $(\Omega, \mathcal{B})$ *if:*

i) $E(\Omega) = I$ (resp. i)' $E(\emptyset)=0$)

ii) $E(\bigcup_{n\geq 1} A_n) = \sum_{n\geq 1} E(A_n)$ *whenever* $\{A_n; n\geq 1\}$ *is a sequence in* $\mathcal{B}$ *whose elements are disjoint*.

The convergence in ii) has to be understood in the sense of the strong convergence of operators. This means that for each fixed f in **H**, the series $\sum_{n\geq 1} E(A_n)f$ converges for the norm of **H** to $E(\bigcup_{n\geq 1} A_n)f$. Hence, for each fixed f in **H** the function $\mathcal{B} \ni A \dashrightarrow E(A)f \in \mathbf{H}$ is countably additive. It is what is usually called a **vector valued measure**. Since the norm of a projection is 0 or 1, the series in ii) cannot converge in the norm of $L(\mathbf{H})$, unless all but finitely many $E(A_n)$ are zero. Consequently, except for some trivial cases, E is not countably additive as a function in $L(\mathbf{H})$. In other words, E is not a $L(\mathbf{H})$-valued measure. Nevertheless one easily checks that:

a projection valued function E *on* $\mathcal{B}$ *is a resolution (resp. subresolution) of the identity of* **H** *if and only if:*

i) $E(\Omega) = I$ (resp. i)' $E(\emptyset)=0$)

ii) *for all* f *and* g *in* **H** *the complex function* $E_{f,g}$ *defined on* $\mathcal{B}$ *by* $E_{f,g}(A) = \langle E(A)f,g\rangle$ *is a complex measure.*

Remark I.3:
If E is a subresolution of the identity of **H** and if $\underline{\mathbf{H}}$ denotes the range of the projection

$E(\Omega)$, then the function $\underline{E}$ defined on $\mathcal{B}$ by setting $\underline{E}(A)$ equal to the restriction of $E(A)$ to $\underline{H}$ is a resolution of the identity of $\underline{H}$.

Example I.4:
Let μ be a nonnegative measure on $(\Omega,\mathcal{B})$ and let us assume that the Hilbert space $H=L^2(\Omega,\mathcal{B},\mu)$ is separable, and let us set $E(A)f = \mathbf{1}_A f$ for all f in H and A in $\mathcal{B}$. Then, E is a resolution of the identity of H and we will see later that every resolution of the identity on $(\mathbb{R},\mathcal{B}_{\mathbb{R}})$ is unitarily equivalent to a resolution of the identity of this form or to a direct sum of them.

The following theorem is the main result of this section. It gives the definition and the main properties of functions $\phi(E)$ of a resolution of the identity E. These properties are easy to check when the functions ϕ are simple, more involved when the functions ϕ are merely bounded and measurable and rather subtle in the general case of measurable functions ϕ.

Theorem I.5:
Let E *be a resolution of the identity of* H *on* $(\Omega,\mathcal{B})$.

i) *for each measurable function* $\phi:\Omega \to \mathbb{C}$, *the subspace*

$$\mathcal{D}_\phi = \{f\in H;\ \int_\Omega |\phi(\omega)|^2\, E_{f,f}(d\omega) < \infty\} \tag{I.6}$$

is dense in H *and there is a unique operator* $\phi(E)$ *with domain* $\mathcal{D}_\phi$ *and such that:*

$$\langle \phi(E) f,g\rangle = \int_\Omega \phi(\omega)\, E_{f,g}(d\omega) \tag{I.7}$$

for all f *in* $\mathcal{D}_\phi$ *and* g *in* H. *Moreover, for each* f *in* $\mathcal{D}_\phi$ *we have:*

$$\|\phi(E) f\|^2 = \int_\Omega |\phi(\omega)|^2\, E_{f,f}(d\omega) \tag{I.8}$$

ii) *if* ϕ *and* ψ *are measurable functions on* Ω *we have:*

$$\mathcal{D}(\phi(E)\psi(E)) = \mathcal{D}_\psi \cap \mathcal{D}_{\phi\psi} \quad\text{and}\quad \phi(E)\psi(E) \subset [\phi\psi](E) \tag{I.9}$$

and consequently $\phi(E)\psi(E) = [\phi\psi](E)$ *if and only if* $\mathcal{D}_{\phi\psi} = \mathcal{D}_\psi$.

iii) *for each measurable function* ϕ *on* Ω *we have:*

$$\phi(E)^* = \bar{\phi}(E) \quad\text{and}\quad \phi(E)\phi(E)^* = |\phi|^2(E) = \phi(E)^*\phi(E) \tag{I.10}$$

__Example I.6:__
If $\varnothing$ is a measurable function on $(\Omega,\mathcal{B})$ and if E is the resolution of the identity introduced in Example I.4 above one easily checks that $\varnothing(E)=M_{\varnothing}$.

The above example shows that the operator $\varnothing(E)$ can be viewed as the function $\varnothing$ of the operator $H=\varnothing_0(E)$ whenever Ω is a Borel subset of $\mathbf{C}$ and $\mathcal{B}$ is the trace σ-field (recall that $\varnothing_0(w)=w$ by definition). Indeed, (I.9) shows that this interpretation is accurate for polynomial fuctions $\varnothing$. Hence Theorem I.5 provides us with a functional calculus for the self adjoint operators H of the form $\varnothing_0(E)$ and we will see in Section I.4 below that every self adjoint operator is of this form.

Numerous concepts from classical scalar valued measure theory remain meaningful for projection valued measures. For example, if E is a resolution of the identity of $\mathbf{H}$ we define the **essential range** of a measurable function $\varnothing$ as the complement of the largest open subset U of $\mathbf{C}$ which satisfies $E(\varnothing^{-1}(U))=0$ and the function $\varnothing$ is said to be **essentially bounded** if its essential range is bounded, the **essential supremum** of $\varnothing$, which we will denote $||\varnothing||_\infty$ being then defined as the supremum of the $|z|$ for z in the essential range. It is then easy to check that $\Sigma(\varnothing(E))$ is nothing but the essential range of $\varnothing$.

If μ is a nonnegative -finite measure on $(\Omega,\mathcal{B})$ and if E is a subresolution of the identity of $\mathbf{H}$ on $(\Omega,\mathcal{B})$, by the **μ-essential support of E** we will refer to any set A $\mathcal{B}$ satisfying:

i) $\exists A'\in\mathcal{B},\ \mu(A')=0$ and $E((A\cup A')^c)=0$
ii) $\forall A''\in\mathcal{B},\quad E(A''\cap A)=0 \iff \mu(A\cap A'')=0.$

Note that all the μ-essential supports differ only by μ-negligeable sets and we will freely talk of **the** μ-essential support even though the latter is not uniquely defined as a set. Note also that this support does not change if one replaces μ by an equivalent measure, finite for example. Finally, we will simply talk about the essential support without even mentioning μ whenever Ω is the real line and μ is the Lebesgue measure.

The result contained in the following problem will be of practical importance in what follows.

__Problem I.7:__
Let $\{f_k;k\geq 1\}$ be an orthonormal basis in $\mathbf{H}$, E a subresolution of the identity of $\mathbf{H}$ and let us define the measure σ by:

$$\sigma(A) = \Sigma_{k\geq 1} 2^{-k} \langle E(A)f_k,f_k\rangle \qquad\qquad A\in\mathcal{B}$$

a) Check that $\sigma(A)=0$ if and only if $E(A)=0$
b) Conclude that the μ-essential supports of σ and of the projection valued measure E coincide.

If Ω is a topological space and if $\mathcal{B}$ is its Borel σ-field, the **topological support** of E is defined in the usual way as the complement of the largest open set U for which $E(U)=0$. It will be denoted **suppE**.

We end this section with the Lebesgue decomposition of a subresolution of the identity and of the associated operators. But first we state the following important remark.

<u>Remark 1.8:</u>
Let E_1 and E_2 be subresolutions of the identity of H on $(\Omega,\mathcal{B})$ and let us assume that H is the orthogonal direct sum of $H_1=E_1(\Omega)H$ and $H_2=E_2(\Omega)H$. Then, the function E defined by $E(A)=E_1(A)+E_2(A)$ for all A in $\mathcal{B}$ is a resolution of the identity of H on $(\Omega,\mathcal{B})$ and the functional calculus of Theorem I.5 gives, for each measurable function ϕ on Ω and for $j=1,2$:

$$E_j(\Omega)\mathfrak{D}(\phi(E))\subset\mathfrak{D}(\phi(E)) \quad \text{and} \quad \phi(E)H_j\subset H_j$$

which is summarized by saying that *the decomposition reduces the operator* since $\phi(E)$ commutes with the projections $E_j(\Omega)$ on the H_j and:

$$\mathfrak{D}(\phi(E_j)) =\mathfrak{D}(\phi(E))\cap H_j \quad \text{and} \quad \phi(E_j)f =\phi(E)f \quad \text{if} \quad f\in H_j$$

which is summarized by saying that *the components of* $\phi(E)$ *on* H_1 *and* H_2 *are the operators* $\phi(E_1)$ *and* $\phi(E_2)$ if one recalls the notations of Remark I.3.

Recall that each complex measure μ on $(\mathbf{R},\mathcal{B}_{\mathbf{R}})$ has a decomposition:

$$\mu = \mu_{pp} + \mu_{ac} + \mu_{sc} \tag{I.11}$$

where μ_{pp} is a **pure point measure** (i.e. a sum of Dirac point masses), where μ_{ac} is **absolutely continuous** with respect to Lebesgue measure and where μ_{sc} is a **singular continuous measure** (**continuous** means that it has no point masses, i.e. $\mu_{sc}\{x\}=0$ for all x in $\mathbf{R}$ and **singular** means that it is carried by a set of zero Lebesgue measure). This **Lebesgue decomposition** of the scalar measures μ suggests that, given a resolution of the identity of H on $(\mathbf{R},\mathcal{B}_{\mathbf{R}})$, say E, we set:

$$H_{pp}(E) = H_{pp} = \{f\in H;\ E_{f,f} \text{ is pure point}\}$$
$$H_{ac}(E) = H_{ac} = \{f\in H;\ E_{f,f} \text{ is absolutely continuous}\}$$
$$H_{sc}(E) = H_{sc} = \{f\in H;\ E_{f,f} \text{ is singular continuous}\}$$

It is easy to see that the H_a's are orthogonal closed subspaces of H (note that here and in the sequel, a stands for pp, ac or sc) and that:

$$H = H_{pp} \oplus H_{ac} \oplus H_{sc} \tag{I.12}$$

In fact, if f belongs to H and $f=g+h+k$ in this decomposition and if $E_{f,f}= \mu_1+ \mu_2+ \mu_3$ is the

Lebesgue decomposition of $E_{f,f}$ then we necessarily have $\mu_1=E_{g,g}$, $\mu_2=E_{h,h}$, and $\mu_3=E_{k,k}$.

If, a being again either pp, or ac or sc, we denote by Π_a the projection of **H** onto $\mathbf{H}_a$ and if for f and g in **H** and A in $\mathcal{B}_{\mathbf{R}}$ we set:

$$\langle E_a(A)f,g\rangle = [E_{f,g}]_a(A)$$

where the right hand side is defined as the left hand one, then one easily checks that the $\{E_a(A);A\in \mathcal{B}_{\mathbf{R}}\}$'s are resolutions of the identity of

$$E_{pp}(\mathbf{R})\mathbf{H} = \mathbf{H}_{pp}, \qquad E_{ac}(\mathbf{R})\mathbf{H} = \mathbf{H}_{ac} \qquad \text{and} \qquad E_{sc}(\mathbf{R})\mathbf{H} = \mathbf{H}_{sc}$$

respectively, and that:

$$E = E_{pp} + E_{ac} + E_{sc}. \tag{I.13}$$

E_{pp} (resp. E_{ac}, E_{sc}) is a **pure point** (resp. **absolutely continuous , singular continuous**) **subresolution of the identity** of **H** in the sense that as f varies in **H**, all the scalar measures $\langle E_{pp}(\,.\,)f,f\rangle$ (resp. $\langle E_{ac}(\,.\,)f,f\rangle$, $\langle E_{sc}(\,.\,)f,f\rangle$) are pure point (resp. absolutely continuous, singular continuous). For this reason the decomposition (I.13) is called the **Lebesgue decomposition of the resolution of the identity E.**

According to Remarks I.3 and I.6, for each measurable function ϕ, the operator $\phi(E)$ decomposes in three parts $\phi(E_{pp})$, $\phi(E_{ac})$ and $\phi(E_{sc})$. In the particular case of the function ϕ_0 we have three self adjoint operators $H_{pp}=\phi_0(E_{pp})$, $H_{ac}=\phi_0(E_{ac})$ and $H_{sc}=\phi_0(E_{sc})$ on the Hilbert spaces $\mathbf{H}_{pp}$, $\mathbf{H}_{ac}$ and $\mathbf{H}_{sc}$ respectively. They are called the **pure point part, the absolute continuous part and the singular continuous part of the operator** H.Their spectra $\Sigma(H_{pp})$, $\Sigma(H_{ac})$ and $\Sigma(H_{sc})$ are denoted $\Sigma_{pp}(H)$, $\Sigma_{ac}(H)$ and $\Sigma_{sc}(H)$ and are called **the pure point spectrum, the absolutely continuous spectrum** and **the singular continuous spectrum** of H. These spectra are the topological supports of the resolutions of the identity E_{pp}, E_{ac} and E_{sc}. Notice that:

$$\Sigma(H) = \Sigma(H_{pp}) \cup \Sigma(H_{ac}) \cup \Sigma(H_{sc})$$

but these sets may not be disjoint. Also, $\Sigma(H_{ac})\cup\Sigma(H_{sc})$ is denoted $\Sigma_c(H)$ and called the **continuous spectrum** of H. Finally we warn the reader that in many texts, $\Sigma_{pp}(H)$ stands for the set of eigenvalues of H while for us, it is actually the closure of this set.

Example I.9:

Let $\{f_k;k\geq 1\}$ be an orthonormal basis in **H**, let $\{e(k);k\geq 1\}$ be a sequence in **R** and let us set:

$$E(A) = \sum_{e(k)\in A} f_k \otimes f_k \qquad\qquad A\in\mathcal{B}_{\mathbf{R}},$$

defines a resolution of the identity of H on $(\mathbf{R},\mathcal{B}_{\mathbf{R}})$ such that $E=E_{pp}$ and hence $H=H_{pp}$. The operator $H=\phi_0(E)$ is such that:

$$\mathcal{D}(H) = \{ f\in H;\ \sum_{k\geq 1} e(k)^2|\langle f,f_k\rangle|^2 < \infty \}$$

and

$$Hf = \sum_{k\geq 1} e(k)\ \langle f,f_k\rangle\ f_k$$

whenever $f\in\mathcal{D}(H)$. Moreover, $\Sigma(H)=\Sigma_{pp}(H)=\overline{\{e(1),e(2),\ \dots\}}$ and $\Sigma_c(H)=\emptyset$.

At some point in what follows we will consider the **essential spectrum** of H which will be denoted $\Sigma_{ess}(H)$ and which is defined as the (closed) set whose elements are the real numbers e for which the rank of the spectral projection $E((e-\varepsilon,e+\varepsilon))$ is infinite for all $\varepsilon>0$. We will also say a few words about the **discrete spectrum** $\Sigma_{disc}(H)$ which is defined as the set of isolated eigenvalues of finite multitplicity. Notice that $\Sigma_{ess}(H)=\Sigma(H)\setminus\Sigma_{disc}(H)$.

I.3. Representation theorems:

For later purposes, we present now classical results on some classes of harmonic and analytic functions. We will use the following notations:

$$\begin{aligned} \mathbf{D} &= \{ z\in\mathbf{C};\ |z|<1 \} \\ \Pi_+ &= \{ z\in\mathbf{C};\ \mathrm{Im}\, z>0 \} \\ \mathbf{T} &= \{ z\in\mathbf{C};\ |z|=1 \} \end{aligned}$$

and

$$p_r(\theta) = [1-r^2]/[1-2r\cos\theta+r^2] \qquad\qquad r\in[0,1),\ \theta\in\mathbf{R}$$

for the Poisson kernel. $\mathbf{T}$ will be often identified to intervals in $\mathbf{R}$ and an integral on $\mathbf{T}$ with respect to a measure $\mu\in M(\mathbf{T})$ will be denoted $\int_{[a,a+2\pi)} -\!-d\mu$ while the notation $\int_a^{a+2\pi} -\!-$ will be used only for Lebesgue integrals (i.e. with respect to the Haar measure). We use the notation $M(S)$ for the space of finite (signed) measures on the Borel σ-field of the topological space S.For μ in $M(\mathbf{T})$ we set:

$$P_\mu(z) = (2\pi)^{-1} \int_{[-\pi,+\pi)} P_r(\theta-t)\,\mu(dt) \qquad z=re^{i\theta}\in \mathbf{D}$$

for the Poisson integral of μ. $P_\mu(z)$ is harmonic in $\mathbf{D}$ and the absolutely continuous measures $P_\mu(re^{i\theta})d\theta$ converge weakly to μ as $r\to 1$. In fact we have:

a nonnegative function f on $\mathbf{D}$ is harmonic if and only if it is the Poisson integral of some nonnegative measure μ on $\mathbf{T}$ (i.e. $f=P_\mu$ for some μ $M_+(\mathbf{T})$) **and μ is uniquely determined by f,**

where we use the obvious notation $M_+(S)$ for the subset of nonnegative elements of $M(S)$. Wether or not the measure μ is nonnegative we have the famous Fatou's theorem:

any Poisson integral $f=P_\mu$ has nontangential limits at almost every point of $\mathbf{T}$, the limit is equal to the Radon-Nikodym derivative of the absolute continuous part of μ with respect to the normalized Lebesgue measure of the circle, and its singular part is carried by the set of points where the limit exists and is ∞.

The last statement is often refered to as the "de La Vallée Poussin theorem". Using the fact that the real and imaginary parts of analytic functions are harmonic we obtain:

a function f on $\mathbf{D}$ is analytic and has a nonnegative real part if and only if it has a representation:

$$f(z) = i\beta + \int_{[-\pi,+\pi)} [e^{it}+z]/[e^{it}-z]\,\mu(dt) \qquad z\in\mathbf{D},$$

for some real number β and some nonnegative measure μ on $\mathbf{T}$ which are uniquely determined by the function f.

This representation theorem, together with Fatou's theorem makes possible the study of the fine structure of the measure μ from the investigation of the boundary behavior of the analytic function f. The classical results which we just recalled are not exactly in the form we will need them. Indeed the measures we intend to study are on $\mathbf{R}$ instead of $\mathbf{T}$. Using the bijection:

$$w = i\,[1+z]/[1-z]$$

from $\mathbf{D}$ onto Π_+ the above representation theorem is transformed in the more convenient form:

a function f on Π_+ is analytic and has a nonnegative imaginary part if and only if it has a representation:

$$f(w) = a + bw + \int_{\mathbf{R}} [1+ew]/[e-w]\,\sigma(de) \qquad w\ \Pi_+$$

for some real numbers a and $b\geq 0$ and for a nonnegative finite measure σ on $\mathbf{R}$ which are uniquely determined by the function f.

The functions having such a representation are often called Pick functions. Modulo an extra assumption, this reprentation can still be improved:

a function f on Π_+ has a representation:

$$f(w) = \int_{\mathbf{R}} 1/[e-w]\sigma(de) \qquad w \in \Pi_+ \qquad (1.14)$$

for a nonnegative finite measure σ on $\mathbf{R}$ if and only if f is analytic, its imaginary part is nonnegative and

$$\sup_{y>0} y\,|f(iy)| < \infty. \qquad (1.15)$$

Such functions f are often called **Herglotz functions** . As before Fatou's and de la Vallé Poussin's theorems are transported to the present setting so that the function $\mathrm{Im} f$ has nontangential limits at almost every point of $\mathbf{R}$ (which is the boundary of Π_+) and these limits are equal to π times the density of the absolutely continuous component σ_{ac} of σ. Moreover, the singular component σ_s of σ is carried by the set of points where the limit exists and equals $+\infty$.

I.4. The spectral theorem:

We saw in Section I.2 that we could associate a self adjoint operator $\emptyset_0(E)$ to each resolution of the identity E of H on $(\mathbf{R}, \mathcal{B}_{\mathbf{R}})$. The present section is devoted to the converse of this result.

<u>Theorem I.10:</u>
A densely defined operator H *on* H *is self adjoint if and only if there exists a resolution of the identity of* H *on* $(\mathbf{R}, \mathcal{B}_{\mathbf{R}})$, *say* E, *such that* $H=\emptyset_0(E)$.

The proof of the converse is as follows. For each f in H, the function $f_f(z) = -\langle [H-zI]^{-1} f, f\rangle$ is defined and analytic outside the spectrum and in particular on Π_+. The resolvent identity (I.2) gives:

$$-2\,\mathrm{Im} z\,\|[H-zI]^{-1} f\|^2 = 2i\,\mathrm{Im}\,\langle [H-zI]^{-1} f, f\rangle$$

which shows that the imaginary part of f_f is nonnegative in the upper half plane. We conclude that f_f is a Herglotz function by remarking that (I.3) implies (I.15). Then the representation theorem gives the existence of a finite nonnegative measure μ_f on $\mathbf{R}$ such that:

$$\langle [zI-H]^{-1} f, f\rangle = \int_{\mathbf{R}} 1/[z-e]\ \mu_f(de) \qquad z \in \Pi_+ \qquad (1.16)$$

Now, for f and g in **H** we set:

$$\mu_{f,g} = [\mu_{f+g} - \mu_{f-g} + i(\mu_{f+ig} - \mu_{f-ig})]/4$$

By polarization of (I.16) we get:

$$\langle [zI-H]^{-1}f, g\rangle = \int_{\mathbf{R}} 1/[z-e] \;\; \mu_{f,g}(de) \qquad z\in\Pi_+ \qquad (I.17)$$

The uniqueness part of the representation theorem gives that for each Borel set A in **R** the function $(f,g) \dashrightarrow \mu_{f,g}(A)$ is linear in f and antilinear in g and hence there is a linear map E(A) from H into itself such that $\langle E(A)f, g\rangle = \mu_{f,g}(A)$ for all f and g in H. It is then a routine exercise to show that E(A) is an orthogonal projection and that $\{E(A); A\in\mathcal{B}_{\mathbf{R}}\}$ is a resolution of the identity of H such that $\emptyset_0(E)=H$ because of (I.17). □

We now come to the important problems of **canonical forms** and **multiplicity** of self adjoint operators.

A subspace K of H is said to be **generator** for E if the set $\{E(A)k; A\in\mathcal{B}_{\mathbf{R}}, k\in K\}$ is total in H. The **multiplicity** of E is then defined as the infimum of the dimensions of the generating subspaces of E. If E is the resolution of a self adjoint operator H, the **spectral multiplicity** of H is defined as the multiplicity of E. This definition can be localized to the components of H in its Lebesgue decomposition or to a part of the spectrum of H by considering the restriction of E to the Borel subset in question. If the spectral multiplicity of H is one, any nonzero element of a generating subspace is called a **cyclic vector** for H. Our interest in cyclic vectors relies in the following:

Proposition I.11:
If g *is a cyclic vector for* H, *the formula:*

$$Vf = \int_{\mathbf{R}} f(e)\, E(de)g \qquad (I.18)$$

for f in $L^2(\mathbf{R}, \mu_g)$, *defines a unitary isomorphism of the Hilbert spaces* H *and* $L^2(\mathbf{R},\mu_g)$. V *maps the domain of the operator* $M_{\emptyset o}$ *in* $L^2(\mathbf{R}, \mu_g)$ *onto* $\mathcal{D}(H)$ *and* $H(Vf)=V(\emptyset_0 f)$ *for* f *in* $\mathcal{D}(M_{\emptyset o})$.

Proof:
By definition of the integral in (I.18) with respect to a measure with values in H, we have, for A in $\mathcal{B}_{\mathbf{R}}$ and f in $L^2(\mathbf{R}, \mu_g)$:

$$\langle E(A)g, Vf\rangle = \int_{\mathbf{R}} f(e) \langle E(de)g, E(A)g\rangle$$

and consequently:

$$||Vf||^2 = \int_{\mathbf{R}} |f(e)|^2 \langle E(de)g, g\rangle$$

which shows that V preserves the norm and hence the inner product by polarization. V is onto because g is cyclic and the end of the proof is straightforward. □

The above result easily generalizes to the finite multiplicity case if one uses the theory of L^2-spaces with respect to matrix measures. The above proof can be adapted and we obtain:

Proposition I.11':
If H *is of finite multiplicity, if* $\{g(1), \dots ,g(n)\}$ *is an orthonormal basis of a generating subspace and if* M *is the matrix measure* $[\mu_{g(j),g(k)}]_{j,k=1,\dots,n}$ *then the formula:*

$$VF = \Sigma_{1\le j\le n} \int_{\mathbf{R}} f_j(e)\, E(de)g(j) \qquad F=(f_1,\dots,f_n)\in L^2(\mathbf{R},dM)$$

defines a unitary isomorphism of the Hilbert spaces $L^2(\mathbf{R}, M)$ *and* H *which also satisfies* $V\mathbf{D}(M_{\emptyset o})=\mathbf{D}(H)$ *and* $H(VF)=V(\emptyset_o F)$ *for* F *in* $\mathbf{D}(M_{\emptyset o})$.

The isomorphism V is said to realize the canonical form of H. In the case of finite multiplicity, many properties of the operator H can be obtained from the knowledge of the nonnegative measure $\mu = \Sigma_{1\le j\le n} M_{jj}$ trace of the matrix measure M. In fact it is easy to see that:

$\Sigma(H)$ *is the topological support of the measure* μ. *Moreover,* μ *is the least upper bound (for the order given by the absolute continuity of measures) of the family* $\{E_{f,f};f\in\mathbf{H}\}$.

In the general case an induction argument leads to:

Proposition I.11":
For each self adjoint operator H *there exist a (possibly infinite) integer* N *and nonnegative measures* $\mu_1, \dots ,\mu_N$ *on* $\mathbf{R}$ *and a unitary isomorphism* V *of* $\oplus_{1\le j\le N} L^2(\mathbf{R},\mu_j)$ *onto* H *such that* $V\mathbf{D}(M_{\emptyset o})=\mathbf{D}(H)$ *and* $H(VF)=V(\emptyset_o F)$ *for* $F=(f_1,\dots,f_N)$ *in* $\mathbf{D}(M_{\emptyset o})$ *and with* $[\emptyset_o F](e)= (e f_1(e), \dots ,e f_N(e))$.

I.5. Perturbations and quadratic forms:

The first problem we want to consider is the sum of self adjoint operators: given two self adjoint operators H_0 and $\mathbf{q}$ we want to investigate the properties of the operator $H=H_0+\mathbf{q}$. We can think of H as the energy Hamiltonian of a quantum system, H_0 being the kinetic energy (a multiple of the Laplacian as we will see in Section III.1) and the operator $\mathbf{q}$ (which will be for us the operator of multiplication by a measurable function q) representing some potential energy. When the operator $\mathbf{q}$ is bounded (in our case when the function q will be bounded) the operator H is self adjoint as we already noticed in Section I.1. Its spectral properties can be investigated from the characteristics of the unitary group $\{e^{itH}; t\in\mathbf{R}\}$ generated by H. These operators can be computed thanks to the famous Trotter product formula which we now recall in the particular case we are intersted in.

Proposition I.12:
Let $H=H_0+\mathbf{q}$ *with* H_0 *self adjoint and* $\mathbf{q}$ *self adjoint and bounded. Then:*

$$e^{itH} f = \lim_{n\to\infty} [e^{i(t/n)H_0}\, e^{i(t/n)\mathbf{q}}]^n f$$

for all t *in* $\mathbf{R}$ *and* f *in* H.

Unfortunately the potential functions we will have to deal with are most of the time unbounded and the self adjointness problem is usually difficult. We will not dwell on this now but we will recall in Chapter III a result by Faris and Lavine which will be appropriate to our study of Schrödinger operators. Even though we try to avoid abstract perturbation results, we need to introduce a minimum from the theory of quadratic forms (see Chapter III below for the applications to the Dirichlet and Neumann Laplacians).

If $\mathbf{Q}$ is a dense subspace of the Hilbert space $\mathbf{H}$ and if $q: \mathbf{Q}\times\mathbf{Q} \to \mathbf{C}$ is a sesquilinear form, the function $\mathbf{Q}\ni f \to q(f)=q(f,f)\in\mathbf{C}$ is called a **quadratic form**, and $\mathbf{Q}$ is denoted $\mathbf{Q}(q)$ and called the **quadratic form domain** of q. q is said to be **symmetric** if $q(f,g)=q(g,f)$ for all f and g in $\mathbf{Q}(q)$ and **nonnegative** if $q(f)\geq 0$ for all f in $\mathbf{Q}(q)$. More generally we will say that a sesquilinear for is **bounded below** if

$$q(f,f) \geq m\,||f||^2$$

for some finite constant m and all f in $\mathbf{Q}(q)$. We will then say that q is **closed** if furthermore the vector space $\mathbf{Q}(q)$ is closed for the norm

$$||f||_{(q)} = [\, q(f,f) + (m+1)\,||f||^2\,]^{1/2}.$$

If H is a self adjoint operator such that $\Sigma(H)\subset\mathbf{R}_+$, then the quadratic form q defined by $\mathbf{Q}(q)=\mathbf{D}(H^{1/2})$ and

$$q(f,g) = \langle H^{1/2}f, H^{1/2}g\rangle \qquad (1.19)$$

for f and g in $\mathcal{D}(H^{1/2})$ is a closed nonnegative form, and we say that it is the form of the operator H. Note that the right hand side of (I.19) will very often be denoted $\langle Hf,g\rangle$ even though f may not belong to the domain of H. Conversely, and this is a very important result, **every closed nonnegative quadratic form is associated with a unique self adjoint operator whose spectrum is contained in $\mathbf{R}_+$.**

I.6. Self adjoint extensions of symmetric operators:

The **unbounded operators** to be considered below are differential or partial differential operators or finite difference operators which are only **given formally**. It will be usually easy to choose domains (made of test functions) on which these operators become **symmetric**, but it will sometimes be difficult to see if they are **self adjoint**, or if they have **self adjoint extensions.** And even if this is the case, we would like to be able to classify the various self adjoint extensions in order to choose among them those which are relevent to the physics of the problem. The theory of **boundary conditions** which we now outline gives a satisfactory answer.

If the operator H is closed and symmetric, the function $z \to \dim \mathrm{Ker}[zI-H^*]$ is equal to a constant $n_+(H)$ in the upper half plane Π_+ and to a constant $n_-(H)$ in the lower half plane Π_-. The integers $n_+(H)$ and $n_-(H)$ are called the **deficiency indices** of H. Moreover $\Sigma(H)$ is

either the closed upper half plane $\overline{\Pi}_+$

or the closed lower half plane $\overline{\Pi}_-$

or the whole complex plane $\mathbf{C}$

or a closed subset of $\mathbf{R}$.

Finally, one can show that H is **self adjoint** if and only if $\Sigma(H)\subset\mathbf{R}$, or equivalently, $n_+(H)=n_-(H)=0$. Consequently, extending closed symmetric operators in self adjoint ones amounts to finding extensions with smaller and smaller deficiency indices until they finally vanish.

Let H be a symmetric operator on $\mathcal{H}$. We define the inner product $\langle\, .\, ,\, .\, \rangle_H$ on $\mathcal{D}(H^*)$ by:

$$\langle f, g\rangle_H = \langle f, g\rangle + \langle H^*f, H^*g\rangle \qquad f,g\in\mathcal{D}(H^*).$$

$\mathcal{D}(H^*)$ is a Hilbert space for this inner product and "H-closure", "H-orthogonality", "H-continuity",.... will be understood in the sense of this inner product. An H-continuous linear form B on $\mathcal{D}(H^*)$ which vanishes on $\mathcal{D}(H)$ will be called a **boundary value** for H and the equation

$$B(f) = 0 \qquad f\in\mathcal{D}(H)$$

will be called a **boundary condition** for H. A set $\{B_j(f)=0;\ 1\le j\le k\}$ of boundary conditions is said to be **symmetric** if :

$$(B_j(f)=B_j(g)\ \ j=1,\dots,k) \quad \Rightarrow \langle H^*f,g\rangle = \langle f,H^*g\rangle.$$

The importance of these definitions relies in the following important result:

> *if* H *is a closed symmetric operator with equal deficiency indices* $n=n_+(H)=n_-(H)$, *the self adjoint extensions of* H *are the restrictions of* H^* *to the subspaces of* $\mathfrak{D}(H^*)$ *determined by symmetric families of* n *linearly independent boundary conditions*

(i.e. to subspaces of the form $\{f\in\mathfrak{D}(H^*);\ B_j(f)=0;\ 1\le j\le n\}$ where $\{B_j(f)=0;\ 1\le j\le n\}$ is a symmetric family of n linearly independent boundary conditions).

I.7. References:

The notations and the terminology of this Chapter are standard. They are used in the basic texts like [Kato(1966)] and [Reed-Simon(1981)]. Our presentation of resolutions of the identity and of the projection valued measures and of the associated functional calculus is patterned after [Halmos(1951)] and [Rudin(1973)], the only novelty being the part dealing with the Lebesgue decomposition.

The representation theorems for analytic and harmonic functions are classical. They can be found in [Hoffman(1965)] or in[Donoghue(1965)]. The de La Vallée Poussin's Theorem is certainly less well known than all the other ones. It can be found for example in [Rudin(1974)] where we have to put together Theorem 11.12 and Theorem 8.10.

The details of the proof of the spectral theorem can be found for example in [Fuhrmann(1981)] where the interested reader will also find some complements on the canonical representations. The theory of multiplicity is very subtle. [Halmos(1951)] gives a nice account of it while the intuitive ideas are enlightened in the case of Schrödinger operators in Section C.5 of [Simon(1982b)].

The Trotter product formula and the results on quadratic forms which we quoted are proven in [Kato(1966)] and [Reed-Simon(1981)].

Finally the reader interested in the theory of the self adjoint extensions of symmetric operators may consult [Reed-Simon(1981)] and especially [Dunford-Schwarz(1963)] for an approach based on the concepts of boundary values and boundary conditions.

II. MEASURABILITY AND ERGODICITY OF SELF ADJOINT OPERATORS.

Throughout this chapter $(\Omega, \mathcal{B})$ will be a **fixed measurable space** and H will be a **separable Hilbert space** on which all the operators to be considered will be defined. We want to investigate the **measurability properties of functions** :

$$\omega \longrightarrow H(\omega)$$

from Ω into the space of operators on H. We are mainly interested in self adjoint operators and, OF COURSE, possibly unbounded ones. Having in mind the approach we chose to define and study the convergence of these operators, we will rely on the functional calculus for self adjoint operators and on the properties of bounded functions of our operators $H(\omega)$.

The second half of the chapter deals with the various **spectra** of the operators $H(\omega)$ when a **ergodic flow** acts on $(\Omega, \mathcal{B})$ and when the function $H(\omega)$ is **stationary** in a natural sense to be defined later.

II.1. Measurability:

The Hilbert space H being separable, its σ-field $\mathcal{B}_H$ is equal to the smallest σ-field for which all the functions

$$H \ni f \longrightarrow \langle f, g \rangle \in \mathbf{C}$$

for g in H are measurable. Hence, a function

$$\Omega \ni \omega \longrightarrow f(\omega) \in H$$

is measurable (for the Borel σ-field $\mathcal{B}_H$) if and only if all the functions

$$\Omega \ni \omega \longrightarrow \langle f(\omega), g \rangle \in \mathbf{C}$$

are measurable. Consequently, for a function $\omega \longrightarrow T(\omega)$ in the space $L(H)$, the following properties are equivalent:

i) $\forall f \in H, \forall g \in H, \quad \Omega \ni \omega \longrightarrow \langle f, T(\omega) g \rangle \in \mathbf{C}$ is measurable,

ii) $\forall f \in H$, $\Omega \ni \omega \longrightarrow T(\omega)f \in H$ is measurable,
iii) $\Omega \ni \omega \longrightarrow T(\omega) \in L(H)$ is measurable for the Borel σ-field $\mathcal{B}_{L(H)}$ for the uniform norm.

A function $\Omega \ni \omega \longrightarrow T(\omega) \in L(H)$ will be said **measurable** if it satisfies any one of the above three properties. The functional calculus for self adjoint operators and especially the following proposition will make possible the extension of this definition to the unbounded operators.

__Proposition II.1:__
If for each ω *in* Ω, $H(\omega)$ *is a self adjoint operator on* H *and if* $\{E(A,\omega): A \in \mathcal{B}_{\mathbf{R}}\}$ *denotes its resolution of the identity, the following three properties are equivalent:*

i) $\Omega \ni \omega \longrightarrow E(A,\omega) \in L(H)$ *is measurable for all* $A \in \mathcal{B}_{\mathbf{R}}$,

ii) $\Omega \ni \omega \longrightarrow e^{itH(\omega)} \in L(H)$ *is measurable for all* $t \in \mathbf{R}$,

iii) $\Omega \ni \omega \longrightarrow [H(\omega)-zI]^{-1} \in L(H)$ *is measurable for all* $z \in \mathbf{C}\setminus\mathbf{R}$.

Notice that a simple monotone class argument shows that i) is in fact equivalent to the measurability of the function $\Omega \ni \omega \longrightarrow E(A,\omega) \in L(H)$ for all Borel sets A of the form $(-\infty,a]$ for a in $\mathbf{R}$, or even for a in any dense subset of $\mathbf{R}$.

__Proof:__
For each ω in Ω and f and g in H the functional calculus gives:

$$\langle f, e^{itH(\omega)} g \rangle = \int_{\mathbf{R}} e^{it\varepsilon} \langle E(d\varepsilon, \omega)f, g \rangle \qquad \text{(II.1)}$$

and i) $\Rightarrow$ ii) follows by approximating the right hand side by integrals of step functions. If ε is real and if $\operatorname{Im} z > 0$, we have:

$$[z-\varepsilon]^{-1} = i^{-1} \int_0^\infty e^{izt} e^{-i\varepsilon t}\, dt$$

(and the same formula with $-i$ instead of i if $\operatorname{Im} z < 0$). Then, the functional calculus gives, for each ω Ω:

$$\langle f, [H(\omega)-zI]^{-1}, g \rangle = i^{-1} \int_0^\infty e^{izt} \langle f, e^{-itH(\omega)} g \rangle\, dt \qquad \text{(II.2)}$$

(and the corresponding formula in the case $\operatorname{Im} z < 0$). This proves ii) $\Rightarrow$ iii).
Let us now assume that a and b are fixed real numbers such that $-\infty < a < b < +\infty$ and for each $\alpha > 0$ let us define the function $F_{a,b,\alpha}$ by:

$$F_{a,b,\alpha}(\varepsilon) = (2i\pi)^{-1} \int_a^b [(t-i\alpha-\varepsilon)^{-1} - (t+i\alpha-\varepsilon)^{-1}]\, dt.$$

If we let α tend to 0, then, $F_{a,b,\alpha}(\varepsilon)$ converges to 0 if $\varepsilon \notin [a,b]$, to $1/2$ if $\varepsilon=a$ or $\varepsilon=b$ and to 1 if $\varepsilon \in (a,b)$. Consequently, $\lim_{\alpha \to 0} F_{a-\beta,b+\beta,\alpha}(\varepsilon)$ converges monotonically to the characteristic function of the interval (a,b) when β tends to 0. Since $F_{a,b,\alpha}(\varepsilon)$ is uniformly bounded as $\alpha \searrow 0$ we conclude that (using again the functional calculus for self adjoint operators):

$$\langle E((a,b),\omega)f,g\rangle$$

$$= \lim_{\beta\searrow 0} \lim_{\alpha\searrow 0} (2i\pi)^{-1}\int_{a-\beta}^{b+\beta} \langle f,([(t-i\alpha)I-H(\omega)]^{-1}-[(t+i\alpha)I-H(\omega)]^{-1})g\rangle\, dt \qquad (II.3)$$

which proves the measurability of $\omega \longrightarrow E(A,\omega)$ for $A=(a,b)$. The general form of iii) $\Rightarrow$ i) is then obtained via a monotone class argument. □

This proposition justifies the following choice for the definition of measurability for (possibly unbounded) operator valued functions.

Definition II.2:
A function $\omega \to H(\omega)$ *with values in the set of self adjoint operators on* $\mathcal{H}$ *will be said to be measurable if any of the three equivalent properties of Proposition II.1 is satisfied.*

Even though very elementary the following criteria will be of practical importance in the sequel.

Proposition II.3:
Let $\{H_n;\ n\geq 1\}$ *be a sequence of functions from* Ω *into the set of self adjoint operators on* $\mathcal{H}$ *and let* H *be another such function. If the sequence* $\{H_n(\omega); n\geq 1\}$ *converges to* $H(\omega)$ *in the strong resolvent sense for each* ω *in* Ω *and if each function* $H_n(\omega)$ *is measurable, then the function* $H(\omega)$ *is also measurable.*

Proof:
For each z in $\mathbf{C}\setminus\mathbf{R}$, ω in Ω and f and g in $\mathcal{H}$ we have (recall the definition of the strong resolvent convergence):

$$\lim_{n\to\infty} \langle f,[H_n(\omega)-zI]^{-1}g\rangle = \langle f,[H(\omega)-zI]^{-1}g\rangle$$

which shows that the right hand side is measurable as limit of measurable functions. □

Eventhough it is not directly related to the measurability problems, the following remark is in order. It will be used in Section IV.3.

Let $\{H_n;n\geq 0\}$ be a sequence of self adjoint operators on H and let $\{E_n;n\geq 0\}$ be the corresponding sequence of resolutions of the identity of H. A proof along the same lines as above, and especially using the formulae (II.1), (II.2) and (II.3), easily shows that **the convergence of the sequence $\{H_n;n\geq 1\}$ to H_0 in the strong resolvent sense** is equivalent to any of the three properties:

i) **for each** z **in** $\mathbf{C}\setminus\mathbf{R}$, $[zI-H_n]^{-1}$ **converges strongly to** $[zI-H_0]^{-1}$

ii) **for each** t **in** $\mathbf{R}$, e^{itH_n} **converges strongly to** e^{itH_0}

iii) **for each** f **in** H**, the measure** $\langle E_n(.)f,f\rangle$ **converges vaguely to the measure** $\langle E_n(.)f,f\rangle$.

We could have considered as well the narrow or weak convergence of measures since all the measures in iii) are nonnegative and have the same total mass $||f||^2$. In fact, one can restate iii) in the following equivalent form obtained by polarization:

iii)' **for each** f **and** g **in** H**, the measure** $\langle E_n(.)f,g\rangle$ **converges vaguely to the measure** $\langle E_n(.)f,g\rangle$.

Also, when all the spectra of the H_n for $n\geq 0$ are contained in the same half line $[a,\infty)$ (i.e. when the operators are uniformly bounded below) the Laplace transform can be used instead of the Fourier transform in the characterization of the weak convergence of measures and ii) becomes equivalent to:

ii)' **for each** $t>0$, e^{-tH_n} **converges strongly to** e^{-tH_0}.

We now consider the following problem: **does the Lebesgue decomposition of self adjoint operators preserve the measurability?** Because of our definition of the measurability this problem rewrites in the simpler way: **does the Lebesgue decomposition of a measurable resolution of the identity give measurable components?** The answer is YES without any doubt, but in order to check the mathematical details we need some definitions and an elementary lemma from classical measure theory.

We will denote by $\mathcal{B}_M$ the smallest σ-field of subsets of the space $M(\mathbf{R})$ of complex (finite) measures on $\mathbf{R}$, for which all the functions $M(\mathbf{R})\ni\mu \dashrightarrow \mu(A)\in\mathbf{C}$ are measurable. According to this definition property i) of Proposition II.1 says that for each f and g in H the complex function:

$$\omega \dashrightarrow \langle f, E(\,.\,,\omega)g\rangle$$

from Ω into $M(\mathbf{R})$ is measurable. The nice thing about our choice for this σ-field is that it gives the measurability of the components of the Lebesgue decomposition of scalar measures. Indeed:

Lemma II.4:

The three applications $\mu \dashrightarrow \mu_{pp}$, $\mu \dashrightarrow \mu_{ac}$ *and* $\mu \dashrightarrow \mu_{sc}$ *from* $M(\mathbf{R})$ *into itself are measurable for the* σ*-field* $\mathcal{B}_M$.

Proof:

Since the sum of two measures and the product of a measure by a scalar are measurable operations, it is enough to show that $\mu \dashrightarrow \mu_s = \mu_{pp} + \mu_{sc}$ and $\mu \dashrightarrow \mu_c = \mu_{ac} + \mu_{sc}$ are measurable. Moreover we can first decompose μ as the sum of its real part and its imaginary part, and further decompose each of them into positive and negative parts. All these operations are measurable so that we can restrict to nonnegative measures μ and show the measurability of the restrictions of these maps to $M_+(\mathbf{R})$.

Let $\mathfrak{I}$ be the set of all finite unions of real intervals with rational endpoints. $\mathfrak{I}$ is countable and generates $\mathcal{B}_{\mathbf{R}}$. Also, it is easy to check that for each A in $\mathcal{B}_{\mathbf{R}}$ we have:

$$\mu_s(A) = \lim_{n \to \infty} \sup_{I \in \mathfrak{I},\ |I| < 1/n} \mu(A \cap I)$$

where we used the notation $|\ .\ |$ for the Lebesgue measure. This proves the measurability of the function $\mu \dashrightarrow \mu_s(A)$ for all A in $\mathcal{B}_{\mathbf{R}}$, and hence of $\mu \dashrightarrow \mu_s$.

To prove the measurability of $\mu \dashrightarrow \mu_c$ it suffices to show the measurability of the function $\mu \dashrightarrow \mu_c(A)$ for each fixed Borel set A of the form $A = [a,b)$ with $-\infty < a < b < +\infty$. This follows from:

$$\mu_c(A) = \lim_{\epsilon \searrow 0} f_\epsilon(\mu)$$

where the limit is monotone nonincreasing and where $f_\epsilon(\mu)$ is defined as the monotone limit of the nondecreasing family of $f_{\epsilon,n}$ where:

$$f_{\epsilon,n}(\mu) = \Sigma_{0 \le k < 2^n - 1}\ \mu(I_{n,k})\ 1_{[0,\epsilon)}(\ \mu(I_{n,k})\ )$$

and

$$I_{n,k} = [a + k2^{-n}(b-a),\ a + (k+1)2^{-n}(b-a))$$

for $k = 0, 1, \ldots, 2^n - 1$. This completes the proof because the measurability of $f_{\epsilon,n}(\mu)$ is obvious.□

Now we set:

Definition II.5:

We will say that a subresolution E *of the identity of* H *is measurable on* $(\Omega, \mathcal{B})$ *if for each* ω *in* Ω, $E(\omega)$ *is a subresolution of the identity of* H *on* $(\mathbf{R}, \mathcal{B}_{\mathbf{R}})$ *such that the function:*

$$\Omega \ni \omega \longrightarrow \langle E(\omega)f, g\rangle \in M(\mathbf{R})$$

is measurable for all f *and* g *in* H.

Hence, Definition II.2 simply says that a function $\omega \longrightarrow H(\omega)$ with values in the set of self adjoint operators on **H** is measurable if and only if the corresponding resolution of the identity is measurable in the sense of Definition II.5 above. Moreover, Lemma II.4 gives:

Proposition II.6:
If $\omega \longrightarrow E(\omega)$ *is a measurable resolution of the identity of* H *and if for each* ω *in* Ω *we denote by* $E_{pp}(\omega)$, $E_{ac}(\omega)$ *and* $E_{sc}(\omega)$ *its components in its Lebesgue decomposition then* E_{pp}, E_{ac} *and* E_{sc} *are measurable subresolutions of the identity of* H.

II.2. Spectra in the ergodic case:

The second half of this chapter deals with the study of the various spectra of measurable self adjoint operators $H(\omega)$ defined on an ergodic dynamical system $(\Omega, \mathcal{B}, \mathbf{P}, \{T_x; x \in \Theta_0\})$. From now on $(\Omega, \mathcal{B}, \mathbf{P})$ will be a fixed complete probability space and Θ_0 a subsemigroup of the additive group $\Theta = \mathbf{R}^d$. The reader can think of Θ_0 as of $\mathbf{R}^d$ itself, or $\mathbf{Z}^d$, or else, $\mathbf{R}_+$ or $\mathbf{Z}_+$ in the one dimensional case. **H** will again be a fixed separable Hilbert space and $\{U_x; x \in \Theta_0\}$ will be assumed to be a unitary representation of Θ_0 on **H** in the following sense: $U_0 = I$ and for each x in Θ_0, U_x is a unitary operator on **H** such that:

$$U_{x+y} = U_x U_y$$

for all x and y in Θ_0. We will say that **the representation is total** if the set F of all the f of **H** for which there exists an infinite set $\{x_k(f); k \geq 1\}$ in Θ_0 such that

$$\langle U_{x_k(f)} f, U_{x_j(f)} f \rangle = \delta_{j,k} \qquad j,k \geq 1 \qquad (II.4)$$

is **total in** H. The classical example (which is at the origin of this definition) deals with $H = L^2(\mathbf{R}^d, dx)$ and $U_x f = f(\,.\,-x)$, in which case F contains the set of normalized functions with compact supports. Finally, we assume that $\{T_x; x \in \Theta_0\}$ is a fixed **ergodic subsemigroup** of the group of all the automorphisms of the measurable space $(\Omega, \mathcal{B})$ which preserve the probability measure **P**. The ergodicity assumption gives measure 0 or 1 to all the elements of $\mathcal{B}$ left invariant, up to **P**-null sets, by all the T_x's for x in Θ_0.

Now that the decor is set we can start. The first result we prove is very elementary but it will turn out to be very useful in the sequel.

Lemma II.7:
If $\omega \dashrightarrow E(\omega)$ *is a measurable function in the set of orthogonal projections of* H *which satisfies:*

$$E(T_x\omega) = U_x^* E(\omega) U_x \tag{II.5}$$

for all ω *in* Ω *and* x *in* Θ_0*, then there exists a (possibly infinite) nonnegative integer* r *such that for* **P**-*almost all* ω *in* Ω *the rank of* $E(\omega)$ *is equal to* r. *Moreover, if the representation* $\{U_x; x\in\Theta_0\}$ *is total, we must have* $r=0$ *or* $r=\infty$, *and in fact:*

$$r=0 \iff \forall f\in F,\ \mathbf{E}\{\langle Ef, f\rangle\}=0 \iff E=0 \ \mathbf{P}\text{-a.s.}$$

and consequently:

$$r=\infty \iff \forall f\in F,\ \mathbf{E}\{\langle Ef, f\rangle\} > 0.$$

Here and in what follows we use the notation $\mathbf{E}\{\,.\,\}$ for the expectation with respect to the probability measure **P** and a.s. as an abbreviation for almost surely.

Proof:
Let $\{f_k; k\geq 1\}$ be an orthonormal basis for H and for each ω in Ω let us set:

$$r(\omega) = \operatorname{trace} E(\omega) = \Sigma_{k\geq 1} \langle E(\omega) f_k, f_k\rangle.$$

r is a nonnegative random variable and for each x in Θ_0 we have:

$$\begin{aligned} r(T_x\omega) &= \Sigma_{k\geq 1} \langle E(T_x\omega) f_k, f_k\rangle \\ &= \Sigma_{k\geq 1} \langle E(\omega) U_x f_k, U_x f_k\rangle \\ &= \operatorname{trace} E(\omega) \\ &= r(\omega) \end{aligned}$$

where we used the assumption (II.5) and the fact that U_x is unitary. The ergodicity implies that r is a constant **P**-a.s. Let $f\in F$ and let $\{x_k(f); k\geq 1\}$ satisfy (II.4). Since $\{U_{x_k(f)} f; k\geq 1\}$ is an orthonormal system in H, for each ω in Ω we have:

$$r = r(\omega) \geq = \Sigma_{k\geq 1} \langle E(\omega) U_{x_k(f)} f_k, U_{x_k(f)} f_k\rangle$$

$$= \sum_{k\geq 1} \langle E(T_{x_k(r)}\omega)f\ , f\rangle$$

and taking expectations of both sides we obtain:

$$r \geq \sum_{k\geq 1} \mathbf{E}\{\langle Ef, f\rangle\}$$

$$= \begin{cases} 0 & \text{if } \mathbf{E}\{\langle Ef, f\rangle\} = 0 \\ \infty & \text{if } \mathbf{E}\{\langle Ef, f\rangle\} > 0. \end{cases}$$

If $r < \infty$, for each $f \in F$ we must have $\mathbf{E}\{\langle Ef, f\rangle\} = 0$, and hence $\langle Ef, f\rangle = 0$ $\mathbf{P}$-a.s. Since we can assume without any loss of generality that F is countable, we can interchange the "for each $f \in F$" and the "$\mathbf{P}$-a.s.". So, for $\mathbf{P}$-almost all ω in Ω we have:

$$\forall f \in F, \quad \langle E(\omega)f, f\rangle = 0$$

and hence:

$$\forall f \in H, \quad \langle E(\omega)f, f\rangle = 0$$

since F is total. This gives $E(\omega) = 0$ by polarization and hence $r=0$. This also completes the proof. □

The first application of the above lemma is the following.

Proposition II.8:
Let $\omega \dashrightarrow E(\omega)$ *be a measurable subresolution of the identity such that* (II.5) *is satisfied by the function* $\omega \dashrightarrow E(A,\omega)$ *for each* A *in* $\mathcal{B}_{\mathbf{R}}$. *Then there exists a closed subset* Σ *of* $\mathbf{R}$ *such that the topological support of the projection valued measure* $E(\,.\,,\omega)$ *is* Σ *for* $\mathbf{P}$-*almost all* ω *in* Ω.

Proof:
The topological support of $E(\omega)$ is the set of ε in $\mathbf{R}$ such that for each neighborhood A of ε we have $E(A,\omega) \neq 0$ or equivalently, $\operatorname{trace} E(A,\omega) \neq 0$. But since this trace is independent of ω we may think that this set of ε's is also independent of ω. Let us prove it. For each rational ε' and ε'' such that $\varepsilon' < \varepsilon''$ we let $\Omega_{\varepsilon',\varepsilon''}$ be a set in $\mathcal{B}$ of full $\mathbf{P}$-measure and $r_{\varepsilon',\varepsilon''}$ a nonnegative (possibly infinite) integer such that:

$$\forall \omega \in \Omega_{\varepsilon',\varepsilon''}, \quad \operatorname{trace} E((\varepsilon',\varepsilon''),\omega) = r_{\varepsilon',\varepsilon''}.$$

We then set:

$$\Sigma = \{\, \varepsilon \in \mathbf{R}\,;\ \forall \varepsilon',\varepsilon'' \in \mathbf{Q},\ \ \varepsilon' < \varepsilon < \varepsilon'' \Rightarrow r_{\varepsilon',\varepsilon''} \neq 0 \,\}$$

and:

$$\Omega' = \bigcap_{\substack{\epsilon',\epsilon'' \in \mathbf{Q} \\ \epsilon' < \epsilon''}} \Omega_{\epsilon',\epsilon''}$$

Then, Ω' is in $\mathcal{B}$, $\mathbf{P}(\Omega')=1$ and supp $E(\omega)=\Sigma$ for all $\omega \in \Omega'$. □

We now discuss an easy consequence of the above result. In order to state and prove it we need an extra assumption:

> $\mathcal{B}$ *is the Borel* σ*-field for a Polish structure on* Ω *and the function* $\omega \dashrightarrow E(\omega)$ *is continuous in the following sense: the measures* $\langle E(\,.\,,\omega_n)f, f\rangle$ *converge weakly to* $\langle E(\,.\,,\omega)f, f\rangle$ *whenever* ω_n *converges to* ω *in* Ω *and* $f \in H$.

Note that the measurability of $\omega \dashrightarrow E(\omega)$ follows. Let now $\mathbf{P}_1$ and $\mathbf{P}_2$ be two probability measures on $(\Omega,\mathcal{B})$ which are invariant and ergodic for the action of $\{T_x; x \in \Theta_0\}$. Proposition II.8 gives the existence of two closed subsets Σ_1 and Σ_2 of $\mathbf{R}$ such that the topological support of $E(.,\omega)$ (which we will denote supp$E(.,\omega)$) is equal to Σ_1 for $\mathbf{P}_1$-almost all ω in Ω and equal to Σ_2 for $\mathbf{P}_2$-almost all ω in Ω. The following proposition is elementary. It shows that the sets Σ's are nondecreasing functions of the topological supports of the probability measures $\mathbf{P}$.

__Corollary II.9:__
Under the above assumptions we have:

$$\text{supp}\,\mathbf{P}_1 \subset \text{supp}\,\mathbf{P}_2 \quad \Rightarrow \quad \Sigma_1 \subset \Sigma_2$$

__Proof:__
As we just noticed, according to Proposition II.8 we have the existence of Ω_j in $\mathcal{B}$ such that $\mathbf{P}_j(\Omega_j)=1$ and such that supp $E(.,\omega)=\Sigma_j$ for all the ω's in Ω_j for $j=1,2$. Let ω be chosen in $\Omega_1 \cap \text{supp}\,\mathbf{P}_1$ which is not empty because it has $\mathbf{P}_1$-measure 1. We have supp $E(.,\omega)=\Sigma_1$ but also $\omega \in \text{supp}\,\mathbf{P}_2$ and hence $\omega \in \overline{\Omega}_2$ since the topological support of $\mathbf{P}_2$ is the smallest closed set with $\mathbf{P}_2$-measure 1. Hence, the exists a sequence $\{\omega_n; n \geq 1\}$ in Ω_2 which converges to ω in Ω. Our continuity assumption implies the convergence of the measure $\langle E(.,\omega_n)f, f\rangle$ to $\langle E(.,\omega)f, f\rangle$ for each f in H. If A is an open subset of $\mathbf{R}$ contained in the complement of Σ_2, we must have $\langle E(A,\omega_n)f,f\rangle=0$ for all $n \geq 1$ since $\omega_n \in \Omega_2$ and hence $\langle E(A,\omega)f,f\rangle=0$ in the limit. Since f was arbitrary, this proves that $E(.,\omega)$ vanishes on the complement of Σ_2 and consequently that its topological support is contained in Σ_2. □

We now apply the above results to the particular case of the Lebesgue decomposition of random self adjoint operators.

Proposition II.19:

Let $\omega \dashrightarrow H(\omega)$ *be a measurable function in the set of self adjoint operators on* $\mathbf{H}$ *and let us assume that for each* A *in* $\mathcal{B}_{\mathbf{R}}$ *the function* $\omega \dashrightarrow E(A,\omega)$ *satisfies* (II.5) *in the sense that:*

$$E(A,T_x\omega) = U_x^* E(A,\omega) U_x. \tag{II.5)'$$

Then, there exist closed subsets Σ, Σ_{pp}, Σ_{ac} *and* Σ_{sc} *of* $\mathbf{R}$ *such that for* $\mathbf{P}$-*almost all* ω *in* Ω *we have* $\Sigma(H(\omega))=\Sigma$, $\Sigma_{pp}(H(\omega))=\Sigma_{pp}$, $\Sigma_{ac}(H(\omega))=\Sigma_{ac}$ *and* $\Sigma_{sc}(H(\omega))=\Sigma_{sc}$.

Proof:

For each ω in Ω the definitions of the subresolutions of the identity $E_{pp}(.,\omega)$, $E_{ac}(.,\omega)$ and $E_{sc}(.,\omega)$, and the uniqueness in the Lebesgue decomposition of measures on $(\mathbf{R},\mathcal{B}_{\mathbf{R}})$ imply that the functions $\omega \dashrightarrow E(.,\omega)$, $\omega \dashrightarrow E_{pp}(.,\omega)$, $\omega \dashrightarrow E_{ac}(.,\omega)$ and $\omega \dashrightarrow E_{sc}(.,\omega)$ satisfy the assumptions of Proposition II.8. □

Remark II.11:

The results and the proofs of Propositions II.8 and II.10 above give also the existence of a closed subset Σ_c of $\mathbf{R}$ and of two subsets Σ_{ess} and Σ_{disc} of $\mathbf{R}$, the union of which is Σ, and such that we have $\mathbf{P}$-almost surely:

$$\Sigma_c(H(\omega)) = \Sigma_c,\ \Sigma_{ess}(H(\omega)) = \Sigma_{ess} \text{ and } \Sigma_{disc}(H(\omega)) = \Sigma_{disc}.$$

Remark II. 12:

Using formulae (II.1), (II.2) and (II.3) in the same way as in the proof of Proposition II.1 one easily checks that for a function $\omega \dashrightarrow H(\omega)$ in the set of self adjoint operators on $\mathbf{H}$ the folowing properties are equivalent:

i) *for each* A *in* $\mathcal{B}_{\mathbf{R}}$ *the function* $\omega \dashrightarrow E(A,\omega)$ *is measurable and satisfies*

$$E(A,T_x\omega) = U_x^* E(A,\omega) U_x \tag{II.6.i}$$

for all x *in* $\mathbf{G}_0$.

ii) *for each* t *in* $\mathbf{R}$ *the function* $\omega \dashrightarrow e^{itH(\omega)}$ *is measurable and satisfies*

$$e^{itH(T_x\omega)} = U_x^* \, e^{itH(\omega)} \, U_x \tag{II.6.ii}$$

for all x *in* Θ_0.

ii) *for each* z *in* $\mathbb{C}\backslash\mathbb{R}$ *the function* $\omega \dashrightarrow [zI-H(\omega)]^{-1}$ *is measurable and satisfies*

$$[zI-H(T_x\omega)]^{-1} = U_x^* \, [zI-H(\omega)]^{-1} \, U_x \tag{II.6.iii}$$

for all x *in* Θ_0.

When either one of the above properties is satisfied we say that the function $\omega \dashrightarrow H(\omega)$ (which is measurable according to Definition II.2) satisfies:

$$H(T_x\omega) = U_x^* \, H(\omega) \, U_x \tag{II.7}$$

for all x in Θ_0. The property (II.7) can also be checked directly. Indeed:

Proposition II.13:

Let D *be a dense subspace of* $\mathcal{H}$ *which is a common core for all the operators* $H(\omega)$ *and such that:*

$$H(T_x\omega)f = U_x^* \, H(\omega) \, U_x \, f$$

for all f *in* D, x *in* Θ_0 *and* ω *in* Ω. *Under these conditions the function* $\omega \dashrightarrow H(\omega)$ *satisfies* (II.7).

Proof:

Let us fix z in $\mathbb{C}\backslash\mathbb{R}$, x in Θ_0 and ω in Ω and let us check (II.6.iii). If f belongs to the domain of $H(T_x\omega)$ there exists a sequence $\{f_n; n\geq 1\}$ in D which converges to f and such that $H(T_x\omega)f_n$ converges to $H(T_x\omega)f$ as $n \to \infty$. Consequently we must have:

$$U_x^* [zI-H(\omega)]^{-1} U_x \; [zI-H(T_x\omega)]f = \lim_{n\to\infty} U_x^* [zI-H(\omega)]^{-1} U_x \; [zI-H(T_x\omega)]f_n$$

$$= \lim_{n\to\infty} U_x^* [zI-H(\omega)]^{-1} [zI-H(\omega)] U_x f$$

$$= f. \; \square$$

Proposition II.14:

Let us assume that the function $\omega \dashrightarrow H(\omega)$ *in the set of self adjoint operators on* $\mathcal{H}$ *is measurable and satisfies* (II.7) *and that the representation* $\{U_x; x\in\Theta_0\}$ *is total. Then:*

i) $\Sigma_{disc} = \emptyset$

ii) *If the spectral multipicity of* $H(\omega)$ *is* $\mathbf{P}$*-almost surely finite, then, for each* ε *in* Ω, ε *is* $\mathbf{P}$*-almost surely not an eigenvalue of* $H(\omega)$.

Proof:

Both i) and ii) are consequences of the following remark. When $\varepsilon \in \mathbf{R}$ is fixed, the rank of the projection $E(\{\varepsilon\},\omega)$ is finite either because of the definition of Σ_{disc} or because of the assumption on the spectral multiplicity of $H(\omega)$. Consequently this rank has to be zero according to Lemma II.7.□

Remark II.15:

We will see later that the above proposition applies to one dimensional Schrödinger operators. Nevertheless, ii) should not be understood as the **absence of eigenvalues** , since as we will see in Chapter VII, that the spectrum is expected to be (and actually is) very often **pure point**. ii) simply means that the eigenvalues are very much dependent on ω and when the latter varies, they move so fast that a fixed real number does not see them almost surely.

Proposition II.14 above and especially its part ii) raises the question of the **spectral multiplicity of the random self adjoint operators** . It seems that many investigators in the field of random Schrödinger operators think that under very mild conditions (for example the usual stationarity and ergodicity assumptions and some assumption insuring some actual randomness in the problem) the spectral multiplicity of the random Schrödinger operators should be almost surely equal to 1. In fact I personally believe that under very general conditions the following result should be true (at least for $H_{pp}(\omega)$ if not for $H(\omega)$):

for each f *in* $\mathbf{H}$ *there exists a set* Ω_f *in* $\mathcal{B}$ *with probability* 1 *and such that* f *is a cyclic vector for the operator* $H(\omega)$ *for all* ω *in* Ω_f.

This conjecture is formulated without any reference to Schrödinger operators. Since it cannot be true in general for stationary ergodic self adjoint operators satisfying (II.7), some assumption on the type of randomness is in order as we already pointed out.

In fact, it is commonly believed that **the pure point spectrum only should be of spectral multiplicity** 1 (i.e. the spectral multiplicity of $H_{pp}(\omega)$ should be almost surely equal to 1)while **the absolutely continuous spectrum should be almost surely of infinite multiplicity** (i.e. the spectral multiplicity of $H_{ac}(\omega)$ should almost surely be infinite). See the discussion of the recent results on the existence of pure point spectrum for the multidimensional lattice case at the end of Chapter VII.

None of these conjectures has been settled in general!

We now investigate the essential support of our stationary (i.e. satisfying (II.7)) random self adjoint operators. But first we prove a technical lemma.

Lemma II.16:

Let $\mathcal{B}'$ *be the Borel* σ*-field of a Polish space* Ω' *and let* μ' *be a nonnegative finite measure on* $(\Omega',\mathcal{B}')$. *Then, the functions:*

$$M_1(\Omega',\mathcal{B}') \ni \mu \dashrightarrow d\mu_{ac} / d\mu' \in L^1(\Omega',\mathcal{B}',\mu')$$

and

$$L^1(\Omega',\mathcal{B}',\mu') \ni f \dashrightarrow \mathbf{1}_{\{f\neq 0\}} \in L^1(\Omega',\mathcal{B}',\mu')$$

are measurable where $M_1(\Omega',\mathcal{B}')$ *denotes the set of probability measures and* μ_{ac} *the absolute continuous component of* μ *in the Lebesgue decomposition of* μ *with respect to* μ'.

Proof:

The function $L^1(\Omega',\mathcal{B}',\mu') \ni f \dashrightarrow f.\mu' \in M(\Omega',\mathcal{B}')$ defined by $[f.\mu'](A) = \int_A f\, d\mu'$ for A in $\mathcal{B}'$ is measurable because for each $A \in \mathcal{B}'$ the function $f \dashrightarrow [f.\mu'](A)$ is continuous, hence measurable. Moreover it is one-to-one. Since $L^1(\Omega',\mathcal{B}',\mu')$ and $M(\Omega',\mathcal{B}')$ are Polish spaces, its range is a Borel subset of $M(\Omega',\mathcal{B}')$ and the inverse function is also measurable. If one compose the latter with the function $\mu \dashrightarrow \mu_{ac}$ (which is easily seen to be measurable by a proof along the same lines as in Lemma II.4) the first claim of the lemma follows.

To prove the second claim, for each integer $n \geq 1$ we consider first, the continuous even function $\int_n$ on $\mathbb{R}$ defined by $\int_n(t)=1$ if $t>1/n$, $\int_n(0)=0$, and which is linear between 0 and $1/n$, and second the function F_n defined by:

$$L^1(\Omega',\mathcal{B}',\mu') \ni f \dashrightarrow F_n(f) = \int_n \circ f \in L^1(\Omega',\mathcal{B}',\mu').$$

We then notice that:

$$\| F_n(f) - F_n(g) \|_{L^1} \leq n \| f-g \|_{L^1}$$

which proves that F_n is continuous and hence measurable. Furthermore, if $f \in L^1(\Omega',\mathcal{B}',\mu')$ we have:

$$\| F_n(f) - \mathbf{1}_{\{f\neq 0\}} \|_{L^1} \leq \int_{\{0<|f|<1/n\}} |f|\, d\mu'$$

which tends to zero as n tends to infinity. This shows that the function $\mathbf{1}_{\{f\neq 0\}}$ is the pointwise limit of the F_n. Its meassurability follows. □

Proposition II.17:
Let $\omega \longrightarrow E(\omega)$ *be a measurable subresolution of the identity of* H *which satisfies* (II.6.i) *and let* μ *be a* σ*-finite nonnegative measure on* $\mathbf{R}$. *Then there exists a Borel subset* Σ_μ *of* $\mathbf{R}$ *such that for each* $\omega \in \Omega$ *the* μ*-essential support of the projection valued measure* $E(\omega)$ *is equal to* Σ_μ *up to sets of* μ*-measure* 0.

The result above and the proof below apply as well to the case of a measurable projection valued measure $E(\omega)$ on a Polish space $(\Omega',\mathcal{B}')$. Note also that since the μ-essential support of $E(\omega)$ depends only on the sets of μ-measure 0, we can assume without any loss of generality that the measure μ is actually finite.

Proof:
Let $\{f_k;k\geq 1\}$ be an orthonormal basis for H and for each ω in Ω let us define the probability measure μ^ω by:

$$\mu^\omega = \Sigma_{1\leq k<\infty}\ 2^{-k} \langle E(.,\omega)f_k, f_k\rangle.$$

Now, if $x\in\Theta_0$ we have:

$$\mu^{T_x\omega} = \Sigma_{1\leq k<\infty}\ 2^{-k} \langle E(.,\omega)U_x f_k, U_x f_k\rangle.$$

This shows that $\mu^{T_x\omega}$ can be obtained from $E(.,\omega)$ and $\{U_x f_k;k\geq 1\}$ in the same way as μ^ω is obtained from $E(.,\omega)$ and $\{f_k;k\geq 1\}$. Consequently, $\mu^{T_x\omega}$ and μ^ω have the same null sets (namely those of $E(.,\omega)$). $\omega \longrightarrow \mu^\omega$ is measurable because $\omega \longrightarrow E(.,\omega)$ is measurable and thus,

$$\Omega\ni\omega \longrightarrow \mathbf{1}_{\{d(\mu^\omega)_{ac}/d\mu \neq 0\}} \in L^1(\mathbf{R},d\mu)$$

is measurable because of the preceeding lemma. This function depends only on the null sets of the measure μ^ω and, according to what we saw, it must be invariant and hence constant $\mathbf{P}$-almost surely. The proof is complete because $E(.,\omega)$ and μ^ω have the same μ-essential support. □

The above proposition has been designed for the following application.

If $\omega \longrightarrow H(\omega)$ ***is a measurable function in the set of self adjoint operators on*** H ***which satisfies*** (II.7), ***there exists a set*** Σ_{Leb} ***defined up to a set of Lebesgue measure*** 0, ***which carries the absolutely continuous component*** $E_{ac}(.,\omega)$ ***of the resolution of the identity of*** $H(\omega)$.

The absolute continuous spectrum of $H(\omega)$ is concentrated on Σ_{Leb} in the sense that if $A\in\mathcal{B}_{\mathbf{R}}$ we must have:

$$A \cap \Sigma_{Leb} = \emptyset \;\Rightarrow\; E_{ac}(A,\omega) = 0.$$

We will see in Chapter VI a result analog to Corollary II.9 for the essential supports of the absolutely continuous spectra of Schrödinger operators in one dimension: **these essential supports are nonincreasing functions of the topologcal supports of the probability measures $\mathbf{P}$**. It would be nice to know wether or not this result is general or if it actually depends on the special characteristics of the dimension 1.

We end this section with an improvement of Lemma II.7 which will apply to the case of the lattice $\mathbb{Z}^d$ and to the continuous one dimensional case. We will assume that $\Theta_0=\Theta$, that $\mathbf{H}=\mathbf{L}^2(\Theta,m)$ where m is the Haar measure of Θ, and that U_x is given by $U_x f=f(.-x)$.

Lemma II.18:
On the top of the assumptions of Lemma II.7 *we assume that for each* $\omega\in\Omega$ *the projection* $E(\omega)$ *has a weak kernel* $E(x,y,\omega)$ *which is a jointly continuous function of* $(x,y)\in\Theta\times\Theta$. *Then:*

$$E(\,.\,)=0 \quad \mathbf{P}\text{-a.s.} \quad\Leftrightarrow\quad \mathbf{E}\{\,E(0,0,.\,)\,\}=0,$$

and hence:

$$\text{rank}\,E(\,.\,)=\infty \quad \mathbf{P}\text{-a.s.} \quad\Leftrightarrow\quad \mathbf{E}\{\,E(0,0,.\,)\,\}>0.$$

Recall that a bounded operator T is said to have a **weak kernel** $T(x,y)$ if

$$\langle Tf,g\rangle = \int\int T(x,y)\,f(y)\,\overline{g(x)}\,dy\,dx$$

for all continuous functions f and g with compact supports.

Proof:
The kernel has to be nonnegative on the diagonal because of its continuity. Consequently,

$$\mathbf{E}\{\,E(0,0,.\,)\,\}=0 \quad\Leftrightarrow\quad E(0,0,\omega)=0 \quad \mathbf{P}\text{-a.s. in } \omega\in\Omega$$

$$\Rightarrow\; [\,\forall k\geq 1,\, E(0,0,T_{x(k)}\omega)=0] \qquad \mathbf{P}\text{-a.s. in } \omega\in\Omega$$

where $\{x(k);k\geq 1\}$ is any countable dense subset of Θ,

$$\Rightarrow\; [\,\forall k\geq 1,\; E(x(k),x(k),\omega)=0] \qquad \mathbf{P}\text{-a.s. in } \omega\in\Omega,$$

$$\Rightarrow [\ \forall x \in \Theta,\ E(x,x,\omega)=0] \qquad \mathbf{P}\text{-a.s. in } \omega\in\Omega,$$

because of the continuity of the kernel. Using Schwarz inequality we obtain:

$$\Rightarrow [\ \forall x,y \in \Theta,\ E(x,y,\omega)=0] \qquad \mathbf{P}\text{-a.s. in } \omega\in\Omega,$$

which implies that $E(\omega)=0$ $\mathbf{P}$-a.s. in $\omega\in\Omega$, and the proof is complete because the converse is obvious. □

II.3. Comments:

We already mentioned the open problems in the text so that we restrict this short paragraph to some bibliographical comments.

The main ideas of this chapter are due to Pastur. They can be traced back to his fundamental paper [Pastur(1974)]. They appeared in English only 6 years later. See [Pastur(1980)]. They have been used and generalized in many papers. See for example [Kunz-Souillard(1980)] and [Kirsch-Martinelli(1982a),(1982b)] for example. We tried to present them with the necessary care to check the numerous measurability problems which we believe have been ignored too often (see nevertheless [Kunz-Souillard(1980)] for the first serious efforts in this direction. It seems that the situation is very simple once the right definitions have been chosen. Note that essentially nothing but the definition of the ergodicity has been used.

Our contribution reduces to the introduction of the Lebesgue decomposition of the resolutions of the identity and to its systematic use. Also Proposition II.17 is published here for the first time. It answers a question raised in [Deift-Simon(1983)]. Finally one will notice that Corollary II.9 is nothing but Theorem 1 of [Kotani(1984a)].

III. SCHRÖDINGER OPERATORS.

We present in this chapter the deterministic Schrödinger operators which we will *randomize* in the sequel. We illustrate the abstract theory of the first chapter by concrete examples like the **Laplacian operators** (see Section III.1 below) and the **perturbed Laplacians** (Section III.2) which are actually our **Schrödinger operators**. We then concentrate on the continuous one dimensional case and we present the time honored approach to their spectral study based on the theory of ordinary differential equations (O.D.E. for short). In Section III.3 below, we spend a lot of time introducing the notations and proving various technical results. This is fully justified for we will use this preparatory work intensively in the chapters VI and VII when we study the absolute continuous spectrum and the exponential localization in one dimension.

It turns out that, once properly defined as self adjoint operators, Schrödinger operators generate semigroups (when they are bounded below) and unitary groups (when they are defined on the lattice $\mathbf{Z}^d$) which can be expressed via **path-space integral formulae**. These formulae are presented in the last section of this chapter. They are very useful in the investigation of the so-called integrated density of states of random Schrödinger operators as we will see in Chapter V, but their importance is quite general and should not be limited to the random case.

III.1. The free Laplacians:

In classical mechanics, observables are measurable functions on the state space of the system. For example, the momentum is represented by a vector $\vec{p} = (p_1, p_2, p_3)$ in $\mathbf{R}^3$ and the kinetic energy is then given by $e = m\,|\vec{p}|^2/2$.

Throughout the rest of these lectures we will assume that all the physical constants (masses, Planck constant, ...) have been set to be 1. *This convention does not change the mathematics and will make our life easier.*

In quantum mechanics, the observables are operators (hopefully self adjoint) on some infinite dimensional Hilbert space. For example, the components p_j of the momentum are transformed into the partial differential operators $i\,\partial/\partial x_j$ on $L^2(\mathbf{R}^3, dx)$ and consequently, the kinetic energy which was :

$$e = (1/2)[p_1^2 + p_2^2 + p_3^2]$$

is transformed into the operator:

$$H_0 = (1/2)\,[\,-\partial^2/\partial x_1^2 - \partial^2/\partial x_2^2 - \partial^2/\partial x_3^2\,].$$

H_0 is thus equal to one half of the negative Laplacian in $\mathbf{R}^3$. It is usually called the **free Hamiltonian** because it does not take into account any interaction energy. We would like to define

H_0 as a self adjoint operator on the Hilbert space $L^2(\mathbf{R}^3, dx)$. But first we need to fix some notations.

If f is an element of the Schwartz space $\mathcal{S}(\mathbf{R}^d)$ of rapidly decreasing test functions, its Fourier transform is denoted by $\hat{f}$ and defined by:

$$\hat{f}(\lambda) = (2\pi)^{-d/2} \int_{\mathbf{R}^d} e^{-i\lambda.x} f(x)\, dx \tag{III.1}$$

The Fourier transform of distributions is defined by duality. If T is a tempered distribution, its Fourier transform $\hat{T}$ is characterized by:

$$\hat{T}(f) = T(\hat{f}) \qquad\qquad f \in \mathcal{S}(\mathbf{R}^d).$$

The Fourier transform is a bijection of $\mathcal{S}(\mathbf{R}^d)$ onto itself. Its inverse transform is given by:

$$\check{f}(x) = (2\pi)^{-d/2} \int_{\mathbf{R}^d} e^{ix.\lambda} f(\lambda)\, d\lambda \tag{III.2}$$

and similarly for the distributions. The Fourier transform is the most efficient tool in the study the partial differential operators with constant coefficients. The following is a clear illustration of this fact.

There are several candidates for the definition of the free Hamiltonian. The first one is the operator H_0^{min} defined by $\mathfrak{D}(H_0^{min}) = C_c^\infty(\mathbf{R}^d)$ the space of C^∞-functions with compact supports in $\mathbf{R}^d$ and by:

$$H_0^{min} f = -(1/2)\, \Delta f \tag{III.3}$$

whenever $f \in C_c^\infty(\mathbf{R}^d)$. If instead of trying a minimal domain for our candidate, we want the largest possible domain, we will consider the operator H_0^{max} whose domain $\mathfrak{D}(H_0^{max})$ is the set of all elements of $H = L^2(\mathbf{R}^d, dx)$ whose Laplacian in the sense of distributions is still a function in $L^2(\mathbf{R}^d, dx)$ and we will set:

$$H_0^{max} f = -(1/2)\, \Delta f \tag{III.4}$$

whenever f is in $\mathfrak{D}(H_0^{max})$ so defined. It is then easy to see that:

H_0^{min} is essentially self adjoint , H_0^{max} is self adjoint and $\overline{H_0^{min}} = H_0^{max}$.

The proof relies on the following simple facts:

- for each tempered distribution T we have $\widehat{-\Delta T} = |\lambda|^2 \hat{T}$.
- $f \in \mathfrak{D}(H_0^{max})$ if and only if the function $\lambda \dashrightarrow |\lambda|^2 \hat{f}(\lambda)$ is square integrable in which case $H_0^{max} f(\cdot) = \widecheck{|\lambda|^2 \hat{f}(\lambda)/2}$

The reader is refered to the Section IX.7 of [Reed-Simon(1975)] for more details.

An important part of the spectral theory of partial differential operators deals with discretization techniques. In the case of the Laplacian this gives the free Hamiltonian on the lattice which we introduce in the following problem.

Problem:

Let $\mathbf{H}=\ell^2(\mathbf{Z}^d)$ be the set of sequences $\mathbf{x}=\{x(\underline{n})\}_{\underline{n}\in\mathbf{Z}^d}$ of complex numbers for which the quantity:

$$||\mathbf{x}|| = [\sum_{\underline{n}\in\mathbf{Z}^d} |x(\underline{n})|^2]^{1/2}$$

is finite. $\mathbf{H}$ is a Hilbert space for the inner product obtained by polarization of the above norm. If $\mathbf{x}$ is any element of $\mathbf{H}$ we set:

$$[H_0\mathbf{x}](\underline{n}) = -(1/2d) \sum_{|\underline{m}-\underline{n}|=1} x(\underline{m}) + x(\underline{n}) \qquad (III.5)$$

where $|\underline{m}-\underline{n}|= \Sigma_{1\leq j\leq d}|m_j-n_j|$ if $\underline{m}=(m_1,\dots,m_d)$ and $n=(n_1,\dots,n_d)$. Show that H_0 is a **bounded self adjoint operator the spectrum of which is** $[0,2]$.

We now come to the case of the Laplacian in a bounded domain instead of $\mathbf{R}^d$. Such "an operator" is formally symmetric and consequently we have to choose boundary conditions. We will concentrate on **Dirichlet and Neumann boundary conditions** and we rely on the theory of quadratic forms presented in the first chapter. More general boundary conditions will be considered in the one dimensional case (see Section III.3 below).

Let D be a bounded open subset of $\mathbf{R}^d$ and let us set $H=L^2(D,dx)$ for a while.

- Let $q_0^{(N)}$ be the sesquilinear form defined by:

$$q_0^{(N)}(f,g) = (1/2) \int_D \nabla f(x)\cdot\overline{\nabla g(x)}\, dx$$

on its domain

$\mathbf{Q}(q_0^{(N)}) = \{f\in\mathbf{H};$ the gradient ∇f of f (in the sense of distributions) is square integrable in $D\}$.

$q_0^{(N)}$ is closed and nonnegative. The corresponding self adjoint operator is called the **Neumann Laplacian** in D and is denoted $H_0^{(N)}$. The terminology can be justified by checking that if the boundary of D is smooth and if f is C^∞ in a neighborhood of the closure of D, then the solution of the classicasl Neumann problem is given by the function

$$u = [zI- H_0^{(N)}]^{-1}f$$

in the sense that u satisfies $(1/2)\Delta u + zu = f$ in D, and the normal derivative of u vanishes on the boundary of D.

- The sesquilinear form defined by:

$$(f,g) \dashrightarrow (1/2) \int_D \nabla f(x) \cdot \overline{\nabla g(x)}\, dx \qquad \text{(III.6)}$$

wit domain the space $C_c^{\infty}(D)$ of C^{∞}-functions with compact supports contained in D is closable. Its closure $q_0^{(D)}$ is a nonnegative closed sesquilinear form and the corresponding self adjoint operator is denoted by $H_0^{(D)}$ and called the **Dirichlet Laplacian** in D because its resolvent operator gives the solution of the classical Dirichlet problem in D.

The interested reader may look at [Simon(1978)] for more details.

III.2. The perturbed Laplacians:

Let D be a **bounded open** subset of $\mathbf{R}^d$ and let H be the Hilbert space $L^2(D,dx)$.
If the measurable function $\mathbf{R}^d \ni x \dashrightarrow q(x) \in \mathbf{R}$ is locally bounded, then the multiplication operator by the function q is a bounded operator on H and the perturbed operator $H = H_0^{(.)} + q$ is self adjoint wether the dot stands for D in the case of Dirichlet boundary condition or N in the case of Neumann boundary condition. In some cases, the function is merely locally L^p for some $p > d/2$ instead of locally bounded and in order to check the self adjointness we have to proceed in the following way:

a) we first split q into the sum of a bounded part which is harmless as we just saw, and a second piece with a very small L^p-norm

b) we then use the classical Sobolev inequality to check that this last part is a small perturbation of the free Laplacian (and hence of the Laplacian perturbed by the bounded part of q)

c) we finally appeal to simple perturbation results to conclude that H is self adjoint.

This argument actually works for more general potential functions q. The right class seems to be the class K_d as explained in [Simon(1982b)]. q is said to be in the class K_d if:

$$\lim_{a \searrow 0} \sup_{x \in \mathbf{R}^d} \int_{|x-y| \le a} |x-y|^{-(d-2)} |q(y)|\, dy = 0$$

when $d \ge 3$, when d=2 if:

$$\lim_{a \searrow 0} \sup_{x \in \mathbf{R}^d} \int_{|x-y| \le a} \mathrm{Log}(1/|x-y|)\, |q(y)|\, dy = 0$$

and, when d=1, if:

$$\sup_{x \in \mathbf{R}^d} \int_{|x-y| \le 1} |q(y)|\, dy < \infty.$$

The function q will be said to be in $K_{d,loc}$ when $q\mathbf{1}_D$ is in K_d for all bounded domains D.

As we just explained the self adjointness of H in bounded domains is not really a problem. Moreover, most of the potential functions we will deal with are locally bounded. So, we will not dwell on this special case. We will rather elaborte on the case of the operator $H=H_0+q$ in the whole space $\mathbf{R}^d$ and try to prove self adjointness on the Hilbert space $\mathcal{H}= L^2(\mathbf{R}^d,dx)$. We should be prepared to settle for essential self adjointness on $C_c^\infty(\mathbf{R}^d)$. Indeed, if the potential function q is locally square integrable, for every function f in $C_c^\infty(\mathbf{R}^d)$ the Laplacian $H_0 f$ is still smooth with compact support while qf is square integrable so that

$$Hf = H_0 f + qf \tag{III.7}$$

is an element of $\mathcal{H}$. One can thus define the operator H on the natural domain $C_c^\infty(\mathbf{R}^d)$. Moreover this operator is easily seen to be symmetric. Whenever it is essentially self adjoint it has a **unique self adjoint extension** which we will still denote H, and in many cases it is possible to show that its domain $\mathcal{D}(H)$ is actually the set of elements f of $L^2(\mathbf{R}^d,dx)$ for which the quantity (III.7) computed in the sense of distributions is a square integrable function (which is the value of Hf). This discussion screams for essential self adjointness criteria. The following one turns out to be very convenient. It is due to Faris and Lavine (the curious reader may consult [Reed-Simon(1975)] for a proof and complements).

But first we introduce some notations which we will use throughout these notes.

$$C(\underline{0}) = \{x=(x_1,\ldots,x_d)\in\mathbf{R}^d;\ |x_j|<1/2\ \ j=1,\ldots,d\} \quad \text{and} \quad C(\underline{n}) = \underline{n} + C(\underline{0}) \quad \text{for } \underline{n}\in\mathbf{Z}^d.$$

Also, $L^p_{loc,unif}(\mathbf{R}^d,dx)$ will denote the space of measurable functions on $\mathbf{R}^d$ which satisfy

$$\sup_{\underline{n}\in\mathbf{Z}^d} \int_{C(\underline{n})} |f(x)|^p\,dx < \infty$$

and $\ell^1(L^p(\mathbf{R}^d,dx))$ the space of measurable functions on $\mathbf{R}^d$ which satisfy

$$\Sigma_{\underline{n}} [\int_{C(\underline{n})} |f(x)|^p\,dx\,]^{1/p} < \infty.$$

__Proposition III.1:__
Let q_1 *and* q_2 *be real valued measurable functions on* $\mathbf{R}^d$ *such that:*

i) $q_1\in L^2_{loc}(\mathbf{R}^d,dx)$ *and* $q_1(x)\geq -a(|x|^2+1)$ *for some* $a>0$,

ii) $q_2\in L^p_{loc,unif}(\mathbf{R}^d,dx)$ *with* $p\geq 2$ *if* $d\leq 3$, $p>2$ *if* $d=4$ *and* $p\geq d/2$ *if* $d\geq 5$.

Then $H=H_0+q_1+q_2$ *is essentially self adjoint on* $C_c^\infty(\mathbf{R}^d)$.

III.3. The one dimensional case:

Throughout this section q will denote a **real valued measurable locally square integrable function** on $\mathbf{R}$ which satisfies:

$$q(t) \geq -a_0\,(t^2+1) \qquad \text{(III.8)}$$

for some $a_0 > 0$ and $|t|$ large enough.

Whenever we are dealing with the one dimensional case we will use the letter "t" rather than "x" for the space variable. This abuse of notation is usual in the O.D.E. literature and will be very convenient when we are dealing with random potential functions in which case "t" will become the parameter of a stochastic process.

The assumptions on q and especially (III.8) give the *essential self adjointness* of the operator H defined on $C_c^\infty(\mathbf{R})$ by:

$$Hf = -f'' + qf$$

We will denote by the same symbol H its unique self adjoint extension. Our aim is to investigate the spectral characteristics of this operator but we first proceed to a review of some typical examples to show the diversity in the possible spectral behaviors of these otherwise innocent looking operators.

Note that we use a prime to denote the derivation with respect to the variable t *and that the coefficient* $1/2$ *in front of the Laplacian disappeared. This is the second departure from our conventions so far.*

Examples:

1. If $\lim_{|t|\to\infty} q(t)=+\infty$, the resolvent of H is a compact operator from which it follows that $\Sigma(H)$ is a countable set without finite accumulation points, that $\Sigma(H)=\Sigma_{pp}(H)$ and even $\Sigma(H)=\Sigma_{disc}(H)$, and that $E=E_{pp}$. See for example Section 14 of Chapter XIII of [Reed-Simon(1978)].

2. If $q(t)$ is piecewise continuous and **periodic** $\Sigma(H)$ is of the form $\bigcup_{n\geq 1}[a_n,b_n]$ with $b_n \leq a_{n+1}$. The spectrum is **purely absolutely continuous** in the sense that $\Sigma(H)=\Sigma_{ac}(H)$ and $E=E_{ac}$. The intervals $[a_n,b_n]$ are called the **bands** of the spectrum and (b_n,a_{n+1}) the **gaps**. The latter are said to be **open** if $b_n < a_{n+1}$. See for example the Section 16 of Chapter XIII of [Reed-Simon(1978)].

2'. If f is a function on $\mathbf{R}^d$ (resp. $\mathbf{Z}^d$) and if $x \in \mathbf{R}^d$ (resp. $\mathbf{Z}^d$) we will use the notation $U_x f$ for the translated function:

$$[U_x f](y) = f(y-x) \qquad x \in \mathbf{R}^d \ (\text{resp.}\ \mathbf{Z}^d). \qquad (III.9)$$

An element f of the Banach space $C_b(\mathbf{R})$ of bounded continuous functions on R is said to be **almost periodic** if the set $\{U_x f;\ x \in \mathbf{R}\}$ is relatively compact in $C_b(\mathbf{R})$. Similar definitions exist for almost periodic functions on $\mathbf{Z}$, $\mathbf{R}^d$, or $\mathbf{Z}^d$. We will not need them.

The concept of almost periodicity is a natural generalization of the concept of periodicity. Nevertheless, the spectral characteristics of Schrödinger operators with almost periodic potentials are by far much more complicated than those of periodic ones. Let us simply note that the spectrum has a tendency to become a **Cantor set** (see for example [Moser(1981)], [Chulaevski(1981)], [Avron-Simon(1982a)], [Johnson-Moser(1982)] and [Bellissard-Simon(1982)]), that it can still be *absolutely continuous* (see [Dinaburg-Sinai(1976)]), but also *purely singular continuous* (see [Avron-Simon(1982)]), *pure point* (see [Craig(1983)], [Pöschel(1983)], and [Bellissard et al.(1983)], or even of *mixed type* (see [Bellissard et al.(1983)]. For a comprehensive review the reader is referred to [Simon(1982a)].

Note also that very often in the sequel, we will regard them as particular cases of **random** Schrödinger operators.

3. Examples of potential functions q giving rise to **purely singular continuous spectrum** (i.e. $\Sigma(H)=\Sigma_{sc}(H)$ and $E=E_{sc}$) were known to people like Aronzjan but the first systematic study is to be found in [Pearson(1978)].

4. In the context of the diffusion theory in a **constant electric field** it is natural to consider potential functions of the form:

$$q(t) = Ft + v(t)$$

where F is a real constant and v is a bounded function with two bounded derivatives. In this case, **the spectrum is the whole real line** and it is **purely absolutely continuous**, i.e. $\Sigma(H)=\Sigma_{ac}(H)=\mathbf{R}$ and $E=E_{ac}$. See for example [Bentosela et al.(1983)], [Ben Artzi(1985)] or [Carmona(1983)].

5. If $q(t)$ is integrable in a neighborhood of $+\infty$ <u>OR</u> $-\infty$, then $[0,\infty) \subset \Sigma(H)$ and $E=E_{ac}$ on $[0,\infty)$. Moreover, if $q(t)$ is integrable near <u>BOTH</u> $+\infty$ and $-\infty$ then $\Sigma(H)$ equals $[0,\infty)$ possibly augmented with an at most countable set of points which can accumulate only near 0, and $E=E_{ac}$ on $[0,\infty)$ while obviously $E=E_{pp}$ on $(-\infty,0]$. See [Carmona(1983)] for details.

6. There are many examples, among which the famous Wigner-Von Neumann's one, of potential functions giving rise to **eigenvalues embedded in the continuous spectrum**. We simply refer to the Section 13 of Chapter XIII of [Reed-Simon(1978)].

We now come to the **O.D.E. approach** to the spectral theory of these operators.

For each real numbers a and b such that $-\infty < a < 0 < b < +\infty$ and each α and β in $[0,\pi)$ we define the operator $H_{a,b,\alpha,\beta}$ on the domain

$$\mathcal{D}(H_{a,b,\alpha,\beta}) = \{f \in L^2([a,b],dt);\ f \text{ is differentiable, } f' \text{ is absolutely continuous and } -f''+qf \text{ is square integrable on } [a,b], \text{ and } f(a)\cos\alpha + f'(a)\sin\alpha = f(b)\cos\beta + f'(b)\sin\beta = 0\}$$

by:

$$H_{a,b,\alpha,\beta} f = -f'' + qf.$$

It is a self adjoint operator on the Hilbert space $\mathbf{H} = \mathbf{L}^2([a,b],dt)$ and the resolvent operator is compact. Indeed we will see below that it has a jointly continuous kernel and thus is Hilbert-Schmidt. Hence, the spectrum of H is a bounded below discrete set. Let us call

$$\lambda(0) < \lambda(1) \le \lambda(2) \le \ldots$$

the eigenvalues and

$$f_0, f_1, f_2, \ldots$$

the corresponding complete orthonormal system of eigenfunctions. We then set:

$$\begin{aligned} \xi_{a,b,\alpha,\beta} &= \textstyle\sum_{j\ge 0} \beta_j^2 \delta_{\lambda(j)} \\ \zeta_{a,b,\alpha,\beta} &= \textstyle\sum_{j\ge 0} \gamma_j^2 \delta_{\lambda(j)} \\ \eta_{a,b,\alpha,\beta} &= \textstyle\sum_{j\ge 0} \beta_j \gamma_j \delta_{\lambda(j)} \end{aligned} \tag{III.10}$$

where we use the notation δ_λ for the Dirac measure at λ, and where, for each $j \ge 0$, the coefficients β_j and γ_j are given by the expansion

$$f_j = \beta_j y_1(\,.\,,\lambda(j)) + \gamma_j y_2(\,.\,,\lambda(j)) \tag{III.11)'}$$

where for each λ in $\mathbf{C}$ the function $t \dashrightarrow y_1(t,\lambda)$ (resp. $t \dashrightarrow y_2(t,\lambda)$) is the unique solution of the equation:

$$-y''(t) + [q(t) - \lambda]\, y(t) = 0 \tag{E.V.E.}$$

which satisfies:

$$y_1(0,\lambda) = 1 \quad \text{and} \quad y_1'(0,\lambda) = 0$$

$$(\text{resp. } y_2(0,\lambda) = 0 \quad \text{and} \quad y_2'(0,\lambda) = 1\,). \tag{III.11)''}$$

If f is in $L^2([a,b],dt)$ we also set:

$$[U_{a,b,\mathfrak{a},\mathfrak{b}} f](\varepsilon) = \left(\int_a^b f(t)y_1(t,\varepsilon)\,dt, \int_a^b f(t)y_2(t,\varepsilon)\,dt\right) \tag{III.12}$$

$U_{a,b,\mathfrak{a},\mathfrak{b}}$ so defined is a unitary isomorphism of $L^2([a,b],dt)$ onto $L^2(\mathbf{R}, M_{a,b,\mathfrak{a},\mathfrak{b}}(d\varepsilon))$ where $M_{a,b,\mathfrak{a},\mathfrak{b}}$ is the matrix measure:

$$M_{a,b,\mathfrak{a},\mathfrak{b}} = \begin{bmatrix} \xi_{a,b,\mathfrak{a},\mathfrak{b}} & \eta_{a,b,\mathfrak{a},\mathfrak{b}} \\ \eta_{a,b,\mathfrak{a},\mathfrak{b}} & \zeta_{a,b,\mathfrak{a},\mathfrak{b}} \end{bmatrix}.$$

$U_{a,b,\mathfrak{a},\mathfrak{b}}$ transforms the operator $H_{a,b,\mathfrak{a},\mathfrak{b}}$ into the operator of multiplication by the variable ε. Consequently $U_{a,b,\mathfrak{a},\mathfrak{b}}$ realizes the canonical form of $H_{a,b,\mathfrak{a},\mathfrak{b}}$.

In the limit $a \to -\infty$ and $b \to +\infty$ the measures $\xi_{a,b,\mathfrak{a},\mathfrak{b}}$, $\zeta_{a,b,\mathfrak{a},\mathfrak{b}}$ and $\eta_{a,b,\mathfrak{a},\mathfrak{b}}$ **converge** vaguely to measures ξ, ζ, and η **independent of the boundary conditions** $\mathfrak{a}$ and $\mathfrak{b}$. Here, the assumption (III.8) plays a crucial role. If one defines the matrix measure M by:

$$M = \begin{bmatrix} \xi & \eta \\ \eta & \zeta \end{bmatrix}$$

and if one sets:

$$[U f](\varepsilon) = \left(\text{"}\int_{-\infty}^{+\infty}\text{"} f(t)y_1(t,\varepsilon)\,dt, \text{"}\int_{-\infty}^{+\infty}\text{"} f(t)y_2(t,\varepsilon)\,dt\right) \tag{III.13}$$

one defines a unitary isomorphism of $L^2(\mathbf{R},dt)$ onto $L^2(\mathbf{R}, M(d\varepsilon))$ which transform the operator H in the operator of multiplication by the independent variable ε. Note that the integrals in (III.13) are in quotation marks because they are meaningful in the L^2-sense only. Hence:

the spectral information on the operator is contained in the nonnegative measure

$$\mu(d\varepsilon) = \operatorname{trace} M(d\varepsilon) = \xi(d\varepsilon) + \zeta(d\varepsilon) \tag{III.14}$$

Indeed, for each Borel subset A of $\mathbf{R}$ we have:

$$(\forall f \in L^2(\mathbf{R},dt),\quad \langle E(A)f, f\rangle = 0) \Leftrightarrow \mu(A) = 0.$$

We will see later on that it is possible to control the above limit in many cases. But first, we introduce the so-called Titchmarsch-Weyl functions $\mathbf{m}$.

Our assumption (III.8) implies that for each complex number $z \notin \mathbf{R}$ there exists a unique complex number $\mathbf{m}_{+(-)}(z)$ such that the solution

$$t \longrightarrow \mathbf{f}_{+(-)}(t,z) = y_1(t,z) +(-)\, \mathbf{m}_{+(-)}(z)\, y_2(t,z)$$

of (E.V.E.) with $\epsilon = z$ is square integrable near $t = +(-)\infty$. Then the negative resolvent $-R(z,H)$ has an integral kernel given by:

$$g_z(s,t) = -[\, m_+(z) + m_-(z)\,]^{-1} f_+(t,z)\, f_-(s,z) \qquad \text{(III.15)}$$

for $s<t$. This integral kernel will be called the **Green's function** of the problem (E.V.E.). Now, if we choose an interval $(\epsilon_1,\epsilon_2) \in \mathbf{R}$ for the Borel set A, we have:

$$\xi((\epsilon_1,\epsilon_2)) + [\xi(\{\epsilon_1\}) + \xi(\{\epsilon_2\})]/2 = (1/\pi)\lim_{a \searrow 0} \pi^{-1} \int_{\epsilon_1}^{\epsilon_2} \mathrm{Im}\, [m_-(\epsilon + ia) + m_+(\epsilon + ia)]^{-1}\, d\epsilon$$

and (III.16)

$$\zeta((\epsilon_1,\epsilon_2)) + [\zeta(\{\epsilon_1\}) + \zeta(\{\epsilon_2\})]/2 = -(1/\pi)\lim_{a \searrow 0} \pi^{-1} \int_{\epsilon_1}^{\epsilon_2} \mathrm{Im}\, \big(m_-(\epsilon + ia)\, m_+(\epsilon + ia)\, [m_-(\epsilon + ia) + m_+(\epsilon + ia)]^{-1}\big)\, d\epsilon$$

Wether or not one plans on using the functions m, the knowledge of the asymptotic behavior of the solutions of (E.V.E.) seems to be a necessary step in the study of the measure $\mu(d\epsilon)$. We rewrite (E.V.E.) as:

$$Y'(t) = \mathbf{A}_\epsilon(t)\, Y(t) \qquad \text{(III.17)}$$

a first order differential system with:

$$\mathbf{A}_\epsilon(t) = \begin{bmatrix} 0 & 1 \\ q(t) - \epsilon & 0 \end{bmatrix} \qquad \text{(III.18)}$$

The solutions are constructed with the propagator $\{\mathbf{U}_\epsilon(t,s) : t,s \in \mathbf{R}\}$ which is the flow of unimodular 2×2 matrices satisfying:

$$\mathbf{U}_\epsilon(t,s)\, Y(s) = Y(t).$$

Also, setting $Y(t) = \begin{bmatrix} y(t) \\ y'(t) \end{bmatrix}$ and:

$$\begin{cases} y(t) = r(t) \sin \theta(t) \\ y'(t) = r(t) \cos \theta(t) \end{cases} \qquad \text{(III.19)}$$

with $r(t) \geq 0$ (in which case $r(t) = \|Y(t)\| = [|y(t)|^2 + |y'(t)|^2]^{1/2}$), we can switch to polar coordinates. This operation is called the **Prüffer transform**, and the functions $r(t)$ and $\theta(t)$ are called the **amplitude** and the **phase** of the solution $y(t)$. With these notations, $y(t)$ is a solution of (E.V.E.) if and only if $r(t)$ and $\theta(t)$ are solutions of the system:

$$\theta'(t) = \cos^2\theta(t) + [\boldsymbol{\varepsilon}-q(t)] \sin^2\theta(t) \qquad (III.20)$$

$$r'(t) = -(1/2)\, r(t)\, [\boldsymbol{\varepsilon}-q(t)-1] \sin 2\theta(t) \qquad (III.21)$$

This system is nonlinear so we may think that we did not gain in simplicity. Nevertheless this is not really the case because the second equation can be integrated directly if one assume the knowledge of the solution of the first one. It gives:

$$r(t) = r(s) \exp[-(1/2)\int_s^t [\boldsymbol{\varepsilon}-q(u)-1]\sin 2\theta(u)\, du] \qquad (III.22)$$

The choice of the "initial condition $r(s)$" is arbitrary especially because it amounts to fixing the amplitude of the solution at the point s while we are only interested in the asymptotic behavior. For similar reasons we may think of $\theta(t)$ as defined only modulo π instead of modulo 2π. From time to time we will also use the notation

$\theta(t,a,\boldsymbol{\alpha},\boldsymbol{\varepsilon})$ for the value of the phase computed at t, knowing that it equals $\boldsymbol{\alpha}$ for $t=a$,

and

$r(t,a,\boldsymbol{\alpha},\boldsymbol{\varepsilon})$ for the value of the amplitude computed at t, knowing that for $t=a$, the amplitude equals 1 and the phase equals $\boldsymbol{\alpha}$. If we recall (III.22) we get:

$$r(t,a,\boldsymbol{\alpha},\boldsymbol{\varepsilon}) = \exp[-(1/2)\int_a^t [\boldsymbol{\varepsilon}-q(u)-1]\sin 2\theta(u,a,\boldsymbol{\alpha},\boldsymbol{\varepsilon})\, du]. \qquad (III.23)$$

We will see below that it is sometimes very useful to consider the operator on random intervals $[a,b]$ or with random boundary conditions $\boldsymbol{\alpha}$ and $\boldsymbol{\beta}$. Indeed we will use the average spectral measure

$$\mu_{a,b,\boldsymbol{\beta}} = \pi^{-1}\int_{[-\pi/2,+\pi/2)} \mu_{a,b,\boldsymbol{\alpha},\boldsymbol{\beta}}\, d\boldsymbol{\alpha} \qquad (III.24)$$

where $\mu_{a,b,\boldsymbol{\alpha},\boldsymbol{\beta}} = \xi_{a,b,\boldsymbol{\alpha},\boldsymbol{\beta}} + \zeta_{a,b,\boldsymbol{\alpha},\boldsymbol{\beta}}$. It happens that $\mu_{a,b,\boldsymbol{\beta}}$ **is absolutely continuous with respect to the Lebesgue measure** and that its density $\partial\mu_{a,b,\boldsymbol{\beta}}/\partial\boldsymbol{\varepsilon}$ can be explicitely computed:

$$\partial\mu_{a,b,\boldsymbol{\beta}}/\partial\boldsymbol{\varepsilon} = r(a,0,\theta(0,b,\boldsymbol{\beta},\boldsymbol{\varepsilon}),\boldsymbol{\varepsilon})^{-2} \qquad (III.25)$$

I personally regarded this formula as very strange at first, miraculous may be. In fact, it has been extended to the case of the half line in [Kotani(1984b)] and can be understood in the abstract framework of averaging over rank one perturbations. See [Simon-Wolff(1985)].

This formula is very interesting because it separates the contributions of $\{q(t);t\geq 0\}$ and $\{q(t);t<0\}$. Indeed, the values of $q(t)$ for $t>0$ (and the boundary condition $\boldsymbol{\beta}$ at b) enter only in the computation of the angle $\theta=\theta(0,b,\boldsymbol{\beta},\boldsymbol{\varepsilon})$ while the values of $q(t)$ for $t<0$ enter in the computation of $r(a,0,\theta,\boldsymbol{\varepsilon})^{-2}$ only. Hence, if one can show that as a tends to $-\infty$, the amplitude $r(a,0,\theta,\boldsymbol{\varepsilon})$ remains bounded away from zero independently of θ in $[-\pi,+\pi)$ and e restricted to some set, then the limiting **spectral measure is absolutely continuous** on this set of energies $\boldsymbol{\varepsilon}$, and has a bounded density there. This is the way to prove the results stated in the Example 5 above. This remark will also have other applications in the sequel.

Except for this last result which is proved in [Carmona(1983)], all the results we discussed in this section are classical and can be found for example in the texts [Coddington-Levinson(1955)] and [Levitan-Sargsjan(1976)].

There is still one more result which we will need in the sequel. It deals with the "**eigenfunction expansions**". The result we will use is not explicitely statefd in the texts quoted above. It can be found in the paper [Simon(1982b)].

We already noticed that whenever th operator H has discrete spectrum $\{\varepsilon(1),\varepsilon(2),\ldots\}$ and cooresponding orthonormal basis $\{f_j;j\geq 1\}$ of eigenfunctions, the resolution of the identity of H, say E, could be written in the form:

$$E(A) = \Sigma_{\varepsilon(j)\in A}\ f_j \otimes f_j$$

At least for bounded Borel subsets of $\mathbf{R}$, such an operator has the integral kernel:

$$E(A,s,t) = \Sigma_{\varepsilon(j)\ A}\ f_j(s)\, f_j(t)$$

which can be written in the form $E(A,s,t) = \int_A E(\varepsilon,s,t)\,\mu(d\varepsilon)$ where $\mu=\Sigma_{j\geq 1}\delta_{\varepsilon(j)}$. Hence,

> *for* μ-*almost all* ε *in* $\mathbf{R}$ (i.e. outside a E-negligeable set) *there exists a projection* $E(\varepsilon)$ *such that the projection valued measure* $E(d\varepsilon)$ *desintegrates in the form* $E(\varepsilon)\mu(d\varepsilon)$. *These projections have an integral kernel* $E(\varepsilon,s,t)$ *and for* μ-*almost every* ε *and almost every* s *the function* $E(\varepsilon,s,\,.\,)$ (which is nothing but $f_j(s)f_j(.)$ if $\varepsilon=\varepsilon(j)$) *is an eigenfunction of* H.

In the general case, on can still desintegrate the projection valued measure $E(d\varepsilon)$ with respect to the measure $\mu(d\varepsilon)$ defined in (III.14), or on any other nonnegative measure having the same null sets as E. The fiber operators $E(\varepsilon)$ are now defined from a smaller space than $\mathbf{L}^2(\mathbf{R},dt)$ into a larger space than $\mathbf{L}^2(\mathbf{R},dt)$. These spaces can actually be taken to be weighted $\mathbf{L}^2$-spaces. The $E(\varepsilon)$'s are still integral operators but now, for each bounded Borel subset A of $\mathbf{R}$, the function

$$\int_A E(\varepsilon,s,t)\,\mu(d\varepsilon)$$

of s and t is simply a **weak kernel** for the bounded operator $E(A)$ on $\mathbf{L}^2(\mathbf{R},dt)$ (see the proof of Lemma II.18 for the definition of weak kernel). Nevertheless, the $E(\varepsilon,s,\,.\,)$'s are still solutions of the equation (E.V.E.) for the value ε. They are called **generalized eigenfunctions**. They are no longer in $\mathbf{L}^2(\mathbf{R},dt)$ but only in a weighted $\mathbf{L}^2$-space. The weight provides some information on the asymptotic behavior of the generalized eigenfunctions at plus and minus infinity. In the present situation it is possible to show that for each $\varepsilon>0$, the weight $(1+|t|)^{-(1+\varepsilon)}$ can be used and this implies that the absolute value of the generalized eigenfunctions are bounded above by multiples of $(1+|t|)^{(1+\varepsilon)/2}$. In particular, *except for a set of* $\varepsilon \in \mathbf{R}$ *of E-measure zero, there exists a generalized eigenfunction* (i.e. a solution of (E.V.E.) for ε) *which increases at most polynomially in absolute value as* $t \to \infty$.

III.4. Complements: the Feynman- Kac and Molcanov formulae.

We present in this last section some probabilistic formulae which will be very useful in what follows. They give exponential functions of the Schrödinger operator as path integrals.

The first one deals with the continuous case of $\mathbf{R}^d$. It is the famous **Feynman-Kac formula.** If f $\mathbf{L}^2(\mathbf{R}^d,dx)$ and if $H=H_0+q$, then we have:

$$[e^{-tH} f](x) = \mathbf{E}_{0,x}\{f(X_t)\exp[-\int_0^t q(X_s)ds]\} \tag{III.26}$$

where we used the following notations on the right hand side: $\mathbf{W}$ is the space $\mathbf{C}(\mathbf{R}_+,\mathbf{R}^d)$ of continuous functions on $\mathbf{R}_+$ with values in $\mathbf{R}^d$. For each $t \geq 0$, X_t denotes the coordinate function $\mathbf{W} \ni w \to X_t(w)=w(t) \in \mathbf{R}^d$. $\mathfrak{W}$ denotes the smallest σ-field of subsets of $\mathbf{W}$ for which all the X_t's are measurable. Finally, for each $s \geq 0$ and x in $\mathbf{R}^d$, $\mathbf{W}_{s,x}$ denotes the **Wiener measure** on $(\mathbf{W},\mathfrak{W})$ of the **Brownian motion starting from** x **at time** s. The corresponding expectation will be denoted by $\mathbf{E}_{s,x}$.

First we remark that the right hand side of (III.26) makes sense for spaces of functions much larger than $\mathbf{L}^2(\mathbf{R}^d,dx)$, and that this formula is often used to define Schrödinger semigroups (and hence the Schrödinger operators themselves as infinitesimal generators of these semigroups) on function spaces other than $\mathbf{L}^2(\mathbf{R}^d,dx)$.

See for example [Carmona(1979)] and [Simon(1982b)].

Second we remark that formula (III.26) cannot be true without some assumptions on the potential function q. We refer the interested reader to the paper [Simon(1982b)] for a complete discussion of these assumptions. Essentially, the positive part of q should be locally in the space $\mathbf{K}_d$ while the negative part should be in $\mathbf{K}_d$ itself.

Actually we will need the analog of formula (III.26) for the operator in a bounded open domain D. So let T_D be the first exit of D for the Brownian path, i.e.

$$T_D(w) = \inf\{ t>0 \,;\, X_t(w) \notin D \}.$$

One then shows that:

$$(\exp[-tH_0^{(D)}]f)(x) = \mathbf{E}_{0,x}\{f(X_t);\, t<T_D\} \tag{III.27}$$

for f in $L^2(D,dx)$, $t\geq 0$ and x in D. This formula is well known. By a simple perturbation argument one obtains easily:

$$(\exp[-t(H_0^{(D)}+q)]f)(x) = \mathbf{E}_{0,x}\{f(X_t)\exp[-\int_0^t q(X_s)ds];\, t<T_D\} \tag{III.28}$$

whenever q is bounded. In fact one can show that this formula remains true when q is in K_d. For more details see [Aizenman-Simon(1981)].

We now consider simultaneously the case $D=\mathbf{R}^d$ (in which case $T_D=\infty$ identically) and D a bounded open domain. In both cases one has:

$$(\exp[-t(H_0^{(D)}+q)]f)(x)=(2\pi t)^{-d/2}\int_{\mathbf{R}^d}\mathbf{E}_{0,x}\{\exp[-\int_0^t q(X_s)ds];\, t<T_D|X_t=y\}\, e^{-|x-y|^2/2t} f(y)dy$$

which shows that the operators have **integral kernels** given by:

$$\exp[-t(H_0^{(D)}+q)](x,y) = (2\pi t)^{-d/2}\, \mathbf{E}_{0,x,t,y}\{\exp[-\int_0^t q(X_s)ds];\, t<T_D\}\, e^{-|x-y|^2/2t}$$

where we used the notation $\mathbf{E}_{0,x,t,y}\{\,.\,\}$ for the conditional expectation $\mathbf{E}_{0,x}\{\,.\,|X_t=y\}$ (i.e. the expectation with respect to the probability measure $\mathbf{W}_{0,x,t,y}$ of the Brownian motion starting from x at time 0 and conditioned to be at y at time t). Under the usual assumptions, that is for q^+ in $K_{d,loc}$ and q^- being the sum of a bounded function and a function in K_d, **these kernels are bounded and jointly coninuous in** (x,y) **for each fixed** $t>0$ **and in** t **for** $t>0$.

Note that when D is bounded, the operators $\exp[-t(H_0^{(D)}+q)]$ are Hilbert-Schmidt on $L^2(D,dx)$ and consequently their spectra must be of the form $\{e^{-t\kappa(j)};j\geq 1\}$ for some sequence $\{\kappa(j);j\geq 1\}$ increasing to $+\infty$, which is nothing but the spectrum of the operator $H_0^{(D)}+q$. We will use these facts in Chapter V to construct the distribution of states for random Schrödinger operators.

We now concentrate on the case of the lattice $\mathbf{Z}^d$, and we try to write the **unitary group** $\{e^{itH};t\in\mathbf{R}\}$ by means of paths integrals (see Proposition III.2 below). Such a formula, if any, should be called the **Feynman formula**. Indeed we know that Feynman proposed such a formula for the continuous case of $\mathbf{R}^d$ but we also know that, in this case, it cannot be given by an integral with respect to a regular measure on path space. Fortunately the lattice case is much simpler and such a path space representation of the unitary group can be proven rigorously. This fact is not well known. We learned it in 1981 during a private discussion with S.A.Molcanov and this is the reason why we will call it the **Molcanov formula**. We will use it in Chapter V.

Let $\mathbf{W}^{(d)}$ be the space of right continuous functions with left limits from $\mathbf{R}$ to $\mathbf{Z}^d$, and for each

$t \geq 0$, let us still denote by X_t the coordinate function $W^{(d)} \ni w \dashrightarrow X_t(w)=w(t) \in \mathbf{Z}^d$. We will use the notation $\mathcal{W}_t^{(d)}$ for the smallest σ-field of subsets of $W^{(d)}$ for which all the X_s's for $0 \leq s \leq t$ are measurable, $\mathcal{W}^{(d)}$ corresponding to the case $t=\infty$. For each $s \geq 0$ and $\underline{n}$ in $\mathbf{Z}^d$ we denote by $\mathbf{W}_{s,\underline{n}}^{(d)}$ the probability measure on $(W^{(d)}, \mathcal{W}^{(d)})$ of the Brownian motion on $\mathbf{Z}^d$ starting from $\underline{n}$ at time s. The corresponding expectation will be denoted $\mathbf{E}_{s,\underline{n}}^{(d)}$. Remember that under $\mathbf{W}_{s,\underline{n}}^{(d)}$, $\{X_t;t \geq s\}$ is at $\underline{n}$ at time s, stays there for an exponential holding time, and then jumps at one of the 2d neighbors of $\underline{n}$ in $\mathbf{Z}^d$ with equal probabilities. It stays there an exponential holding time independent of everything else and again jumps to one of the 2d neighbors with equal probabilities, ...

For each $t \geq 0$ and w in $W^{(d)}$ we will denote by $N(t,w)$ the number of jumps of the path w before time t. In fact, $\{N(t);t \geq 0\}$ is a **Poisson process with parameter 1 independent of the sites of $\mathbf{Z}^d$ visited** by the path.

Proposition III.2:

Let q *be a real function on* $\mathbf{Z}^d$ *and let* H *be the self adjoint operator* $H=H_0+q$ *on* $\mathbf{H}=\ell^2(\mathbf{Z}^d)$ *where* H_0 *has been defined in* (III.5). *Then for each* $\mathbf{x}=(x(\underline{n}))_{\underline{n}\in\mathbf{Z}^d}$ *in* $\mathbf{H}$ *and* $\underline{n}$ *in* $\mathbf{Z}^d$ *we have:*

$$[e^{itH}\mathbf{x}](\underline{n}) = e^{(1-i)t}\,\mathbf{E}_{0,\underline{n}}^{(d)}\{x(X_t)\; i^{N(t)} \exp[i\textstyle\int_0^t q(X_s)ds]\} \qquad \text{(III.29)}$$

Proof:

The operator H_0 is self adjoint and bounded and the multiplication operator by the function q is self adjoint. Consequently (recall Chapter I) the operator H is also self adjoint. Setting $u(t,\underline{n})$ for the right hand side of (III.29) we need only to show that the function u satisfies

$$\partial u/\partial t = i\,Hu \qquad \text{(III.30)}$$

and $u(0,\underline{n})=x(\underline{n})$ for all $\underline{n}$ in $\mathbf{Z}^d$. The latter is obvious. Note also that $\mathbf{x}$, as a function of $\underline{n}$ is bounded. Now:

$$u(t+\Delta t,\underline{n}) = e^{(1-i)t}e^{(1-i)\Delta t}\,\mathbf{E}_{0,\underline{n}}^{(d)}\{x(X_{t+\Delta t})\; i^{N(t+\Delta t)} \exp[i\textstyle\int_0^{t+\Delta t} q(X_s)ds]\}$$

$$= e^{(1-i)\Delta t}\,\mathbf{E}_{0,\underline{n}}^{(d)}\{u(t,X_{\Delta t})\; i^{N(\Delta t)} \exp[i\textstyle\int_0^{\Delta t} q(X_s)ds]\} \qquad \text{(III.31)}$$

where we used the simple Markov property at time Δt. Moreover,

$$\mathbf{E}_{0,\underline{n}}^{(d)}\{u(t,X_{\Delta t})\; i^{N(\Delta t)} \exp[i\textstyle\int_0^{\Delta t} q(X_s)ds]\}$$

$$\begin{aligned}&= u(t,\underline{n})\, e^{i\Delta t q(\underline{n})}\, \mathbf{W}_{0,\underline{n}}^{(d)}\{N(\Delta t)=0\}\\ &\quad + i\,\mathbf{E}_{0,\underline{n}}^{(d)}\{u(t,X_{\Delta t}) \exp[i\textstyle\int_0^{\Delta t} q(X_s)ds]\,;\, N(\Delta t)=1\}\\ &\quad + \mathbf{E}_{0,\underline{n}}^{(d)}\{u(t,X_{\Delta t})\; i^{N(\Delta t)} \exp[i\textstyle\int_0^{\Delta t} q(X_s)ds]\,;\, N(\Delta t)>1\}\end{aligned}$$

u is bounded because x is so. Since the function q is real valued, the last expectation is at most of the order of the probability of more than two jumps in less than Δt, which is of the order of $(\Delta t)^2$. Hence, its product with $e^{(1-i)\Delta t}$ converges to 0 as $\Delta t \to 0$, even after dividing by Δt.

After multiplication by $e^{(1-i)\Delta t}$ the first term equals:

$$e^{(1-i)\Delta t}\, u(t,\underline{n})\, e^{i\Delta t q(n)}\, e^{-\Delta t} = u(t,\underline{n}) - i\Delta t u(t,\underline{n}) + i\Delta t\, q(\underline{n})u(t,\underline{n}) + o(\Delta t^2).$$

Finally, the second terms gives:

$$i\,\mathbf{E}_{0,\underline{n}}^{(d)}\{u(t,X_{\Delta t});\ N(\Delta t)=1\} + i\,\mathbf{E}_{0,\underline{n}}^{(d)}\{u(t,X_{\Delta t})\,(\exp[i\textstyle\int_0^{\Delta t} q(X_s)ds]-1);\ N(\Delta t)=1\}$$

$$= i\,[\Sigma_{|\underline{m}-\underline{n}|=1}\,(2d)^{-1}u(t,\underline{m})]\ \mathbf{W}_{0,\underline{n}}^{(d)}\{N(\Delta t)=1\} + o(\Delta t^2)$$

$$= i\ (2d)^{-1}\Sigma_{|\underline{m}-\underline{n}|=1}\ u(t,\underline{m}) + o(\Delta t^2).$$

Collecting all the terms we obtain:

$$[u(t+\Delta t,\underline{n})-u(t,\underline{n})]/\Delta t = -iu(t,\underline{n}) + iq(\underline{n})u(t,\underline{n}) + i\ (2d)^{-1}\Sigma_{|\underline{m}-\underline{n}|=1}\ u(t,\underline{m}) + o(\Delta t)$$

which is exactly what we want, namely (III.30), if we notice that the convergence is not only pointwise but also in the $\ell^2(\mathbf{Z}^d)$ sense. □

IV. RANDOM SCHRÖDINGER OPERATORS.

The purpose of this chapter is to present the Schrödinger operators

$$H(\omega) = -(1/2)\,\Delta + q(\,.\,,\omega) \tag{IV.1}$$

with a potential function depending on a parameter ω running through a probability space. We give sufficient conditions of essential self adjointness on $C_c^\infty(\mathbf{R}^d)$ and of measurability in the sense of Chapter II. We then illustrate the results on the examples the spectral properties of which will be investigated in details later on.

Most of the corresponding problems in the lattice case of $\mathbf{Z}^d$ are trivial and we will hardly consider them here.

IV.1. Essential self adjointness and measurability:

Proposition IV.1:

Let $\{q(x);\ x \in \mathbf{R}^d\}$ be a real valued measurable stochastic process on a complete probability space $(\Omega,\mathcal{B},\mathbf{P})$, let us assume for each ω in Ω that the function $x \dashrightarrow q(x,\omega)$ is locally square integrable and that the operator (IV.1) is essentially self adjoint on $C_c^\infty(\mathbf{R}^d)$. Then $\omega \dashrightarrow H(\omega)$ is measurable.

Proof:

Let us first assume that for each ω, the path or the function :

$$\mathbf{R}^d \ni x \dashrightarrow q(x,\omega) \in \mathbf{R}$$

is bounded. For each t in $\mathbf{R}$, the Trotter product formula (recall Proposition I.12) gives:

$$\langle e^{itH(\omega)}f, g\rangle = \lim_{n\to\infty} \langle [e^{i(t/n)H_0}\, e^{i(t/n)q(\,.\,,\omega)}]^n f, g\rangle$$

for all f and g in $H=L^2(\mathbf{R}^d,dx)$. Using the measurability of the process q one easily shows the measurability of the right hand side. This concludes the proof in the case of bounded paths. In the general case, for each integer $k\geq 1$, we define the potential process $q^{(k)} = \{q^{(k)}(x);\ x\in \mathbf{R}^d\}$ by;

$$q^{(k)}(x) = [\,q(x)\vee(-k)]\wedge k.$$

It has bounded paths and then, the function

$$\omega \dashrightarrow H^{(k)}(\omega) = H_0 + q^{(k)}(\,.\,,\omega)$$

is measurable. Since $D = C_c^{\infty}(\mathbf{R}^d)$ is a common core for (almost) all the $H^{(k)}(\omega)$ and since, for each f in D we have:

$$H(\omega)f = H_0 f + q(\,.\,,\omega)f$$

$$= \lim_{k \to \infty} H_0 f + q^{(k)}(\,.\,,\omega)f$$

in $\mathbf{H}$, we can conclude that $H^{(k)}(\omega)$ converges to $H(\omega)$ in the strong resolvent sense (recall (I.5)) and the measurability of the function $\omega \dashrightarrow H(\omega)$ follows Proposition II.3. □

The above proof applies to the lattice case and gives (recall that the free Hamiltonian H_0 is now defined by (III.5)):

if $\{q(n);\ n \in \mathbf{Z}^d\}$ *is any family of real valued random variables on a complete probability space* $(\Omega, \mathcal{B}, \mathbf{P})$ *then the function* $\omega \dashrightarrow H(\omega) = H_0 + q(\,.\,,\omega)$ *in the space of self adjoint operators on* $\mathbf{H} = \ell^2(\mathbf{Z}^d)$ *is measurable.*

The following proposition gives a sufficient condition for the assumptions of Proposition IV.1 to be satisfied. Recall the notations introduced before Proposition III.1.

Proposition IV.2:

Let $\{q(x);\ x \in \mathbf{R}^d\}$ *be a real valued measurable stochastic process on a complete probability space* $(\Omega, \mathcal{B}, \mathbf{P})$, *let us assume that*

$$\sup_{n \in \mathbf{Z}^d} \mathbf{E}\{(\,[\int_{C(n)} |q(x)|^p dx]^{1/p})^r\} < \infty \qquad \text{(IV.2)}$$

for some $p > p(d)$ *and some* $r > dp/2[p - p(d)]$ *where* $p(d) = 2$ *if* $d \leq 3$ *and* $p(d) = d/2$ *if* $d \geq 4$. *Then, the operator* $H(\omega) = H_0 + q(\,.\,,\omega)$ *is* $\mathbf{P}$-*almost surely essentially self adjoint on* $C_c^{\infty}(\mathbf{R}^d)$.

Proof:

We check that the assumptions of the Faris-Lavine criteria (recall Proposition III.1) are satisfied. For each x in $\mathbf{R}^d$ let $n(x)$ be the unique element of $\mathbf{Z}^d$ for which $x \in C_{n(x)}$ and for each ω in Ω we define the functions $x \dashrightarrow q_1(x,\omega)$ and $x \dashrightarrow q_2(x,\omega)$ by:

$$q_1(x,\omega) = q(x,\omega)\, \mathbf{1}_{\{q(.,\omega) \geq -|n(.)|^2\}}(x)$$

and

$$q_2(x,\omega) = q(x,\omega) - q_1(x,\omega).$$

The proof reduces to showing that $q_2(.,\omega)$ is in $L^p_{loc,unif}(\mathbf{R}^d,dx)$ for $p=p(d)$ and $\mathbf{P}$-almost all ω.

$$[\int_{C(\underline{n})} |q_2(x)|^p\,dx]^{1/p} \le [\int_{C(\underline{n})} |q(x)|^p\, \mathbf{1}_{\{|q(.,\omega)|\ge|\underline{n}|^2\}}(x)\,dx]^{1/p}$$

$$\le [\int_{C(\underline{n})} |q(x)|^{\mathbf{p}}\,dx]^{1/\mathbf{p}} [\int_{C(\underline{n})} \mathbf{1}_{\{|q(.,\omega)|\ge|\underline{n}|^2\}}(x)\,dx]^{1/p-1/\mathbf{p}}$$

$$\le [\int_{C(\underline{n})} |q(x)|^{\mathbf{p}}\,dx]^{1/\mathbf{p}} [\int_{C(\underline{n})} |q(x)|^{\mathbf{p}}\,dx]^{1/p-1/\mathbf{p}}\, |\underline{n}|^{-2\mathbf{p}(1/p-1/\mathbf{p})}$$

where we used $\mathbf{p} > p$ to obtain the second inequality. Consequently,

$$\mathbf{P}\{ [\int_{C(\underline{n})} |q_2(x)|^p\,dx]^{1/p} > 1\} \le \mathbf{P}\{ [\int_{C(\underline{n})} |q(x)|^{\mathbf{p}}\,dx]^{1/\mathbf{p}} > |\underline{n}|^{2\mathbf{p}(1/p-1/\mathbf{p})} \}$$

$$\le |\underline{n}|^{-2k\mathbf{p}(1/p-1/\mathbf{p})}\, \mathbf{E}\{([\int_{C(\underline{n})} |q(x)|^{\mathbf{p}}\,dx]^{1/\mathbf{p}})^k \}$$

Now, setting $k=rp/\mathbf{p}$ the exponent of $|n|$ is smaller than $-d$ and the right hand side is consequently the general term of a convergent series because of the assumption (IV.2). We conclude by using Borel Cantelli lemma. □

The following corollaries illustrate the usefulness of the above proposition. The proof of the first one of them is obvious and we omit it.

Corollary IV.3:
Let $\{q(x);\ x \in \mathbf{R}^d\}$ *be a real valued measurable stochastic process on a complete probability space* $(\Omega,\mathcal{B},\mathbf{P})$ *and let us assume that it is stationary and that* $\mathbf{E}\{|q(0)|^k\} < \infty$ *for some* $k > p(d)+d/2$, *or that* $q(x)=\Sigma_{\underline{n}\in\mathbf{Z}^d}\, q_{\underline{n}}\, f(x-\underline{n})$ *for some* f *in* $\ell^1(L^p(\mathbf{R}^d,dx))$ *and a stationary sequence* $\{q_{\underline{n}};\, \underline{n}\in\mathbf{Z}^d\}$ *such that* $\mathbf{E}\{|q_{\underline{0}}|^r\} < \infty$ *with* p *and* r *as in Proposition IV.2. Then, the operator* $H(\omega) = H_0+q(\,.\,,\omega)$ *is* $\mathbf{P}$-*almost surely essentially self adjoint on* $C_c^\infty(\mathbf{R}^d)$.

Corollary IV.4:
Let us assume that:

$$q(x,\omega) = \int_{\mathbf{R}^d} f(x-y)\,\mu(dy,\omega) \tag{IV.3}$$

for a measurable function $\omega \dashrightarrow \mu(\,.\,,\omega)$ *from* Ω *into* $M(\mathbf{R}^d)$ *and for some* f *in* $\ell^1(L^{\mathbf{p}}(\mathbf{R}^d,dx))$ *and let us assume furthermore that:*

$$\sup_{\underline{n}\in \mathbf{Z}^d} \mathbf{E}\{|\mu(\,.\,,\omega)|(C(\underline{n}))^r\} < \infty \tag{IV.4}$$

with $p=p(d)$, p *and* r *as in Proposition IV.2. Then, the operator* $H(\omega) = H_0+q(\,.\,,\omega)$ *is* $\mathbf{P}$-*almost surely essentially self adjoint on* $C_c^\infty(\mathbf{R}^d)$.

Proof:

We show that the stochastic process $\{q(x);\, x\in \mathbf{R}^d\}$ defined by (IV.3) satisfies the assumptions of Proposition IV.2. Jensen's inequality gives:

$$[\int_{C(\underline{n})} |q(x)|^p dx]^{1/p} \le \int_{\mathbf{R}^d} [\int_{C(\underline{n})} |f(x-y)|^p dx]^{1/p} |\mu(\,.\,,\omega)|(dy)$$

$$= \Sigma_{\underline{m}\in \mathbf{Z}^d} [\int_{C(\underline{n})-C(\underline{m})} |f(x)|^p dx]^{1/p} |\mu(\,.\,,\omega)|(C(\underline{m}))$$

and hence:

$$\mathbf{E}\{([\int_{C(\underline{n})} |q(x)|^p dx]^{1/p})^r\}^{1/r} \le \Sigma_{\underline{m}\in \mathbf{Z}^d} [\int_{C(\underline{n})-C(\underline{m})} |f(x)|^p dx]^{1/p} \mathbf{E}\{|\mu|(C(\underline{m}))^r\}^{1/r}$$

$$\le c\, \Sigma_{\underline{m}\in \mathbf{Z}^d} [\int_{C(\underline{n})-C(\underline{m})} |f(x)|^p dx]^{1/p}$$

$$\le c\, ||f||_{\ell^1(L^p(\mathbf{R}^d,dx))}$$

which proves (IV.2) since the above right hand side does not depend on $\underline{n}$. □

IV.2. Examples:

We now present the particular random potentials $\{q(x);x\in \mathbf{R}^d\}$ which we will use in the sequel. Except for the 5th example below, for which we will be able to investigate the asymptotics of the distribution of states, all of them are one-dimensional.

IV.2.1. Let q be an almost periodic function on $\mathbf{R}$ (recall Example 2' in Chapter III) and let Ω be the closure in $C_b(\mathbf{R})$ of the relatively compact set $\{U_t q;\, t\in \mathbf{R}\}$. The shift extends to Ω and Ω becomes a compact metrizable group. Let $\mathcal{B}$ be its Borel σ-field and let $\mathbf{P}$ be the normalized Haar measure. Then $(\Omega,\mathcal{B},\mathbf{P},\{U_t;t\in\mathbf{R}\})$ is an **ergodic dynamical system** and $\{q(t);t\in\mathbf{R}\}$ can be regarded as an ergodic stationary stochastic process on the probability space $(\Omega,\mathcal{B},\mathbf{P})$ with **bounded continuous paths** and such that $U_t q(o)=q(t)$.

IV.2.2. Let $\{X_t;t\in\mathbf{R}\}$ be a **stationary Gaussian process continuous in quadratic mean** and let us set:

$$q(t) = F(X_t) \tag{IV.5}$$

where F is a measurable real valued function on $\mathbf{R}$. Then $\{q(t);t\in\mathbf{R}\}$ is a measurable stationary real valued process . It is **ergodic** whenever the spectral measure of the process $\{X_t;t\in\mathbf{R}\}$ is **continuous** .

V.2.3. Let χ be a bounded measurable real valued function on $\mathbf{R}$ which vanishes outside the interval $[0,1]$ and let $\{\xi_n;n\in\mathbf{Z}\}$ be an independent identically distributed (**i.i.d.** for short) sequence of real valued random variables on a complete probability space $(\Omega,\mathcal{B},\mathbf{P})$ and let us set:

$$q(t,\omega) = \Sigma_{n\in\mathbf{Z}}\, \xi_n(\omega)\, \chi(t-n) \tag{IV.6}$$

for ω in Ω and t in $\mathbf{R}$. χ can be regarded as a potential well and the potential function $q(\,.\,,\omega)$ as a sequence of wells of the same type but with random depth $\xi_n(\omega)$.

IV.2.4. Let $\{X_t;t\in\mathbf{R}\}$ be the **stationary Brownian motion** process on a **connected compact Riemmannian manifold** M (i.e. the stationary Markov process whose infinitesimal generator is the Laplace Beltrami operator of the manifold), let F:M ---> [0,1] be a **Morse function** (we will assume that $\inf_{m\in M} F(m)=0$ and $\sup_{m\in M} F(m)=1$ for simplicity), and we define the random potential $\{q(t);t\in\mathbf{R}\}$ by formula (IV.5) as before.

The above examples give stationary ergodic potential functions, but they are more or less random. Indeed, the almost periodic functions (and especially the periodic ones) give **deterministic processes** while the last two examples give **strongly mixing** ones. The third example is interesting because its randomness can be varied by adjusting the covariance of the Gaussian process.

Let us also remark that except for some of the cases from Example IV.2.3, all of the potential functions presented so far can be regarded via formula (IV.5), as the images of simple Markov processes by a real valued function on the state space of the process. We will often say that these **random potential functions** are of the **Markovian type**.

Notice also that the first and the third examples above have natural multidimensional generalizations which are of some physical interest while the two other ones are purely academic. Their interest relies in our capability of proving mathematical results which have been predicted by the physicists.

Finally we dwell on a particular case of the class of examples treated in Corollary IV.4 above and which we will investigate in full detail in Chapter V, Section V.2, when we study the asymptotic behavior of the distribution of states.

IV.2,5. Let μ **be a infinitely divisible random measure with independent increments** on $\mathbf{R}^d$. This is a random variable μ on some complete probability space $(\Omega,\mathcal{B},\mathbf{P})$ with values in the space $M_+(\mathbf{R}^d)$ of nonnegative measures on $\mathbf{R}^d$ such that for each integer n there exist similar random variables $\mu_1, \ldots, \mu_n$ such that μ and $\mu_1 + \ldots + \mu_n$ have the same distribution on $M_+(\mathbf{R}^d)$ and such that for each finite collection $A_1, \ldots A_m$ of mutually exclusive Borel subsets of $\mathbf{R}^d$ the nonnegative random variables $\mu(A_1), \ldots \mu(A_m)$ are independent.

These random measures are characterized by their Laplace transforms which has to be given for each function f in the space $C_c(\mathbf{R}^d)$ of continuos functions on $\mathbf{R}^d$ with compact supports, by:

$$\mathbf{E}\{e^{-\langle f,\mu\rangle}\} = \exp[-\langle f,\boldsymbol{\alpha}\rangle - \int_{\mathbf{R}+}\int_{\mathbf{R}^d} [1-e^{-tf(x)}]\, \boldsymbol{\gamma}(dt,dx)] \qquad (IV.7)$$

for some nonnegative measures $\boldsymbol{\alpha}$ on $\mathbf{R}^d$ and $\boldsymbol{\gamma}$ on $\mathbf{R}_+ \times \mathbf{R}^d$, where we used the notation $\langle . , . \rangle$ for the duality between functions and measures, i.e.

$$\langle f, \mu \rangle = \int f(x)\, \mu(dx).$$

We will concentrate on the Poisson case which can be characterized by $\boldsymbol{\alpha} = 0$ and the fact that the measure $\boldsymbol{\gamma}$ is concentrated on $\{1\}\times\mathbf{R}^d$. In this case, μ is a point measure with probability one and formula (IV.7) rewrites:

$$\mathbf{E}\{e^{-\langle f,\mu\rangle}\} = \exp[-\int_{\mathbf{R}^d} [1-e^{-tf(x)}]\, m(dx)] \qquad (IV.8)$$

where m is a (nonrandom) nonnegative measure on $\mathbf{R}^d$ called the **intensity** of the random measure μ. Note that the measure m can be obtained via the formula:

$$m(A) = \mathbf{E}\{\mu(A)\}$$

where A is any Borel subset of $\mathbf{R}^d$. Note also that for each A in $\mathcal{B}_{\mathbf{R}^d}$, $\mu(A)$ is a Poisson random variable with parameter $\mu(A)$. Hence, condition (IV.4) rewrites:

$$\sup_{\underline{n}\in\mathbf{Z}^d} e^{-m(C(\underline{n}))} \sum_{0\le k<\infty} k^r (k!)^{-1} m(C(\underline{n}))^k < \infty \qquad (IV.9)$$

Finally we remark that the random measure μ is stationary (for the appropriate action of the translations of $\mathbf{R}^d$) if and only if the intensity measure m is translation invariant. Consequently, we will concentrate on the case where m is the usual Lebesgue measure and this will force the potential process $\{q(x);\, x\in\mathbf{R}^d\}$ defined by (IV.3) to be stationary. It will also be ergodic because there is enough independence in Poisson measures.

We close this section with the checking of the **almost sure essential self adjointness on** $C_c^\infty(\mathbf{R}^d)$ (**and** hence of the **measurability** according to Proposition IV.1) of the random Schrödinger operators constructed from the random potential functions we presented in the above examples.

The examples 1 and 4 are trivial since the sample paths of the potential function are bounded and hence, for each ω in Ω, the operator $H(\omega)$ is self adjoint on the domain of the free Hamiltonian H_0.

If the absolute value of the function F in the third example is bounded above by an exponential function, then it is easy to see that for each $k>0$ we have:

$$\mathbf{E}\{ |q(0)|^k \} < \infty$$

and Corollary IV.3 gives the desired essential self adjointness.

The existence of the r-th moment for the common distribution of the ξ_n gives the same estimate as above and one concludes in the same way.

Finally, if μ is a stationary Poisson random measure like in the fifth example, the expectation in formula (IV.4) is independent of $\underline{n}$. It is in fact the r-th moment of a Poisson random variable with parameter $|C(\underline{0})|=1$, and consequently, the supremum is finite for all r's. Hence, the assumptions of Corollary IV.4 are satisfied whenever f is in $\mathbf{L}^1(L^p(\mathbf{R}^d,dx))$ for some $p > p(d)$.

IV.3. Complements:

While Proposition IV.1 is a simple exercise, we borrowed Proposition IV.2 from [Kirsch-Martinelli(1983b)]. We simply corrected the arithmetic.

Let us now comment on the closed set Σ which is the spectrum of almost all the operators $H(\omega)$. According to Corollary II.9, this set is **increasing** with respect to the **topological support of the distribution of the potential function** $\{q(x);x\in \mathbf{R}^d\}$. This result is from [Kotani(1984a)]. In the same paper the **essential support of the absolute continuous spectrum is** shown to be **decreasing** in this toppological support of the distribution of the potential function **in the one dimensional case** and as we already pointed out it would be nice to have a proof of this result in the general multidimensional case.

The characterization of Σ as a set gave rise to various publications.

In the lattice case of $\mathbf{Z}^d$ it is shown in [Kunz-Souillard(1980)] that Σ is equal to the sum of the interval [0,2] and the support of the common distribution of the random variables $q(\underline{n})$ when the potential function is given by an i.i.d. sequence $\{q(\underline{n});\underline{n}\in\mathbf{Z}^d\}$.

In some continuous cases it can be shown that Σ is the union of the spectra of the deterministic Schrödinger operators with periodic potential chosen from the topological support of the distribution of the random potential function $\{q(x);x\in \mathbf{R}^d\}$ in path space. See [Kirsch-Martinelli(1982b)].

Finally we note that the case of **almost periodic potentials** has been investigatedf by many functional analysts without appealing to any probability theory. Roughly speaking one can say that the further it is from a periodic function, the more frequencies one needs to approximate it by trgonometric polynomials and the **more gaps** will appear in the spectrum Σ. In fact Σ has a strong tendency to become a **Cantor set**. For more details we refer to the fundamental works of [Dinaburg-Sinai(1976)], [Chulaevsky(1981)], [Moser(1981)], [Avron-Simon(1982a)] and [Johnson-Moser(1982)], to the review article [Simon(1982a)] and also to [Bellissard-Simon(1982)].

V. THE INTEGRATED DENSITY OF STATES.

Let $H=H_0+q$ be a Schrödinger operator for a continuous system in $\mathbf{R}^d$ and let us try to count the possible energy levels of the system. Two problems arise immediately. The first is of a mathematical nature. The spectrum of H is continuous most of the time, at least above some energy, and the problem does not make sense as stated. The second one relates to the model. Studying a system in the whole space $\mathbf{R}^d$ is in fact an approximation because the real systems are actually finite, even though large.

A possible way out seems to be the sudy of the system in a bounded domain D of $\mathbf{R}^d$. We will say that the system is put in a **box**. Indeed if the operator H is suitably defined on $L^2(D,dx)$ - note that this will involve the choice of boundary conditions - its spectrum is very likely to be discrete and it will be feasible to count the eingenvalues which we already agreed to call energy levels. To account for the fact that real systems are large, we can divide this number by the volume of the domain D and try to see if the result converges when D is very large to some sort of **average number of energy levels by unit volume**. Such a quantity has been experimentally measured for crystals whose mathematical models are built out of periodic potential functions. Many physicists called this quantity the **density of states**. Even though this terminology is very natural it is very misleading from the mathematical point of view. Indeed, this quantity happens to be the **distribution function** of a limiting measure in the mathematical model we are working with, and , would we know this measure is **absolutely continuous** , its density would be different. Fortunately most of the theoretical physicists and mathematical physicists avoid this possible confusion and use the term "**integrated density of states**" instead. Even though we believe that "**distribution function of states**" should be more appropriate, we will try to stick to the most comonly used terminology.

We prove its existence for stationary ergodic random potential functions. For the continuous case of $\mathbf{R}^d$ in Section V.1 and for the discrete case of the lattice $\mathbf{Z}^d$ in section V.3. We consider the one dimensional continuous case in Section V.4 and we comment on related existence results and asymptotic properties in Section V.5.

We did not choose our proofs to be the most general ones. Rather, we wanted them to be suited to our study of the asymptotic behavior of the integrated density of states and to the subsequent study of the one dimensional case.

We will show that its topological support is the almost sure spectrum of the operators H. Consequently, this distribution function has to vanish at the left hand of this spectrum and we may ask ourselves the following question: how fast does it vanish, or in other words, how many low energy states are there? Using nonrigorous heuristic arguments, the physicist I.M.Lifschitz, showed in some cases that the distribution function was vanishing exponentially in some powers of the energy and these powers are called since then, **Lifschitz exponents** . In Section V.2 below, we study two models for which the existence and the computation of these exponents can be carried out rigorously.

Proving that the integrated density of states is absolutely continuous (and hence, has an actual density) is a difficult problem which has no general answer yet. See nevertheless Sections V.4 and V.3 on the continuous one dimensional case and on the lattice case for which satisfactory answers are available. Note also that the existence and the analyticity of the density proved in [Constantinescu et al.(1983)] are given here a very simple proof based on what we called Molcanov formula.

V.1. Existence in the continuous case:

Let $(\Omega, \mathcal{B}, \mathbf{P})$ be a complete probability space and let $\{T_x\,;\, x\in\mathbf{R}^d\}$ be an ergodic flow for $\mathbf{P}$. Coming back to the notations of Chapter II, this means that we are considering the case $\Theta_0 = \mathbf{R}^d$. We will say a few words at the end of this paragraph on how to handle the case $\Theta_0 = \mathbf{Z}^d$. For the time being, let $q:\Omega \dashrightarrow \mathbf{R}$ be a measurable function and let us set:

$$q(x,\omega) = q(T_x\omega) \qquad\qquad x\in\mathbf{R}^d,\ \omega\in\Omega. \qquad (V.1)$$

We assume that the fuction $x \dashrightarrow q(x,\omega)$ is in $K_{d,loc}$ for each ω Ω. Recall Section III.2 for the definition of $K_{d,loc}$. The reader can think this function is bounded if he wants to postpone until later (or never) the treatment of local singularities in the potential. We already noticed in Section III.4 that for each bounded domain D, the operator $H^{(D)}(\omega)=H_0^{(D)}+q(.,\omega)$ has a discrete spectrum, say

$$-\infty < \varepsilon_0(D,\omega) < \varepsilon_1(D,\omega) \leq \varepsilon_2(D,\omega) \leq \dots$$

and we define the measure $n_{D,\omega}$ by:

$$n_{D,\omega} = |D|^{-1} \sum_{j\geq 0} \delta_{\varepsilon_j(D,\omega)} \qquad (V.2)$$

the distribution function of which is:

$$N_{D,\omega}(\varepsilon) = |D|^{-1}\, \#\{\text{eigenvalues of } H^{(D)}(\omega) \leq \varepsilon\} \qquad (V.3)$$

We now let the domain D tends to the whole of $\mathbf{R}^d$ in a certain sense and we try to show that the quantity $N_{D,\omega}(\varepsilon)$ converges to a limit $N_\omega(\varepsilon)$ (presumably independent of ω because of our ergodicity assumption), which should be interpreted as the **number of possible states of energy at most ε per unit of volume**. We already said that this quantity could be experimentaly measured and this explains its importance for the physics of the problem. From a mathematical point of view, we will see that it enters in the proofs of Chapter VI for example, and also, the study of its asymptotics is interesting for its own sake as the reader will probably see below.

We will restrict ourselves to rectangle domains D of the form $\prod(-a_j,a_j)$ with $a_j>0$, and we will say that $D \dashrightarrow \mathbf{R}^d$ when all the a_j converge to $+\infty$ simultaneously. We aim at showing that $\mathbf{P}$-almost surely in ω Ω, the family of measures $n_{D,\omega}$ converges vaguely to a limiting measure n which we will call the **distribution of states** and whose distribution function

$$\mathbf{N}(\varepsilon) = n((-\infty,\varepsilon])$$

will be called the **integrated density of states.**

<u>Theorem V.1:</u>

Let us assume furthermore that the function:

$$W\times\Omega \ni (w,\omega) \dashrightarrow \exp[\int_0^t q^-(X_s(w),\omega)ds]$$

is in $L^r(W\times\Omega, W_{0,0}\times P)$ *for some* $r>2$ *and all* $t>0$. *Then the measures* $n_{D,\omega}$ *converge vaguely to a nonnegative measure* n *(independent of* ω*) as* $D \dashrightarrow \mathbf{R}^d$ *for* $\mathbf{P}$*-almost all* ω *in* Ω. *Moreover the Laplace transform of* n *is given by:*

$$\mathcal{L}(n,t) = (2\pi t)^{-d/2}\, \mathbf{E}_{0,0,t,0} \times \mathbf{E}\{\exp[-\int_0^t q(X_s(w),\omega)ds\,]\} \tag{V.4}$$

<u>Proof:</u>

It is sufficient to show that for ω in some set Ω_1 of $\mathbf{P}$-probability one, the Laplace transform $\mathcal{L}(n_{D,\omega},t)$ converges pointwise to the right hand side of (V.4). In fact, for each $t>0$ we have:

$$\mathcal{L}(n_{D,\omega},t) = \int_{\mathbf{R}} e^{-\varepsilon t}\, n_{D,\omega}(d\varepsilon)$$

$$= |D|^{-1} \sum_{j\geq 0} e^{-\varepsilon_j(D,\omega)\,t}$$

$$= |D|^{-1}\, \mathrm{trace}[e^{-t H^{(D)}(\omega)}]$$

$$= |D|^{-1} (2\pi t)^{-d/2} \int_D \mathbf{E}_{0,x,t,x}\{\exp[-\int_0^t q(X_s,\omega)ds\,]; T_D>t\}\, dx \tag{V.5}$$

because the kernel of the operator $e^{-t H^{(D)}(\omega)}$ is continuous,

$$= (A_D) - (B_D)$$

with:

$$(A_D) = |D|^{-1} (2\pi t)^{-d/2} \int_D \mathbf{E}_{0,x,t,x}\{\exp[-\int_0^t q(X_s,\omega)ds\,]\}\, dx$$

and:

$$(B_D) = |D|^{-1} (2\pi t)^{-d/2} \int_D \mathbf{E}_{0,x,t,x}\{\exp[-\int_0^t q(X_s,\omega)ds\,]; T_D\leq t\}\, dx.$$

For each $t>0$ fixed we can apply the ergodic theorem to (A_D) and the limit is exactly the right hand side of (V.4). Consequently we need only to show that (B_D) converges to 0.

First we remark that:

$$(B_D) \le \left(|D|^{-1}(2\pi t)^{-d/2}\int_D \mathbf{E}_{0,x,t,x}\{\exp[-(r'/2)\int_0^t q(X_s,\boldsymbol{\omega})ds\,]\}\,dx\right)^{2/r'}$$
$$\left(|D|^{-1}(2\pi t)^{-d/2}\int_D \mathbf{W}_{0,x,t,x}\{\,T_D \le t\,\}\,dx\right)^{(r'-2)/r'}$$

for each r' such that $2<r'<r$. Second :

$$\lim_{D\to\mathbf{R}^d} |D|^{-1}\int_D \mathbf{W}_{0,x,t,x}\{\,T_D\le t\,\}\,dx = 0,$$

and finally that the supremum over D of the first factor is an integrable (and hence finite almost everywhere) random variable by a multidimensional maximal ergodic lemma. This proves that (B_D) tends to 0 for $\mathbf{P}$-almost all $\boldsymbol{\omega}$ in Ω and hence that for each $t>0$, there exists $\Omega_t\in\mathcal{B}$ such that $\mathbf{P}(\Omega_t)=1$ and

$$\lim_{D\to\mathbf{R}^d} \mathcal{L}(\mathbf{n}_{D,\boldsymbol{\omega}},t) = (2\pi t)^{-d/2}\,\mathbf{E}_{0,0,t,0}\times\mathbf{E}\{\exp[-\int_0^t q^-(X_s(w),\boldsymbol{\omega})ds\,]\}$$

for each $\boldsymbol{\omega}$ in Ω_t. Making the $\mathbf{P}$-almost sure independent of t requires some extra work. In fact, it is sufficient to show that there exists a set of full probability on which the family $\{\mathcal{L}(\mathbf{n}_{D,\boldsymbol{\omega}},.)\}_D$ of functions is equicontinuous. This follows from the existence of an increasing process $\{\boldsymbol{\alpha}(\boldsymbol{\varepsilon});\,\boldsymbol{\varepsilon}\in\mathbf{R}\}$ such that:

$$\sup_D \mathbf{n}_{D,\boldsymbol{\omega}},(\boldsymbol{\varepsilon}) \le \boldsymbol{\alpha}(\boldsymbol{\varepsilon},\boldsymbol{\omega})$$

and

$$\int_{\mathbf{R}} e^{-\boldsymbol{\varepsilon}t}\,\boldsymbol{\alpha}(\boldsymbol{\varepsilon},\boldsymbol{\omega})\,d\boldsymbol{\varepsilon} < +\infty$$

for all $t>0$, $\boldsymbol{\varepsilon}$ in $\mathbf{R}$ and $\boldsymbol{\omega}$ in Ω. Recall that a simple integration by parts gives

$$\int_{\mathbf{R}} e^{-\boldsymbol{\varepsilon}t}\,d\boldsymbol{\alpha}(\boldsymbol{\varepsilon}) = t\int_{\mathbf{R}} e^{-\boldsymbol{\varepsilon}t}\,\boldsymbol{\alpha}(\boldsymbol{\varepsilon})\,d\boldsymbol{\varepsilon}.$$

Since $r/2>1$, the multidimensional maximal ergodic lemma already mentionned gives:

$$\mathbf{E}\{|\sup_D |D|^{-1}(\pi t)^{-d/2}\int_D \mathbf{E}_{0,x}\{\exp[2\int_0^{t/2} q^-(X_s,\boldsymbol{\omega})ds\,]\}\,dx|^{r/2}\}$$
$$\le 4^{rd/2}(\pi t)^{-rd/4}[r/(r-2)]^d\,\mathbf{E}\times\mathbf{E}_{0,0}\{\exp[r\int_0^{t/2} q^-(X_s,\boldsymbol{\omega})ds\,]\}. \qquad (V.6)$$

Moreover (V.5) implies:

$$\mathcal{L}(\mathbf{n}_{D,\boldsymbol{\omega}},t) \le |D|^{-1}(2\pi t)^{-d/2}\int_D \mathbf{E}_{0,x,t,x}\{\exp[-\int_0^t q(X_s,\boldsymbol{\omega})ds\,]\}\,dx$$

$$\leq |D|^{-1}(\pi t)^{-d/2}\int_D \mathbf{E}_{0,x}\{\exp[2\int_0^{t/2} q(X_s,\omega)ds]\}\, dx \tag{V.7}$$

because we have the following estimate for the conditioned Wiener measures:

$$\mathbf{W}_{0,x,t,x}(A) \leq [t/(t-u)]^{d/2}\mathbf{W}_{0,x}(A)$$

for all $0 \leq u \leq t$ and $A \in \sigma\{X_s;\, 0 \leq s \leq u\}$, and this implies:

$$\mathbf{E}_{0,x,t,x}\{\exp[-\int_0^t q(X_s,\omega)ds]\} \leq 2^{d/2}\mathbf{E}_{0,x}\{\exp[-2\int_0^{t/2} q(X_s,\omega)ds]\}.$$

Hence, for each $t \geq 1$ and $c>0$ we have:

$$\mathbf{P}\{\sup_D \pounds(n_{D,\cdot},t) > c\} \leq h(t)\, c^{-r/2} \tag{V.8}$$

where $t \dashrightarrow h(t)$ is a strictly increasing continuous function on $[1,\infty)$ which is an upper bound for (V.6) and such that $h(\infty)=\infty$. Setting $c=h(t)^{6/r}$ in (V.8) we obtain:

$$\mathbf{P}\{\sup_D \pounds(n_{D,\cdot},t) > h(t)^{6/r}\} \leq h(t)^{-2}$$

which gives:

$$\sum_{n \geq h(1)} \mathbf{P}\{\sup_D \pounds(n_{D,\cdot},h^{-1}(n)) > n^{6/r}\} \leq \sum_{n \geq h(1)} n^{-2} < +\infty$$

and consequently we have with $\mathbf{P}$-probability 1:

$$\sup_D \pounds(n_{D,\cdot},h^{-1}(n)) \leq n^{6/r}$$

eventualy. This implies that:

$$\sup_D n_{D,\cdot}(n^{6/r}) \leq n^{6/r}\exp[-h^{-1}(n)\, n^{6/r}]$$

eventually if one remarks that $\alpha(-a) \leq a\, e^{-at}$ whenever $\int_{\mathbf{R}} e^{-\varepsilon t} d\alpha(\varepsilon) \leq a$. Consequently, we have:

$$\int_{-\infty} e^{-\varepsilon t}\alpha(\varepsilon,\omega)\, d\varepsilon < +\infty$$

for all $t>0$ if we set:

$$\alpha(\varepsilon,\omega) = \sup_D n_{D,\omega}(\varepsilon).$$

Next we note that for each $0<t\leq 1$ we have by (V.7):

$$\pounds(n_{D,\omega},t) \leq K(\omega)\, t^{-d/2} \tag{V.9}$$

provided we set:

$$K(\omega) = \pi^{-d/2} \sup_D |D|^{-1} \int_D \mathbf{E}_{0,0}\{ \exp[2\int_0^{1/2} q^-(X_s, T_x\omega) ds] \} dx$$

which is finite $\mathbf{P}$-almost surely, since integrable, by the multidimensional maximal ergodic lemma. Note also that:

$$\sup_{0<t\leq 1} t^{d/2} \int_{\mathbf{R}} e^{-\varepsilon t} d\alpha(\varepsilon) \leq K(\omega)$$

implies:

$$\sup_{\varepsilon\geq 1} \alpha(\varepsilon) \leq K(\omega)\, \varepsilon,$$

from which we conclude that:

$$\int^{+\infty} e^{-\varepsilon t} \alpha(\varepsilon,\omega)\, d\varepsilon < +\infty$$

for all $t>0$ because of (V.9). This completes the proof. □

The above proof is borrowed from [Nakao(1977)]. It has in common with most of the functional analytic proofs (see for example [Avron-Simon(1983)] or Subin's work on partial differential operators with almost periodic coefficients [Subin(1982)]) to rely heavily on the regularity properties of the semigroup generated by the operator in question and this is a **major drawback** if one plans to study **operators which are unbounded from below**. This existence proof should be viewed as the result of the efforts of L.A.Pastur in U.S.S.R. and of M.Fukushima and his collaborators in Japan. See [Pastur(1972),(1973)], [Fukushima(1974)], [Fukushima et al.(1975)] and [Nakao(1977)].

The first existence results dealt with the one-dimensional case $d=1$. They did not use the properties of the semigroup. They relied on the Sturm oscillation theory to compare the number of eigenvalues less than e say, to the same quantity for a different choice of boundary conditions, and show that these numbers were additive up to controlable errors. The Birkhoff additive ergodic theorem could then be applied to conclude the proof. See [Benderskii-Pastur(1969),(1970)] and Section V.4 below. This proof has been adapted to the case of the multidimensional lattice $\mathbf{Z}^d$ in [Fukushima(1981)] and to the multidimensional continuous case in [Kirsch-Martinelli(1982b)] where a Dirichlet-Neumann bracketing argument is used instead of the Sturm oscillation theory and a superadditive ergodic theorem instead of the Birkhoff ergodic theorem.

In the above proof we made two arbitrary choices which do not have any physical justification. The first concerns the choice of **rectangles for domains** D. We could have decided on other geometric shapes and other ways to converge to $\mathbf{R}^d$. The existence of the limit and the actual value of **this limit should not depend on these choices**. We will not elaborate on that but the reader familiar with the mathematics of statistical mechanics will easily convince himself that this problem can easily be fixed. A more subtle difficulty come from the choice of the boundary conditions for the operators in the bounded domains D. We picked Dirichlet boundary condition, but the choice of **Neumann boundary condition** is at least as natural. Our proof can be adapted to handle this case by using **reflected Brownian motion** instead of the killed one, but the estimates are more involved. Also, we would have to check that the limit are the same. These two types of boundary conditions are in some sense extremal, but there are many other ones (we can think for example of various mixed boundary conditions, or periodic ones, or...) and the existence of the limit and their independence from all these choices are subtle problems which we will not tackle here.

Remark:

The above proof applies to the case of a random potential which is merely $\mathbf{Z}^d$-stationary instead of $\mathbf{R}^d$ stationary. The only modifications consist in picking domains D which are union of elementary cubes C_n and replacing the integrals by sums over the cubes which constitute D.

We now try to relate the measure $\mathbf{n}$ which we just constructed to the spectral measures of the operators $H(\omega)$ and we will show that the topological support of $\mathbf{n}$ coincides with the almost sure spectrum of the $H(\omega)$'s.

Proposition V.2:

For each strictly positive continuous function f *on* $\mathbf{R}^d$ *such that* $\int f(x)^2 dx = 1$, *and for each bounded Borel set* A *in* $\mathbf{R}^d$, *the operator* $M_f E(A,\omega) M_f$ is trace class $\mathbf{P}$-*almost surely in* ω. *Furthermore, if the potential is uniformly bounded below we also have that:*

$$\mathbf{n}(A) = \mathbf{E}\{\mathrm{trace}[\, M_f E(A,\omega) M_f \,]\} \tag{V.10}$$

In any case, the topological support of the distribution of states equals the almost sure spectrum, i.e. $\mathrm{supp}\, \mathbf{n} = \Sigma$.

Proof:

Let us first assume that the potential is uniformly bounded below in the sense that $q(x,\omega) \geq c$ for some finite constant c and all x in $\mathbf{R}^d$ and ω in Ω. Note that according to Section III.4, the operators $e^{-tH(\omega)}$ are bounded and have a jointly continuous kernel given by the Feynman-Kac formula. Let $\{f_k; k \geq 1\}$ be any orthonormal basis in $\mathbf{H}$ and c' and t be positive constants such that $\mathbf{1}_A(\varepsilon) \leq c' e^{-t\varepsilon}$ for all ε. Then we have:

$$\mathbf{E}\{\textstyle\sum_k \langle [\, M_f E(A,\omega) M_f \,] f_k, f_k \rangle\} \leq c'\, \mathbf{E}\{\textstyle\sum_k \langle e^{-tH(\omega)} (ff_k), (ff_k) \rangle\}$$

$$= c'\, \mathbf{E}\{\mathrm{trace}[\, M_f\, e^{-tH(\omega)} M_f \,]\}$$

$$= c'\, \mathbf{E}\{\textstyle\int f(x)^2\; e^{-tH(\omega)}(x,x)\, dx\}$$

$$= c'\, (2\pi t)^{-d/2}\, \mathbf{E}\{\textstyle\int f(x)^2\; \mathbf{E}_{0,0,t,0}\{\exp[-\int_0^t q(x+X_s,\omega) ds]\, dx\}\}$$

$$= c'\, (2\pi t)^{-d/2}\, \mathbf{E} \times \mathbf{E}_{0,0,t,0}\{\exp[-\textstyle\int_0^t q(X_s,\omega) ds]\,\}$$

which is finite by the main assumption of Theorem V.1. This proves not only that $M_f E(A,\omega)M_f$ is trace class for $\mathbf{P}$-almost all ω, but also that the right hand side of (V.10) defines a nonnegative Radon measure on the real line. If we compute its Laplace transform we obtain:

$$\mathcal{L}(\,\mathbf{E}\{\mathrm{trace}[\,M_f E(\,.\,,\omega)M_f\,]\}\,,t) = \mathbf{E}\{\mathrm{trace}[\,M_f\, e^{-tH(\omega)}M_f\,]\}$$

by the spectral theorem, and this is nothing but $\mathcal{L}(n,t)$ if one follows the above computations and if one recalls (V.4). We thus proved the first half of the present proposition when the potential is uniformly bounded from below. Note also that the above computations actually show that we have:

$$\int_{\mathbf{R}} F(\varepsilon)\, n(d\varepsilon) = \mathbf{E}\{\mathrm{trace}[\,M_f F(H(\omega))M_f\,]\} \qquad (V.12)$$

for each continuous function F with compact support. To study the general case we set $q_{(n)}(x,\omega)=q(x,\omega)v(-n)$. For each integer n, the random self adjoint operator $H_{(n)}(\omega)=H_0+q_{(n)}(x,\omega)$ (whose resolution of ther identity will be denoted $E_{(n)}(\,.\,,\omega)$) is stationary and ergodic so that its spectrum is almost surely equal to a nonrandom closed subset $\Sigma_{(n)}$ of $\mathbf{R}$. The main assumption of Theorem V.1 is obviously satisfied and consequently the distribution of states, say $n_{(n)}$, exists. Finally we remark that the first half of the present proposition applies to $H_{(n)}(\omega)$. Formula (V.4) shows that the Laplace transform of $n_{(n)}$ converges pointwise to the Laplace transform of n and from this we conclude that $n_{(n)}$ converges vaguely to n. Hence, if F is a continuous function with compact support as above we must have:

$$\int_{\mathbf{R}} F(\varepsilon)\, n(d\varepsilon) = \lim_{n\to\infty} \int_{\mathbf{R}} F(\varepsilon)\, n_{(n)}(d\varepsilon)$$

$$= \lim_{n\to\infty} \mathbf{E}\{\textstyle\sum_k \langle [\,M_f F(\,H_{(n)}(\omega)\,)M_f\,]f_k,f_k\rangle\}$$

by applying (V.12) to the operator $H_{(n)}(\omega)$, and:

$$\geq \mathbf{E}\{\textstyle\sum_k \langle [\,M_f F(\,H(\omega)\,)M_f\,]f_k,f_k\rangle\} \qquad (V.13)$$

by Fatou's lemma and the strong resolvent convergence of $H_{(n)}(\omega)$ to $H(\omega)$. This shows that the operator $M_f F(\,H(\omega)\,)M_f$ is $\mathbf{P}$-almost surely trace class and that:

$$\mathbf{E}\{\mathrm{trace}[\,M_f F(H(\omega))M_f\,]\} \leq \int_{\mathbf{R}} F(\varepsilon)\, n(d\varepsilon). \qquad (V.14)$$

If A is a bounded Borel subset of $\mathbf{R}$ we can bound its characteristic function $\mathbf{1}_A$ from above by such a function F and (V.14) shows that the operator $M_f E(A,\omega)M_f$ is almost surely trace class. We now come back to the general case of nonnecessarily bounded below potential to prove the last claim concerning the topological support of the distribution of states. If U is any open set contained in Σ^c we have:

$$n(U) \le \liminf_{n \to \infty} n_{(n)}(U) = 0$$

because U is necessarily contained in $\Sigma_{(n)}{}^c$ for large enough n. This proves that the topological support of n is contained in Σ. To prove the other inclusion we let e_0 be any point in Σ and pick any nonnegative continuous function F with compact support such that $F(e_0)>0$. Since $\Sigma=\Sigma(H(\omega))$ $\mathbf{P}$-almost surely, we must have $F(H(\omega))\neq 0$ $\mathbf{P}$-almost surely. Hence,

$$\int_{\mathbf{R}} F(\varepsilon)\, n(d\varepsilon) = 0$$

would imply that $M_f F(H(\omega)) M_f = 0$ $\mathbf{P}$-almost surely by (V.14) and thus $F(H(\omega))= 0$ $\mathbf{P}$-almost surely which is a contradiction. The proof is now complete.□

The above ideas are borrowed from [Simon(1982a)] where the same results are proved for uniformly bounded below almost periodic potentials.

Remark V.3:

If for each ω and each bounded Borel set A the spectral projection $E(A,\omega)$ has a **jointly continuous kernel** $E(A,\omega,x,y)$ then it is easy to show that:

$$n(d\varepsilon) = \mathbf{E}\{ E(d\varepsilon,\omega,0,0) \} \tag{V.15}$$

by using the above ideas and a simple approximation argument. The reader may also want to take a look at Sections C6 and C7 of [Simon(1982b)] for results in the same spirit.

V.2. Asymptotic behavior:

We showed that under very general conditions, the topological support of the distribution of states was equal to the almost sure spectrum Σ. Consequently, the distribution of states vanishes at the left hand edge of Σ which can very well be $-\infty$ as we will see in some examples, and according to the heuristic arguments of M.I.Lifschitz, it should decay with a specific exponential behavior. The aim of this section is to show that this is perfectly sound.

Proposition V.4:

Let $\{q(x);\ x\in \mathbf{R}^d\}$ *be a stationary ergodic Gaussian process whose covariance function, say* $\gamma(x-y)$, *is continuous. Then, the distribution of states satisfies:*

$$\lim_{\varepsilon \to -\infty} \varepsilon^{-2} \operatorname{Log} n(\varepsilon) = -1/[2\gamma(0)] \tag{V.16}$$

Proof.

We can assume without any loss of generality that the process is centered. Using Jensen's inequality we obtain:

$$\mathbf{E}\times\mathbf{E}_{0,0}\{\exp[r\int_0^t q(X_s,\omega)ds]\} \le t^{-1}\int_0^t \mathbf{E}\times\mathbf{E}_{0,0}\{\exp[rt\,q(X_s,\omega)]\}ds$$

$$= t^{-1}\int_0^t \mathbf{E}_{0,0}\{\exp[t^2r^2\gamma(0)/2]\}ds$$

$$= \exp[t^2r^2\gamma(0)/2]$$

$$< +\infty$$

which shows that we can use Theorem V.1. In the present situation it is easy to see that $\Sigma = \mathbf{R}$ and hence that the topological support of n is the whole real line. Moreover:

$$\mathbf{E}\times\mathbf{E}_{0,0,t,0}\{\exp[-\int_0^t q(X_s,\omega)ds]\} = \mathbf{E}_{0,0,t,0}\{\exp[(1/2)\int_0^t\int_0^t \gamma(X_s-X_u)duds]\}$$

$$\le \exp[\gamma(0)t^2/2]$$

which, together with (V.4) implies:

$$\limsup_{t\to\infty} t^{-2}\operatorname{Log}\mathcal{L}(n,t) \le r(0)/2 \qquad \text{(V.17)}$$

On the other hand we have:

$$\liminf_{t\to\infty} t^{-2}\operatorname{Log}\mathcal{L}(n,t)$$

$$= \liminf_{t\to\infty} t^{-2}\operatorname{Log}\mathbf{E}\times\mathbf{E}_{0,0,t,0}\{\exp[-\int_0^t q(X_s,\omega)ds]\}$$

$$\ge \liminf_{t\to\infty} t^{-2}\operatorname{Log}\mathbf{E}\times\mathbf{E}_{0,0,t,0}\{\exp[-\int_0^t q(X_s,\omega)ds]\ ;\ \sup_{0\le s\le t}|X_s|<a\}$$

$$\ge \liminf_{t\to\infty} t^{-2}\operatorname{Log}\left(\exp([\inf_{|x|\le 2a}\gamma(x)]t^2/2)\,\mathbf{W}_{0,0,t,0}\{\sup_{0\le s\le t}|X_s|<a\}\right)$$

$$\ge \inf_{|x|\le 2a}\gamma(x)/2 + \lim_{t\to\infty} t^{-2}\operatorname{Log}\mathbf{W}_{0,0,1,0}\{\sup_{0\le s\le 1}|X_s|<a\,t^{-1/2}\}$$

$$= \gamma(0)/2$$

since a is arbitrary, γ is continuous and the above probability is of the order $t^{d/2}e^{-c(a)t}$ for each fixed a. Recalling (V.17), this proves:

$$\lim_{t\to\infty} t^{-2}\operatorname{Log}\mathcal{L}(n,t) = \gamma(0)/2$$

from which we get (V.16) by a simple Tauberian argument.□

We now come back to Example IV.2.5 for which the random potential $\{q(x);\ x\in \mathbf{R}^d\}$ is given by:

$$q(x) = \int_{\mathbf{R}^d} \mathfrak{f}(x-y)\ \mu(dy) \tag{V.18}$$

for a **Poisson random measure** μ on $\mathbf{R}^d$ the **intensity of which is the Lebesgue measure**, and for a **nonnegative continuous function** f on $\mathbf{R}^d$ which is **not identically zero and such that** $\mathfrak{f}(x)=o(|x|^{-(d+\varepsilon)})$ when $|x|\text{--->}\infty$ for some $\varepsilon>0$. Such a function f is trivially in $\mathcal{L}^1(L^p(\mathbf{R}^d,dx))$ for all $p>1$ and consequently for each $\omega\in\Omega$, the operator $H(\omega) = H_0 + q(.,\omega)$ is essentially self adjoint on $C_c^\infty(\mathbf{R}^d)$, the function $\omega \text{--->} H(\omega)$ is measurable and the condition (II.7) is satisfied. The nonnegativity of the potential function gives $\Sigma\subset[0,\infty)$ and it is not difficult to show that the decay of $\mathfrak{f}$ implies that there is actually equality. The assumptions of Theorem V.1 are satisfied because q is nonnegative, and the topological support of the distribution of states is also equal to $[0,\infty)$ according to Proposition V.2.

We try to find an equivalent of $N(\varepsilon)$ as $\varepsilon\searrow 0$. To do so, we first find the asymptotic of the Laplace transform of $\mathfrak{n}$ as t tends to infinity and we use a Tauberian argument to obtain the desired result for $N(\varepsilon)$. In the present situation we have for each w in **W**:

$$\mathbf{E}\{\exp[-\int_0^t q(X_s(w),\omega)ds]\} = E\{e^{-\langle f,\,\mu\rangle}\}$$

if we set $f(y) = \int_0^t \mathfrak{f}(X_s(w)-y)ds$, and consequently,

$$\mathbf{E}\{\exp[-\int_0^t q(X_s(w),\omega)ds]\} = \exp[-\int_{\mathbf{R}^d}[1-e^{-\int_0 \mathfrak{f}(X_s-x)ds}]\,dx$$

according to formula (IV.9). If we recall (V.4) we obtain:

$$\mathcal{L}(\mathfrak{n},t) = (2\pi t)^{-d/2}\,\mathbf{E}_{0,0,t,0}\{\exp[-\int_{\mathbf{R}^d}[1-e^{-\int_0 \mathfrak{f}(X_s-x)ds}]\,dx\} \tag{V.19}$$

Let us remark that, for each positive ε, if we let the function $\mathfrak{f}$ be equal to $+\infty$ on the ball of radius ε centered at the origin and to 0 outside this ball, then for each fixed Brownian path, the function $x\text{--->}1 - \exp[-\int_0^t \mathfrak{f}(X_s-x)ds]$ equals the characteristic function of the famous "**Wiener saussage**" of width ε. Motivated by the problem of the Lifschitz exponents, Donsker and Varadhan studied its asymptotic in their celebrated article [Donsker-Varadhan(1975)]. Among other things they proved that if we assume furthermore that the function f is actually $o(|x|^{-(d+2)})$ then:

$$\lim_{t\text{-->}\infty} t^{-d/(d+2)}\operatorname{Log}\mathbf{E}_{0,0}\{\exp[-\int_{\mathbf{R}^d}[1-e^{-\int_0 \mathfrak{f}(X_s-x)ds}]\,dx\} = -(d+2)(2\mathfrak{g}_d/d)^{d/(d+2)}/2 \tag{V.20}$$

where $\mathfrak{g}_d$ is the smallest eigenvalue of $H_0^{(D)}$ when D is the unit ball of $\mathbf{R}^d$. It is easy to see that the result remains unchanged if we replace the expectation $\mathbf{E}_{0,0}$ by the expectation $\mathbf{E}_{0,0,t,0}$ and that consequently, putting together (V.19) and (V.20) on obtains:

$$\lim_{t\to\infty} t^{-d/(d+2)} \mathrm{Log}\, \mathcal{L}(n,t) = -(d+2)(2\mathbf{g}_d/d)^{d/(d+2)}/2,$$

and using a Tauberian argument one gets:

Proposition V.5:
If μ is a Poisson random measure on $\mathbf{R}^d$ the intensity of which is Lebesgue measure, if $\mathbf{f}$ is a continuous nonnegative function on $\mathbf{R}^d$ and if $\mathbf{f}(x)=o(|x|^{-(d+2)})$ when $|x|\to\infty$ then, the distribution function of states $\mathbf{N}$ satisfies:

$$\lim_{\varepsilon\to 0} \varepsilon^{d/2} \mathrm{Log}\mathbf{N}(\varepsilon) = -\mathbf{g}_d{}^{d/2} \qquad \text{(V.21)}$$

Formula (V.21) above tells us that, at least in the sense of the exponential equivalents, we have

$$\mathbf{N}(\varepsilon) \sim e^{-c\varepsilon^{-\mathbf{d}}} \qquad \text{(V.22)}$$

for some positive constants c and **d**. **d** is often called the **Lifschitz exponent** of the model. The existence of a formula similar to (V.22) is usually difficult to prove(see nevertheless next section on the lattice case). The reader may want to look at [Kirsch-Martinelli(1983a)] for an attempt to use large deviation techniques in the study of weakened forms of (V.22).

V.3. The lattice case:

Let $\{q(\underline{n});\ \underline{n}\in\mathbf{Z}^d\}$ be a real stochastic process of the form $q(\underline{n},\omega) = q(T_{\underline{n}}\omega)$ where $(\Omega,\mathcal{B},\mathbf{P},\{T_{\underline{n}};\ \underline{n}\in\mathbf{Z}^d\})$ is an ergodic dynamical system and where $q:\Omega\to\mathbf{R}$ is measurable. For each ω in Ω we consider the self adjoint operator $H(\omega) = H_0+q(.,\omega)$ on the Hilbert space $\mathbf{H}=\ell^2(\mathbf{Z}^d)$ where H_0 is the discrete Laplacian defined in (III.5) and where $q(.,\omega)$ is the operator of multiplication by the real function $\underline{n}\to q(\underline{n},\omega)$. Under these conditions it is not difficult to see that the distribution of states exists without further assumption. See for example [Fukushima(1981)]. Moreover, since the "delta functions" are in the Hilbert space $\mathbf{H}$, each bounded operator has a kernel, and this kernel is necessarily continuous. Consequently, a trivial adaptation of Remark V.3 to the lattice case gives:

$$\mathbf{n}(.) = \mathbf{E}\{E(.,0,0,\omega)\} = \mathbf{E}\{E(.,\delta_0,\delta_0,\omega)\} \qquad \text{(V.23)}$$

This formula remains true if one substitutes any $\underline{n}$ to 0 because of the stationarity of the potential, and this immediately implies that the topological support of the distribution of states is equal to the almost sure spectrum. Finally, if $\hat{\mathbf{n}}$ denotes the Fourier transform of $\mathbf{n}$, we have:

$$\hat{\mathbf{n}}(t) = \mathbf{E}\{\int_{\mathbf{R}} e^{it\epsilon} \langle E(d\epsilon,\omega)\delta_0,\delta_0\rangle\}$$

$$= \mathbf{E}\{\langle e^{itH(\omega)}\delta_0,\delta_0\rangle\}$$

which we can compute using Molcanov formula:

$$= e^{(1-i)t}\, \mathbf{E}\times\mathbf{E}_{0,0}{}^{(d)}\, \{\delta_0(X_t)\, i^{N(t)} \exp[i\int_0^t q(X_s,\omega)ds]\}$$

$$= e^{(1-i)t}\, \mathbf{E}_{0,0}{}^{(d)}\, \{\delta_0(X_t)\, i^{N(t)}\, \mathbf{E}\,\{\exp[i\int_0^t q(X_s,\omega)ds]\}\} \qquad (V.24)$$

Proposition V.6:

Let $\{q(\underline{n});\ \underline{n}\in\mathbf{Z}^d\}$ *be i.i.d., and let us assume that the Fourier transform* $\chi(t)$ *of the common distribution decays exponentially in the sense that:*

$$|\chi(t)| \le c' e^{-c|t|} \qquad (V.25)$$

for some positive constants c' *and* c *such that* $c' < c$, *and all real numbers* t. *Then the distribution function of states has an analytic extension to a strip* $\{z\in\mathbf{C};\ |\mathrm{Im}z| < a\}$ *for some* $a>0$.

Let us remark that the assumption means that the distribution of $q(0)$ has a density which posseses an analytic extension to a similar strip and that proving the desired result amounts to showing that the Fourier transform of $\mathbf{n}$ decays exponentially as in (V.25). See for example the Section IX of [Reed-Simon(1975)] where the relation between the constants c and a is given.

Proof:

If the Brownian path w is fixed we have:

$$\int_0^t q(X_s(w),\omega)ds = \sum_{y\in\mathbf{Z}^d} q(y,\omega)\, l(t,y,w) \qquad (V.26)$$

where $l(t,y,w)$ is the total time spent at the site y by the Brownian path w before time t:

$$l(t,y,w) = \int_0^t \mathbf{1}_{\{y\}}(X_s(w))\, ds.$$

The right hand side of (V.26) is a sum of independent random variables on the probability space $(\Omega,\mathcal{B},\mathbf{P})$, whence:

$$\mathbf{E}\left\{\exp[i\int_0^t q(X_s,\omega)ds]\right\} = \prod_{y\in\mathbf{Z}^d} \chi(\,l(t,y)\,)$$

and consequently:

$$|\hat{n}(t)| \le e^t\, \mathbf{E}_{0,0}^{(d)}\left\{\prod_{y\in\mathbf{Z}^d} |\chi(\,l(t,y)\,)|\right\}$$

$$\le e^t\, \mathbf{E}_{0,0}^{(d)}\left\{\prod_{y\in\mathbf{Z}^d,\, l(t,y)>0} c'\, e^{-c\, l(t,y)}\right\}$$

$$= e^t\, \mathbf{E}_{0,0}^{(d)}\left\{c'^{\,\#(t)}\right\}$$

where $\#(t)$ deenotes the number of different sites visited by the Brownian path before time t. Obviously we have $\#(t)\le N(t)$, and since $\mathbf{E}_{0,0}^{(d)}\{c'^{\,N(t)}\} = e^{(c'-1)t}$ we must also have:

$$\le e^{(c'-c)t}. \; \square$$

Remark V.7:

1. Let us assume that each $q(\underline{n})$ is Gaussian with density $(2\pi s^2)^{-1/2}\exp[-x^2/2s^2]$. Then $\chi(t)=\exp[-s^2t^2/2]$ and for each $a>0$, the inequality (V.25) holds with $c=a$ and $c'=\exp[a^2/2s^2]$. This shows that for **large disorder** (i.e. for large s) we can find a so that $c>c'$ which gives the existence and the analyticity of the density of states. This is a result of [Constantinescu et al.(1984)] for which we announced a simple proof. In a "Note added in proof" to their paper, the authors mention this short proof due to Molcanov and a similar one based on a Duhamel expansion of the unitary group which is due to Simon. For **small disorder** (i.e. for small s) the answer is not known. In fact, there is a controversy in the physicist community regarding this problem and it would be nice to find a mathematical settlement of this case.

2. We beleive that the indepence of the potential at each site should not be needed for the existence and analycity of the density of states in the Gaussian case. If $\{q(\underline{n});\, \underline{n}\in\mathbf{Z}^d\}$ is a mean zero stationary Gaussian process, simple properties of its spectral measure should force the density of states to exist and to be analytic. Note that the ergodicity of the process is implied by the continuity of this spectral measure.

3. The estimations above become exact calculations in the case of the **Cauchy distribution**, the density of which is $a/\pi(x^2+a^2)$ for some $a>0$. In this case we have $\chi(t)=e^{-a|t|}$. Then, formula (V.24) gives:

$$\hat{n}(t) = e^{(1-i)t}\, \mathbf{E}_{0,0}^{(d)}\left\{\delta_0(X_t)\, i^{N(t)}\right\} e^{-a|t|}$$

$$= \langle e^{itH_0}\delta_0, \delta_0\rangle\, e^{-a|t|}$$

$$= J_0^{(d)}(t)\, e^{-a|t|}$$

where the first factor $J_0^{(d)}(t)$ is the Fourier transform of the integrated density of states of the free discrete Laplacian $H_0^{(d)}$; it is a known Bessel function. This proves that in the Cauchy case, n is the **convolution of the Cauchy distribution and the integrated density of states of the free Hamiltonian.** This calculation was originally made by Lloyd. See also [Simon(1983a)].

4. When $\{q(n);\ n\in\mathbf{Z}\}$ **is i.i.d. and** $q(0)$ **has a bounded density**, a functional analytic argument shows that the integrated density of states is absolutely continuous, namely that **the distribution of states has a density**. See [Wegner(1981)].

5. In general, the integrated density of states is **continuous**. See for example [Delyon-Souillard (1983)]. Also, it is possible to show that **it is Log-Hölder continuous**. See [Craig-Simon(1983b)] and Remark V.13 below for the continuous one dimensional case. It can even be C^∞ as shown in [Simon et al.(1984)] and [Simon-Taylor(1985)]. Nevertheless, an argument due to B. Halperin shows that one can construct i.i.d. potentials on Z taking only a finite number of values and for which the integrated density of states is not Hölder continuous of any pre-assigned order. I learnt this fact from a private discussion with Barry Simon.

We now come back to the study of the **Lifschitz exponents** to show that the situation is much simpler in the lattice case. The bounded domain D to be used are now of the form:

$$D = \prod_{1\le j\le d} \{-a_j, -a_j+1, \dots, 0, \dots, a_j-1, a_j\}$$

for some integers $a_1, \dots, a_d$ and $\partial D = \{n\in D;\ n'\in\mathbf{Z}^d\setminus D,\ |n-n'|=1\}$ and the operator $H^{(D)}(\omega)$ is the only self adjoint operator on $\ell^2(D)$ which we can identify with $\mathbf{C}^D$, defined by its domain $\mathcal{D}(H^{(D)}(\omega))$ of the functions x which vanish on the boundary ∂D of D by the formula:

$$[H^{(D)}(\omega)\, x](n) = -(2d)^{-1} \sum_{|m-n|=1,\ m\in D} x(m) + [q(n)+1]\, x(n)$$

We first remark that the integrated density of states is the distribution function of a **probability measure** in the lattice case. Next, we notice that the eigenvalue problem:

$$\begin{cases} M_q u = e\, u & \text{on } \tilde{D} = D\setminus\partial D \\ u(a) = 0 & \text{on } \partial D \end{cases}$$

has (whatever the function q is) $\{q(a);\ a\in\tilde{D}\}$ as eigenvalues and $\{\delta_a;\ a\in\tilde{D}\}$ as corresponding eigenfunctions. Hence, for each u in $\ell^2(D)$ we have:

$$\langle M_q u,u\rangle \le \langle H^{(D)}(\omega)u,u\rangle \le \langle M_{q+2}u,u\rangle$$

where $\langle .,.\rangle$ denotes the $\ell^2(D)$ -scalar product. The min-max principle gives:

$$\#\{\text{eigenvalues of } M_{q+2} \le \varepsilon\} \le \#\{\text{eigenvalues of } H^{(D)}(\omega) \le \varepsilon\} \le \#\{\text{eigenvalues of } M_q \le \varepsilon\}$$

which rewrites:

$$\sum_{a\in \tilde{D}} \mathbf{1}_{(-\infty,\varepsilon]}(q(a)+2) \le \#\{\text{eigenvalues of } H^{(D)}(\omega) \le \varepsilon\} \le \sum_{a\in \tilde{D}} \mathbf{1}_{(-\infty,\varepsilon]}(q(a)) \qquad (V.27)$$

Now, for each fixed ε, the ergodic theorem gives:

$$\lim_{D\to \mathbf{Z}^d} (\#D)^{-1} \sum_{a\in \tilde{D}} \mathbf{1}_{(-\infty,\varepsilon]}(q(a)) = E\{ \mathbf{1}_{(-\infty,\varepsilon]}(q(0)) \} \qquad (V.28)$$

and the conjunction of (V.27), (V.28) and the definition of the distribution of states implies:

Proposition V.8:
Let $\{q(n); n\in\mathbf{Z}\}$ *be a real stationary process and let us set:*

$$F(\varepsilon) = \mathbf{P}\{ q(0) \le \varepsilon\} \qquad \varepsilon\in\mathbf{R},$$

for the distribution function of the potential at one site. Then the distribution function of states satisfies for each $\varepsilon\in\mathbf{R}$:

$$F(\varepsilon-2) \le \mathcal{N}(\varepsilon) \le F(\varepsilon) \qquad (V.29)$$

Formula (V.29) above has the following immediate consequences:

Corollary V.9:
For all positive constants a *and* C *we have:*

i) $\lim_{\varepsilon\to-\infty} |\varepsilon|^{a} \mathcal{N}(\varepsilon) = C \iff \lim_{\varepsilon\to-\infty} |\varepsilon|^{a} F(\varepsilon) = C.$

ii) $\lim_{\varepsilon\to-\infty} |\varepsilon|^{-a} \mathcal{N}(\varepsilon) = C \iff \lim_{\varepsilon\to-\infty} |\varepsilon|^{-a} F(\varepsilon) = C.$

iii) $\lim_{\varepsilon\to\infty} |\varepsilon|^{a} [1-\mathcal{N}(\varepsilon)] = C \iff \lim_{\varepsilon\to\infty} |\varepsilon|^{a} [1-F(\varepsilon)] = C.$

iv) $\lim_{\varepsilon\to\infty} |\varepsilon|^{-a} \mathrm{Log}[1-\mathcal{N}(\varepsilon)] = C \iff \lim_{\varepsilon\to\infty} |\varepsilon|^{a} \mathrm{Log}[1-F(\varepsilon)] = C.$

This last result is taken from [Fukushima(1981)]. It should be looked at as the outcome of the works of L.A.Pastur, M.Fukushima and some of his collaborators which are quoted in the bibliography. See also [Romiero-Wreszinski(1979)].

An important case which is not covered by the above corollary is the situation where the distribution of the potential at each site is concentrated on a bounded interval. This problem is certainly more subtle and the interested reader will find a good account in [Simon(1984)].

V.4. The one dimensional case:

Let us see rapidly how one can go about proving the existence of the distribution of states in the one dimensional case. Let $N_{a,b,\alpha,\beta}(E)$ be the number of eigenvalues of the operator $H_{a,b,\alpha,\beta}$ which are in $(-\infty, E]$. We use the notations of Section III.3 and we skip the dependence in ω. Note that a real number E is an eigenvalue of $H_{a,b,\alpha,\beta}$ if and only if :

$$\theta(b,a,\alpha,E) = \beta + k\pi$$

for some $k \in \mathbb{Z}$. Using various monotonicity properties of θ with respect to some of its arguments (the so-called Sturm oscillation theory) it is easy to check that:

i) $$N_{a,b,0,0}(E) \leq N_{a,b,\alpha,\beta}(E) \leq N_{a,b,\pi/2,\pi/2}(E)$$

for most of the angles α and β. This shows in which sense the **Dirichlet and Neumann boundary conditions** (corresponding to $\alpha=\beta=0$ and $\alpha=\beta=\pi/2$ respectively) **are** regarded as **extreme** . Moreover:

ii) $$0 \leq N_{a,b,\pi/2,\pi/2}(E) - N_{a,b,0,0}(E) \leq 2$$

and if $a< b< c$ we also have:

iii) $$N_{a,c,\pi/2,\pi/2}(E) \leq N_{a,b,\pi/2,\pi/2}(E) + N_{b,c,\pi/2,\pi/2}(E)$$

and

$$N_{a,b,0,0}(E) + N_{b,c,0,0}(E) \leq N_{a,c,0,0}(E)$$

For those boundary conditions α and β satisfying i), one uses iii) to bound each quantity of the form $(2hk)^{-1}N_{-hk,+hk,\alpha,\beta}(E)$ where h is a fixed positive integer by quantities of the form

$$(2hk)^{-1} \sum_{-k\leq j\leq k-1} N_{jh,(j+1)h,\pi/2,\pi/2}(E)$$

from above, and from below by quantities of the form:

$$(2hk)^{-1} \sum_{-k\le j\le k-1} N_{jh,(j+1)h,0,0}(\varepsilon)$$

which are additive and to which one can apply the individual Birkhoff ergodic theorem. We obtain:

$$h^{-1}\mathbf{E}\{ N_{0,h,0,0}(\varepsilon)\} \le \liminf_{k\to\infty} (2hk)^{-1}N_{-hk,+hk,\alpha,\beta}(\varepsilon)$$

$$\le \limsup_{k\to\infty} (2hk)^{-1}N_{-hk,+hk,\alpha,\beta}(\varepsilon) \le h^{-1}\mathbf{E}\{ N_{0,h,\pi/2,\pi/2}(\varepsilon)\}.$$

One then concludes the proof by using the fact that the right most term and the left most one differ at most by $2h^{-1}$ and that h can be chosen arbitrarily large.

Note also that a modern way to conclude the proof is to apply a subadditive ergodic theorem to $N_{.,.,0,0}(\varepsilon)$ and a superadditive ergodic theorem to $N_{.,.,\pi/2,\pi/2}(\varepsilon)$ and to check that the limits are the same by using ii).

In both cases one needs an extra argument to show that the set of full $\mathbf{P}$-measure guaranteed by the ergodic theorem can actually be chosen independent of ε. Also the integrability of $N_{a,b,\alpha,\beta}(\varepsilon)$ in ω has to be justified. This is obvious once we remark that:

$$N_{a,b,\alpha,\beta}(\varepsilon) \le \pi^{-1}|\, \theta(b,a,\alpha,\varepsilon)\, | \qquad \text{(V.30)}$$

and that:

$$\theta(t) = \alpha + \int_0^t [\cos^2\theta(s)+ (\varepsilon-q(s))\sin^2\theta(s)]\, ds. \qquad \text{(V.31)}$$

The above proof has been borrowed from [Benderskii-Pastur(1969),(1970)]. It has been extended to the lattice $\mathbf{Z}^d$ in [Fukushima(1981)] where there is no need of the main assumption of Theorem V.1. See also the earlier work [Slivnyak(1966)].

Integrating (V.30), using the stationarity and setting $l=b-a$, we obtain:

$$(b-a)^{-1}\mathbf{E}\{\theta(b,a,\alpha,\varepsilon)\} = l^{-1}\mathbf{E}\{\theta(l,0,\alpha,\varepsilon)\}$$

$$= \alpha l^{-1} + l^{-1}\int_0^l \mathbf{E}\{\cos^2\theta(s,0,\alpha,\varepsilon) + [e-q(s)]\sin^2\theta(s,0,\alpha,\varepsilon)\}\, ds$$

$$= \alpha l^{-1} + \iint_{\mathbf{R}\times\mathbf{S}} F_\varepsilon(q,\theta)\, \mu_{\alpha,\varepsilon}^{(l)}(dq,d\theta)$$

where $\mathbf{S}$ is the torus $\mathbf{R}/\pi\mathbf{Z}$, which we will identify with $[0,\pi)$, where $F_\varepsilon(q,\theta)=\cos^2\theta+[\varepsilon-q]\sin^2\theta$ and where:

$$\mu_{\alpha,\varepsilon}^{(l)}(dq,d\theta) = l^{-1}\int_0^l \mathbf{P}\{\, q(s)\in dq,\ \theta(s,0,\alpha,\varepsilon)\in d\theta\,\}\, ds.$$

Hence, if the joint distribution of the potential and the phase has a limit in the Cesaro sense, say

$\pi_{\boldsymbol{e}}$, for $\boldsymbol{e} \in \mathbf{R}$ and $\boldsymbol{a} \in \mathbf{S}$ fixed, the distribution function of states has the representation:

$$\mathbf{N}(\boldsymbol{e}) = \pi^{-1} \iint_{\mathbf{R}\times\mathbf{S}} F_{\boldsymbol{e}}(q,\theta)\ \pi_{\boldsymbol{e}}(dq,d\theta). \tag{V.32}$$

We didn't write the dependence in $\boldsymbol{a}$ of the limiting measure because we expect this limit is independent of the choice of the boundary condition. This formula will be very useful in the Markovian cases for we will be able of computing or at least estimating this limiting measure.

The existence of limit points $\pi_{\boldsymbol{e},\boldsymbol{a}}$ for which formula (V.32) holds is quite general. Indeed the torus $\mathbf{S}$ is compact and this implies:

$$\mu_{\boldsymbol{a},e}^{(l)}(\ [-n,+n]\times\mathbf{S}\ ) = l^{-1}\int_0^l \mathbf{P}\{\ |q(s)| \leq n\ \}\, ds$$

$$= \mathbf{P}\{\ |q(0)| \leq n\ \}$$

which shows that the family $\{\ \mu_{\boldsymbol{a},e}^{(l)};\ l>0\}$ is tight and hence, the existence of the limit points for which (V.32) holds. When the family is merely tight without being convergent, the limit points depend on the choices of the boundary condition $\boldsymbol{a}$ and formula (V.32) is less useful.

The so-called density of states has been investigated for operators slightly different from ours. Either because the second order term is not the second derivative any more, or because the perturbation is a random measure rather than a potential function. These operators have to be defined by means of their quadratic forms, but the results are essentially the same. The interested reader is refered to [Kotani(1976)], [Fukushima-Nakao(1977)], [Okura(1979)] and [Thompson(1983)] for example.

Sturm oscillation theory also tells us that $N_{a,b,\boldsymbol{a},\boldsymbol{b}}(e)$ is equal to (or differs at most by one from) π times the number of zeros of the function $t \longrightarrow \theta(t,a,\boldsymbol{a},e)$ on the interval $[a,b]$. Hence, and we already made this remark in the derivation of formula (V.31), the distribution function of states is equal to the almost sure limit of $t^{-1}\theta(t,a,\boldsymbol{a},e)$ up to a factor π. This limit does not depend on a or $\boldsymbol{a}$. It is called the **rotation number** for obvious reasons. It was extensively studied in [Johnson-Moser(1982)] to **label the gaps** of Schrödinger operators with almost periodic potentials. This work has been extended to the lattice case in [Herman(1983)] and [Delyon-Souillard(1983)].

Let us now come back to the case $\mathbf{G}_0=\mathbf{R}$ and let us consider the **m-functions** $m_{+(-)}(z)$ whose definition was given in Section III.3, along the trajectories of our dynamical system $\{T_t;\ t\in\mathbf{G}_0\}$. We set:

$$z_{+(-)}(t,z,\boldsymbol{w}) = m_{+(-)}(z,T_t\boldsymbol{w}).$$

For each fixed z and $\boldsymbol{w}$, the function $t \longrightarrow z_{+(-)}(t,z,\boldsymbol{w})$ satisfies a Ricatti equation:

$$z_{+(-)}'(t,z,\boldsymbol{w}) = +(-)[\ q(t,\boldsymbol{w}) - z - z_{+(-)}(t,z,\boldsymbol{w})^2\ ] \tag{V.33}$$

Note also that:

$$z_{+(-)}(t,z,\omega) = +(-)\, f_{+(-)}'(t,z,\omega)/\, f_{+(-)}(t,z,\omega)$$

since:

$$f_{+(-)}(t,z,\omega) = \exp[\,+(-)\int_0^t z_{+(-)}(s,z,\omega)\, ds\,]. \qquad (V.34)$$

We now add the assumption " Σ **is bounded below** " which is satisfied if, for example:

* **the potential function is uniformly bounded below**, i.e. if $q(t,\omega) \geq c$ for some constant c, for all t in $\mathbf{R}$ and $\mathbf{P}$-almost surely in ω.

or if, and this is less obvious:

* **the potential is locally uniformly bounded below** in the sense that there is a constant c such that $\int_t^{t+1} q^-(s,\omega)\, ds \leq c$, $\mathbf{P}$-almost surely in ω.

Note that a simpler set of assumptions under which the considerations below hold can be found in the appendix of [Kirsch et al.(1985)] which appeared after the completion of this set of notes.

Anyway, under either of these conditions one can check by a deterministic argument for ω fixed that for each compact subset K of Π_+ there is a constant $c(K)>0$ such that we have with probability 1:

$$\begin{aligned} c(K)^{-1} &\leq |m_{+(-)}(z,\omega)| \leq c(K) \\ c(K)^{-1} &\leq |\mathrm{Im}\, m_{+(-)}(z,\omega)| \leq c(K) \\ c(K)^{-1} &\leq |g_z(t,t,\omega)| \leq c(K) \end{aligned} \qquad (V.35)$$

for all z in K and all t in $\mathbf{R}$. See [Kotani(1983),(1984)]. Let us define the function $\mathbf{w}$ by:

$$\mathbf{w}(z) = (1/2)\, \mathbf{E}\{m_+(z) + m_-(z)\} = -(1/2)\, \mathbf{E}\{g_z(0,0)^{-1}\} \qquad (V.36)$$

Since $m_+(z)$ and $m_-(z)$ are analytic on Π_+ with positive imaginary parts, one concludes that $\mathbf{w}$ has the same property: it is also a Herglotz function. The importance of this function lies in the fact that **the real part and the imaginary part of its boundary values on the real axis determine the Ljapunov exponent** (which we introduce in the next chapter) and **the integrated density of states**. See Proposition VI.1. We prepare for the proof of this fundamental result by some technical lemmas.

<u>**Lemma V.10:**</u>

$\mathbf{w}$ *<u>and its derivative are Herglotz functions. Moreover we have:</u>*

$$\mathbf{w}(z) = \mathbf{E}\{m_+(z)\} = \mathbf{E}\{m_-(z)\} \tag{V.37}$$

$$\mathbf{w}'(z) = \mathbf{E}\{g_z(0,0)\} \tag{V.38}$$

Proof:

One easily checks that $m_+(z,T_t\omega) - m_-(z,T_t\omega) = [\mathrm{Log}\, g_z(t,t,\omega)]'$ which implies:

$$\int_a^b [\, m_+(z,T_t\omega) - m_-(z,T_t\omega)\,]\, dt = \mathrm{Log}\, g_z(0,0,T_b\omega) - \mathrm{Log}\, g_z(0,0,T_a\omega)$$

for all $a<b$ if one remarks that $g_z(t,t,\omega) = g_z(0,0,T_t\omega)$. One obtains (V.37) by taking the expectation of both sides. In order to show (V.38) we first notice that the function h defined by:

$$h(\omega) = (1/2)\, g_z(0,0,\omega)\Big[\int_0^{\infty} f_+(t,z,\omega)^2\, dt - \int_{-\infty}^{0} f_-(t,z,\omega)^2\, dt\Big]$$

satisfies:

$$-\partial[2\, g_z(t,t,\omega)]^{-1}/\partial z = g_z(t,t,\omega) + \partial h(T_t\omega)/\partial t$$

which gives, for all $a<b$:

$$(b-a)\,\mathbf{E}\{\, -\partial[2\, g_z(0,0)]^{-1}/\partial z\,\} = (b-a)\,\mathbf{E}\{\, g_z(0,0)\}$$

which concludes the proof since $\mathbf{w}'(z) = \partial\mathbf{E}\{-[2\, g_z(0,0)]^{-1}\}/\partial z$. Note that we interchanged freely integration signs and derivations without special care. In fact everything can be justified by using (V.35).□

Lemma V.11:

We have:

$$\mathbf{E}\{[\mathrm{Im}\, m_{+(-)}(z)]^{-1}\} = -2\mathrm{Re}\, \mathbf{w}(z)/\mathrm{Im}\, z$$

and, in particular $\mathrm{Re}\, \mathbf{w}(z) < 0$. *Also:*

$$-(\mathrm{Im}\, z)^{-1}\, \mathrm{Re}\, \mathbf{w}(z) - \mathrm{Im}\, \mathbf{w}'(z) > 0$$

$$-2\, \mathrm{Re}\, \mathbf{w}(z)\, \mathrm{Im}\, \mathbf{w}(z) \geq \mathrm{Im}\, z.$$

Proof:

Equating imaginary parts of both sides of the Ricatti equation (V.33) we obtain:

$$[\mathrm{Im}\, z_+(t,z,\omega)]' = -2\,[\mathrm{Re}\, z_+(t,z,\omega)]\,[\mathrm{Im}\, z_+(t,z,\omega)] - \mathrm{Im}\, z$$

and this gives:

$$(d/dt)\,\mathrm{Log}[\mathrm{Im}\, z_+(t,z,\omega)] + \mathrm{Im}\, z / [\mathrm{Im}\, z_+(t,z,\omega)] = -2\,\mathrm{Re}\; z_+(t,z,\omega)$$

that is:

$$(d/dt)\,\mathrm{Log}[\mathrm{Im}\, m_+(z,T_t\omega)] + \mathrm{Im}\, z / [\mathrm{Im}\, m_+(z,T_t\omega)] = -2\,\mathrm{Re}\; m_+(z,T_t\omega).$$

Taking expectations of both sides gives the desired result for $m_+(z)$. The cooresponding result for $m_-(z)$ is shown in the same way. The two inequalities can be proven directly. Indeed,

$$-(\mathrm{Im}\, z)^{-1}\,\mathrm{Re}\, w(z) - \mathrm{Im}\, w'(z)$$

$$= (1/4)\mathbf{E}\{(\mathrm{Im}\, m_+(z))^{-1} + (\mathrm{Im}\, m_-(z))^{-1}\} + \mathbf{E}\{\mathrm{Im}\,[m_+(z)+m_-(z)]^{-1}\}$$

$$=4\,\mathbf{E}\{[(\mathrm{Im}\, m_+(z))^{-1} + (\mathrm{Im}\, m_-(z))^{-1}]([\mathrm{Re}\, m_+(z) + \mathrm{Re}\, m_-(z)]^2 + [\mathrm{Im}\, m_+(z) - \mathrm{Im}\, m_-(z)]^2) / |m_+(z) + m_-(z)|^2\} \qquad (V.39)$$

$$> 0.$$

Moreover,

$$\mathrm{Im}\, w(z) / \mathrm{Im}\, z = (\mathrm{Im}\, z)^{-1} E\{\mathrm{Im}\, m_+(z)\}$$

$$= \int_0^\infty \mathbf{E}\{\exp[2\int_0^t \mathrm{Re}\, m_+(z,T_u\omega)\, du]\}\, dt$$

(if we integrate the relation $\int_0^\infty |f_+(t,z,\omega)|^2 dt = \mathrm{Im}\, m_+(z,\omega) / \mathrm{Im}\, z$)

$$\geq \int_0^\infty \exp[2\int_0^t \mathbf{E}\{\mathrm{Re}\, m_+(z)\}\, du]\}\, dt$$

by Jensen's inequality,

$$= \int_0^\infty e^{2t\,\mathrm{Re}\, w(z)}\, dt$$

$$= -(2\,\mathrm{Re}\, w(z))^{-1}. \square$$

Coming back to formula (V.38) we see that:

$$w'(z) = \mathbf{E}\{g_z(0,0)\}$$

$$= \mathbf{E}\{\int_\Sigma (e-z)^{-1}\, E(de,0,0)\}$$

$$= \int_{\mathbf{R}} (\varepsilon - z)^{-1}\, n(d\varepsilon), \qquad (V.40)$$

and a simple integration by parts gives:

$$= \int_{\mathbf{R}} N(\varepsilon)(\varepsilon - z)^{-2}\, d\varepsilon$$

because $\int (|\varepsilon| + 1)^{-1} n(d\varepsilon) < \infty$. Integrating both sides of the relation we just obtained, we get:

$$w(z) = \alpha + \int_{\mathbf{R}} (1+\varepsilon z) N(\varepsilon)/(\varepsilon - z)(1+\varepsilon^2)\; d\varepsilon \qquad (V.41)$$

for some $\alpha \in \mathbf{C}$. If z is chosen real and outside the spectrum Σ, then $w(z) = -(1/2)\mathbf{E}\{g_z(0,0)^{-1}\}$ has to be real and this forces $\operatorname{Im} \alpha = 0$. Hence, if $\varepsilon \in \mathbf{R}$, we have for each $a>0$:

$$\operatorname{Im} w(\varepsilon + ia) = a \int_{\mathbf{R}} N(\varepsilon')/[(\varepsilon - \varepsilon')^2 + a^2]\, d\varepsilon'$$

$$= \int_{\mathbf{R}} N(\varepsilon + au)/[1+u^2]\, du.$$

We already noticed that the distribution of states was continuous because any given real number was not an eigenvalue with probability 1, in the one dimensional case at least. Hence, passing to the limit a-->0 in the above equality we obtain the first part of the following proposition.

Proposition V.12:
The Herglotz function $w(z)$ *satisfies:*

$$\pi N(\varepsilon) = w(\varepsilon + i0) \qquad (V.42)$$

for all ε *in* $\mathbf{R}$, *and if we define the function* $\dot{\alpha}(\varepsilon)$ *by* $\dot{\alpha}(\varepsilon) = -\operatorname{Re} w(\varepsilon + i0)$, *we have:*

$$\dot{\alpha}(\varepsilon) = -\alpha + \int_{\mathbf{R}} \operatorname{Log}|(\varepsilon - x)/(x - i)|\, dN(x) \qquad (V.43)$$

for some real constant α.

Proof:
Formula (V.41) gives:

$$-\operatorname{Re} w(z) = -\alpha + \int_{\mathbf{R}} (1+\varepsilon z) N(\varepsilon)/(z-\varepsilon)(1+\varepsilon^2)\; d\varepsilon$$

$$= -\alpha + \int_{\mathbf{R}} \operatorname{Log}|(x-z)/(x-i)|\, dN(x)$$

by integration by parts. The first inequality of Lemma V.11 gives:

$$\int_{\mathbf{R}} \mathrm{Log}|(x-z)/(x-i)|\, d\mathbf{N}(x) < a \tag{V.44}$$

but if $z=-i\varepsilon$ with ε in $(0,1)$, we have:

$$-\mathrm{Log}|(x-i\varepsilon)/(x-i)| = \mathrm{Log}(1+x^2)-(1/2)\mathrm{Log}[(1+x^2)\varepsilon^2+x^2+x^4]$$

which is positive and increases to $-\mathrm{Log}|x/(x-i)|$ as ε decreases to 0. This shows that $\mathrm{Log}|x|$ is locally integrable with respect to $d\mathbf{N}(x)$. Choosing now $z=\varepsilon+ia$ with $a>0$ we get:

$$-\mathrm{Re}\, \mathbf{w}(\varepsilon+ia) = -\alpha + \int_{\mathbf{R}} \mathrm{Log}|(x-\varepsilon-ia)/(x-\varepsilon)|\, d\mathbf{N}(x) + \int_{\mathbf{R}} \mathrm{Log}|(x-\varepsilon)/(x-i)|\, d\mathbf{N}(x).$$

Using the fact that (V.44) implies that

$$-\int_{|\varepsilon-x|\leq 1} \mathrm{Log}|\varepsilon-x|\, d\mathbf{N}(\varepsilon)$$

is bounded from above by a constant independent of ε whenever the latter is restricted to a compact set, and this justifies passing to the limit $a\to 0$ in (V.45) which gives (V.43) and concludes the proof. □

Remark V.13:

1. We will see in Chapter VI that $\dot{\alpha}(\varepsilon)$ is in fact the upper Ljapunov exponent of the problem. Keeping that in mind, formula (V.43) is usually called **Thouless formula** even though it appeared in a paper by Herbert and Jones one year prior to Thouless' work. It can be very useful as shown in [Avron-Simon(1983)].

2. The above calculations show also that for each compact set K, there is a positive constant $c'(K)$ such that:

$$|\mathbf{N}(\varepsilon_1)-\mathbf{N}(\varepsilon_2)| \leq c'(K)/\mathrm{Log}|\varepsilon_1-\varepsilon_2|$$

for all ε_1 and ε_2 in K. This is the **"Hölder continuity" of the integrated density of states** which we already mentioned in the lattice case. The above derivation is borrowed from [Kotani(1983)], but the result was first proven in [Craig-Simon(1983b)].

We close this Section with the following lemma, also taken from [Kotani(1983)], which will be fundamental in the investigation of the absolute continuous spectrum.

Lemma V.14:
If $\dot{\alpha}(\varepsilon)=0$ *for almost all* ε *in a compact subset* K *of* $\mathbf{R}$ *we have:*

$$-\lim_{a\to 0} a^{-1}\int_K \mathrm{Re}\, \mathbf{w}(\varepsilon+ia)\, d\varepsilon = \lim_{a\to 0} a^{-1}\int_K \mathrm{Im}\, \mathbf{w}'(\varepsilon+ia)\, d\varepsilon$$

$$= \mathbf{n}_{ac}(K). \qquad (V.46)$$

Proof:
For each $a>0$, we define the measure $\mathbf{n}_a$ by:

$$\mathbf{n}_a(d\varepsilon) = a^{-1}[-\mathrm{Re}\, \mathbf{w}(\varepsilon+ia) + \mathrm{Re}\, \mathbf{w}(\varepsilon+i0)]\, d\varepsilon$$

For each continuous function f with compact support in $\mathbf{R}$ we have:

$$\int f(\varepsilon)\, \mathbf{n}_a(d\varepsilon) = (2a\pi)^{-1}\iint f(\varepsilon)\, \mathrm{Log}[(x-\varepsilon)^2+a^2]/(x-\varepsilon)^2\, \mathbf{n}(dx)\, d\varepsilon$$

(if one remembers the formula

$$-\mathrm{Re}\, \mathbf{w}(\varepsilon+ia) + \mathrm{Re}\, \mathbf{w}(\varepsilon+i0) = (1/2)\int \mathrm{Log}[(x-\varepsilon)^2+a^2]/(x-\varepsilon)^2\, \mathbf{n}(dx)$$

which we obtained in the proof of the preceding proposition)

$$= (1/2)\int y^{-2}\mathrm{Log}(1+y^2)[\int f(x-a/y)\mathbf{n}(dx)]\, dy$$

which converges to $\pi\int f(\varepsilon)\mathbf{n}(d\varepsilon)$ since $\int_0^\infty y^{-2}\mathrm{Log}(1+y^2)\, dy = \pi$. This proves that the measure $\mathbf{n}_a$ converges vaguely to $\pi\mathbf{n}$ when $a\to 0$. The function $z \to \mathrm{Re}\, \mathbf{w}(z)$ is harmonic and can be reconstructed from its boundary values by means of the Poisson kernel:

$$-\mathrm{Re}\, \mathbf{w}(\varepsilon+ia) = -\pi^{-1}\int_{\mathbf{R}} a\, \mathrm{Re}\, \mathbf{w}(x+i0)/[(\varepsilon-x)^2+a^2]\, dx.$$

By integrating both sides over a closed subset F of K, we obtain:

$$\mathbf{n}_a(F) = \pi^{-1}\int_{\mathbf{R}}\int_F \dot{\alpha}(x)\, /[(\varepsilon-x)^2+a^2]\, dx\, d\varepsilon$$

and using the monotone convergence of the integrals and the vague convergence of the measures we have:

$$\lim_{a\to 0} \mathbf{n}_a(F) = \pi^{-1}\int_F\Big(\int_{\mathbf{R}} \dot{\alpha}(x)\, /(\varepsilon-x)^2\, dx\Big)\, d\varepsilon \le \pi\, \mathbf{n}(F).$$

This last inequality extends to the trace of the Borel σ-field $\mathcal{B}_{\mathbf{R}}$ on K, and we conclude from that:

$$\pi^{-2}\int_{\mathbf{R}} \dot{\alpha}(x)\, /(\varepsilon-x)^2\, dx \le [d\, \mathbf{n}_{ac}/d\varepsilon](\varepsilon)$$

for almost every ε in K. Actually equality holds because we already saw that:

$$-a^{-1}\,\mathrm{Re}\,\mathbf{w}(\varepsilon+ia) > \mathrm{Im}\,\mathbf{w}'(\varepsilon+ia) = \int a/[(x-\varepsilon)^2+a^2]\,\mathbf{n}(dx)$$

and the right hand side converges to $[\,d\mathbf{n}_{ac}/d\varepsilon](\varepsilon)$ for almost all ε because of Fatou's theorem (recall Section I.3) and because formula (V.40) implies:

$$[\,d\mathbf{n}_{ac}/d\varepsilon](\varepsilon) = \mathrm{Im}\,w'(\varepsilon+i0)$$

for almost all ε.□

V.5. Complements and comments:

We proved the existence of the integrated density of states using a version of the additive ergodic theorem:

$$\mathcal{N}(\varepsilon) = \lim_{D\to\mathbf{R}^d} |D|^{-1} \sum_{k\geq 0} \mathbf{1}_{(-\infty,\,\varepsilon]}(\varepsilon_k(D))$$

where the $\varepsilon_k(D)$ are the eigenvalues of our random operators restricted to bounded domains D. The natural question to ask is that of a corresponding **Central Limit Theorem** , namely, do the random variables:

$$\tilde{\mathcal{N}}_D(\varepsilon) = |D|^{1/2}(\mathcal{N}_D(\varepsilon) - E\{\mathcal{N}_D(\varepsilon)\})$$

where $\mathcal{N}_D(\varepsilon)$ has been defined in (V.3), converge in distribution when $D\to\mathbf{R}^d$, and can we identify the limit? A positive answer has been given for the case of Example IV.2.4 when the manifold M is chosen to be the unit circle. In fact, in this case a **functional form** of the central limit theorem holds: for each compact interval I of $\mathbf{R}$, the stochastic process $\{\tilde{\mathcal{N}}_{[-k,+k]}(\varepsilon)\ ;\ \varepsilon\in I\}$ converges in law to a mean zero Gaussian process with continuous sample paths and the covariance function of which is computable. See [Reznikova(1981)] for details. This work has been extended to the lattice $\mathbf{Z}$ when the potential is an i.i.d. sequence in [Le Page(1983)].

The fact that the topological support of the integrated density of states is the almost sure spectrum Σ is an interesting result by itself since it tells us that the information in the set Σ is contained in the integrated density of states. Nevertheless the reader should not be misslead. The integrated density of states is a **very poor predictor of the spectral properties of the operators** $H(\omega)$. Indeed, there are simple examples of random potentials depending on some parameter, for which the integrated density of states is independent of the parameter while the spectra are almost surely pure point for some value of the parameter and almost surely singular continuos for other values of the parameter. See [Simon(1983a)]. Also we expect that some models which have non trivial absolutely continuous and pure point components in their spectrum will also have an analytic density of states. See [Kunz-Souillard(1983)].

The asymptotic behavior of the disribution function of states at the edge of the spectrum has been studied for more general operators. Indeed, formally symmetric differential operators with singular coefficients can be defined as self adjoint operators by the method of quadratic forms. As an example one can consider Schrödinger operators for which the potential $\{q(t);\ t\in\mathbf{R}\}$ is **white noise**. The interested reader may want to look at [Kotani(1976)], [Fukushima-Nakao(1977)], [Nakao(1977)], [Fukushima(1981)] or [Thompson(1983)] for details.

We close these complements with some words on the important problem of the **attraction** (or **repulsion**) **of the energy levels.** Many incorrect claims can be found in the physics litterature on this problem and it is very difficult to understand what is actually happening on a rigorous mathematical level. Consequently we will restrict to a rather loose discussion of some mathematical results which have been proven in one dimension for the very reccurent random potentials constructed from Brownian motion processes as in example IV.2.4. and we will ask the reader to see [Molcanov(1981)] and [Genkova et al.(1983)] for details.

We want to investigate the asymptotic behavior of the normalized spectral interval lengths:

$$I_k(a) = \left(\mathbf{E}\{\varepsilon_{k+1}(a)-\varepsilon_k(a)\}\right)^{-1}[\ \varepsilon_{k+1}(a)-\varepsilon_k(a)\ ]$$

between the k-th and (k+1)-th eigenvalues of (E.V.E.) on $[-a,+a]$ with Dirichlet boundary conditions for example. We want to answer the following questions: does $I_k(a)$ converge as $a\to\infty$? If yes, in which sense? What is the limit? Can one estimate the speed of convergence? Can one compute large deviation probabilities? .....

We will assume that $I_k(a)$ converges in distribution to some probability measure μ_k concentrated on $[0,\infty)$ and we will say that there is:

* **repulsion of the energy levels** if $\lim_{\varepsilon\to 0}\ \varepsilon\mu_k([0,\varepsilon])=0$.
* **no interaction between the energy levels** if $\lim_{\varepsilon\to 0}\ \mu_k([0,\varepsilon])=c_k\neq 0$.
* **attraction of the energy levels** if $\lim_{\varepsilon\to 0}\ \varepsilon^{-1}\mu_k([0,\varepsilon])=\infty$.

Also we may want to study the joint distribution of several consecutive normalized spectral intervals, and their limit correlation. In fact the mathematical problem we tried to define above is meaningless because the k-th energy level $\varepsilon_k(a)$ converges to the infimum of the spectrum Σ when $a\to\infty$. This is the reason why we chose k as a function of a in such a way that $\varepsilon_{k(a)}(a)$ converges to a number ε when $a\to\infty$ and we will talk about the attraction-repulsion of the energy level at energy ε.

For the random potential of Example IV.2.4. it has been shown that for each ε_0 in Σ, the marginals of the process $\{n_\varepsilon{}^*(a);\ \varepsilon\geq 0\}$ defined by:

$$n_\varepsilon{}^*(a) = \#\{\text{eigenvalues of } H_{-a,+a,0,0}(\omega) \text{ in } (\varepsilon_0,\varepsilon_0+\varepsilon/2a]\}$$

converge as $a\to\infty$ to those of a Poisson process with parameter $[dn/d\varepsilon](\varepsilon_0)$. In this case it is easy to check that the density of states $dn/d\varepsilon$ exists and is smooth. More precisely it is possible to see that if $0\leq a_1<b_1\leq a_2<b_2\leq \ldots\ldots \leq a_n<b_n$, $\varepsilon_0\ \Sigma$ and $n,k(1),k(2),\ldots,k(n)\ \mathbf{N}$, then:

$\lim_{a\to\infty} \mathbf{P}\{$ #eigenvalues of $H_{-a,+a,0,0}$ in $I_1=k(1)$, ... ,#eigenvalues of $H_{-a,+a,0,0}$ in $I_n=k(n)\}$

$$=e^{[dn/de](\epsilon_0)(b_1-a_1)}([dn/de](\epsilon_0)(b_1-a_1))^{k(1)}/k(1)! \;\ldots\ldots\ldots\ldots\ldots$$

$$e^{[dn/de](\epsilon_0)(b_n-a_n)}([dn/de](\epsilon_0)(b_n-a_n))^{k(n)}/k(n)!$$

where we set $I_j=(\epsilon_0+a_j/(2a),\epsilon_0+b_j/(2a)]$ for $j=1, \ldots ,n$. This **"Poissonian" asymptotic behavior** , and hence the **nonrepulsion** of the energy levels, has been proven for other models, for eamples for potentials made out of a periodic sequence of Dirac masses with i.i.d. coefficients with an exponential distribution. Note also that some operators of the form:

$$H(\omega) = -(d/dt)a(t,\omega)(d/dt)$$

have been shown to exhibit the phenomenum of **repulsion** of the energy levels.

An intuitive explanation of the above results was given to us by B.Souillard and we try to present it now. The interaction between two energy levels should be understood in perturbation when they are very close to each other, and it should be proportional to the overlap of the corresponding wave functions. Repulsion should occur when the latter is important, and this will be the case when the states are extended while, when the states are localized, the interaction is smaller and the repulsion should not occur. Indeed, for random potentials giving exponential localization (see Chapter VII), it is possible to compute numerically the eigenfunctions in bounded intervals and check that they are localized in further and further regions when the corresponding energies are closer and closer. This is also responsible of the insulator behavior of the system exhibited in [Kunz-Souillard(1980)] where it is shown thar the electric conductivity of these system is vanishing. Nevertheless, the localization length depends on the energy and it happens that it is not small enough to be responsible for this phenomenum to occur at certain energies, and this is the cause of some of the apparent paradoxes which lead to the incorrect statements we already alluded to.

Undoubtedly, these problems are more subtle than expected! But after all, this may be one of the main attractions of the subject.

VI. THE ABSOLUTELY CONTINUOUS SPECTRUM IN ONE DIMENSION.

In this chapter we will assume that we are given a *complete dynamical system* $(\Omega,\mathcal{B},\mathbf{P},\{T_t;\, t\in\Theta_0\})$ with Θ_0 equal to $\mathbf{R}$ or $\mathbf{Z}$. This means that $(\Omega,\mathcal{B},\mathbf{P})$ is a complete probability space and $\{T_t;t\in\Theta_0\}$ an ergodic group of automorphisms of the measurable space $(\Omega,\mathcal{B})$ which leave invariant the measure $\mathbf{P}$. We will also assume that the function:

$$q:\ \Omega \dashrightarrow \mathbf{R}$$

is measurable, integrable and such that the stochastic process $\{q(t);t\in\mathbf{R}\}$ defined by $q(t,\omega)=q(T_t\omega)$ for each real t and ω in Ω, satisfies:

> *there exists a constant* $a>0$ *such that* $q(t)\geq -a(t^2+1)$ *for* $|t|$ *large enough and the function* $t \dashrightarrow q(t)$ *is locally square integrable,*

$\mathbf{P}$-almost surely. Hence, the operator:

$$H(\omega) = -(d^2/dt^2) + q(t,\omega)$$

is essentially self adjoint on $C_c^\infty(\mathbf{R})$ $\mathbf{P}$-almost surely and is a measurable function of ω. The aim of this chapter is to study the absolute continuous spectrum of this random self adjoint operator, and more precisely, the essential support of the absolutely continuous component $E_{ac}(.,\omega)$ of the resolution of the identity of $H(\omega)$. We know already that this essential support is independent of ω in Ω. We will show that it actually equals the set $\{\varepsilon\in\mathbf{R};\ \alpha(\varepsilon)=0\}$, where $\alpha(\varepsilon)$ is the upper Ljapunov exponent for (E.V.E.) which we introduce in the first paragraph.

VI.1. The Ljapunov exponent:

Following the notations of Section III.3, we denote by $\{U_\varepsilon^{\omega}(t,s);\ t,s\in\mathbf{R}\}$ the propagator of (E.V.E.) for $\varepsilon\in\mathbf{R}$ and $\omega\in\Omega$. By existence and uniqueness of such a propagator we have:

$$U_\varepsilon^{T_h\omega}(t,s) = U_\varepsilon^{\omega}(t+h,s+h) \qquad \text{(VI.1)}$$

Thus, Kingman's subadditive ergodic theorem gives, for each $\varepsilon\in\mathbf{R}$, a set $\Omega_\varepsilon\in\mathcal{B}$ with $\mathbf{P}$-probability 1 and such that the limits

$$\alpha(\varepsilon) = \lim_{t \to +(-)\infty} |t|^{-1} \operatorname{Log} \|U_\varepsilon^\omega(t,0)\| \qquad \text{(VI.2)}$$

exist for ω in Ω_ε. Indeed we have:

$$\mathbf{E}\{\sup_{0\le t\le 1} \operatorname{Log}^+ \|U_\varepsilon^\omega(t,0)\|\} \le \mathbf{E}\{\int_0^1 \|A_\varepsilon(s)\|\, ds\}$$

$$\le \mathbf{E}\{|q(0)|\} + |\varepsilon|$$

$$< \infty.$$

The (nonnegative) number $\alpha(\varepsilon)$ defined by (VI.2) is called the **Ljapunov exponent** of (E.V.E.). The above estimate and the unimodularity of the matrices $U_\varepsilon^\omega(t,s)$ make possible the use of the deterministic part of the Oceledec multiplicative ergodic theorem and, for each ω in Ω_ε there exist unit vectors in $\mathbf{R}^2$, say $\Theta^+_{\varepsilon,\omega}$ and $\Theta^-_{\varepsilon,\omega}$ such that:

$$\lim_{t \to +\infty} t^{-1} \operatorname{Log} \|U_\varepsilon^\omega(t,0)\, \Theta^+_{\varepsilon,\omega}\| = -\alpha(\varepsilon)$$

and:

$$\lim_{t \to +\infty} t^{-1} \operatorname{Log} \|U_\varepsilon^\omega(t,0)\, \Theta\| = \alpha(\varepsilon)$$

whenever Θ is linearly independent of $\Theta^+_{\varepsilon,\omega}$ in $\mathbf{R}^2$, and:

$$\lim_{t \to -\infty} |t|^{-1} \operatorname{Log} \|U_\varepsilon^\omega(t,0)\, \Theta^-_{\varepsilon,\omega}\| = -\alpha(\varepsilon)$$

and:

$$\lim_{t \to -\infty} |t|^{-1} \operatorname{Log} \|U_\varepsilon^\omega(t,0)\, \Theta\| = \alpha(\varepsilon)$$

whenever Θ is linearly independent of $\Theta^-_{\varepsilon,\omega}$ in $\mathbf{R}^2$. The above relations will be very useful when we know that the Ljapunov exponent $\alpha(\varepsilon)$ is strictly positif. We summarize these facts in a picture which the reader should keep in mind from now on.

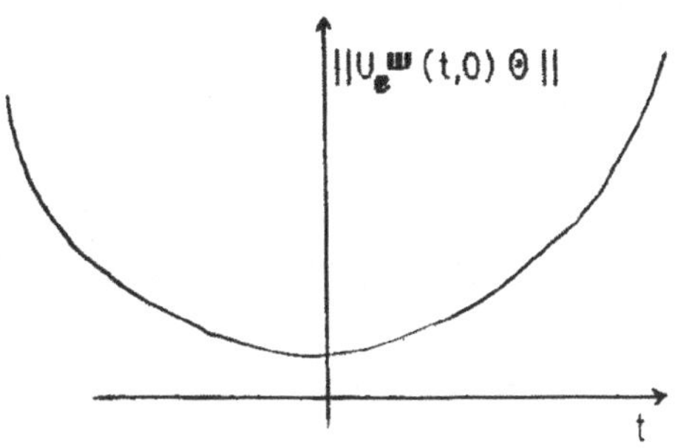

Θ and $\Theta^+_{\varepsilon,\omega}$ linearly independent

Θ and $\Theta^-_{\varepsilon,\omega}$ linearly independent

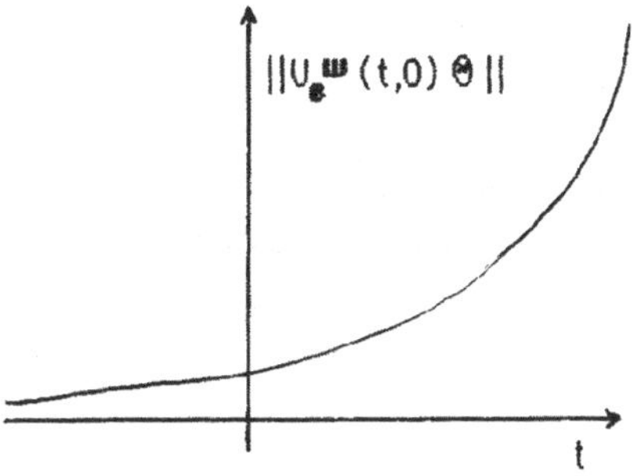

Θ and $\Theta^+_{\varepsilon,\omega}$ linearly independent

Θ and $\Theta^-_{\varepsilon,\omega}$ linearly dependent

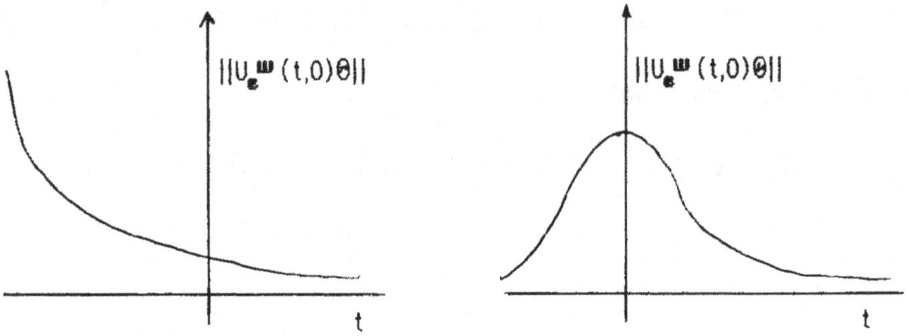

θ and $\theta^{+}_{\varepsilon,\omega}$ linearly dependent
θ and $\theta^{-}_{\varepsilon,\omega}$ linearly independent

θ and $\theta^{+}_{\varepsilon,\omega}$ linearly dependent
θ and $\theta^{-}_{\varepsilon,\omega}$ linearly dependent

Figure 1.

We conclude this section with the identification of the Ljapunov exponent with quantities which we introduced in Chapter III.

Using the Oceledec multiplicative theorem and Fubini's theorem one sees that the set Ω_1 defined by:

$$\Omega_1 = \{\omega\in\Omega;\ \lim_{t\to+(-)\infty} t^{-1}\,\mathrm{Log}\,\|U_\varepsilon^\omega(t,0)\left[\begin{smallmatrix}1\\0\end{smallmatrix}\right]\| \text{ and}$$

$$\lim_{t\to+(-)\infty} t^{-1}\,\mathrm{Log}\,\|U_\varepsilon^\omega(t,0)\left[\begin{smallmatrix}0\\1\end{smallmatrix}\right]\| \text{ exist and are finite for almost every } \varepsilon\in\mathbf{R}\}$$

is of probability 1. Also, using a uniform version of the additive ergodic theorem and the definition (recall (V.36)) of the function $\mathbf{w}$, one easily sees that the set:

$$\Omega_2 = \{\omega\in\Omega;\ \forall z\in\Pi_+\ \lim_{t\to+(-)\infty} t^{-1}\int_0^t \mathbf{m}_{+(-)}(z,T_s\omega)ds = \mathbf{w}(z)\}$$

is also of full probability.

Also there exists another full set, say Ω_3, on which the limits $\lim_{t\to+(-)\infty} t^{-1}\theta(t,a,\alpha,\varepsilon)$ exist and equal $\pi\mathbf{N}(\varepsilon)$ for all a,α and ε. Recall Section V.4.

Proposition VI.1:
For each $\omega\in\Omega_0 = \Omega_1\cap\Omega_2\cap\Omega_3$ *and for almost every* ε *in* $\mathbf{R}$, *the four limits in the definition of* Ω_2 *are equal to the Ljapunov exponent* $\alpha(\varepsilon)$ *and to the number* $\dot\alpha(\varepsilon)$ *defined in Proposition V.12 by* $\dot\alpha(\varepsilon) = -\,\mathrm{Re}\,\mathbf{w}(e+i0)$.

Proof:
For each z in Π_+, t in $\mathbf{R}$ and ω in Ω_0 we set:

$$\eta_z(t,\omega) = y_1(t,z,\omega) - i\,y_1'(t,z,\omega).$$

$\eta_z(t,\omega)$ never vanishes so that we can choose a determination of the logarithm for which the function:

$$\mathbf{R} \times \Pi_+ \ni (t,z) \longrightarrow \operatorname{Log} \eta_z(t,\omega) \in \mathbf{C}$$

is continuous and analytic in z for each ω in Ω_0. (Note $\operatorname{Log} \eta_i(t,\omega) = \operatorname{Log}(-i) = 3i\pi/2$).

Since $\omega \in \Omega_2$, $t^{-1}\int_0^t \mathbf{m}_+(z,T_s\omega)ds$ converges to $\mathbf{w}(z)$ whose real part is strictly negative (recall Lemma V.11). This implies that $|f_+(t,z,\omega)|$ decays exponentially as $t \to \infty$. Recall (V.34) and the definition of the function $z_+(t,z,\omega)$. This forces

$$f_+'(t,z,\omega) = \mathbf{m}_+(z,T_t\omega)\, f_+(t,z,\omega)$$

to decay exponentially. Using again the definition of the functions $f_+(t,z,\omega)$ and $f_-(t,z,\omega)$ one easily sees that

$$\begin{aligned}\eta_z(t,\omega) = & -i\, \mathbf{m}_-(z,\omega)[\mathbf{m}_+(z,\omega)+\mathbf{m}_-(z,\omega)]^{-1}[\, f_+(t,z,\omega)+i\, f_+'(t,z,\omega)] \\ & -i\, \mathbf{m}_+(z,\omega)[\,\mathbf{m}_+(z,\omega)+\,\mathbf{m}_+(z,\omega)]^{-1}[\, f_-(t,z,\omega)+i\, f_-'(t,z,\omega)]\end{aligned}$$

so that we conclude that:

$$\begin{aligned}\lim_{t\to\infty} t^{-1} \operatorname{Log} \eta_z(t,\omega) &= \lim_{t\to\infty} t^{-1} \operatorname{Log} f_-(t,z,\omega) \\ &= \lim_{t\to\infty} -t^{-1}\int_0^t \mathbf{m}_+(z,T_s\omega)ds \\ &= -\mathbf{w}(z)\end{aligned}$$

since $\omega \in \Omega_2$. For each $t \in \mathbf{R}$, $z \in \Pi_+$ and ω in Ω_0 we set:

$$k_t(z,\omega) = t^{-1} \operatorname{Log} \eta_z(t,\omega) - i\, t^{-1} \operatorname{Log} n_i(t,\omega)$$

Note that $k_t(\,.\,,\omega)$ is analytic in Π_+ and continuous in $\overline{\Pi}_+$. Moreover:

$$\lim_{t\to\infty} \operatorname{Im} k_t(z,\omega) = -\pi\mathbf{N}(z) + \operatorname{Im} \mathbf{w}(i)$$

if $z \in \mathbf{R}$ and

$$\lim_{t\to\infty} k_t(z,\omega) = -\,\mathbf{w}(z) + i \operatorname{Im} \mathbf{w}(i) =: k_\infty(z)$$

exists if $z \in \Pi_+$. Note also that:

$$\operatorname{Re} k_t(z,\omega) = t^{-1} \operatorname{Log}[\,|y_1(t,z,\omega)|^2 + |y_1'(t,z,\omega)|^2\,]^{1/2}$$

has a limit because $\omega \in \Omega_1$. Now we know that $\lim_{t\to\infty} \operatorname{Im} k_t(\varepsilon + i0,\omega) = \operatorname{Im} k_\infty(\varepsilon + i0)$ in $\mathbf{L}^p$

for any p in (1,2) so that we can conclude that $\lim_{t \to \infty}$ Re $k_t(\varepsilon + i0, \omega)$ also exists in the same L^p. Using the Poisson kernel we check that this limit is the boundary value of some harmonic function, say $F(z)$, and necessarily we have $F(\varepsilon + i0) = \mathrm{Re}\, k_\infty(\varepsilon + i0)$. Consequently:

$$\dot{\alpha}(\varepsilon) = -\mathrm{Re}\, \mathbf{w}(\varepsilon + i0)$$

$$= \mathrm{Re}\, k_\infty(\varepsilon + i0)$$

$$= \lim_{t \to \infty} \mathrm{Re}\, k_t(\varepsilon + i0, \omega)$$

in L^p and hence for almost every real ε,

$$= \lim_{t \to \infty} t^{-1} \mathrm{Log}[\,|y_1(t,z,\omega)|^2 + |y_1'(t,z,\omega)|^2\,]^{1/2}$$

$$= \alpha(\varepsilon). \ \square$$

VI.2. The Ishii-Pastur result:

We present an argument due to L.A.Pastur (see [Pastur(1974),(1980)]), but note that a similar result was previously proven by K.Ishii in [Ishii(1973)]. Note also that partial results in the same direction were obtained earlier for example in [Casher-Lebowitz(1971)] and [Yoshioka(1973)].

Proposition VI.1:
Let A *be a Borel subset of* $\mathbb{R}$ *such that* $\alpha(\varepsilon)>0$ *for almost every* ε *in* A. *Then* $\mathbb{P}$- *almost surely we have:*

$$E_{ac}(A) = 0.$$

Proof:
Let $A^\sim$ be the set of couples (ω,ε) in ΩxA for which

$$\lim_{|t| \to \infty} |t|^{-1} \mathrm{Log}\|U_\varepsilon^{\omega}(t,0)\| \qquad \text{(VI.3)}$$

exists and is strictly positive. The function $(\omega,\varepsilon) \dashrightarrow U_\varepsilon^{\omega}(t,0)$ is measurable in ω when ε is fixed and continuous in ε for ω fixed. Consequently, it is jointly measurable, and this implies the measurability of the set $A^\sim$. The latter is of full measure for the product $\mathbf{P}\times|.|$ of the probability $\mathbf{P}$ and the Lebesgue measure $|.|$ restricted to A, and by Fubini's theorem this implies that for $\mathbf{P}$-almost all ω we have:

$$|\{\varepsilon\in A; (\omega,\varepsilon)\in A^{\sim}\}^c| = 0 \tag{VI.4}$$

For each of these ω, we know that (recall the end of Section III.3) for $E(.,\omega)$-almost all ε in A there exists a generalized eigenfunction (i.e. a solution of (E.V.E.) for ε) the amplitude of which is polynomially bounded. If $E_{ac}(A) \neq 0$, the set of ε's for which there exists such a polynomially bounded solution of (E.V.E.) is of nonzero Lebesgue's measure. This measure remains strictly positive even after removing from this set the L^2-eigenvalues since the latter are at most countable. Now, we can use the argument leading to Figure 1 in the preceeding paragraph to get a contradiction, showing that we must actually have $E_{ac}(A) = 0$. □

Remarks:

1. The proof above actually proves that **for any -finite measure μ on the real line, the spectral measures are almost surely orthogonal to μ on the set where the Ljapunov exponent vanishes**. Indeed, the only special property of the Lebesgue measure we used was its continuity when we claimed it did not charge the set of L^2-eigenvalues. Proposition II.14 tells us that any given real number is not an L^2-eigenvalue with probability 1. Thus, by throwing away a negligeable set of ω for each point mass of the measure μ, we can assume that μ is continuous without any loss of generality, and the proof of Proposition VI.1 above applies.
2. In fact, the above remark can be reformulated in the following way: if one picks two independent ω in Ω according to the distribution $\mathbf{P}$, say ω_1 and ω_2, then the spectral measures $E(.,\omega_1)$ and $E(.,\omega_2)$ are mutually singular in the sense that they are carried by disjoint Borel subsets of A.

VI.3. Kotani's converse:

Proposition VI.2:

Let A *be a Borel subset of the real line such that* $\alpha(\varepsilon)=0$ *for almost every* ε *in* A *with respect to the Lebesgue measure. Then for* ω *in a set of probability one we have:*

$$[d\mu_{ac}(\,.\,,\omega)/d\varepsilon](\varepsilon) > 0$$

for almost every $\varepsilon\in A$. *In this case we also have:*

$$m_+(\varepsilon+i0,\omega) = -\overline{m_-(\varepsilon+i0,\omega)} \tag{VI.5}$$

for almost every $\varepsilon\in A$.

Proof:

Without any loss of generality we can assume that the Lebesgue measure of A is positive. Let K be a compact set contained in A with positive Lebesgue measure. According to Lemma V.11 we have:

$$\mathbf{E}\{[\operatorname{Im} \mathbf{m}_{+(-)}(\mathbf{e}+i\varepsilon)]^{-1}\} = -2^{-1} \operatorname{Re} \mathbf{w}(\mathbf{e}+i\varepsilon) \tag{VI.6}$$

so that, integrating over K and using the classical Fatou's lemma of integration theory in the limit $\varepsilon \to 0$ we obtain:

$$\mathbf{E}\{\int_K [\operatorname{Im} \mathbf{m}_{+(-)}(\mathbf{e}+i0)]^{-1} d\mathbf{e}\} \le \lim_{\varepsilon \downarrow 0} \mathbf{E}\{\int_K [\operatorname{Im} \mathbf{m}_{+(-)}(\mathbf{e}+i\varepsilon)]^{-1} d\mathbf{e}\}$$

$$= 2\, \mathbf{n}_{ac}(K)$$

according to (VI.6) and Lemma V.13. In particular this implies that $\operatorname{Im} \mathbf{m}_{+(-)}(\mathbf{e}+i0)>0$ almost everywhere in $\mathbf{e} \in A$ with probability 1. Recall that for $\mathbf{e} \in \mathbf{R}$ we have:

$$\operatorname{Im} \mathbf{g}_{\mathbf{e}+i\varepsilon}(0,0,\boldsymbol{\omega}) = (\operatorname{Im} \mathbf{m}_+(\mathbf{e}+i\varepsilon)+ \operatorname{Im} \mathbf{m}_-(\mathbf{e}+i\varepsilon))/|\ \mathbf{m}_+(\mathbf{e}+i\varepsilon)+ \mathbf{m}_-(\mathbf{e}+i\varepsilon)|$$

and by the Fatou's theorem which we recalled in Section I.3, we get:

$$[d\mu_{ac}(\ .\ ,\boldsymbol{\omega})/d\mathbf{e}](\mathbf{e}) = \operatorname{Im} \mathbf{g}_{\mathbf{e}+i0}(0,0,\boldsymbol{\omega})$$

$$> 0$$

which proves the first claim of the proposition. In order to prove (VI.5) we first notice that:

$$0 = \lim_{\varepsilon \downarrow 0} \left(- \int_K \operatorname{Re} \mathbf{w}(\mathbf{e}+i\varepsilon)\, d\mathbf{e} - \int_K \operatorname{Im} \mathbf{w}(\mathbf{e}+i\varepsilon)\, d\mathbf{e} \right)$$

because of Lemma V.13 and if we recall (V.39) this gives:

$$0 = \mathbf{E}\Big\{\int_K [(\operatorname{Im} \mathbf{m}_+(\mathbf{e}+i0))^{-1}+(\operatorname{Im} \mathbf{m}_-(\mathbf{e}+i0))^{-1}]\Big([\operatorname{Re} \mathbf{m}_+(\mathbf{e}+i0) + \operatorname{Re} \mathbf{m}_-(\mathbf{e}+i0)\ ]^2+$$
$$[\operatorname{Im} \mathbf{m}_+(\mathbf{e}+i0) - \operatorname{Im} \mathbf{m}_-(\mathbf{e}+i0)\ ]^2\Big) /|\ \mathbf{m}_+(\mathbf{e}+i0) - \mathbf{m}_-(\mathbf{e}+i0)\ |^2\, d\mathbf{e}\Big\}$$

by the classical Fatou's lemma of integration theory. Consequently, with probability 1 the integrand is 0 for almost every $\mathbf{e}$ in K and this implies:

$$\begin{aligned} \operatorname{Re} \mathbf{m}_+(\mathbf{e}+i0) &= -\operatorname{Re} \mathbf{m}_-(\mathbf{e}+i0) \\ \operatorname{Im} \mathbf{m}_+(\mathbf{e}+i0) &= \operatorname{Im} \mathbf{m}_-(\mathbf{e}+i0) \end{aligned} \tag{VI.7}$$

Taking an increasing sequence of compact sets whose union is of full measure in A shows that (VI.7) actually holds almost everywhere in A. □

Putting together the Ishii-Pastur result and Kotani's converse we obtain:

Corollary VI.3:
The essential support of the resolution of the identity of $H(\omega)$ *is almost surely equal to* $\{\varepsilon \in \mathbf{R}; \alpha(\varepsilon)=0\}$.

But we also have:

Corollary VI.4:
If $\alpha(\varepsilon)=0$ *for almost every* ε *of an open interval* I *then* $H(\omega)$ *has purely absolute continuous spectrum in* I.

Proof:
According to formula (VI.5) the real part of $m_+(\varepsilon+i0,\omega) + m_-(\varepsilon+i0,\omega)$ vanishes almost everywhere on the interval I. Since the function $m_+(\ .\ ,\omega) + m_-(\ .\ ,\omega)$ is analytic and has a positive imaginary part in the upper half plane Π_+, it can be analytically continued through the interval I by the Schwarz reflection principle, and the extension does not vanish on I. This shows that:

$$g_z(0,0,\omega) = -[\ m_+(z,\omega) + m_-(z,\omega)\ ]^{-1}$$

is analytic on I. This concludes the proof because $E(\ .\ ,0,0,\omega)$ is recovered from the boundary values of the imaginary part of the Green's function $g_z(0,0,\omega)$. □

The last application of Proposition VI.2 shows that the stochastic potential giving absolute continuous spectrum cannot be very random.

Corollary VI.5:
If $H(\omega)$ *has a nontrivial absolute continuous component for a set of* ω *with positive probability, the stochastic process* $\{q(t);t\ \mathbf{R}\}$ *is deterministic in the sense that the* σ*-field* $\mathcal{B}_{-\infty}$ *intersection of all the* σ*-fields generated by* $\{q(s);s\leq t\}$ *is equal to the* σ*-field* $\mathcal{B}_{\infty}$ *generated by* $\{q(t);t\in\mathbf{R}\}$.

Proof:
The knowledge of $\mathcal{B}_{-\infty}$ gives the asymptotic behavior of the potential function $q(t,\omega)$ as $t \to -\infty$. Now, the knowledge of the potential function $q(t,\omega)$ on the negative t-axis determines the m-function $m_-(z,\omega)$ in the upper half plane and its boundary value on the real axis. If the absolute continuous component of $H(\omega)$ is nontrivial, formula (VI.5) gives us the boundary value of $m_+(z,\omega)$ on a set of strictly positive Lebesgue measure. This is enough to reconstruct the analytic function $m_+(z,\omega)$ in the whole upper half plane from which we can obtain the potential function $q(t,\omega)$ on the positive t-axis (recall Section III.3). Using the stationarity of the potential function,

one can determine the potential function $q(t,\omega)$ on the positive t-axis from its knowledge on arbitrary left hand intervals, and thus from the knowledge of $\mathcal{B}_{-\infty}$. Note that we have to use different m-functions, namely the ones appropriate for the half intervals in question. This completes the proof. □

We would like to thank B.Simon for correcting an error in the first version of our presentation of the above proof.

VI.4. Complements and bibliographical comments:

The existence of the Ljapunov exponents is a consequence of a result in [Furstenberg-Kesten(1960)] on the products of random matrices. The right way to understand this result is via the Kingman's subadditive ergodic theorem (see for example [Kingman(1976)]). The interested reader will find simpler proofs of this theorem in [Derriennic(1975)], [Katznelson-Weiss(1982)] and [Ledrappier(1985a)].

The multiplicative ergodic theorem is due to Oceledec, but the original proof is so involved that we rather refer to [Ruelle(1979)] for a proof suited to our needs.

The use of the ideas and the results of the theory of products of random matrices in the study of random differential equations goes back to the early seventies, but one had to wait almost a decade for the converse of the Ishii-Pastur result. In this respect, Kotani's contribution to the understanding of the spectral properties of one-dimensional random Schrödinger operators is remarkable. Essentially we learn that ***the more randomness in the potential, the less absolute continuous spectrum we may expect.*** [Deift-Simon(1983)] contains an interesting result in this direction: in the one dimensional lattice case, the Lebesgue meas8ure of the essential support of the absolute continuous spectrum is always at most less than 2 and is only equal to 2 when the potential is constant. Also, recall Corollary VI.5, the search for a non trivial absolutely continuous component in the spectrum has to be restricted to those operators with **"deterministic"** potentials, like for example the analytic ones studied in [Herbst-Howland(1981)]. See [Kirsch et al.(1985)] for more information on this subject.

Kotani's theory has been extended to the lattice case in [Simon(1983b)]. See also [Ledrappier(1985b)].More recently it has also been extended to general random Sturm-Liouville operators, the framework of which including both the lattice and the continuous cases. See [Minami(1985)].

Finally we recall a result of Kotani which we already quoted in Chapter II: ***the essential support of the absolutely continuous spectrum is a nonincreasing function of the topological support of the distribution of the potential process.*** This can be conveniently used to exhibit examples of deterministic random potentials giving rise to Schrödinger operators without absolute continuous spectrum. See [Kotani(1984a)] and [Kirsch et al.(1985)]. In particular, if the function F used in the definition of the random potential in Example IV.2.2 is bonded, Lipschitz and nonconstant, then the absolute continuous component of the spectrum is empty. In the case the Gaussian process has analytic sample paths and the function F is analytic we obtain ***analytic potential functions giving rise to Schrödinger operators without absolute continuous spectrum***.

We conclude this chapter by a discussion of the similar problems in the multidimensional case.

The first result in this direction deals with the **band** $\{1,\ldots,n\}\times\mathbb{Z}$ on which an analog of our discrete Laplacian can be defined and on which the random potential is usually chosen to be i.i.d. In this case, the spectrum is almost surely pure point. This result is due to Goldsheid. See [Goldsheid(1981)] and [Lacroix(1983),(1984)] for a detailed proof. The various approaches used so far are one dimensional in spirit.

Another model of multidimensional disordered system has been extensively investigated. The state space is the so-called Bethe lattice(called Caley tree in mathematics) and it is usually, at least in statistical mechanics, consider as an infinite dimensional lattice. The results announced in [Kunz-Souillard(1983)] will be discussed in the comments to the following chapter.

The first genuinely multidimensional result was obtained by Frölich and Spencer in their remarkable work [Frölich-Spencer(1983)]. They studied the case of the lattice $\mathbb{Z}^d$ with a Gaussian i.i.d. sequence for random potential and they proved, for large disorder or low energy, the ***absence of diffusion*** in a sense similar but slightly weaker than the one discussed in the introduction. As remarked in [Martinelli-Scoppola(1984)], their main estimate implies the absence of absolutely continuous spectrum. Here is a simple proof from a forthcoming paper of B.Simon and T.Wolff and which we learned from Barry Simon in a private discussion. Applying Borel-Cantelli lemma to the Frölich-Spencer estimate shows that for each real number $\mathbf{e}$ in some interval $\mathbf{I}$, there exists a positive number $m(\mathbf{e})$ such that $\mathbf{P}$-almost surely in $\boldsymbol{\omega}\in\Omega$ there exists a positive constant $c(\mathbf{e},\boldsymbol{\omega})$ such that:

$$|\langle [H(\boldsymbol{\omega})-(\mathbf{e}+i\varepsilon)I]^{-1}\delta_{\underline{0}},\delta_{\underline{n}}\rangle| \leq c(\mathbf{e},\boldsymbol{\omega})\, e^{-m(\mathbf{e})|\underline{n}|} \qquad \text{(VI.8)}$$

for all ε in $(0,1)$. The resolvent equation gives:

$$\operatorname{Im}\langle [H(\boldsymbol{\omega})-(\mathbf{e}+i\varepsilon)I]^{-1}\delta_{\underline{0}},\delta_{\underline{0}}\rangle = \varepsilon\sum_{\underline{n}\in\mathbb{Z}^d} |\langle [H(\boldsymbol{\omega})-(\mathbf{e}+i\varepsilon)I]^{-1}\delta_{\underline{0}},\delta_{\underline{n}}\rangle|^2$$

and the estimate (VI.8) gives:

$$\lim_{\varepsilon\to 0} \operatorname{Im}\langle [H(\boldsymbol{\omega})-(\mathbf{e}+i\varepsilon)]^{-1}\delta_{\underline{0}},\delta_{\underline{0}}\rangle = 0. \qquad \text{(VI.9)}$$

Using Fubini's theorem we can conclude that $\mathbf{P}$-almost surely in $\boldsymbol{\omega}\in\Omega$, (VI.9) holds for almost every $\mathbf{e}$ in $\mathbf{I}$. We then learn from Fatou's theorem that the absolute continuous component of the spectral measure $\langle E(\,.\,,\boldsymbol{\omega})\,\delta_{\underline{0}},\delta_{\underline{0}}\rangle$ vanishes on the interval $\mathbf{I}$. Using the stationarity of the potential one easily shows that not only $\langle E(\,.\,,\boldsymbol{\omega})\,\delta_{\underline{0}},\delta_{\underline{0}}\rangle$, but all the spectral measures have no absolute continuous components in $\mathbf{I}$.

VII. EXPONENTIAL LOCALIZATION IN ONE DIMENSION.

We discovered in the last chapter that one dimensional random Schrödinger operators could have some absolutely continuous spectrum only when the stochastic potential was not very random as in the case of some almost periodic potentials. Even though we will not dwell on that, let us point out that there also exist examples of almost periodic potentials on the lattice $\mathbb{Z}$ giving rise to random operators for which $E=E_{sc}$ almost surely, see [Avron-Simon(1982b),(1983)]. Even more surprinsingly, it has been recently remarked that the same random potential functions, for almost every choice of a boundary condition at the origin, give rise to pure point spectrum when the operators are considered on $L^2([0,\infty),dt)$. See [Kotani(1984b)].

Nevertheless, since the pioneering work of Ph. Anderson (see [Anderson(1958)]) and the numerous contributions of his followers, we expect that the following occurs almost surely in dimension one (at least when the potential is random enough):

i) $E = E_{pp}$ (and hence $E_{ac} = E_{sc} = 0$)

ii) the set of eigenvalues is dense in Σ and the amplitudes of the corresponding eigenfunctions decay exponentially.

iii) the rate of exponential falloff of the eigenfunctions is given by the corresponding Ljapunov exponent.

This is called the *exponential localization phenomenum* . Note that except may be for the exponential decay of the eigenfunctions, the same thing is expected in dimension two. The Russian school should be credited for the construction of the first rigorous mathematical models. After a partial succes in [Goldsheid-Molcanov(1976)], the first example of random Schrödinger operator with dense pure point spectrum was announced in [Goldsheid- Molcanov-Pastur(1977)]. Details and a proof of the point ii) above appeared one year later in [Molcanov(1978)]. The third result was conjectured in this latter article. The upper bound was proved in [Carmona(1982)] and the lower bound in [Craig-Simon(1983a)].

The works of the Russian school initiated a wave of interest, especially in France where the lattice case as well as the continuous case were studied (see [Kunz-Souillard(1980)], [Carmona(1982)] and [Royer(1982)]). Despite these efforts, and the clever improvment given in [Kotani(1984b)], the situation is still unsatisfactory, especially in the continuous case. The main reason is that the techniques used in the one-dimensional case cannot be extended to higher dimensional ones. See nevertheless the discussion at the end of this chapter where we comment on recent progresses in the lattice case.. Even worst, the existing proofs require artificial assumptions which are obviously unnecessary but which have not been bypassed so far. We will try to emphasize this last remark throughout this chapter.

VII.1. Informal discussion:

Let us assume that ***the spectrum is pure point and that the eingenfunctions decay exponentially with probability one***. So, $\mathbf{P}$-almost surely in ω there exists a Borel subset A_ω of $\mathbf{R}$ such that $E(A_\omega,\omega)=0$ and such that there is a function $f_{\varepsilon,\omega}$ satisfying

$$- f''_{\varepsilon,\omega}(t) + [q(t,\omega)-\varepsilon]\, f_{\varepsilon,\omega}(t) = 0$$

and:

$$(\,|f_{\varepsilon,\omega}(t)|^2 + |f'_{\varepsilon,\omega}(t)|^2\,)^{1/2} \leq k(\varepsilon,\omega)\, e^{-\alpha(\varepsilon,\omega)|t|} \qquad \text{(VII.1)}$$

for each real t and for some positive constants $k(\varepsilon,\omega)$ and $\alpha(\varepsilon,\omega)$ independent of t, whenever ε does not belong to A_ω. If we set:

$$Y_{\varepsilon,\omega}(t) = \begin{bmatrix} f_{\varepsilon,\omega}(t) \\ \\ f'_{\varepsilon,\omega}(t) \end{bmatrix}$$

then (VII.1) rewrites:

$$\|\, Y_{\varepsilon,\omega}(t)\,\| \leq k(\varepsilon,\omega)\, e^{-\alpha(\varepsilon,\omega)|t|} \qquad \text{(VII.2)}$$

Now, let Θ be a unit vector in $\mathbf{R}^2$ which is linearly independent of $Y_{\varepsilon,\omega}(0)$ and let us set

$$Z(t) = U_\varepsilon^{\omega}(t,0)\,\Theta.$$

$Z(t)$ is a solution of the system (III.17) which is linearly independent of $Y_{\varepsilon,\omega}(t)$. Hence, the Wronskian of these solutions, say $c_{\varepsilon,\omega}$, is a nonzero constant independent of t, and we have:

$$0 < |\, c_{\varepsilon,\omega}\,| < \|\, Y_{\varepsilon,\omega}(t)\,\|\ \|Z(t)\|$$

by Schwarz inequality, and this implies:

$$\liminf_{|t|\to\infty} |t|^{-1}\mathrm{Log}\|\, U_\varepsilon^{\omega}(t,0)\,\| > 0 \qquad \text{(VII.3)}$$

if one recalls (VII.2). For each ω in Ω we choose in a measurable way, a measure μ^ω such that μ^ω has the same sets of measure zero as $E(\,.\,,\omega)$. We then define the measure $\mathbf{P}^*$ on the product space $\mathbf{R}\times\Omega$ by:

$$\mathbf{P}^*(\,A\times B\,) = \int_B \mu^{\omega}(\,A\,)\, d\mathbf{P}(\omega) \qquad \text{(VII.4)}$$

Consequently, the above argument tells us that a **necessary condition for the exponential localization is that (VII.3) is satisfied $\mathbf{P}^*$-almost everywhere**. We now show that **this condition is "essentially sufficient"**. We used "essentially" because we will have to assume a little bit more in the sense that we now assume that for $\mathbf{P}^*$-almost every couple (ε,ω) in the product $\mathbf{R}\times\Omega$ the quantity $|t|^{-1}\mathrm{Log}\|U_\varepsilon^\omega(t,0)\|$ has a limit, say $\alpha(\varepsilon,\omega)$ which is strictly positive. If we fix such a couple (ε,ω), we can apply the deterministic part of the proof of the multiplicative ergodic theorem and obtain the existence of unit vectors $\Theta^+_{\varepsilon,\omega}$ and $\Theta^-_{\varepsilon,\omega}$ with the same properties as in Section VI.1, and the only possible asymptotic behaviors of $\|U_\varepsilon^\omega(t,0)\Theta\|$ are discussed in Fig. 1. We can appeal once more to the deterministic result on the existence of polynomially bounded eigenfunctions to conclude that the only possibility is to have these generalized eigenfunctions decay exponentially at both $+\infty$ and $-\infty$. Consequently, for $\mathbf{P}$-amost all ω in Ω, μ^ω almost all ε in $\mathbf{R}$ is actually an L^2-eigenvalue, the fall off of the corresponding eigenfunction being exponential and the rate being given by the Ljapunov exponent. This is precisely the result we were after.

It appears that the solution to our problem reduces to finding the asymptotic behavior of the norm of a cocycle of random matrices. Unfortunately, the usual techniques to tackle this sort of problem cannot be used here because we cannot obtain much information on the measure $\mathbf{P}^*$ without proving already the desired result. A possible way out coming first to mind, is to try to show that $\mathbf{P}$-almost surely in ω, the limit of $|t|^{-1}\mathrm{Log}\|U_\varepsilon^\omega(t,0)\|$ exists and is strictly positive for all ε instead of for μ^ω-almost all ε. Indeed, for each ε in Σ a theorem "à la Furstenberg" could very well give us the existence of a set Ω_ε of probability 1 and of a strictly positive number $\alpha(\varepsilon)$ such that:

$$\forall\omega\in\Omega_\varepsilon,\qquad \lim_{|t|\to\infty} |t|^{-1}\mathrm{Log}\|U_\varepsilon^\omega(t,0)\| = \alpha(\varepsilon) \tag{VII.5}$$

We know from the asymptotic theory of products of random matrices that this occurs quite often. Consequently one might wonder if:

$$\mathbf{P}\Big(\bigcup_{\varepsilon\in\Sigma}\Omega_\varepsilon^c\Big) = 0. \tag{VII.6}$$

In order to obtain such a result one could take the union over a countable dense set of ε's first, and then try to use perturbation arguments. Indeed it is sometimes possible to change a little bit the coefficients of a differential equation with a so-called ***exponential dichotomy of the solutions*** and preserves the latter. Unfortunately, **(VII.6) is false**. In fact, (VII.5) is much weaker than what is usually called an exponential dichotomy. It merely says that the asymptotic behavior of the amplitudes of the solutions is controled by exponential functions rather than being exponential itself. This property is not as strong and is very unstable, even under very small perturbations. We now explain why (VII.6) does not hold in the present situation.

Let $\Theta = \begin{bmatrix}-\sin\theta\\ \cos\theta\end{bmatrix}$ and $\Theta' = \begin{bmatrix}-\sin\theta'\\ \cos\theta'\end{bmatrix}$ be two unit vectors in $\mathbf{R}^2$ and for each $\omega\in\Omega$, let us consider the Schrödinger operators $H_\theta(\omega)$ and $H_{\theta'}(\omega)$ on the half axis $[0,\infty)$ which are defined in the usual way and whose domains are determined by the boundary conditions $f(0)\cos\theta+f'(0)\sin\theta=0$, and $f(0)\cos\theta'+f'(0)\sin\theta'=0$ respectively. Let $\mathbf{I}$ be an interval contained in both spectra Σ_θ and $\Sigma_{\theta'}$ and let us assume that:

$$\mathbf{P}\Big(\bigcup_{\varepsilon \in \mathbf{I}} \Omega_\varepsilon^c \Big) = 0.$$

With the argument we used earlier in this section for the operator on the whole real line one can show that the spectra are dense pure point in both cases. This means that $\mathbf{P}$-almost surely in ω there exist a dense set of ε in $\mathbf{I}$ for which $\Theta^+_{\varepsilon,\omega} = \Theta$, and another set of ε in $\mathbf{I}$ for which $\Theta^+_{\varepsilon,\omega} = \Theta'$, and this forces the function $\mathbf{I} \ni \varepsilon \dashrightarrow \Theta^+_{\varepsilon,\omega}$ to *oscillate wildly*. This is impossible because if one recalls the construction of $\Theta^+_{\varepsilon,\omega}$ (see for example [Ruelle(1979)]) one sees that this function is the pointwise limit of the sequence of functions $\varepsilon \dashrightarrow \tilde{\Theta}^{(n)}_{\varepsilon,\omega}$ where $\tilde{\Theta}^{(n)}_{\varepsilon,\omega}$ is the normalized eigenvector of $|U_\varepsilon^\omega(n,0)|^{1/n}$ associated to its largest eigenvalue. These functions are continuous and this implies that the set of discontinuity points of the limit should be of the second Baire category and this is the desired contradiction.

This argument appeared for the first time in [Goldsheid(1980)]. It has been independently rediscovered several times since then. The reader interested in the Furstenberg theorem can consult the original article [Furstenberg(1963)] or the lectures [Guivarch'(1979)]. He can also look at [Virtser(1979)], [Royer(1980)], [Ledrappier-Royer(1980)], [Guivarch'(1981)] and [Virtser(1983)] for complements on the nonindependent cases.

VII.2. A proof:

The following hypotheses will be in force throughout this paragraph.

> $\{q(t);\ t \in \mathbf{R}\}$ *is a real stationary ergodic process of the form* $q(t,\omega) = q(T_t\omega)$ *for some* $\mathcal{B}$*-measurable function* $q:\Omega \dashrightarrow \mathbf{R}$ *where* $(\Omega, \mathcal{B}, \mathbf{P}, \{T_t; t \in \mathbf{R}\})$ *is an ergodic dynamical system. We will also assume that for each* ω *the path* $t \dashrightarrow q(t,\omega)$ *is locally square integrable and satisfies* $q(t,\omega) \geq -a_\omega(t^2+1)$ *for all* t *in* $\mathbf{R}$ *and for some* $a_\omega > 0$.

and on the top oif them we will add the assumptions (A1) and (A2) which will be found in the text.

We know that the function:

$$\omega \dashrightarrow H(\omega) = -d^2/dt^2 + q(.,\omega)$$

with values in the self-adjoint operators on $H = L^2(\mathbf{R}, dt)$ is measurable and satisfies (II.7) and hence that there exists a closed subset of $\mathbf{R}$, say Σ, which is equal to the spectrum of $H(\omega)$ for $\mathbf{P}$-almost all ω. We choose an open interval $\mathbf{I}$ contained in Σ and we try to proved that exponential localization holds in $\mathbf{I}$. Note that the fact that $\mathbf{I}$ is closed or open is irrelevant since we know that a given point is almost surely not an eigenvalue. Our first assumption will be:

(A1). **for each ε in $\mathbf{I}$ we have $\alpha(\varepsilon)>0$.**

We knew that the Ljapunov existed but we demand now that it is **strictly positive** in $\mathbf{I}$. According to the preceeding section the problem reduces to the existence for $\mathbf{P}$-almost all ω and outside a set of ε's in $\mathbf{I}$ which is negligible for the resolution of the identity $E(.,\omega)$, of a solution of (E.V.E.) for ε which decays exponentially at + and - infinity, and to the identification of the rate of exponential fall off. Note that we can replace $E(.,\omega)$ by any nonnegative measure having the same sets of measure zero. We choose to use the measure μ^{ω} which was introduced in Section III.3. Next we remark that we need only construct a solution which decays exponentially at $+\infty$. Indeed, by symmetry and by throwing away various sets of measures zero, we can assume that we can also construct a solution-which is a priori different- which decays exponentially at $-\infty$. Then, because of the Wronskian argument, the situation has to be given by Fig.1 in Section VI.1, and we conclude as before by using the deterministic result on the existence of polynomially bounded generalized eigenfunctions.

Consequently, a sufficient condition is the existence of $\gamma>0$ and $\gamma'>0$ such that, $\mathbf{P}$-amost surely in $\omega\in\Omega$ we have, for μ^{ω}-almost all ε in $\mathbf{I}$:

$$\inf_{\|\Theta\|=1} \int_0^{\infty} \|U_{\varepsilon}^{\omega}(t,0)\Theta\|^{\gamma'} e^{\gamma t}\, dt < \infty.$$

Without any loss of generality we can assume that $\mathbf{I}$ is bounded. In fact we will try to prove:

$$\mathbf{E}\{\int_{\mathbf{I}} \mu^{\omega}(d\varepsilon) \left(\inf_{\|\Theta\|=1} \int_0^{\infty} \|U_{\varepsilon}^{\omega}(t,0)\,\Theta\|^{\gamma'} e^{\gamma t}\, dt \right)\} < \infty. \tag{VII.7}$$

This is the sufficient condition we will actually work with. The integral between 0 and $+\infty$ is the limit of the same integral between 0 and T when T tends to ∞. So, by Fatou's lemma it suffices to prove the existence of a constant C such that:

$$\mathbf{E}\{\int_{\mathbf{I}} \mu^{\omega}(d\varepsilon) \left(\inf_{\|\Theta\|=1} \int_0^{T} \|U_{\varepsilon}^{\omega}(t,0)\Theta\|^{\gamma'} e^{\gamma t}\, dt \right)\} \leq C. \tag{VII.8}$$

Let us assume that $\omega\in\Omega$ is fixed for a little while. Using the regularity properties of the propagators and the fact that $B_1=\{\Theta\in\mathbf{R}^2;\ \|\Theta\|=1\}$ is compact, one easily checks that the function:

$$\mathbf{I}\times B_1 \ni (\varepsilon,\Theta) \longrightarrow \int_0^{T} \|U_{\varepsilon}^{\omega}(t,0)\Theta\|^{\gamma'} e^{\gamma t}\, dt$$

is jointly continuous, and using again the compactness of B_1 one shows that the function:

$$\mathbf{I}\ni\varepsilon \longrightarrow \inf_{\|\Theta\|=1} \int_0^{T} \|U_{\varepsilon}^{\omega}(t,0)\Theta\|^{\gamma'} e^{\gamma t}\, dt$$

is also continuous. Hence, if μ_n^{ω} is any sequence of nonnegative measures which converge vaguely to μ^{ω}, we must have:

$$\lim_{n\to\infty} \int_{\mathbf{I}} \mu_n^{\boldsymbol{\omega}}(d\boldsymbol{e}) \left(\inf_{\|\Theta\|=1} \int_0^T \|U_{\boldsymbol{e}}^{\boldsymbol{\omega}}(t,0)\Theta\|^{\gamma'} e^{\gamma t}\, dt \right)$$

$$= \int_{\mathbf{I}} \mu^{\boldsymbol{\omega}}(d\boldsymbol{e}) \left(\inf_{\|\Theta\|=1} \int_0^T \|U_{\boldsymbol{e}}^{\boldsymbol{\omega}}(t,0)\Theta\|^{\gamma'} e^{\gamma t}\, dt \right)$$

since the endpoints of $\mathbf{I}$ are not charged by the measure $\mu^{\boldsymbol{\omega}}$. Using again Fatou's lemma we can replace the measure $\mu^{\boldsymbol{\omega}}$ by the measure $\mu_n^{\boldsymbol{\omega}}$ in the sufficient condition (VII.8) as long as the constant C does not depend on n, OF COURSE! According to the results of Section III.3, we can choose:

$$\mu_n^{\boldsymbol{\omega}} = \xi_{a,b,\boldsymbol{a},\boldsymbol{b}}^{\boldsymbol{\omega}} + \zeta_{a,b,\boldsymbol{a},\boldsymbol{b}}^{\boldsymbol{\omega}}$$

where $\xi_{a,b,\boldsymbol{a},\boldsymbol{b}}^{\boldsymbol{\omega}}$ and $\zeta_{a,b,\boldsymbol{a},\boldsymbol{b}}^{\boldsymbol{\omega}}$ have been defined in (III.10), and where a tends to $-\infty$ through a sequence $\{a_m; m\geq 1\}$ and b tends to $+\infty$ through a sequence $\{b_n; n\geq 1\}$. We will implicitely assume that $a<0<T<b$. Also, we may as well consider that $\boldsymbol{a},\boldsymbol{b},a_m$ and b_n are random variables such that a_m and b_n converge almost surely to $-\infty$ and $+\infty$ respectively. To make our life easier we will assume that they are independent and that they are independent of the potential process $\{q(t); t\in\mathbf{R}\}$, and that they are uniformly distributed in $[0,\pi)$, $[0,\pi)$, $(-m-1,-m]$ and $[n,n+1)$ respectively. Using once more Fatou's lemma this amounts to saying that the sufficient condition (VII.8) can be replaced by:

$$\mathbf{E}^{*}\{\int_{\mathbf{I}} F(\boldsymbol{e})\, \mu_n^{\boldsymbol{\omega}}(d\boldsymbol{e})\} \leq C, \tag{VII.9}$$

where we set:

$$F(\boldsymbol{e}) = \inf_{\|\Theta\|=1} \int_0^T \|U_{\boldsymbol{e}}^{\boldsymbol{\omega}}(t,0)\Theta\|^{\gamma'} e^{\gamma t}\, dt, \tag{VII.10}$$

and where the expectation $\mathbf{E}^{*}$ contains an integration with respect to $d\boldsymbol{a}\, d\boldsymbol{b}\, da\, db$ over $[0,\pi)\times[0,\pi)\times(-m-1,m]\times[n,n+1)$. Since the spectral measure $\mu_n^{\boldsymbol{\omega}}$ charges only the eigenvalues of the operator $H_{a,b,\boldsymbol{a},\boldsymbol{b}}(\boldsymbol{\omega})$, and since the latter are the roots of the equation :

$$\Theta(\boldsymbol{e}) = 0 \quad \mathrm{mod}\,\pi$$

if we set:

$$\Theta(\boldsymbol{e}) = \theta(0,a,\boldsymbol{a},e) - \theta(0,b,\boldsymbol{b},e)$$

We must have (recall formula (III.10)):

$$\int_{\mathbf{I}} F(\boldsymbol{e})\, \mu_n^{\boldsymbol{\omega}}(d\boldsymbol{e}) = \sum_{\boldsymbol{e}(j)\in\mathbf{I}} F(\boldsymbol{e}(j))\, \mu_n^{\boldsymbol{\omega}}(\{\boldsymbol{e}(j)\})$$

$$= \lim_{\varepsilon\to 0} (2\varepsilon)^{-1} \int_{\mathbf{I}} F(\boldsymbol{e})\, \mu_n^{\boldsymbol{\omega}}(\boldsymbol{e})\, |\Theta'(\boldsymbol{e})|\, \mathbf{1}_{[-\varepsilon,+\varepsilon]}(\Theta(\boldsymbol{e}))\, d\boldsymbol{e}$$

where $\mu_n^{\boldsymbol{\omega}}(\boldsymbol{e})$ is any continuous function which equals $\mu_n^{\boldsymbol{\omega}}(\{\boldsymbol{e}(j)\})$ when $\boldsymbol{e}=\boldsymbol{e}(j)$. A simple calculation shows that $|G'(\varepsilon)|$ is such a function. Consequently, using again Fatou's lemma we can rewritte the sufficient condition (VII.9) as:

$$\mathbf{E}^*\{\int_I F(\varepsilon)\,\delta(\,\theta(0,a,\mathbf{a},\varepsilon) - \theta(0,b,\mathbf{b},\varepsilon)\,)\,d\varepsilon\} \le C, \qquad \text{(VII.11)}$$

where we used the suggestive notation of the Dirac delta function to avoid having to write the limit over the approximate delta functions which we had above. We can now use Fubini's theorem to interchange the integral over I and the expectation and our sufficient condition becomes:

$$\int_I \mathbf{E}^*\{F(\varepsilon)\,\delta(\,\theta(0,a,\mathbf{a},\varepsilon)-\theta(0,b,\mathbf{b},\varepsilon)\,)\,\}\,d\varepsilon \le C. \qquad \text{(VII.12)}$$

which rewrites:

$$\int_I \mathbf{E}^*\{F(\varepsilon)\,\mathbf{E}^*\,\{\delta(\,\theta(0,a,\mathbf{a},\varepsilon)-\theta)|\mathcal{B}_{[0,\infty)}\}_{|\theta=\theta(0,b,\,\mathbf{b},\varepsilon)}\}\,d\varepsilon \le C. \qquad \text{(VII.13)}$$

where we used the notation $\mathcal{B}_A$ for the σ-field generated by the random variables $q(t)$ for t in A. At this point we make the following assumption:

(A.2). **for each ω in Ω the conditional distribution of $\theta(0,a,\mathbf{a},\varepsilon)$ given $\mathcal{B}_{[0,\infty)}$ is absolutely continuous and the density is continuous and bounded by a constant K_I independent of ω and ε in I.**

Condition (VII.13) becomes:

$$\int_I \mathbf{E}^*\{F(\varepsilon)\,\}\,d\varepsilon \le C\,K_I. \qquad \text{(VII.14)}$$

To get rid of the infimum in the definition of F we choose for Θ the unit vector determined by the angle $\theta(0,b,\mathbf{b},\varepsilon)$. Then, (VII.14) will be a consequence of:

$$\int_I \int_0^T \mathbf{E}^*\{||U_\varepsilon^{\omega}(t,0)\;\theta(0,b,\mathbf{b},\varepsilon)||^{\gamma'}\}\,e^{\gamma t}\,dt\,d\varepsilon \le C\,K_I\,, \qquad \text{(VII.15)}$$

where we identify the angles and the corresponding unit vectors. If we use the notation

$$A.\theta = ||A\theta||^{-1}\,A\theta$$

for the action of the 2x2 matrices on the unit circle, the left hand side of (VII.15) is equal to (note that with these conventions we have $\theta(0,b,\mathbf{b},\varepsilon) = U_\varepsilon^{\omega}(0,b).\mathbf{b}$):

$$\int_I \int_0^T \mathbf{E}^*\{||U_\varepsilon^{\omega}(t,0)\;\theta(0,b,\mathbf{b},\varepsilon)||^{\gamma'}\}\,e^{\gamma t}\,dt\,d\varepsilon$$

$$= \int_I \int_0^T \mathbf{E}^*\{||U_\varepsilon^{\omega}(t,0)\;U_\varepsilon^{\omega}(0,b).\mathbf{b})||^{\gamma'}\}\,e^{\gamma t}\,dt\,d\varepsilon$$

$$= \int_I \int_0^T \mathbf{E}^*\{||U_\varepsilon^{\omega}(t,0)\;U_\varepsilon^{\omega}(0,t).\;U_\varepsilon^{\omega}(t,b).\mathbf{b})||^{\gamma'}\}\,e^{\gamma t}\,dt\,d\varepsilon$$

$$= \int_I \int_0^T \mathbf{E}^*\{ \mathbf{E}^*\{ \| U_\varepsilon^\omega(t,0)\, U_\varepsilon^\omega(0,t).\, U_\varepsilon^\omega(t,b).b) \|^{\gamma'} | \mathcal{B}_{[0,t)} \} \} e^{\gamma t}\, dt\, d\varepsilon$$

$$\leq K_I \int_{[0,2\pi)} \int_I \int_0^T \mathbf{E}^*\{ \| U_\varepsilon^\omega(t,0)\, U_\varepsilon^\omega(0,t).\theta \|^{\gamma'} \} e^{\gamma t}\, dt\, d\varepsilon\, d\theta$$

where we used the same assumption as (A.2) for the reversed process in order to conclude that the angle had a density which we could bound above by K_I. Notice that:

$$\| U_\varepsilon^\omega(t,0)\, U_\varepsilon^\omega(0,t).\theta \| = \| U_\varepsilon^\omega(0,t)\theta \|^{-1}$$

and consequently the whole proof reduces to showing the ecistence of positive constants c and c' such that:

$$\int_{[0,2\pi)} \int_I \mathbf{E}\{ \| U_\varepsilon^\omega(0,t)\theta \|^{-\gamma'} \}\, d\varepsilon\, d\theta \leq c' e^{-ct}$$

for $t>0$ large enough. Using the stationarity of the potential process one sees that the above expectation equals:

$$\mathbf{E}\{ r(-t,0,0,\varepsilon)^{-\gamma'} \}$$

if one recalls the notation for the amplitude of the solutions of (E.V.E.). According to (III.23) it is equal to:

$$\mathbf{E}\{\exp[(\gamma'/2)\int_0^{-t} [\varepsilon - q(u) - 1] \sin 2\theta(u,0,\theta,\varepsilon)\, du]\}$$

Using a subadditivity argument one shows that the limit

$$\mu(\gamma') = \lim_{t \to \infty} t^{-1} \operatorname{Log} \int_{[0,2\pi)} \mathbf{E}\, \{\exp[(\gamma'/2)\int_0^{-t} [\varepsilon - q(u) - 1] \sin 2\theta(u,0,\theta,\varepsilon)\, du] \}\, d\theta$$

exists and that its derivative at $\gamma'=0$ is actually equal to $-\alpha(\varepsilon)$. One completes the proof by appealing to assumption (A.1). □

We now pause to discuss the assumptions made in the above **sketch of proof**.

As we saw in the previous chapter, assumption (A.1) concerning the positivity of the Ljapunov exponent is essentially necessary to insure the nonexistence of an absolutely continuous component in the spectrum.

The second assumption is much more restrictive and much more difficult to check. It is satisfied in the case of Example IV.2.4 of the random potential constructed from a stationary Brownian motion process on a compact connected Riemmannian manifold. The existence and the regularity of the density of the phase is guaranteed by the **smoothness properties of fundamental solutions of hyppoelliptic partial differential equations**. Appealing to such a result should be compared to the use of a sledgehammer. This is certainly not necessary in the present situation, but it sems that the understanding of the technical difficulties is not sufficient yet. The above presentation is patterned after our adaptation of the ideas of the Russian school. See [Carmona(1982)]. The argument can be adapted to treat the case of Example IV.2.3 when the common distribution of the random variables ξ_n have a continuous bounded density. See [Bentosela et al.(1983)].

Furthermore, it is claimed in [Carmona (1984a)] that it can also be pushed to apply to **the Bernouilli case**. This example has to be understood as the epitome of random potentials taking only a finite number of values. It corresponds, in the description of Example IV.2.3 the the case where the function χ is the characteristic function of the unit interval $[0,1]$, and the random variables ξ_n take only the values 0 and 1 with probability 1/2. It is easy to check that $\Sigma=[0,\infty)$. Assumption (A.1) on the positivity of the Ljapunov exponent is satisfied (see for example [Matsuda-Ishii(1976)]). Consequently, the results of the previous chapter imply that there is **no absolutely continuous spectrum**. Moreover, it is claimed in [Carmona(1984a)] that the exponential phenomenum occurs in $[1,\infty)$ and that **the nature of the spectrum in** $[0,1]$ is the only **remaining open problem**. S.Kotani noticed that some steps of the proof given in [Carmona(1984a)] are incorrect as presented. Nevertheless, we believe in the following cure.

The Bernouilli potential is of the Markovian type, i.e. of the form $F(X_t)$ for some stationary Markov process $\{X_t;t\in \mathbb{R}\}$. Indeed, we can consider the Markov process $X_t = (\xi_{[t]},t-[t])$ in $\{0,1\}\times[0,1)$ for $t\geq 0$, start it with its invariant distribution, and define it for all real parameters t. Its first coordinate, say $F(X_t)$, is nothing but a Bernouilli random potential, with the origin uniformly distributed on the interval $[0,1)$. Their spectral characteristics being identical we may as well work with $F(X_t)$. The absolute continuity assumption for the conditional distributions of the phases can be checked in the following way. Since $a<0<b$, the knowledge of $X_t=(q(t),s(t))$ for t in $[0,\infty)$ tells us that the potential $q(t)$ has to be equal to $q(0)$ on the interval $[-s(0),0]$, but the values of the potential $q(t)$ for $a\leq t<-s(0)$ are independent of the conditioning by $\mathcal{B}_{[0,\infty)}$. So, in order to compute the conditional distribution of $\theta(0,a,\omega,\varepsilon)$, we first compute the unconditional distribution of $\theta(-s(0),a,\omega,\varepsilon)$ and then adjust the phase between $-s(0)$ and 0 according to the conditioning. By stationarity, the distribution of $\theta(-s(0),a,\omega,\varepsilon)$ is the same as that of $\theta(-a,s(0),\omega,\varepsilon)$, and we compute the latter by a simple change of variables, using the fact that $-a$ is uniformly distributed on the interval $[m,m+1)$ and the fact that the function $a \dashrightarrow \theta(a,s(0),\omega,\varepsilon)$ is monotone increasing in this interval if $\varepsilon>1$. Notice also that, if e is restricted to a closed interval contained in $(1,\infty)$, we have uniform bounds on the derivative of this function (recall (III.20)), and assumption (A.2) follows.

An interesting question is the settlement of the problem of the spectral nature of the operator in the interval $[0,1]$.

The above proof has been generalized and extended in [Kotani(1984b)]. But the above mentionned problem still remains. Also, as it is given above, it merely gives an upper bound on the amplitude of the eigenfunctions. In fact, since γ can be chosen arbitrarily close to the Ljapunov exponent, it shows that, for each $\epsilon>0$ and for each ω in a set of probability 1, whenever ε is an eigenvalue, there exists a positive constant $k=k_\gamma(\omega,\epsilon,\varepsilon)$ such that the amplitude $r^{\omega}(t,\varepsilon)$ of the corresponding normalized eigenfunction satisfies:

$$r^{\omega}(t,\varepsilon) \leq k e^{-[\alpha(\varepsilon)-\epsilon]|t|}$$

As argued in [Craig-Simon(1983a)], the corresponding lower bound:

$$r^{\omega}(t,\varepsilon) \geq k' e^{-[\alpha(\varepsilon)+\varepsilon]|t|}$$

also holds for another positive constant $k'=k_1'(\omega,\epsilon,\varepsilon)$. The following simple modification of the above proof gives actually the lower bound. We learned this fact in a private discussion with Jean Brossard. Using a Wronskian argument the reader will easily convince himself that in order to prove the lower bound it is in fact enough to prove that:

$$\mathbf{E}\{\int_0^{\infty} \sup_{\|\Theta\|=1} \|U_{\varepsilon}^{\omega}(t,0)\Theta\|^{\delta} e^{-\delta[\alpha(\varepsilon)-\varepsilon]t}\, dt\} < \infty$$

for some $\delta>0$. Using the fact that

$$\sup_{\|\Theta\|=1} \|M\Theta\| \leq \mathbf{c}\, \left[\int \|M\Theta\|^{\delta} d\Theta\right]^{1/\delta}$$

for some positive constant $\mathbf{c}$ independent of the 2x2 matrix M, we can use the argument of the above proof to get the conclusion.

VII.3. Complements:

The publication of the note [Golsheid-Molcanov-Pastur(1977)] created a new vawe of interest for the mathematical theory of disordered systems. The standard methods of spectral analysis are well suited to the investigation of operators with absolutely continuous spectra and possibly some eigenvalues (which could accumulate here and there), mainly because the generic cases to handle were not expected to be **dense pure point spectra** like in the study of the random systems. In order to understand this new spectral type inverse scattering methods have been used to manufacture examples of these strange mathematical objects. See for example [Craig(1983)], [Pöschel(1983)] or [Bellissard et al.(1983)]. All these examples are dealing with the one dimensional lattice and all the potential functions are almost periodic so that the computations can be explicitely carried out.

A simple proof of points i) and ii) of the exponential localization in the onedimensional lattice case can be found in [Delyon et al.(1983)]. This proof has the usual drawbacks: first, it is written for potenteials which are ***independent and identically distributed at each site*** of the lattice, and second ***the common law has to have a bounded smooth density***. Nevertheless, it has the following interesting byproduct: *one can add any deterministic function to the random potential without changing the exponential localization.* Also, as noticed in [Simon(1982c)], it can be favourably modified to handle some potential functions decaying at infinity. For example, if the potential function $\{q(n);n\in\mathbf{Z}\}$ is of the form $q(n)=\lambda n^{-\alpha}q_n$ for some i.i.d. sequence $\{q_n;n\in\mathbf{Z}\}$ with a bounded density, then the spectrum is purely absolutely continuous if $\alpha>1/2$, pure point with eigenfunctions decaying like $\exp[-c|n|^{1-2\alpha}]$ when $\alpha<1/2$ and for $\alpha=1/2$, the spectrum is pure point with eigenfunctions with a power decay if λ is large and singular continuous for small λ. See [Delyon et al.(1984)] and [Delyon(1984)].

The perturbation result quoted above is typical of the lattice case, and is not true in general for the continuous case of operators on $L^2(\mathbf{R},dt)$ with potential functions of the form:

$$q(t,\omega) = q_0(t) + q_1(t,\omega)$$

with a deterministic function $q_0(t)$ and a stationary stochastic process $\{q_1(t,\omega);\ t\ \mathbf{R}\}$. For example adding the linear term $q_0(t) = Ft$ is a mathematical way to put the system in a constant electric field. In this case an interesting problem arise: we know that the disorder, namely the random part of the potential, has a tendency to localize the states and to create dense pure point spectrum, but on the other hand, we also know that the electric field wants to help the diffusion of the wave packets and create absolute continuous spectrum. So the question is, what kind of trade off will the system find? A first attempt to answer this question can be found in [Herbst-Howland(1981)] where the case of a random potential similar to the one in Proposition V.5 with an analytic function f is studied. It is shown that the spectrum is almost surely equal to the whole real line and that it is purely absolutely continuous in the case $F\neq 0$. Unfortunately, the case $F=0$ is not investigated and this is too bad because the spectrum could as well be already absolutely continuous to start with. Indeed such a random potential is deterministic in the probabilistic sense. A more satisfactory answer was given in [Bentosela et al.(1983)] where random potentials $q_1(t,\omega)$ made out of an **i.i.d. sequence of smoothed sqare wells** are studied. It is shown that **exponential localization holds when** $F=0$ and that the spectrum is pure absolutely continuous as soon as the electric field is turned on (i.e. $\mathbf{F\neq 0}$). After competing the first draft of this set of Notes , we received the preprint [Kirsch et al.(1985)] where this problem is discussed and where it is proven that the Herbst-Howland example does not have absolutely continuous spectrum, at least for most of the functions one can pick. Also, the following related result can be found in [Delyon et al.(1984)]: if $q_1(t,\omega)$ is a sequence of point masses at the points of the lattice $\mathbf{Z}$ multiplied by the elements of an i.i.d. sequence with a smooth bounded density, it is shown that one can go *from the exponential localization when* $F=0$, *to pure absolutely continuous spectrum for large* F *via an intermediate regime of dense pure point spectrum with power localized eigenfunctions*

Let us assume now that $q_0(t)$ is periodic and let us denote by $\Sigma_0=\cup[a_n,b_n]$ the spectrum of the operator $(-d^2/dt^2)+ q_0(t)$. It is proved in [Brossard(1983)] that if $q_1(t,\omega)$ is like in the Example IV.2.4 of the Russian school with a function F with range $[-\alpha,+\alpha]$, then $\Sigma=\cup[a_n-\alpha,b_n+\alpha]$ and the *exponential localization occurs*. Nevertheless, it is proved in [Carmona(1983)] that if $q_1(t,\omega)$ is multiplied by the characteristic function of a half axis (i.e. *if the disorder affects one side of the system only*) then the spectrum remains the same as a set, but *the exponential localization occurs only in* $\Sigma\setminus\Sigma_0$ *while the spectrum is pure absolutely continuous in* Σ_0. The proof of this last result is a subtle combination of the ideas of the present chapter, of formula (III.25) and the strategy defined right after this formula. This very idea has been used in [Carmona(1983)] to exhibit other examples of Schrödinger operators with this **new spectral type**: parts of the spectrum are pure absolutely continuous while the rest is dense pure point with exponential localization of the eigenfunctions. In order to create such models one can multiply the random potential of the Russian school for example by a bounded function which is identically equal to one in a neighborhood of one of the infinities and integrable near the other infinity.

We now conclude with a short discussion of the multidimensional results, and especially of the very recent announcements we heard of, and of the preprints we received since this course was given.

If one considers a Caley tree as a multidimensional example (actually infinite dimensional according to some people!), the exponential localization phenomenum has been announced in [Kunz-Souillard(1983)] for **large disorder** or for **weak disorder and low energies**. Some sort of Anderson transition is also discussed. In this case, like in all of the works in the lattice cases, the potential function is chosen to be an i.i.d. sequence indexed by the sites and the common distribution is assumed to have a nice density.

As we already pointed out in the last section of the previous chapter, the exponential localization has been announced in [Goldsheid(1981)] for the band (i.e. a subset of $\mathbb{Z}^2$ of the form $\{1,\dots,n\}\times\mathbb{Z}$) -see [Lacroix(1984)] for a proof- but the problem of the complete lattice $\mathbb{Z}^d$ was not (yet) the target of the annoncers. The situation changed drastically during these last two months of November and December 1984. After Goldsheid's announcement during a conference in Tashkent, and the rumor that Frölich, Martinelli, Scoppola and Spencer were preparing a paper on this pure point character of the spectrum, we received a written announcement by Simon, Taylor and Wolff. Two weeks later, we got in the mail, THE VERY SAME DAY, two preprints with essentially the same result: one by Delyon, Levy and Souillard and the other one by Frölich, Martinelli, Scoppola and Spencer. Rather than risking ourselves in a touchy discussion of the respective priorities, we will try to compare what we understand of the results.

The three papers give a proof of the exponential localization for the Anderson model in the energy region where Frölich and Spencer proved exponential decay of the Gren's function. The three papers use the fundamental estimate of [Frölich-Spencer(1983)] or an improvement of it. This is the reason why ***this paper has to be regarded as the fundamental breakthrough in the multidimensional case.***

More precisly, [Frölich et al.(1985)] obtain their results by reproving and improving the estimates of [Frölich-Spencer(1983)] while [Delyon et al.(1985)] and [Simon-Wolff(1985)] merely supplement the main estimate of [Frölich-Spencer(1983)] with a Fubini's type argument first used in [Kotani(1984b)], and based on an averaging of the spectral measures in the spirit of (III.25). [Simon-Wolff(1985)] gives a nice abstract account of this trick in the framework of rank one perturbations of self adjoint operators. [Delyon et al.(1985)] studies the asymptotic behavior of the eigenfunctions much in the same spirit as [Martinelli-Scoppola(1984)] and proves the pure point character of the spectrum like in the discussion of Section I. The simpicity of the pure point spectrum is obtained as a byproduct of the proof.

When applied to the one dimensional lattice case, these two proofs simplify greatly the existing ones, and shed some light on the nature of the phenomenum by requiring weaker assumptions. Both demand (on the top of the **positivity of the Ljapunov exponent**) the **existence of a density for the joint conditional distribution of the potential at two sites knowing its values everywhere else**. Eventhough this assumption is rather simple, it still **excludes the random potentials taking only a finite number of values.**

REFERENCES

M.AIZENMAN and **B.SIMON** (1982), Brownian Motion and Harnack Inequality for Schrödinger Operators. Comm. Pure Appl. Math. **35**, 209-271.

N.I.AKHIEZER and **I.M.GLAZMAN** (1978), Teory of Linear Operators in Hilbert Space I,II. Frederick Ungar Publish. Co. 3rd printing.

Ph.ANDERSON (1958), Absence of Diffusion in Certain Random Lattices. Physical Review **109**, 1492-1505.

J.AVRON and **B.SIMON** (1982a), Almost periodic Schrödinger operators, I. Limit periodic potentials. Comm. Math. Phys. **82**, 101-120.

J.AVRON and **B.SIMON** (1982b), Singular continuous spectrum for a class of almost periodic Jacobi matrices. (preprint)

J.AVRON and **B.SIMON** (1983), Almost periodic Schrödinger operators, II. The integrated density of states. Duke Math. J. **50**, 369-391.

J.BELLISSARD, **R.LIMA** and **E.SCOPPOLA** (1983), Localization in -dimensional incommensurate structures. Comm. Math. Phys. **88**, 465-477

J.BELLISSARD, **R.LIMA** and **D.TESTARD** (1983), A metal-insulator transition for the almost Mathieu model. Comm. Math. Phys. **88**, 207-234

J.BELLISSARD and **B.SIMON** (1982), Cantor Spectrum for the Almost Mathieu Equation. J. Functionnal Anal. **48**, 408-419

M.BEN-ARTZI (1983), Remarks on Schrödinger operators with an electric field and a deterministic potential. J. Math. Anal. Appl. (to appear)

M.M.BENDERSKII and **L.A.PASTUR** (1969), Calculation of the average number of states in a model problem. Soviet Phys. JETP **30**, 158-162

M.M.BENDERSKII and **L.A.PASTUR** (1970), On the spectrum of the one dimensional Schrödinger equation with a random potential. Math. Sb. **82**, 245-256

F.BENTOSELA, **R.CARMONA**, **P.DUCLOS**, **B.SIMON**, **B.SOUILLARD** and **R.WEDER** (1983), Schrödinger operators with an electric field and random or deterministic potential. Comm. Math. Phys. **88**, 387-397

J.BROSSARD (1983), Perturbations Aléatoires de Potentiels Périodiques. (preprint).

R.CARMONA (1979), Regularity Properties of Schrödinger and Dirichlet Semigroups. J.Functional Anal. **33**, 259-296

R.CARMONA (1982), Exponential Localization in One Dimensional Disordered Systems. Duke Math. J. **49**, 191-213

R.CARMONA (1983), One Dimensional Schrödinger Operators with Random or Deterministic Potentials: New Spectral Types. J. Functional Anal. **51**, 229-258

R.CARMONA (1984a), One Dimensional Schrödinger Operators with Random Potentials. Physica **124A**, 181-188

R.CARMONA (1984b), The essential support of the absolutely continuous spectral measures of ergodic operators is deterministic. (preprint)

A.CASHER and **J.L.LEBOWITZ** (1971), Heat Flow in Regular and Disordered Harmonic Chains. J. Math. Phys. **12**, 1701-1711.

V.CHULAEVSKY (1981), On perturbations of a Schrödinger operator with periodic potential. Russ. Math. Surveys **36**, (5) 143-144.

E.A.CODDINGTON and **N.LEVINSON** (1955), Theory of Ordinary Differential Equations. Mc Graw Hill, New York

F.CONSTANTINESCU, **J.FRÖLICH** and **T.SPENCER** (1983), Analiticity of the density of states and replica method for random Schrödinger operators on a lattice. J. Stat. Phys. **34**, 571-596

W.CRAIG (1983), Pure point spectrum for discrete Schrödinger operators. Comm. Math. Phys. **88**, 113-13

W.CRAIG and **B.SIMON** (1983a), Subharmonicity of the Ljapunov Index.Duke Math. J. **50**, 551-559.

W.CRAIG and **B.SIMON** (1983b), Log-Hölder Continuity of the Integrated Density of States for Stochastic Jacobi Matrices. Comm. Math. Phys. **90**, 207-218.

P.DEIFT and **B.SIMON** (1983), Almost Periodic Schrödinger Operators III, the Absolute Continuous Spectrum. Comm. Math. Phys. **90**, 389-411.

F.DELYON (1984), Apparition of purely singular continuous spectrum in a class of random Schrödinger operators. (preprint)

F.DELYON, **H.KUNZ** and **B.SOUILLARD** (1983), One dimensional wave equations in random media. J. Phys. **A16**, 25-42.

F. DELYON, **Y.LEVY** and **B.SOUILLARD** (1985), Anderson localization for multidimensional systems at large disorder or low energy. (preprint)

F.DELYON and **B.SOUILLARD** (1983), The rotation number for finite difference operators and its properties. Comm. Math. Phys. **89**, 415-427.

F.DELYON, B.SIMON and **B.SOUILLARD** (1984), From Power Law Localized to Extended States in Disordered Systems. Phys. Rev. Lett. **52**, 2187-2189.

Y.DERRIENNIC (1975), Sur le théorème ergodique sous additif. C.R. Acad. Sci. Paris ser.A **282**, 985-988

E.I.DINABURG and **Ya. G. SINAI** (1976), The one dimensional Schrödinger equation with a quasi-periodic potential. Funct. Anal. Appl. **9**, 279-289

W.F.DONOGHUE Jr (1974), Monotone Matrix Functions and Analytic Continuation. Springer Verlag, New York.

M.DONSKER and **S.R.S. VARADHAN** (1975), Asymptotics for the Wiener saussage. Comm. Pure Appl. Math. **28**, 525-565

N.DUNFORD and J.T.SCHWARZ (1963), Linear Operators II. John Wiley and Sons, New York.

W.G.FARIS (1975), Self Adjoint Operators. Lect. Notes in Math. **#433**, Springer Verlag, New York.

J.FRÖLICH, F.MARTINELLI, E.SCOPPOLA and **T.SPENCER** (1985), Anderson localization for large disorder or low energy. (preprint)

J.FRÖLICH and **T.SPENCER** (1983), Absence of Diffusion in the Anderson Tight Binding Model for Large Disoreder or Low Energy. Comm. Math. Phys. **88**, 151-184.

P.A.FUHRMANN (1981), Linear Systems and Operators in Hilbert Space. Mc Graw Hill, New York.

M.FUKUSHIMA (1974), On the spectral distribution of a disordered system and the range of a random walk. Osaka J. Math. **11**, 73-85.

M.FUKUSHIMA (1981), On asymptotics of spectra of Schrödinger oprators. in "Aspects Statistiques et Aspects Physiques des Processus Gaussiens", ed. C.N.R.S.

M.FUKUSHIMA, H.NAGAI and **S.NAKAO** (1975), On the asymptotic property of spectra of a random difference operator. Proc. Japan Acad. **51**, 100-102

M.FUKUSHIMA and **S.NAKAO** (1977), On spectra of the Schrödinger operator with a white noise potential. Z. Wahrsch. verw. Geb. **37**, 267-274.

H.FURSTENBERG (1963), Non commuting random products. Trans. Amer. Soc. **108**, 377-428.

H.FURSTENBERG and **H.KESTEN** (1960), Products of random matrices. Ann.Math.Stat. **31**, 457-469..

I. Ja. GOLSHEID (1980), Asymptotic Properties of the Product of Random Matrices Depending on a Parameter. in "Multicomponent Systems", Advances in Probability, **8**, 239-283.

I.Ja. GOLSHEID (1981), The structure of the spectrum of the Schrödinger random difference operator. Sov. Math. Dokl. **22**, 670-674.

I. Ja. GOLSHEID and **S.A.MOLCANOV** (1976), On Mott's Problem. Soviet Math. Dokl. **17**, 1369-1373.

I. Ja. GOLSHEID, S.A.MOLCANOV and **L.A.PASTUR** (1977), A pure point spectrum of the one dimensional Schrödinger operator. Funct. Anal. Appl. **11**, 1-10.

L.N.GRENKOVA, S.A.MOLCANOV and **Ju.N.SUDAREV** (1983), On the Basic States of the One Dimensional Disordered Structures. Comm. Math. Phys. **90**, 101-123.

Y.GUIVARCH' (1980), Quelques propriétés asymptotiques des produits de matrices aléatoires. in "Ecole d'Eté de Probabilités de Saint-Flour VIII, 1978, ed. P.L.Hennequin, Lect. Notes in Math. **#774, 177-249.**

Y.GUIVARCH' (1981), Exposants de Liapunov des produits de matrices aléatoires en dépendance markovienne. C.R. Acad. Sci. Paris ser.A **292**, 327-329.

P.R.HALMOS (1951), Introduction to Hilbert Space and the Theory of Spectral Multiplicity. Chelsea Publ. Co. New York.

I.HERBST and **J. HOWLAND** (1981), The Stark ladder and other one dimensional external field problems. Comm. Math. Phys. **80**, 23-42.

M.HERMAN (1983), Une méthode pour minorer les exposants de Ljapunov et quelques exemples montrant le caractère local d'Arnold et de Moser sur le tore de dimension 2. Comment. Math. Helvetici **58**, 453-502.

H.HOELDEN and **F.MARTINELLI** (1984), A remark on the absence of diffusion near the bottom of the spectrum for a random Schrödinger operator on $L^2(R^\nu)$. Comm. Math. Phys. **93**, 197-217.

K.HOFFMAN (1965), Banach Spaces of Analytic Functions. Prentice Hall, Englewood Cliffs, N.J.

K.ISHII (1973), Localization of Eigenstates and Transport Phenomena in the One Dimensional Disordered System. Suppl. Progr. Theor. Phys. **53**, 77-138.

K. ISHII and **H.MATSUDA** (1970), Localization of Normal Modes and Energy Transport in the Disordered Harmonic Chain. Suppl. Progr. Theor. Phys. **45**, 56-86.

R.JOHNSON and **J.MOSER** (1982), The Rotation Number for Almost Periodic Potentials. Comm. Math. Phys. **84**, 403-438.

O.KALLENBERG (1976), Random Measures. Academic Press, London.

T.KATO (1966), Perturbation Theory for Linear Operators. Springer Verlag, New York.

Y.KATZNELSON and B.WEISS (1982), A simple proof of some ergodic theorems. Israel J. Math. **42**, 291-296.

J.F.C.KINGMAN (1976), Subadditive Processes. in "Ecole d'Eté de Probabilités de Saint-Flour V, 1975, ed. P.L.Hennequin, Lect. Notes in Math.#**539**, 167-223.

W. KIRSCH, S. KOTANI and B. SIMON (1985), Absence of absolutely continuous spectrum for some one dimensional random but deterministic potentials. (preprint)

W.KIRSCH and F.MARTINELLI (1982a), On the ergodic properties of the spectrum of general random operators. J. Reine Angew. Math. **334**, 141-156.

W.KIRSCH and F.MARTINELLI (1982b), On the Spectrum of Schrödinger Operators with a Random Potential. Comm. Math. Phys. **85**, 329-350.

W.KIRSCH and F.MARTINELLI (1982c), On the density of states of Schrödinger operators with a random potential. J. Phys. A **15**,2139-2156.

W.KIRSCH and F.MARTINELLI (1983a), Large Deviations and the Lifschitz Singularity of the Integrated Density of States of Random Hamiltonians. Comm. Math. Phys. **89**, 27-40.

W.KIRSCH and F.MARTINELLI (1983b), On the essential self adjointness of stochastic Schrödinger operators. Duke Math. J. **50**, 2155-1260. erratum ibidem.

S.KOTANI (1976), On Asymptotic Behavior of the Spectra of a One Dimensional Hamiltonian with a certain Random Coefficient. Publ. Res. Inst. Math. Sci. Kyoto Univ. **12**, 447-492.

S.KOTANI (1983), Ljapunov indices determine absolute continuous spectra of stationary one dimensional Schrödinger operators. in Proc. Taneguchi Intern. Symp. on Stochastic Analysis. Kataka and Kyoto (1982), ed. K.Ito. North Holland, 225-247.

S.KOTANI (1984a), Support theorems for random Schrödinger operators. (preprint).

S.KOTANI (1984b), Lyapunov exponents and spectra for one-dimensional random Schrödinger operators. (preprint).

H.KUNZ and B.SOUILLARD (1980), Sur le spectre des opérateurs aux différences finies aléatoires. Comm. Math. Phys. **78**, 201-246.

H.KUNZ and B.SOUILLARD (1983), The localization transition on the Bethe lattice. J. Phys. (Paris) Lett. **44**, 411-414.

H.KUNZ and B.SOUILLARD (in prep), Localization Theory: A Review.

J.LACROIX (1983), Singularité du spectre de l'opérateur de Schrödinger aléatoire dans un ruban ou un demi ruban. Ann. Inst. H. Poincaré ser.A **38**, 385-399.

J.LACROIX (1984), Localisation pour l'opérateur de Schrödinger aléatoire dans un ruban. Ann. Inst. H. Poincaré ser.A **40**, 97-116.

F.LEDRAPPIER (1984), Quelques propriétés des exposants caractéristiques. Ecole d'Eté de Probabilités XII, Saint Flour 1982. Lect. Notes in Math. #1097, Springer Verlag.

F.LEDRAPPIER (1985), Positivity of the exponent for stationary sequences of matrices (preprint).

F.LEDRAPPIER et **G.ROYER** (1980), Croissance exponentielle de certains produits aléatoires de matrices. C.R. Acad. Sci. Paris ser.A **290**, 513-514.

E.LEPAGE (1983), Répartition d'etats pour les matrices de Jacobi à coefficients aléatoires. (preprint).

B.M.LEVITAN and **I.S.SARGSJAN** (1976), Introduction to Spectral Theory. Transl. Math. Monographs, #39, Amer. Math. Soc. Providence.

I.M.LIFSHITZ (1965), Energy spectrum structure and quantum statea of disordered condensed systems. Sov. Phys. Usp. **7**, 549-573.

F.MARTINELLI and **E.SCOPPOLA** (1984), Remark on the absence of absolutely continuous spectrum for d-dimensional Schrödinger operators with random potential for large disorder or low energy. (preprint).

N.MINAMI (1985), An extension of Kotani's theorem to random generalized Sturm Liouville operators. (preprint).

S.A.MOLCANOV (1978), The structure of eigenfunctions of one dimensional unordered structures. Math. U.S.S.R. Izvestija **12**, 69-101.

S.A.MOLCANOV (1981), The Local Structure of the Spectrum of the One Dimensional Schrödinger Operator. Comm. Math. Phys. **78**, 429-446.

J.MOSER (1981), An example of a Schrödinger equation with almost periodic potential and nowhere dense spectrum. Comm. Math. Helvetici **56**, 198-224.

S.NAKAO (1977), On the spectral distributionof the Schrödinger operator with a random potential. Japan. J. Math. **3**, 11-139.

H.OKURA (1979), On the spectral distribution of certain integro-differential operators with random potential. Osaka J. Math. **16**, 633-666.

L.A.PASTUR (1972), On the distribution of the eingenvalues of the Schrödinger equation with a random potential. Funct. Anal. Appl. **6**, 163-165.

L.A.PASTUR (1973), Spectra of random self adjoint operators. Russ. Math. Surveys **28**, 1-67.

L.A.PASTUR (1974), On the spectrum of the random Jacobi matrices and the Schrödinger equation on the whole axis with random potential. (preprint), Kharkov, in Russian.

L.A.PASTUR (1980), Spectral Properties of Disordered Systems in the One Body Approximation. Comm. Math. Phys. **75**, 167-196.

D.B.PEARSON (1978), Singular Continuous Measures in Scattering Theory. Comm. Math. Phys. **60**, 13-36.

J.PÖSCHEL (1983), Examples of Discrete Schrödinger Operators with Pure Point Spectrum. Comm. Math. Phys. **88**, 447-463.

M. REED and **B.SIMON** (1975), Methods of Modern Mathematical Physics II. Fourier Analysis, Self Adjointness. Academic Press, New York.

M. REED and **B.SIMON** (1978), Methods of Modern Mathematical Physics IV. Analysis of Operators. Academic Press, New York.

M. REED and **B.SIMON** (1979), Methods of Modern Mathematical Physics III. Scattering Theory. Academic Press, New York.

M. REED and **B.SIMON** (1981), Methods of Modern Mathematical Physics I. Functional Analysis. Academic Press, New York. 2nd printing.

A.Ja.REZNIKOVA (1981), The Central Limit Theorem for the Spectrum of the One Dimensional Schrödinger Operator. J. Stat. Phys. **25**, 291-308.

M.ROMIERO and **W.WRESZINSKI** (1979), On the Lifshitz Singularity and the Tailing in the Density of States for Random Lattice Systems. J. Stat. Phys. **21**, 169-179.

G.ROYER (1980), Croissance exponentielle de produits de matrices aléatoires. Ann. Inst. H.Poincaré ser.B, **16**, 49-62.

G.ROYER (1982), Etude des opérateurs de Schrödinger à potentiel aléatoire en dimension un. Bull. Soc. Math. France **110**, 27-48.

W.RUDIN (1973), Functional Analysis. Mc Graw Hill, New York.

W.RUDIN (1974), Real and Complex Analysis. 2nd edit. Mc Graw Hill, New York.

B.SIMON (1978), Classical boundary conditions as a tool in quantum physics. Adv. in Math. **30**, 268-281.

B.SIMON (1982a), Almost Periodic Schrödinger Operators: A Review. Adv. Appl. Math. **3**, 463-490.

B.SIMON (1982b), Schrödinger Semigroups. Bull. Amer. Math. Soc. **7**, 447-526.

B.SIMON (1982c), Some Jacobi matrices with decaying potential and dense point spectrum. Comm. Math. Phys. **87**, 253-258.

B.SIMON (1983a), Equality of the density of states in a wide class of tight-binding Lorentzian random models. Phys. Rev. B, **27**, 3859-3860.

B.SIMON (1983b), Kotani theory for one dimensional stochastic Jacobi matrices. Comm. Math. Phys. **89**, 227-

B.SIMON (1984), Lifschitz Tails for the Anderson Model. (preprint).

B.SIMON, **M.TAYLOR** and **T.WOLFF** (1984), Some rigorous results for the Anderson model. (preprint).

B.SIMON and **M.TAYLOR** (1985), Harmonic analysis on SL(2,R) and smoothness of the density of states in the Anderson model. (preprint).

B.SIMON and **T.WOLFF** (1985), Singular continuous spectrum under rank one perturbation and localization for random Hamiltonians. (preprint).

I.M.SLIVNYAK (1956), Spectrum of the Schrödinger Operator with Random Potential. Zh. vychist. Mat. mat. Fiz. **6**, 1104-1108.

M.A.SUBIN (1982), The Density of States of Self Adjoint Elliptic Operators with Almost Periodic Coefficients. Amer. Math. Soc. Transl. **118**, 307-339.

M. THOMPSON (1983), The State Density for Second Order Ordinary Differential Equations with White Noise Potential. Bollettino U.M.I. **2-B**, 283-296.

A.D.VIRTSER (1979), On products of random matrices and operators. Theory Proba. Appl. **24**, 367-377.

A.D.VIRTSER (1983), On the simplicity of the spectrum of the Ljapunov characteristic indices of a product of random matrices. Theory Proba. Appl. **26**, 122-136.

F.WEGNER (1981) Bounds on the density of States in Disordered Systems. Z. Phys. B Condensed Matter. **44**, 9-15.

Y.YOSHIOKA (1973), On the singularity of the spectral measures of a semi-infinite random system. Proc. Japan Acad. **49**, 665-668.

NOTATIONS

INDEX

ASPECTS OF FIRST PASSAGE PERCOLATION

Harry KESTEN

1. Introduction

Introduction. First-passage percolation was invented by Hammersley and Welsh in 1963 (cf. [19]; an excellent survey of the early results in the subject was given by Smythe and Wierman in [31]). Roughly speaking, first-passage percolation deals with the collection of points which can be reached within a given time from some fixed starting point, when the network of roads is given, but the passage times of the road are random. A mathematical model for this can be obtained by taking for the roads the edges of some graph $\mathcal{G}$. In these notes we shall only consider the case where this graph is $\mathbb{Z}^d$. Assign to each edge e between adjacent vertices of $\mathbb{Z}^d$ a random variable $t(e)$. Of course $u = (u_1,\dots,u_d)$ and $v = (v_1,\dots,v_d)$ are adjacent on $\mathbb{Z}^d$ if and only if

(1.1) $$\sum_{i=1}^{d} |u_i - v_i| = 1.$$

We interpret $t(e)$ as the passage time of e. Throughout we assume that

(1.2) all $t(e)$, $e \in \mathbb{Z}^d$, are independent and have the same distribution F,

and

(1.3) $$F(0-) = 0.$$

(1.3) implies that all $t(e)$ are positive w.p.1. For vertices u, v of any graph $\mathcal{G}$ a $\mathcal{G}$-path (from u to v) is an alternating sequence $(v_0,e_1,v_1,\dots,e_n,v_n)$ of vertices and edges of $\mathcal{G}$ such that v_i is adjacent on $\mathcal{G}$ to v_{i-1} and e_i is an edge between v_i and v_{i-1} for $i = 1,2,\dots,n$ (and $v_0 = u$, $v_n = v$). For a $\mathbb{Z}^d$-path $r = (v_0,e_1,\dots,e_n,v_n)$ define the passage time of r as

(1.4) $$T(r) = \sum_{i=1}^{n} t(e_i) .$$

The travel time from x to y is defined as

(1.5) $$T(u,v) = \inf\{T(r): r \text{ a path from } u \text{ to } v\}.$$

Finally, with $\underline{0}$ denoting the origin,

(1.6) $$\tilde{B}(t) = \{v \in \mathbb{Z}^d: T(\underline{0},v) \le t\} .$$

As the names indicate we interpret $T(r)$ as the time needed to traverse the path r, and $T(u,v)$ as the minimal time needed to go from u to v along some path. $\tilde{B}(t)$ then consists of all points which can be reached in time t from the origin. One now wishes to make statements about the behavior for large time of $\tilde{B}(t)$. To simplify the statement of the all important Theorem 1.7 it is advantageous to replace $\tilde{B}$ by a somewhat fattened version of $\tilde{B}$, namely the set

$$B(t) = \{v + \overline{U}: v \in \tilde{B}(t)\},$$

where $\overline{U}$ is the closed cube

$$\overline{U} = \{x = (x(1),\dots,x(d)): -\tfrac{1}{2} \le x(i) \le \tfrac{1}{2}, 1 \le i \le d\} .$$

The principal result says that $B(t)$ grows linearly and has an asymptotic shape which is not random. The exact result is the following theorem of Cox and Durrett [7], which was inspired on work of Richardson [28].

(1.7) THEOREM. Assume in addition to (2.2) and (2.3) that

(1.8) $$E \min\{t_1^d,\dots,t_{2d}^d\} < \infty ,$$

where $t_1,\dots,t_{2d}$ are i.i.d. random variables with distribution F. Then there exists a nonrandom convex set $B_0 \subset \mathbb{R}^d$, which is invariant under permutations of the coordinates and under reflections in the coordinate hyperplanes, has nonempty interior, and which is either compact or equals all of $\mathbb{R}^d$, and has the following property:
If B_0 is compact, then for all $\varepsilon > 0$

(1.9) $$(1-\varepsilon)B_0 \subset \frac{1}{t} B(t) \subset (1+\varepsilon)B_0 \quad \text{eventually w.p.1.}$$

If $B_0 = \mathbb{R}^d$, then for all $\varepsilon > 0$

(1.10) $$\{x : |x| \le \varepsilon^{-1}\} \subset \frac{1}{t} B(t) \quad \text{eventually w.p.1.}$$

If (1.8) fails then

(1.11) $$\limsup_{v\to\infty} \frac{1}{|v|} T(\underline{0},v) = \infty \quad \text{w.p.1.}$$ ///

The last statement shows that (1.8) is a necessary condition for the almost sure linear growth of $B(t)$. Nevertheless we shall show in the next section that without any moment condition one can define a modified passage time $\hat{T}(u,v)$ such that the family of random variables

$$\{\hat{T}(u,v) - T(u,v) : u,v \in \mathbb{Z}^d\}$$

is tight (see (2.32) and the lines following it). In particular

$$\frac{1}{|v|}[\hat{T}(\underline{0},v) - T(\underline{0},v)] \to 0 \quad \text{in prob., } |v| \to \infty .$$

Consequently we can view

$$\hat{B}(t) := \{v + \overline{U} : v \in \mathbb{Z}^d , \hat{T}(\underline{0},v) \le t\}$$

as an approximation to $B(t)$. It turns out that the analogues of (1.9) and (1.10) hold for $\hat{B}$ without the assumption (1.8). For $d = 2$ this was first shown by Cox and Durrett [7].

Historically, one began not with the study of all of $B(t)$, but with its rightmost point or its furthest point to the right on the first coordinate axis. Specifically, Hammersley and Welsh [19] introduced

$$a_{0,n} = T(\underline{0},(n,0,\ldots,0))$$

and

$$b_{0,n} = \inf\{T(\underline{0},(n,k_2,\ldots,k_d)) : k_2,\ldots,k_d \in \mathbb{Z}\} .$$

These are called the <u>point to point</u> and <u>point to line passage times</u>, respectively. (For $d > 2$, b_{0n} should be called the "point to hyperplane passage time", but we shall maintain the traditional term "point to line passage time".) If

(1.12) $$E \min\{t_1,\ldots,t_{2d}\} < \infty,$$

with $t_1,\ldots,t_{2d}$ as in (1.8), then there exists a constant $\mu = \mu(F,d) < \infty$, the so-called <u>time constant</u>, such that

(1.13) $$\lim_{n\to\infty} \frac{1}{n} a_{0,n} = \lim_{n\to\infty} \frac{1}{n} b_{0,n} = \mu \quad \text{w.p.1 and in } L^1.$$

It is easy to see ([19], Theorem 4.1.9) that in general $\mu < E\{t(e)\}$. Even without (1.12) one can define a time constant $\hat{\mu} < \infty$, but then one only obtains

(1.14) $$\frac{1}{n} a_{0,n} \to \hat{\mu} \text{ in probability and } \frac{1}{n} b_{0,n} \to \hat{\mu} \text{ w.p.1.}$$

Clearly B_0 must intersect the positive first coordinate axis in the segment from $\underline{0}$ to $(\mu^{-1},0,\ldots,0)$. We therefore see (compare proof of Theorem 1.7) that $B_0 = \mathbb{R}^d$ if and only if $\mu = 0$. We shall give a necessary and sufficient criterion for this.

(1.15) THEOREM. $\mu = 0$ and $B_0 = \mathbb{R}^d$ if and only if

$$F(0) \geq p_T(d),$$

where $p_T(d)$ is a critical probability for Bernoulli (bond) percolation on $\mathbb{Z}^d$ (defined in (5.7)). ///

Thus the distinction between compact and noncompact B_0 (or $\mu > 0$ and $\mu = 0$) depends only on the atom of F at the origin. μ depends continuously on F (using the weak topology on the space of distribution functions); see Theorem 6.9. Some results about the behavior of $\mu(F,d)$ for large d will be discussed in Sect. 8. However, little else is known about μ and even less about B_0. Durrett and Liggett [14] constructed a surprising example in which B_0 has flat edges (see Theorem 6.13). In Sect. 8 we show that for large d and F an exponential distribution B_0 is <u>not</u> a Euclidean ball. This is of some interest because early Monte Carlo simulations (cf. [15], [28]) suggested that B_0 was a disc for $d = 2$ and exponential F.

Theorem 1.7 and (1.13) correspond to laws of large numbers. Are there distributional limit theorems which give the rate of convergence? For instance does $(\gamma(n))^{-1}\{a_{0n}-n\mu\}$ have a limit distribution for some $\gamma(n) \to \infty$, $\gamma(n) = o(n)$, under suitable conditions. No results of this precision are known. However, we show in Sect. 5 that under some moment conditions there exists a $\rho > 0$ such that

(1.16) $$\limsup_{n\to\infty} (\log n)^{\rho}\left|\frac{\theta_{0,n}}{n} - \mu\right| < \infty \quad \text{w.p.1.}$$

for $\theta = a$ or b. Moreover for all $\varepsilon > 0$

(1.17) $P\{|\theta_{0,n} - n\mu| \geq \varepsilon n\} \to 0$ exponentially fast for $\theta = a$ or b.

Various generalizations and other interpretations of the above problems come to mind immediately. One may discuss passage times between two sets of vertices F_0 and F_1, given by

(1.18) $$T(F_0,F_1) = \inf\{T(u,v): u \in F_0, v \in F_1\}.$$

$\mathbb{Z}^d$ can be replaced by any regular (infinite) graph $\mathcal{G}$. Instead of associating the basic passage times to the edges, one can associate passage times to the vertices of $\mathcal{G}$. Of course, one can also do both, i.e., associate (different) passage times to vertices and to edges. One can also take the $t(e)$ not identically distributed; for instance different distributions for the passage times of horizontal and vertical edges of $\mathbb{Z}^2$ could be used. The basic random variables $t(e)$ can be interpreted as costs or amounts of fuel needed to traverse e.

There is an interesting relation between first-passage percolation and Eden's growth model [15]; this was first realized by Richardson [28]. If F, the distribution of $t(e)$, is exponential, i.e.,

(1.19) $$F(x) = \begin{cases} 0 & \text{if } x < 0 \\ 1-e^{-x} & \text{if } x \geq 0, \end{cases}$$

then the shape of $B(\cdot)$ at certain random times has the same distribution as Eden's animal. For simplicity consider the two-dimensional situation. Eden assumed that at time 1 an animal consists of one cell, represented by the unit square with center at the origin. At each integer time a new cell is chosen from the unit squares adjacent to the existing animal, but not yet belonging to it, and added to the animal. Thus at time n+ the animal consists of a connected set A_n

of n unit squares with centers in $\mathbb{Z}^2$. Several methods of choosing the new cell to produce A_{n+1} from A_n are possible. One version mentioned by Eden is to choose a given cell with a probability proportional to the number of edges it has in common with A_n. Thus, in Fig. 1.1 the cell with center at x has probability $2/8$ to be added at time 4 to A_3, while the cells with center at one of the o's have

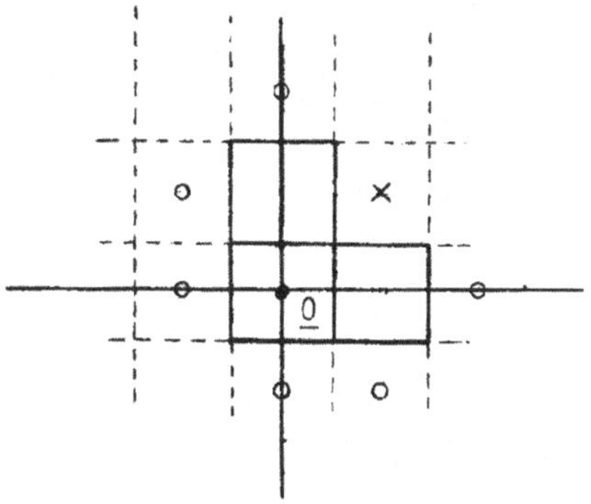

Fig.1.1. A possible configuration of A_3, with the centers of possible additions at time 4 marked by o and $\times$.

probability $1/8$. Using simple properties of the exponential distribution one can show that in this version A_n has the same distribution as $B(t_n+)$, when (1.19) prevails and

$$t_n = \inf\{t : B(t) \text{ contains } n \text{ vertices}\}.$$

Thus, the asymptotic results for $B(t)$ show that $n^{-1/2}A_n$ has an asymptotic shape (see [28]). Similar statements hold in all dimensions. The version in which all cells adjacent to A_n have the same probability of being added at time $(n+1)$ corresponds to a first-passage percolation model in which the travel times are attached to the vertices instead of the edges of $\mathbb{Z}^d$; see [14] for a discrete time analogue.

A rather differently looking class of problems concerns the maximal flow between two vertices v and w - or more generally between two disjoint sets of vertices F_0 and F_1 - of a graph $\mathcal{G}$. When $\mathcal{G}$ is the graph consisting of the restriction of $\mathbb{Z}^2$ to a rectangle these problems have been studied by means of first-passage percolation by Grimmett and Kesten [17]. For $\mathcal{G}$ equal to the restriction of $\mathbb{Z}^d$ with $d \geq 3$ to boxes one needs new generalizations of first-passage percolation A beginning on these for $d = 3$ can be found in [24]. We shall discuss these a little in Sect. 9. For now we only outline the general problem (for unoriented graphs). Let F_0 and F_1 be disjoint sets of vertices of $\mathcal{G}$. A permissable flow from F_0 to F_1 is an assignment of a number $f(e) \geq 0$ and a direction to each edge e such that for each vertex v outside $F_0 \cup F_1$

$$(1.20) \qquad \sum_v^+ f(e) = \sum_v^- f(e) ,$$

where $\sum_v^+$ ($\sum_v^-$) is the sum over all edges incident to v and directed towards v (away from v), as well as

$$(1.21) \qquad 0 \leq f(e) \leq t(e), \quad e \text{ an edge of } \mathcal{G}.$$

One now thinks of each edge as a channel or tube through which fluid can flow; $t(e)$ is the (random) capacity of e and $f(e)$ is the amount of fluid flowing through e per unit time in the direction assigned to e. (1.20) is a conservation law; at vertices outside $F_0 \cup F_1$ the total inflow equals the total outflow, so that no fluid can be gained or lost at such vertices. (1.21) is the reason for calling $t(e)$ the capacity of e; $t(e)$ is an upper bound for the amount of fluid which can flow through e per unit time (in either direction). The amount of fluid leaving F_0 per unit time is

$$\phi(F_0,F_1) = \sum{}' f(e) - \sum{}'' f(e),$$

where $\sum'$ ($\sum''$) is the sum over all edges e with one endpoint, u say, in F_0 and the other endpoint, v say, outside F_0 and directed from u to v (from v to u). The problem is to maximize the flow $\phi(F_0,F_1)$. By using the max flow-min cut theorem (see problem (9.1)) one can sometimes show that this maximal flow equals the minimal passage time $T(F_0^*,F_1^*)$ between vertex sets F_0^* and F_1^* on a dual graph. In Sect. 9.1 we mention some such results from [17] for rectangles in $\mathbb{Z}^2$ and discuss the needed generalization of first-passage percolation to treat similar problems on blocks in $\mathbb{Z}^d$. By and large the problems for these generalizations are still open.

Other open problems appear at the end of Sect. 9. Each section begins with a statement of the principal theorems for that section, thus allowing the reader to browse through the results without proofs.

Finally we list some general notation.

$\underline{0}$ denotes the origin or the vector $(0,\ldots,0)$.

We write the coordinates of $x \in \mathbb{R}^d$ as $(x(1),\ldots,x(d))$ and

$$|x| = \sum_1^d |x(i)|.$$

$\#A$ denotes the number of elements in A.

I_A denotes the indicator function of A.

$\lfloor a \rfloor$ denotes the largest integer $\le a$.

We say that a is positive if $a \ge 0$ and strictly positive if $a > 0$. Similarly a function f is called increasing if $f(x_1) \ge f(x_2)$ when $x_1 > x_2$ and strictly increasing if $f(x_1) > f(x_2)$ when $x_1 > x_2$. Similar conventions are adopted for "negative" and "decreasing".

H_k is the hyperplane $\{x : x(1) = k\}$.

E^c is the complement of E.

A path $r = (v_0, e_1, \ldots, e_n, v_n)$ is called self-avoiding or without double points if $v_i \neq v_j$ for $i \neq j$.

Acknowledgement. The author wishes to thank his colleagues M. Cohen and R. Connelly for help with the topological arguments in Sect. 2.

2. Kingman's subadditive ergodic theorem and the time constant.

Kingman developed his celebrated subadditive ergodic theorem (see [25], [26]) to prove (1.13). This subadditive ergodic theorem generalizes Birkhoff's ergodic theorem and has by now been proved in a variety of ways. Liggett [27] has recently given a useful variant and also gives a complete set of references. We state his result here.

(2.1) THEOREM. Let $\{X_{mn}\}_{0 \leq m < n}$ be a family of random variables which satisfies the conditions (2.2)-(2.5) below.

(2.2) $$X_{0n} \leq X_{0m} + X_{mn} \quad \text{for all} \quad 0 < m < n.$$

(2.3) The distribution of the sequences $\{X_{m+h,m+h+k}\}_{k \geq 1}$ is the same for all $h \geq 0$.

(2.4) For each $k \geq 1$, the sequence $\{X_{nk,(n+1)k}\}_{n \geq 0}$ is stationary.

(2.5) $$EX_{01}^{+} < \infty .$$

Then

(2.6) $$\lim_{n\to\infty} \frac{1}{n} X_{0n} \quad \text{exists and is} \quad < \infty \quad \text{w.p.1.}$$

If

(2.7) the stationary sequences in (2.4) are ergodic,

then the limit in (2.6) is constant w.p.1 and equals

(2.8) $$\lim_{n\to\infty} \frac{1}{n} EX_{0n} = \inf_{n>0} \frac{1}{n} E\, X_{0n} .$$

If, in addition to (2.2)-(2.5) one has

(2.9) $$E\, X_{0n}^{-} \geq -An \quad \text{for some} \quad A < \infty ,$$

then the convergence in (2.6) also holds in L^1. ///

(2.10) REMARK. Kingman requires that the distribution of the whole family $\{X_{m+h},X_{n+h}: m < n\}$ is the same for all h, instead of (2.3) and (2.4). ///

Now define

$$a_{m,n} = T((m,0,\ldots,0),(n,0,\ldots,0)). \tag{2.11}$$

It is clear that $X_{mn} = a_{m,n}$ satisfies (2.2)-(2.4) and (2.9) (with $A = 0$). Also (2.7) holds, since the shift invariant sets in[1] $\sigma\{X_{nk,(n+1)k}: n \geq 0\}$ are also sets in $\sigma\{t(e): e \in \mathbb{Z}^d\}$ which are invariant under the shift by $(k,0,0,\ldots,0)$ (compare [4], Prop. 6.31). The latter sets all have probability zero or one (cf. [4], Prop. 6.32 and Cor. 6.33). If (1.12) holds, then it is not hard to show that also (2.5) holds, because for any $n > 0$ there exist $2d$ disjoint paths $r_1,\ldots,r_{2d}$ between $\underline{0}$ and $(n,0,\ldots,0)$ and

$$a_{0,n} \leq \min\{T(r_1),T(r_2),\ldots,T(r_{2d})\}.$$

In fact, take for r_1 the path along the straight line segment of the first coordinate axis from $\underline{0}$ to $(n,0,\ldots,0)$. For $2 \leq i \leq d$ take for r_{2i-1} the path from $\underline{0}$ to $(0,\ldots,0,1,0,\ldots,0)$, from there parallel to the first coordinate axis to $(n,0,\ldots,0,1,0,\ldots,0)$ and then to $(n,0,\ldots,0)$, with the one on the ith place. For r_{2i} replace the 1 on the ith place by a -1. Finally, let r_2 be the path from $\underline{0}$ through $(-1,0,\ldots,0),(-1,1,0,\ldots,0),(-1,2,0,\ldots,0)$, then parallel to the first coordinate axis to $(n+1,2,0,\ldots,0)$ and further through $(n+1,1,0,\ldots,0),(n+1,0,0,\ldots,0)$ and finally to $(n,0,\ldots,0)$ (see Fig. 2.1 for $d = 2$; cf. [31], Sect. 4.2 and [7] for the proof of (2.5)).

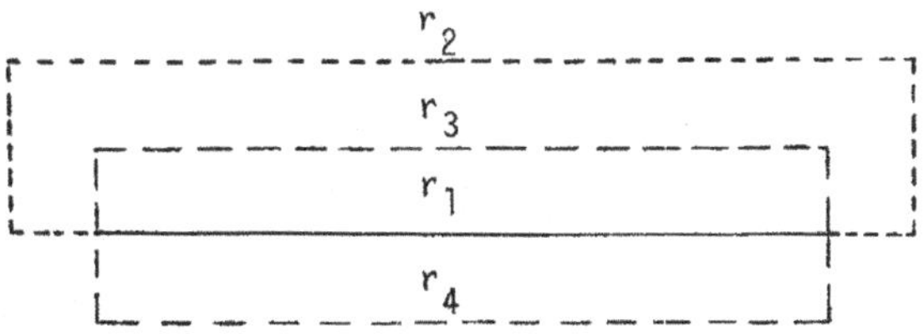

Fig. 2.1. The paths r_1 - r_4.

---

1) For a family $\{Y_\alpha: \alpha \in A\}$ of random variables, $\sigma\{Y_\alpha: \alpha \in A\}$ denotes the σ-field generated by this family.

In view of these observations (1.13) for a_{0n} follows from Theorem 2.1, whenever (1.12) holds. It is much harder to prove (1.13) for b_{0n}, since

(2.12) $$b_{m,n} := \inf\{T(r): r \text{ a path from } (m,0,\dots,0) \text{ to some point in the hyperplane } \{x(1) = n\}\}$$

does not satisfy (2.2). (1.13) for $b_{0,n}$ was first shown by Wierman and Reh [36], see also [31], Sect. 5.3. We shall follow Cox and Durrett [7] to derive this from the convergence of $t^{-1}\hat{B}(t)$ (see Sect. 3), even without (1.12). We remark that Hammersley and Welsh [19] also introduced two further passage times. We use the slightly modified versions of these, as defined in [31]. Let

(2.13) $$H_k = \{x(1) = k\}$$

(a hyperplane in $\mathbb{R}^d$) and set for $m < n$

(2.14) $$t_{m,n} = \inf\{T(r): r \text{ a path from } (m,\dots,0) \text{ to } (n,0,\dots,0) \text{ which, except for its endpoints, lies strictly between } H_m \text{ and } H_n\},$$

(2.15) $$s_{m,n} = \inf\{T(r): r \text{ a path from } (m,\dots,0) \text{ to some point in } H_n \text{ which, except for its endpoint lies strictly between } H_m \text{ and } H_n\}.$$

$t_{0,n}$ and $s_{0,n}$ are called the _cylinder point to point_ and _cylinder point to line passage times_, respectively. It is clear that

(2.16) $$b_{m,n} \le s_{m,n}, a_{m,n} \le t_{m,n},$$

but $s_{m,n}$ and $a_{m,n}$ are not comparable. It is also easy that $t_{m,n}$ satisfies (2.2)-(2.4), (2.7), (2.9). One can show (2.5) for $t_{0,n}$ in the same way as for $a_{0,n}$, but this time one requires

(2.17) $$E \min\{t_1,\dots,t_{2d-1}\} < \infty .$$

Under this condition one therefore obtains from Theorem 2.1 the convergence of $n^{-1}t_{0n}$. It turns out (see [19], Theorem 4.3.7) that its limit is the same μ as in (1.13). Combined with (1.13) and (2.16) this yields the following result.

(2.18) THEOREM. If (2.17) holds then there exists a constant $\mu < \infty$ such that

$$\lim \frac{1}{n}\, \theta_{0,n} = \mu \quad \text{w.p.1 and in } L^1$$

for $\theta = a,b,s$ or t. ///

The moment condition (2.17) was weakened by Wierman [35], at the price of obtaining only convergence in probability for $n^{-1}a_{0n}$ and $n^{-1}t_{0n}$. Instead of discussing this generalization, we shall go directly to the definition of the time constant μ without any moment conditions. For $d = 2$ the argument is due to Cox and Durrett [7]. To motivate the construction note (as in [7]) that

$$(2.19) \qquad \frac{1}{n}\, a_{0,n} \ge \frac{1}{n} \min\{t(e): e \text{ incident to } (n,0,\dots,0)\}.$$

There are $2d$ edges incident to $(n,0,\dots,0)$ and their passage times are independent. Thus the expectation of the right hand side of (2.19) equals the left hand side (1.12). Consequently, if (1.12) fails, then the lim sup (as $n \to \infty$) of the right hand side of (2.19) is ∞ w.p.1. Thus (1.12) is necessary for the a.s. convergence of $n^{-1}a_{0n}$; if (1.12) fails there may be edges with too large a passage time incident to $(n,0,\dots,0)$. The idea is not to include these bad edges in the calculation of a_{0n}. Instead, we construct a "shell" $S(v)$ around each vertex v, such that $S(v)$ contains only edges with passage times $\le M$, for a suitable large M. $a_{0,n}$ will be replaced by the passage time $\hat{a}_{0,n}$ between the shells around $\underline{0}$ and $(n,0,\dots,0)$. It will be the case that the $a_{0,n} - \hat{a}_{0,n}$, $n \ge 1$, are tight, so that $\hat{a}_{0,n}$ is a good approximation to $a_{0,n}$. We proceed with the formal definitions.

Let

$$(2.20) \qquad \pi_0 = 3^{-d}\, 3^{-(d+1)7^d}$$

and M such that

$$(2.21) \qquad P\{t(e) > M\} \le \pi_0 .$$

Assign (random) colors to the vertices of $\mathbb{Z}^d$: x is white if all edges e incident to x have passage time $t(e) \leq M$, and x is black otherwise. Next we need an auxilliary graph $\mathcal{L}$. The vertex set of $\mathcal{L}$ is the same as that of $\mathbb{Z}^d$ (so that the vertices of $\mathcal{L}$ automatically have a color black or white). Two distinct vertices $u = (u(1),\dots,u(d))$ and $v = (v(1),\dots,v(d))$ are adjacent on $\mathcal{L}$ - and hence have an edge of $\mathcal{L}$ between them - if

$$|u(i)-v(i)| \leq 1 \quad \text{for all} \quad 1 \leq i \leq d.$$

E.g. if $d = 3$ this means that $\mathcal{L}$ is obtained from $\mathbb{Z}^3$ by adding body diagonals in each unit cube with corners on $\mathbb{Z}^3$, and diagonals in each face of such a unit cube. For any graph $\mathcal{G}$ we call a set $\mathcal{S}$ of its vertices (edges) connected if for each $u_1,u_2 \in \mathcal{S}$ ($e_1,e_2 \in \mathcal{S}$) there exists a path on $\mathcal{G}$ whose first and last vertex (edge) are u_1 and u_2, respectively (e_1 and e_2, respectively). If A is any $\mathcal{L}$-connected set of vertices we define $\mathcal{C}(A,b)$, the black cluster of A on $\mathcal{L}$ as the union of A and the set of all vertices v_0 of $\mathcal{L}$ for which there exists a path $(v_0,e_1,\dots,e_n,v_n)$ on $\mathcal{L}$ from v_0 to some $v_n \in A$ such that $v_0,\dots,v_{n-1}$ are all outside A and black. Similarly, for a $\mathbb{Z}^d$-connected set of vertices A, $\mathcal{C}(A,w)$, the white cluster of A on $\mathbb{Z}^d$ is the union of A and the set of all vertices v_0 of $\mathbb{Z}^d$ for which there exists a path $(v_0,e_1,\dots,e_n,v_n)$ on $\mathbb{Z}^d$ from v_0 to some $v_n \in A$ such that $v_0,\dots,v_{n-1}$ are all outside A and white. We shall write $\mathcal{C}(v,w)$ for $\mathcal{C}(\{v\},w)$. Note that we defined black clusters only on $\mathcal{L}$ and white clusters only on $\mathbb{Z}^d$. The color of the vertices in A has no influence on $\mathcal{C}(A,b)$. By definition always $A \subset \mathcal{C}(A,b)$. Similarly $A \subset \mathcal{C}(A,w)$.

Unfortunately there seems to be no way of avoiding the use of $\mathbb{Z}^d$- connectedness in some entities and $\mathcal{L}$-connectedness in others. This also shows up in the next definition and Lemma 2.24 below. For an $\mathcal{L}$-connected set of vertices C we define its exterior boundary as

$$\partial_{ext} C = \{v : v \text{ a vertex of } \mathcal{L},\ v \notin C, \text{ but } v \text{ is adjacent on } \mathcal{L} \text{ to some vertex } u \text{ in } C \text{ and there exists a path on } \mathbb{Z}^d \text{ from } v \text{ to } \infty \text{ which is disjoint from } C\}.$$

We next define the shells $S(v)$. For any vertex v of $\mathbb{Z}^d$ (or $\mathcal{L}$), let

$$D_k(v) = [v(1)-k,v(1)+k] \times \dots \times [v(d)-k,v(d)+k],$$

$n = n(v) =$ minimal k for which there exists a vertex $u \in D_k(v)$ with an infinite white cluster $\mathcal{C}(u,w)$ on $\mathbb{Z}^d$,

$$S(v) = \partial_{ext}\mathcal{C}(D_{n(v)}(v),b).$$

$S(v)$ is the exterior boundary of the black cluster of some cube surrounding v. We have chosen this cube large enough to make sure that it contains points of an infinite white cluster. Note first that

(2.22) All vertices in $S(v)$ are white.

This is true because all vertices of $S(v)$ are $\mathcal{L}$-adjacent to a black cluster but do not belong to this black cluster itself. The other properties of $S(v)$ which we need are stated in the next two lemmas. K_i will denote a constant in $(0,\infty)$ whose specific value has no significance for our purposes, and it may have different values at different appearances. Distances will always be L^1 distances, i.e.,

$$|x-y| = \sum|x(i)-y(i)| .$$

(2.23) LEMMA. For any finite $\mathcal{L}$-connected set of vertices C, $\partial_{ext}C$ is $\mathbb{Z}^d$-connected. In particular, if $\mathcal{C}(D_{n(v)}(v),b)$ is finite, then $S(v)$ is $\mathbb{Z}^d$-connected. Moreover, in this case $S(v)$ separates v from ∞, in the sense that any path on $\mathbb{Z}^d$ from v to ∞ must intersect $S(v)$. ///

Note that $\partial_{ext}C$ is $\mathbb{Z}^d$-connected not merely $\mathcal{L}$-connected (despite the definition of ∂_{ext} by means of the adjacency relation on $\mathcal{L}$)

(2.24) LEMMA. For all n

$$P\{\mathcal{C}(D_n(v),b) \text{ is finite}\} = 1$$

and for all $k \geq 0$

$$P\{n(v) > k\} \leq K_1(k+1)3^{-k} ,$$

$$P\{\text{diameter of } S(v) > k\} \leq K_2(k+3)^{d-1}\, 3^{-\frac{k}{4}} . \qquad ///$$

We postpone the proofs of these lemmas and first show how they can be used to define a time constant.

(2.25) DEFINITION. $\hat{T}(u,v) = T(S(u),S(v)) = \inf\{T(r): r \text{ a path from } S(u) \text{ to } S(v)\}$,

$$\hat{a}_{m,n} = \hat{T}((m,0,\dots,0),(n,0,\dots,0)).$$

(2.26) THEOREM. There exists a constant $\hat{\mu} = \hat{\mu}(F,d) < \infty$ such that

$$\lim \frac{1}{n} \hat{a}_{0,n} = \hat{\mu} \quad \text{w.p.1 and in } L^1 .$$

The family $\{a_{0,n} - \hat{a}_{0,n} : n \geq 1\}$ is tight. If (1.12) holds, then $\hat{\mu} = \mu$. ///

REMARK: Because of the last result we shall not distinguish $\hat{\mu}$ from μ, and drop the hat.

<u>Proof</u>: We follow [7] closely. Call a $\mathbb{Z}^d$ path $(v_0, e_1, \dots, v_m)$ <u>white</u> if <u>all</u> its vertices are white. Set

$\tilde{S}(u) = \{v: \exists$ white $\mathbb{Z}^d$-path from v to some point z of $S(u)$ and v has the property that every $\mathbb{Z}^d$-path from v to ∞ must intersect $S(u)\}$.

One should think of $\tilde{S}$ as the white points in "the interior of S" or on S which are connected to S by a white path. We allow the path in the above definition to consist of the point v only, so that

$$S(u) \subset \tilde{S}(u). \tag{2.27}$$

Next we define

$\Delta(u)$ = sum of $t(e)$ over all edges e of $\mathbb{Z}^d$ which have at least one endpoint in $\tilde{S}(u)$,

$$X_{m,n} = \hat{a}_{m,n} + \Delta((n,0,\dots,0)), \quad m < n .$$

The first part of the theorem will follow from Theorem 2.1 if we can show (2.2)-(2.5), (2.7) and (2.9) for this $X_{m,n}$. (2.2) follows from the fact that $S((m,0,\dots))$ is $\mathbb{Z}^d$ connected (by (2.23)). (2.3), (2.4) and (2.9) are obvious, while (2.7) follows in the same way as for $a_{m,n}$ (see the lines after (2.11)). Thus we only have to verify (2.5). This in turn follows from the following three facts:

(2.28) $E\{\Delta^p(u)\} < \infty$ for all $p > 0$,

(2.29) $\tilde{S}(u)$ is $\mathbb{Z}^d$-connected,

(2.30) $\tilde{S}(u) \cap \tilde{S}(v) \neq \emptyset$ whenever u, v are $\mathbb{Z}^d$-adjacent.

Indeed (2.29) and (2.30) together imply that there is a path on $\mathbb{Z}^d$ from $S(\underline{0})$ to $S(1,0,\dots,0)$ consisting of only edges with at least one endpoint in $\tilde{S}(\underline{0}) \cup \tilde{S}(1,0,\dots,0)$. Thus

$$X_{01} \leq \Delta(\underline{0}) + 2\Delta(1,0,\dots,0),$$

and (2.5) is implied by (2.28).

(2.28) is easy since $S(u) \subset D_{d(u)}(u)$, where $d(u)$ = diameter of $S(u)$. Therefore $\tilde{S}(u) \subset D_{d(u)}$. Moreover, all vertices in $\tilde{S}(u)$ are white and any edge e with a white endpoint has $t(e) \leq M$. Thus

$$\Delta(u) \leq \text{sum of } t(e) \text{ over all } e \text{ which have a white endpoint in } D_{d(u)}$$
$$\leq M(2d(u)+1)^d .$$

Now use Lemma 2.24.

Also (2.29) is easy. Indeed, every $v \in \tilde{S}(u)$ is connected by a white $\mathbb{Z}^d$-path $(v,e_1,\dots,e_m,v_m)$ to $S(u)$. If we choose m minimal, i.e., such that $v_i \notin S(u)$ for $i < m$, then $v_1,\dots,v_{m-1}$ must also belong to $\tilde{S}(u)$. Indeed, if any of these vertices could be connected to ∞ without intersecting $S(u)$, then the same would be true for v, in contradiction to $v \in \tilde{S}(u)$. Thus any $v \in \tilde{S}(u)$ is connected inside $\tilde{S}(u)$ to $S(u)$, while $S(u)$ is itself connected by (2.23). Thus (2.29) holds.

Finally, we prove (2.30). Let u and v be adjacent on $\mathbb{Z}^d$ and let ζ be a path on $\mathbb{Z}^d$ from v to ∞. Since we obtain a path from u to ∞ by adjoining u and the edge of $\mathbb{Z}^d$ between u and v at the beginning of ζ, ζ must contain a vertex of $S(v)$ and a vertex in $S(u)$ (by virtue of (2.23) and the fact that $u \notin S(u)$). For the sake of argument, let the first point of ζ in $S(u) \cup S(v)$ belong to $S(u)$ (only trivial changes are needed if this point belongs to $S(v)$). Call this point z. Then any $\mathbb{Z}^d$-path ϕ from z to ∞

must intersect $S(v)$; otherwise we would be able to go from v to z via ζ and then to ∞ via ϕ, without hitting $S(v)$, which contradicts the fact that $S(v)$ separates v from ∞ (see (2.23)). We claim that z can also be connected to a point of $S(v)$ by a white $\mathbb{Z}^d$-path. Once this is proved we know that $z \in S(u) \cap \tilde{S}(v) \subset \tilde{S}(u) \cap \tilde{S}(v)$, and hence that (2.30) holds. To prove this claim note that there exists a vertex $y \in D_{n(u)} \subset \mathcal{C}(D_{n(u)}(u),b)$ for which $\mathcal{C}(y,w)$ is infinite. Take any $\mathbb{Z}^d$-path ψ from y to ∞ in $\mathcal{C}(y,w)$. Since $\mathcal{C}(D_{n(u)}(u),b)$ is finite by (2.24), and since ψ starts in $\mathcal{C}(D_{n(u)}(u),b)$ it must have a last vertex in $\mathcal{C}(D_{n(u)}(u),b)$. The next vertex on ψ, call it x, must belong to $S(u)$. Since $S(u)$ is connected, and since all its vertices are white (see (2.23) and (2.22)) we can now connect z to x on $S(u)$, and then via ψ to ∞. This path on $\mathbb{Z}^d$ from z to ∞ contains only white vertices and must intersect $S(v)$ (by what we already proved). This proves our claim, and hence (2.30). The convergence of $n^{-1}\hat{a}_{0,n}$ to a finite $\hat{\mu}$ follows.

The tightness of the $a_{0,n}-\hat{a}_{0,n}$ follows from the fact that if $v \notin D_{d(u)}(u)$ then v can be connected to ∞ by a path which does not intersect $S(u)$. Therefore v lies "outside" $S(u)$ and any path from u to v must intersect $S(u)$. Similarly, if $u \notin D_{d(v)}(v)$ any path from u to v must intersect $S(v)$. Therefore, as soon as

$$(2.31) \qquad v \notin D_{d(u)}(u) \quad \text{and} \quad u \notin D_{d(v)}(v)$$

any path r from u to v contains a piece r' from some $y \in S(u)$ to some $z \in S(v)$. Thus $T(r) \geq T(r') \geq \hat{T}(u,v)$. Taking the infimum over r yields $T(u,v) \geq \hat{T}(u,v)$. Conversely, let r' be a path from $y \in S(u)$ to $z \in S(v)$ and ξ a path from u to $y' \in S(u)$ and ζ a path from v to $z' \in S(v)$. By (2.23) and (2.27) we can connect y' to y (z to z') by a path on $S(u)$ ($S(v)$) with a passage time of at most $\Delta(u)$ ($\Delta(v)$). Thus

$$T(u,v) \leq T(\xi) + T(\zeta) + \Delta(u) + \Delta(v) + T(r') \,.$$

By taking the inf over r', ξ and ζ we finally obtain

$$(2.32) \qquad 0 \leq T(u,v) - \hat{T}(u,v) \leq \Delta(u) + \Delta(v) + T(u,S(u)) + T(v,S(v)).$$

Since $\Delta(u) + T(u,S(u))$ has the same distribution as $\Delta(\underline{0}) + T(\underline{0},S(\underline{0}))$ for all u, the required tightness follows.

Lastly, if (1.12) holds, then

$$\frac{1}{n} a_{0,n} \to \mu \qquad \text{and} \quad \frac{1}{n} \hat{a}_{0,n} \to \hat{\mu} \quad \text{w.p.1}$$

while

$$\frac{1}{n}(a_{0,n}-\hat{a}_{0,n}) \to 0 \quad \text{in probability.}$$

Thus $\hat{\mu} = \mu$ if (1.12) holds. ///

Proof of Lemma 2.23. This proof should be skipped by anyone not interested in topological details. Some algebraic topology seems unavoidable. We have, however, chosen to use direct arguments instead of the apparatus of topology when a simple direct argument for our special situation was possible. As before we take $\overline{U}$ the closed cube

$$\overline{U} = \{x \in \mathbb{R}^d : |x(i)| \le \frac{1}{2} , 1 \le i \le d\} .$$

We also introduce for some $0 < \varepsilon \le \frac{1}{8}$ the open cube

$$U^{\varepsilon} = \{x \in \mathbb{R}^d : |x(i)| < \frac{1}{2} + \varepsilon, 1 \le i \le d\}$$

and set

$$N = \bigcup_{v \in C} \{v + U^{\varepsilon}\} .$$

If C is finite and $\mathcal{L}$-connected, then N is a bounded open connected set in $\mathbb{R}^d$. In fact it is a union of open cubes and hence of a very nice form. The proof will be carried out in the following steps:

(i) ∂N, the topological boundary of N, is a topological (d-1) manifold (without boundary). One of the components of ∂N, D say, separates N from ∞, in the sense that any path in $\mathbb{R}^d$ from N to ∞ must intersect D.

(ii) If v', v'' are any vertices of $\mathbb{Z}^d$ which are connected by a path ϕ in $\mathbb{R}^d \setminus N$, then v' and v'' can also be connected by a $\mathbb{Z}^d$-path ψ which is disjoint from N, and intersects only such cubes $v + \overline{U}$, $v \in \mathbb{Z}^d$, which also contain a point of ϕ.

(iii) A vertex v of $\mathbb{Z}^d$ belongs to $\partial_{ext} C$ if and only if $v \notin C$ but $v + \overline{U}$ intersects D.

(iv) If v' and v'' belong to $\partial_{ext} C$ then they can be connected by a path on $\mathbb{R}^d$ which lies in $(v'+\overline{U}) \cup (v''+\overline{U}) \cup D$. By the procedure of (ii) this path can be deformed into a path on $\mathbb{Z}^d$ which only contains vertices of $\partial_{ext}C$.

(iv) states that $\partial_{ext}C$ is $\mathbb{Z}^d$-connected. It follows that $S(v) = \partial_{ext}\mathcal{C}(D_{n(v)},b)$ is $\mathbb{Z}^d$-connected, since by definition $\mathcal{C}(D_{n(v)},b)$ is $\mathcal{L}$-connected. It is also clear that $S(v)$ must separate v from ∞, because if ϕ is any $\mathbb{Z}^d$-path from v to ∞, and u is the last vertex on ϕ in $\mathcal{C}(D_{n(v)},b)$, then by definition the next vertex of ϕ lies in $\partial_{ext}\mathcal{C}(D_{n(v)},b) = S(v)$. Thus, the lemma will follow from (i)-(iv)..

Proof of (i). Let $x = (x(1),\dots,x(d)) \in \partial N$. Since $\partial N \subset \bigcup_{v\in C} \{v+\partial U^\varepsilon\}$, where ∂U^ε is the topological boundary of U^ε, some of the $x(i)$ have to equal $n\pm(\frac{1}{2}+\varepsilon)$ for some $n \in \mathbb{Z}$. If $x(i) = n+(\frac{1}{2}+\varepsilon)$ replace $x(i)$ by $x(i)-n$; if $x(i) = n-(\frac{1}{2}+\varepsilon)$ replace $x(i)$ by $-x(i)+n$. By similar transformations we may assume that each $x(i)$ which is not of the form $n\pm(\frac{1}{2}+\varepsilon)$ lies in the interval $[0,\frac{1}{2}]$. Finally, by renumbering the coordinates we may assume that x is of the following form, for some $r \geq 1$,

$$(2.33)\qquad (\tfrac{1}{2}+\varepsilon,\dots,\tfrac{1}{2}+\varepsilon,x(r+1),\dots,x(d)) \text{ with } 0 \leq x(i) \leq \tfrac{1}{2}$$
$$\text{but } x(i) \neq \tfrac{1}{2}-\varepsilon,\ r+1 \leq i \leq d.$$

To show that ∂N is a manifold, we must show that each $x \in \partial N$ of the form (2.33) has a neighborhood V_1 which is homeomorphic to $\mathbb{R}^{d-1}$ (see [32], Sect. 6.2). Denote the indicator function of a set A by $I(A,\cdot)$ for the remainder of this proof. Observe that $I(v+U^\varepsilon,y)$ is constant in $y(i)$ for $y(i) \in (-\frac{1}{2}+\varepsilon,\frac{1}{2}-\varepsilon)$, as well as for $y(i) \in (\frac{1}{2}-\varepsilon,\frac{1}{2}+\varepsilon)$. Therefore, if we choose $\eta > 0$ such that

$$(x(i)-\eta,x(i)+\eta) \subset [0,\tfrac{1}{2}-\varepsilon) \cup (\tfrac{1}{2}-\varepsilon,\tfrac{1}{2}],\ r+1 \leq i \leq d,$$

then the $(y(r+1),\dots,y(d))$-section of ∂N,

$$M(y(r+1),\dots,y(d)) := \{y(1),\dots,y(r):(y(1),\dots,y(d)) \in \partial N\}$$

is the same for all $(y(r+1),\dots,y(d)) \in (x(r+1)-\eta,x(r+1)+\eta) \times(x(r+2)-\eta,x(r+2)+\eta)\times\dots\times(x(d)-\eta,x(d)+\eta)$. It therefore suffices to find a neighborhood V of $(\frac{1}{2}+\varepsilon,\dots,\frac{1}{2}+\varepsilon)$ in $\mathbb{R}^r$ such that $V \cap M(x(r+1),\dots,x(d))$ is homeomorphic to $\mathbb{R}^{r-1}$ (for then we can take $V_1 = V\times(x(r+1)-\eta,x(r+1)+\eta)\times\dots\times(x(d)-\eta,x(d)+\eta)$). Thus for the x of (2.33) the $(r+1)$-th through the d-th coordinate play no role in the construction of V_1 and we therefore restrict ourselves to the case $r = d$, so that $x(i) = \frac{1}{2} + \varepsilon$ for all i.

Now let H be the hyperplane through 0

$$H = \{y \in \mathbb{R}^d : y(1) + \ldots + y(d) = 0\}$$

and let π be the projection on H. Clearly π is continuous. We claim that it is a one to one map from the compact neighborhood

$$V_2 := \{y : \tfrac{1}{2} - \tfrac{\varepsilon}{2} \le y(i) \le \tfrac{1}{2} + 4\varepsilon\} \cap \partial N$$

of x in ∂N onto a compact neighborhood of $\underline{0} = \pi(x)$. This will imply that π is a homeomorphism between V_2 and $\pi(V_2)$ (see [12], Theorem XI.2.1), and hence will complete the proof that ∂N is a $(d-1)$-manifold. To prove this claim we first note that $v + \overline{U}^\varepsilon$ (with $\overline{U}^\varepsilon :=$ closure of U^ε) contains x if and only if all coordinates $v(i)$ equal 0 or 1. If $v_1 := (1,1,\ldots,1) \in C$, then $x \in v_1 + U^\varepsilon$ and x is an interior point. Thus we may assume that $v_1 \notin C$, but that $v \in C$ for some other v with all $v(i) = 0$ or 1. On the other hand, each cube $v + U^\varepsilon$ with $v(i) = 0$ or 1 for all i contains the small closed cube (see Fig. 2.2)

$$G := [\tfrac{1}{2} - \tfrac{\varepsilon}{2}, \tfrac{1}{2} + \tfrac{\varepsilon}{2}] \times \ldots \times [\tfrac{1}{2} - \tfrac{\varepsilon}{2}, \tfrac{1}{2} + \tfrac{\varepsilon}{2}] \subset N .$$

Now denote by L_y the line $\{y + t(1,1,\ldots,1) : t \in \mathbb{R}\}$. For $y \in G$ this line contains the point y of N. But also, for $2\varepsilon < t_0 \le 3\varepsilon$ and $y \in G$ one has $\frac{1}{2} + \varepsilon < y(i) + t_0 \le \frac{1}{2} + 4\varepsilon, 1 \le i \le d$, so that $y + t_0(1,\ldots,1)$ lies in $u + \overline{U}_\varepsilon$ only for $u = v_1$. Therefore $y + t_0(1,\ldots,1) \notin \overline{N}$ and L_y contains a point, $z(y)$ say, of ∂N corresponding to a value of t in $[0,3\varepsilon]$. It is clear that for $y \in G$, $z(y) \in V_2$. Since $\pi(z) = \pi(y)$ for each $z \in L_y$, $\pi(V_2) \supset \pi(G)$, and one easily sees (compare Fig. 2.2) that $\pi(G) \supset H \cap ([-\frac{\varepsilon}{2}, \frac{\varepsilon}{2}] \times \ldots \times [-\frac{\varepsilon}{2}, \frac{\varepsilon}{2}])$. This, and the continuity of π show that $\pi(V_2)$ is a compact neighborhood of the origin in H. To show that π is one to one in V_2 we merely have to show that for any y, L_y can contain only one point of V_2. To see this note first that for each $u \in \mathbb{Z}^d$, $I(u+U^\varepsilon, z)$ is a decreasing function of $z(i)$ on $(\frac{1}{2} - \varepsilon, \frac{1}{2} + 5\varepsilon)$, $1 \le i \le d$. This is then also the case for

$$I(N,z) = \max_u I(u+U^\varepsilon, z).$$

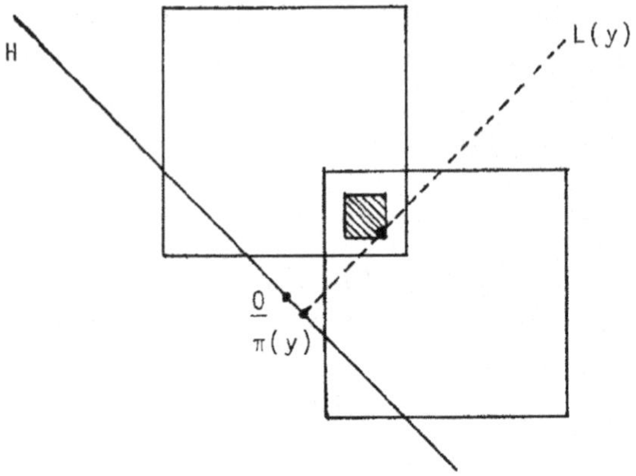

Fig.2.2. G is the small hatched square. y is denoted by a small solid circle in G. The two big squares are $(0,1)+U^{\varepsilon}$ and $(1,0)+U^{\varepsilon}$, respectively.

Now assume $z_j = y + t_j(1,\dots,1) \in V_2$ for $j = 1,2$ and some $t_1 < t_2$. Then $z_1 \in \partial N$ so that $I(N,\xi) = 0$ for some ξ's arbitrarily close to $z_1 \in V_2$. In particular $I(N,\xi) = 0$ for some ξ with $\frac{1}{2} - \varepsilon < \xi_i < z_2(i) < \frac{1}{2} + 5\varepsilon$ for each i (recall $z_2 \in V_2$), and by the above monotonicity $I(N,\zeta) = 0$ for all ζ in some neighborhood of z_2. This precludes $z_2 \in \partial N$ and hence z_1 and z_2 cannot both belong to V_2. This proves that π is one to one on V_2 and that N is a manifold.

We shall now use Poincaré and Alexander duality (see [32], Sect. 6.2) to show that some component of ∂N separates N from ∞. Let $D_1,\dots,D_\ell$ be the path components of ∂N. These are the same as the ordinary components, because ∂N is a manifold. ℓ is finite because ∂N is compact (since N is bounded). Let $N, E_1,\dots,E_k$ denote the components of $\mathbb{R} \setminus \partial N$. These are open and therefore are also the path components of $\mathbb{R} \setminus \partial N$. We shall show that $k = \ell$ and that each D_i separates N from exactly one $E_{j(i)}$, with $j(i_1) \neq j(i_2)$ when $i_1 \neq i_2$. Thus some D_i (which we shall denote by D) will separate N from the (unique) unbounded component among the E_i's, which is exactly what we want to show.

To carry out the details we first observe that the zeroeth singular homology group of a topological space X, with coefficients $\mathbb{Z}_2$, $H_0(X;\mathbb{Z}_2)$, equals the direct sum of n copies of $\mathbb{Z}_2$, where n equals the number of path components of X (see [32], Cor. 4.4.8 with

$\mathbb{Z}$ replaced by $\mathbb{Z}_2$). We now apply the Poincaré duality theorem, or rather Theorem 6.2.17 of [32] with the following choices: $X = A = \partial N$, $B = \emptyset$, $n = d-1$ = dimension of ∂N and $q = 0$. This yields, in the notation of [32]

$$H_0(\partial N;\mathbb{Z}_2) \approx \bar{H}^{d-1}(\partial N;\mathbb{Z}_2) .$$

Note that this theorem can be applied because ∂N is oriented over $\mathbb{Z}_2$ ([32], Theorem 6.2.9). Still in the notation of [32], we have further by the Alexander duality theorem (6.2.16 in [32])

$$\bar{H}^{d-1}(\partial N;\mathbb{Z}_2) \approx \tilde{H}_0(\mathbb{R}^d \setminus \partial N;\mathbb{Z}_2) .$$

But

$$H_0(\mathbb{R}^d \setminus \partial N;\mathbb{Z}_2) \approx \tilde{H}_0(\mathbb{R}^d \setminus \partial N;\mathbb{Z}_2) \oplus \mathbb{Z}_2$$

by [32], Lemma 4.3.1 with $\mathbb{Z}$ replaced by $\mathbb{Z}_2$. Thus, we finally arrived at

$$H_0(\partial N;\mathbb{Z}_2) \oplus \mathbb{Z}_2 \approx H_0(\mathbb{R}^d \setminus \partial N;\mathbb{Z}_2)$$

so that

$$\begin{aligned} k+1 &= \text{number of path components of } \mathbb{R}^d \setminus \partial N \\ &= (\text{number of path components of } \partial N) + 1 = \ell+1. \end{aligned}$$

Thus $k = \ell$ and ∂N separates $\mathbb{R}^d$ into $\ell+1$ components.

If we replace ∂N by one of the D_i everywhere in the last paragraph we obtain also that for each $1 \leq i \leq \ell$ $\mathbb{R}\setminus D_i$ has exactly two components. One of these components must contain N; denote this component by F_i^0. The other component - call it F_i^1 - must be the union of some of the E_j and some of the D_r, $r \neq i$, because $\mathbb{R}\setminus D_i$ is the union of the connected sets $N, E_1, \ldots, E_k, D_1, \ldots, D_\ell$, except for D_i. In fact F_i^1 cannot contain any D_r, because F_i^1 is open, and if it would contain any point of $D_r \subset \partial N$, then it would contain points of N. Thus F_i^1 is the union of some E_j. We claim that actually $F_i^1 = E_{j(i)}$ for a single $j(i)$. Indeed if F_i^1 would contain E_r and E_s for some $r \neq s$, then there would be a path ϕ from E_r to E_s in F_i^1. Such a path would have to contain a point of $\partial N = \cup D_j$, for E_r and E_s are distinct components of $\mathbb{R}^d\setminus\partial N$. But we have just seen that F_i^1 does not contain any point of $\cup D_j$, so that our claim follows.

We have now shown that $F_i^1 = E_{j(i)}$ for a unique i. We finally complete the proof of (i) by showing that $j(i_1) \neq j(i_2)$ whenever $i_1 \neq i_2$. Note that any path ϕ from $E_{j(i)} = F_i^1$ to F_i^0 must intersect D_i, since D_i separates $E_{j(i)}$ from F_i^0. The piece of such a ϕ from its initial point to its first intersection with D_i connects $E_{j(i)}$ to D_i, so that $E_{j(i)} \cup D_i$ is connected. Also $N \cup D_i$ is connected, since D_i is a connected subset of ∂N. Thus, for each i, $E_{j(i)} \cup D_i \cup N$ is connected. If $i_2 \neq i_1$ then $E_{j(i_2)} \cup D_{i_2} \cup N$ is also disjoint from D_{i_1} and therefore must be contained in one component of $\mathbb{R}^d \setminus D_{i_1}$. This component must be $F_{i_1}^0$ by definition of F_i^0. Therefore $j(i_2)$ is not $j(i_1)$ for $i_2 \neq i_1$. ///

Proof of (ii). For $\phi:[0,1] \to \mathbb{R}^d \setminus N$ be a path outside N from v' to v'' for two vertices v', v'' of $\mathbb{Z}^d$. Set $u_0 = v'$. If distinct vertices $u_0,\dots,u_k$ have already been found, and $u_k \neq v''$, then set

$$t_{k+1} = \sup\{t:\phi(t) \in u_k+\bar{U}\} .$$

Then $\phi(t_{k+1})$ lies in the boundary of $u_k+\bar{U}$ and there exists a vertex u_{k+1} of $\mathbb{Z}^d$ such that $\phi(t) \in u_{k+1}+U$ for some $t > t_{k+1}$ arbitrarily close to t_{k+1}. We continue this process until some t_n equals 1 and $u_n = v''$. Note that the above definition of u_{k+1} shows that $t_{k+2} > t_{k+1}$. Thus the sequence of t_k's is strictly increasing and we obtain inductively that $u_{k+1} \notin \{u_0,\dots,u_k\}$. Thus the u_k are distinct and we must have $u_n = v''$ for some finite n. Also

$$\phi(t_{k+1}) \in (u_k+\bar{U}) \cap (u_{k+1}+\bar{U}) \tag{2.34}$$

so that u_k and u_{k+1} must be $\mathcal{L}$-adjacent. Furthermore, no u_k can be in C, because $u_{k+1} \in C$ would imply

$$\phi(t_{k+1}) \in u_{k+1}+\bar{U} \subset N.$$

Unfortunately u_k and u_{k+1} do not have to be $\mathbb{Z}^d$-adjacent. However, we shall be able to insert a number of vertices $v_{k,1},v_{k,2},\dots,v_{k,r}$ (with $r = r(k)$) outside C such that two successive vertices in the sequence $u_k,u_{k,1},\dots,v_{k,r},u_{k+1}$ are $\mathbb{Z}^d$-adjacent and such that

(2.35) $$\phi(t_{k+1}) \in v_{k,j}+\overline{U},\ 1 \le j \le r(k).$$

Indeed, for the sake of argument, let

$$u_k = (s_1,\dots,s_d),\quad s_i \in \mathbb{Z},$$

$$u_{k+1} = (s_1+1,\dots,s_\ell+1,s_{\ell+1},\dots,s_d),\quad \text{and}$$

$$\phi(t_{k+1}) = (s_1+\tfrac{1}{2},\dots,s_\ell+\tfrac{1}{2},\tilde{s}_{\ell+1},\dots,\tilde{s}_d)\quad \text{with}$$

$$\tilde{s}_m \in [s_m-\tfrac{1}{2},s_m+\tfrac{1}{2}]\quad \text{for}\quad \ell+1 \le m \le d.$$

Then we can take $r(k) = \ell-1$ and

$$v_{k,j} = (s_1+1,\dots,s_j+1,s_{j+1},s_{j+2},\dots,s_d),\ 1 \le j \le \ell-1.$$

These points clearly satisfy our requirements. Now any two successive vertices in the full sequence

(2.36) $$v' = u_0,v_{0,1},\dots,v_{0,r(0)},u_1,v_{1,1},\dots,v_{1,r(1)},u_2,\dots,$$
$$v_{n-1,r(n-1)},u_n = v''$$

are $\mathbb{Z}^d$-adjacent, so that there exists a $\mathbb{Z}^d$-path ψ which passes only through these vertices. By (2.34), (2.35) ϕ intersects the cube $v+\overline{U}$ for each vertex v in the sequence (2.36), and as before this means $v \notin C$ for each such vertex. Consequently the path ψ cannot come nearer than distance 1 to the set C. This implies that the distance of ψ to N is at least $1-\frac{1}{2}-\varepsilon > 0$ (since each point of N is within L^1 distance $\frac{1}{2}+\varepsilon$ from some vertex in C). ///

<u>Proof of (iii)</u>. Let $v \in \partial_{ext}C$. Then by definition $v \notin C$ and v is $\mathcal{L}$-adjacent to some vertex $u \in C$, and v can be connected to ∞ via a path on $\mathbb{Z}^d \setminus C$. As at the end of the last step this path lies outside N, so that v lies in the unbounded component of $\mathbb{R}^d \setminus N$. Consequently the straight line segment ξ from v to u must intersect D. Since v and u are $\mathcal{L}$-adjacent

$$\xi \subset (u+\overline{U}) \cup (v+\overline{U}).$$

On the other hand D is disjoint from N and hence disjoint from $u+\bar{U}$, for $u \in C$. Therefore

$$\xi \cap D \subset (v+\bar{U}),$$

so that $v+\bar{U}$ does intersect D.

Conversely, assume $v+\bar{U}$ intersects D, but $v \notin C$. As in (i) we may assume (after some simple transformations) that $(v+\bar{U}) \cap D$ contains a point x of the form $(\frac{1}{2}+\varepsilon,\ldots,\frac{1}{2}+\varepsilon,x(k+1),\ldots,x(d))$ with

$$(2.37) \qquad 0 \le x(j) \le \tfrac{1}{2} \quad \text{but} \quad x(j) \ne \tfrac{1}{2}-\varepsilon,\ k+1 \le j \le d.$$

Then $x \in v+\bar{U}$ forces $v(i) = 1$ for $1 \le i \le k$ and $v(j) = 0$ for those $j \ge k+1$ for which $0 \le x(j) < \frac{1}{2}$. For those j for which $x(j) = \frac{1}{2}$ we can have $v(j) = 0$ or 1. However, if $x(j) = \frac{1}{2}$ and $v(j) = 1$ we can make a simple coordinate change, $y \to y'$ with

$$y'(\ell) = \begin{cases} y(\ell) & \text{if } \ell \ne j \\ 1-y(j) & \text{if } \ell = j \end{cases}$$

Then $x' = x$ while $v(j') = 0$. Thus, after several such changes we may assume that (2.37) holds and that $v = (1,\ldots,1,0,\ldots,0)$, i.e., $v(i) = 1(0)$ for $i \le k$ $(i \ge k+1)$.

By assumption $v \notin C$. However $x \in D$ implies that some point $u = (u(1),\ldots,u(d)) \ne v$ with $u(i) = 0$ or 1 must belong to C, so that v is $\mathcal{L}$-adjacent to some point of C. To show that $v \in \partial_{ext}C$ we merely have to show that there exists a path ϕ in $\mathbb{R}^d \setminus N$ from v to ∞. (ii) will then guarantee that there also exists a path from v to ∞ on $\mathbb{Z}^d \setminus C$. To construct ϕ we first show that the straight line segment ξ from v to x lies outside N. This follows from the monotonicity argument which was already used in (i). The indicator function of N, $I(N,y)$, is decreasing in each $y(i)$ on the intervals $\frac{1}{2}+\frac{\varepsilon}{2} \le y(i) \le 1$ and constant on the intervals $0 \le y(i) < \frac{1}{2}-\varepsilon$ and $\frac{1}{2}-\varepsilon < y(i) \le \frac{1}{2}$ separately. $I(N,y)$ can only jump downwards as $y(i)$ moves from the right to the left of $\frac{1}{2}-\varepsilon$. Now x is a boundary point of N and therefore there exist points y^n with $y^n \to x$, $y^n \notin N$. By the above we may choose $y^n(i) = x(i)+\frac{1}{n}$, $1 \le i \le k$, and $y^n(i) = x(i)$ for $k+1 \le i \le d$. Now

$$1 \geq tx(i) + (1-t)v(i) = t(\tfrac{1}{2} + \varepsilon) + 1 - t \geq y^n(i) = \tfrac{1}{2} + \varepsilon + \tfrac{1}{n},\ 1 \leq i \leq k,$$

$$0 \leq tx(i) + (1-t)v(i) = tx(i) \leq x(i) = y^n(i),\ k+1 \leq i \leq d,\ 0 \leq t \leq 1.$$

Thus, again by the above monotonicity property, $y^n \notin N$ implies $tx + (1-t)v \notin N$. This is true for all $0 \leq t \leq 1$. Therefore ξ is disjoint from N, as claimed. The desired ϕ can now be constructed by continuing ξ by a path ψ from x to ∞ outside N. Such a ψ exists since D is path connected and separates N from ∞. ///

Proof of (iv). Now let $v', v'' \in \partial_{ext}C$. As in step (iii) there then exist $x', x'' \in D$ such that the straight line segments ξ', ξ'' from x' to v' and x'' to v'', respectively, are disjoint from N, and such that $\xi' \subset v' + \bar{U}$, $\xi'' \subset v'' + \bar{U}$. Since D is path connected, there is a path ζ in D from x' to x''. The path ϕ obtained by concatenating ξ', ζ and the reverse of ξ'' connects v' to v'' outside N. Therefore, by (ii) there exists a path ψ on $\mathbb{Z}^d$ from v' to v''. This path ψ only intersects cubes $v + \bar{U}$ which also contain a point of ξ', ζ or ξ''. Since x', x'' and ζ are contained in D all those cubes contain a point of D. Also ψ is disjoint from N so that for any such cube, its center $v \notin C$. By (iii) this implies $v \in \partial_{ext}C$. Consequently all vertices on ψ belong to $\partial_{ext}C$ as desired.

Proof of Lemma 2.24: $C(D_n(v), b)$ can be infinite only if for some vertex $u \in \partial_{ext}D_n(v)$ there exists a black $\mathcal{L}$-path from u to ∞ (a black path is a path all of whose vertices are black). To prove the first statement of the lemma it therefore suffices to show that for each fixed v

(2.38) $$P\{\exists \text{ black } \mathcal{L}\text{-path } (v, e_1, v_2, \ldots, e_m, v_m) \text{ with } m \text{ vertices which are all distinct}\} \to 0 \text{ as } m \to \infty.$$

The proof of (2.38) is standard. The number of paths of m steps on $\mathcal{L}$, starting at v and without double points is at most $(3^d - 1)^m$. The probability that all vertices in such a path are black is at most $(2d\pi_0)^{\alpha m}$, where $\alpha = 3^{-d}$. This is so, because for any vertex z

$$P\{z \text{ is black}\} \leq 2d\pi_0 \quad \text{(see (2.21))},$$

and the colors of any set of vertices $z_1,\dots,z_t$ for which

$$\max_i |z_\ell(i)-z_k(i)| \geq 2,\ \ell \neq k, \tag{2.39}$$

are independent. Any path $(v,e_1,v_1,\dots,e_m,v_m)$ without double points contains a set of at least $3^{-d}m$ vertices such that all pairs of these vertices satisfy (2.39). In view of these observations the left hand side of (2.38) is at most

$$\{(3^d-1)(2d\pi_0)^\alpha\}^m \leq 3^{-m} .$$

Thus the first statement of the lemma holds.

To prove the second statement consider a configuration of colors with $n(v) > k$. Form $\mathcal{C}(D,w)$, the white cluster of $D = D_k(v)$ on $\mathbb{Z}^d$. By definition, $n(v) > k$ means that there cannot be an infinite white path on $\mathbb{Z}^d$ with initial point in or adjacent to $D_k(v)$. Thus, the assumption $n(v) > k$ implies $\mathcal{C}(D_k(v),w)$ is finite. Its exterior boundary, T_k, is therefore a $\mathbb{Z}^d$-connected set (by (2.23); note that $\mathcal{C}(D,w)$ is $\mathbb{Z}^d$-connected, and a fortiori $\mathcal{L}$-connected). If $v \in T_k$, then v is $\mathcal{L}$-adjacent to some $u \in \mathcal{C}(D,w)$, but v itself does not belong to $\mathcal{C}(D,w)$. In particular, this means that if v is even $\mathbb{Z}^d$-adjacent to u, then v must be black (compare (2.22)). If v is not $\mathbb{Z}^d$-adjacent to u, then there exists at least a path $u = v_0,e_1,v_1,\dots,v_{\ell-1},e_\ell,v_\ell = v$ on $\mathbb{Z}^d$ from u to v such that two successive v_j differ at most in one coordinate by 1 (so that $\ell \leq d$). At least one of the vertices v_j must be black (otherwise v itself belongs to $\mathcal{C}(D,w)$. Thus T_k is such that for each $v \in T_k$ there exists a black v' with $\max_i |v(i)-v'(i)| \leq 1$. We can also say something about the size of T_k. Since it separates $\mathcal{C}(D_k(v),w)$ from ∞ it also separates $D_k(v)$ from ∞, and therefore must intersect the first coordinates in points $(\ell_1,0,\dots,0)$ and $(-\ell_2,0,\dots,0)$ with $\ell_1,\ell_2 > k$. To contain these two points the connected set T_k must contain a $\mathbb{Z}^d$-path of at least $\ell_1+\ell_2$ vertices. Therefore

$$\{n(v) > k\} \subset \bigcup_{\ell>k}\ \bigcup_{m>\ell+k} \{\exists\ \mathbb{Z}^d\text{-path } r \text{ of at least } m \text{ vertices,}$$
$$\text{which starts at the vertex } (\ell,0,\dots,0) \text{ and is such that for}$$
$$\text{each } v \in r \text{ there is a black } v' \text{ with } \max_i |v(i)-v'(i)| \leq 1\} .$$

Very much as in the argument for (2.38) this yields

$$(2.40) \qquad P\{n(v) > k\} \le \sum_{m=2k+2}^{\infty} m(2d-1)^{m-1}(3^d\pi_0)^{m7^{-d}}$$

$$\le \sum_{m=2k+2}^{\infty} m\, 3^{-m} ,$$

which implies the second statement of the lemma.

Finally we estimate the tail of the distribution of

$$(2.41) \qquad d(v) := \text{diameter of } S(v).$$

Note that if $d(v) > 2k$, then $S(v)$ must contain points outside $D_k(v)$. If $n(v) = n < k$ and $d(v) > 2k$ there must exist a black path $(u,e_1,\dots,e_m,u_m)$ on $\mathcal{L}$ without double points, which starts at some u adjacent on $\mathcal{L}$ to $D_n(v)$ and ends on the boundary of D_k. This forces $m \ge k-n-1$. Since the number of vertices u adjacent to D_n is at most $2d(2n+3)^{d-1}$ we find (again as in the proof of (2.38))

$$P\{d(v) > 2k\} \le P\{n(v) > \tfrac{k}{2}\} + 2d(k+3)^{d-1}\{(3d-1)(2d\pi_0)^{\alpha}\}^{\frac{k}{2}-1}$$

$$\le K_1 k 3^{-\frac{k}{2}} + 2d(k+3)^{d-1} 3^{-\frac{k}{2}+1} .$$

3. <u>The asymptotic shape of $B(t)$ and $\hat{B}(t)$</u>. In this section we prove (part of) Theorem 1.7 plus the following analogue for $\hat{B}(t)$:

(3.1) THEOREM. There exists a nonrandom convex set $B_0 \subseteq \mathbb{R}^d$, which is invariant under permutations of the coordinates and under reflections in the coordinate hyperplanes, has nonempty interior, and which is either compact or equals all of $\mathbb{R}^d$, and has the following property: If B_0 is compact, then for all $\varepsilon > 0$

(3.2) $$(1-\varepsilon)B_0 \subset \frac{1}{t}\hat{B}(t) \subset (1+\varepsilon)B_0 \quad \text{eventually w.p.1.}$$

If $B_0 = \mathbb{R}^d$, then for all $\varepsilon > 0$

(3.3) $$\{x : |x| \le \varepsilon^{-1}\} \subset \frac{1}{t}\hat{B}(t) \quad \text{eventually w.p.1.}$$ ///

We shall also complete the proof of (1.14).

The idea of the proof of Theorems 1.7 and 3.1 is to first use subadditivity to demonstrate the linear growth of $B(t)$ and $\hat{B}(t)$, respectively, in a fixed (rational) direction. This implies the right growth rate in any finite number of directions simultaneously. To obtain the full result from this one needs some uniform estimate on the different growth rates in two directions which are close together. This is provided by Lemmas 3.5 and 3.6. We only prove Theorem 1.7 under the moment condition

(3.4) $$E\, t^2(e) < \infty,$$

which is stronger than (1.8). However this part of the proof exhibits the main ideas and the reader can consult [7] for the few additional technical steps needed when one works with (1.8) instead of (3.4).

(3.5) LEMMA. If (3.4) holds, then for $\lambda \ge 2d$

$$P\{T(u,v) \ge 2Et(e)(|u-v|+\lambda)\} \le K_3(|u-v|+\lambda)^{-2d}.$$

<u>Proof</u>: Without loss of generality we take $u = \underline{0}$. Observe that (as in the argument following (2.11)) there exist $2d$ edge disjoint paths from $\underline{0}$ to v, each containing at most $|v|+2d$ edges. (Recall that $|u-v| = \sum|u_i-v_i|$.) The $(2i-1)$-th and $2i$-th path, r_{2i-1} and r_{2i},

start with a step along the i-th coordinate axis in the positive and negative direction, respectively ($1 \le i \le d$). Rather than describe these paths formally we refer the reader to Fig. 3.1 which exhibits the paths when $d = 2$. Since the r_i are edge disjoint

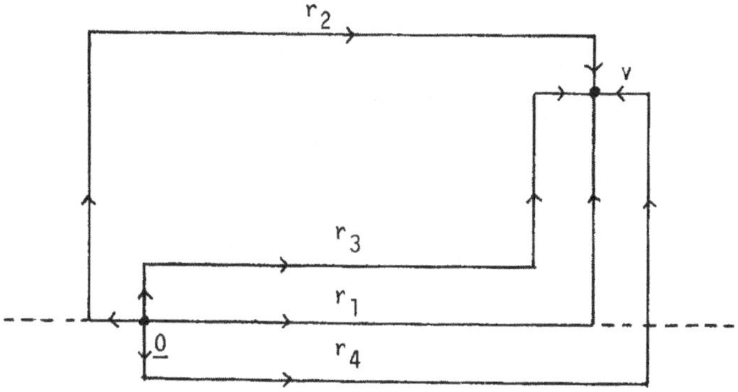

Fig.3.1. Four edge disjoint paths $r_1 - r_4$ from $\underline{0}$ to v of at most $|v| + 4$ edges.

$$P\{T(0,v) \ge 2Et(e)(|v|+\lambda) \le \prod_{i=1}^{2d} P\{T(r_i) \ge 2Et(e)(|v|+\lambda)\} .$$

Now

$$ET(r_i) \le Et(e)\cdot(\#\text{ of edges in } r_i) \le Et(e)(|u-v|+2d),$$

$$\sigma^2(T(r_i)) \le Et^2(e)\cdot(\#\text{ of edges in } r_i) \le Et^2(e)(|u-v|+2d),$$

so that the lemma follows from Chebyshev's inequality.

(3.6) LEMMA. If (3.4) holds then there exists a K_4 such that for all $0 < \varepsilon \le \frac{1}{2}$

$$P\{T(u,v) \le K_4\varepsilon|u| \text{ for all } u,v \in \mathbb{Z}^d \text{ with } |u-v| \le \varepsilon|u|$$
$$\text{and } |u| \text{ sufficiently large}\} = 1.$$

<u>Proof</u>: Fix $0 < \varepsilon \le \frac{1}{2}$. Let $C_k = C_k(\varepsilon)$ be the collection of all vertices $v = (v(1),\ldots,v(d))$ with[1)k]

---

1 $\lfloor a \rfloor$ denotes the largest integer $\le a$.

$$|v| = \lfloor (1+\varepsilon)^k \rfloor \quad \text{and} \quad v(1),\ldots,v(d-1) \quad \text{divisible by} \quad \lfloor \varepsilon(1+\varepsilon)^{k-1} \rfloor .$$

The number of possible values for $v(i)$, with $v \in C_k$, is at most $8/\varepsilon$ $(1 \leq i \leq d-1)$. Moreover, $|v(d)|$ is determined by $v(1),\ldots,v(d-1)$ when $v \in C_k$. Thus

$$|C_k| := \text{number of vertices in } C_k \leq (\tfrac{8}{\varepsilon})^d .$$

Introduce the event

$$E_k = E_k(\varepsilon) = \{T(u,v) \leq K_5\varepsilon(1+\varepsilon)^k \text{ for all pairs of vertices } u,v \text{ with } u \in C_k,\ |u-v| \leq 4d\varepsilon(1+\varepsilon)^k\},$$

where $K_5 = 10dEt(e)$. Then, by (3.5), for sufficiently large k[1)]

$$P\{E_k^c\} \leq \sum_{u \in C_k} \sum_{v:|u-v| \leq 4d\varepsilon(1+\varepsilon)^k} P\{T(u,v) > K_5\varepsilon(1+\varepsilon)^k\}$$

$$\leq (\tfrac{8}{\varepsilon})^d \{8d\varepsilon(1+\varepsilon)^k+1\}^d K_6\{\varepsilon(1+\varepsilon)^k\}^{-2d} \leq \frac{K_7}{\varepsilon^{2d}}(1+\varepsilon)^{-dk} .$$

Thus, by the Borel-Cantelli lemma

$$P\{E_k^c \text{ occurs infinitely often}\} = 0.$$

Now consider any realization of the $t(e)$ for which there exists a $k_0 < \infty$ such that E_k occurs for all $k \geq k_0$. Without loss of generality assume $\varepsilon(1+\varepsilon)^{k_0} \geq 1$. We claim that for such a realization

(3.7) $$T(u,v) \leq 4K_5\varepsilon|u| \quad \text{whenever} \quad |u| \geq (1+\varepsilon)^{k_0} \quad \text{and} \quad |u-v| \leq \varepsilon|u| .$$

To prove (3.7), let $\ell = \ell(u)$ be the unique integer for which

(3.8) $$\lfloor (1+\varepsilon)^{\ell} \rfloor \leq |u| < \lfloor (1+\varepsilon)^{\ell+1} \rfloor ,$$

---

1) E^c denotes the complement of E

and let $z = z(u)$ be a vertex of $C_{\ell+1}$ for which

$$|u-z(u)| = \min_{y \varepsilon C_{\ell+1}} |u-y| .$$

If $|u| \geq (1+\varepsilon)^{k_0}$, then $\ell(u) \geq k_0$, and by definition of $C_{\ell+1}$ there exists a $y \varepsilon C_{\ell+1}$ such that

$$(3.9) \qquad |y(i)| \leq |u(i)|, |y(i)-u(i)| \leq \lfloor \varepsilon(1+\varepsilon)^{\ell} \rfloor, \ 1 \leq i \leq d-1,$$

$$\text{and} \quad \operatorname{sgn} y(d) = \operatorname{sgn} u(d).$$

Then, by virtue of (3.8)

$$|y(d)-u(d)| = |y(d)|-|u(d)| \leq \lfloor (1+\varepsilon)^{\ell+1} \rfloor - \sum_{i=1}^{d-1} |y(i)|$$

$$- \{\lfloor (1+\varepsilon)^{\ell} \rfloor - \sum_{i=1}^{d-1} |u(i)|\} \leq d\varepsilon(1+\varepsilon)^{\ell} + 1,$$

and, by (3.9) and $\varepsilon(1+\varepsilon)^{\ell} \geq \varepsilon(1+\varepsilon)^{k_0} \geq 1$,

$$|y-u| \leq 2d\varepsilon(1+\varepsilon)^{\ell} .$$

A fortiori

$$|u-z(u)| \leq 2d\varepsilon(1+\varepsilon)^{\ell} .$$

Since $E_{\ell+1}$ occurs

$$T(z(u),u) \leq K_5\varepsilon(1+\varepsilon)^{\ell+1} ,$$

and if $|u-v| \leq \varepsilon|u| \leq \varepsilon(1+\varepsilon)^{\ell+1}$, then also

$$|v-z(u)| \leq |v-u| + |u-z(u)| \leq (2d+1)\varepsilon(1+\varepsilon)^{\ell+1}$$

and

$$T(z(u),v) \leq K_5\varepsilon(1+\varepsilon)^{\ell+1} ,$$

$$T(u,v) \leq T(z(u),u) + T(z(u),v) \leq 2K_5\varepsilon(1+\varepsilon)^{\ell+1} \leq 4K_5\varepsilon|u| .$$

This proves (3.7) and the lemma with $K_4 = 4K_5$.

Proof of Theorem 1.7 (under hypothesis (3.4)). We show that if $\mu > 0$, then (1.9) must hold for a compact convex B_0 which is invariant under permutations of the coordinates and under reflections in the coordinate hyperplanes, and which intersects the i-th coordinate axis in $(0,\ldots,0,\pm\mu^{-1},0,\ldots,0)$. We leave it to the reader to prove the (easier) fact that $\mu = 0$ implies (1.10). Let $\mathcal{V}_M$ be the set of all vectors $x = (x(1),\ldots,x(d))$ for which each $x(i)$ is an integer multiple of M^{-1}. For any $x \in \mathcal{V}_M$ the process

$$X_{m,n} = T(mMx,nMx),\ m < n,$$

satisfies (2.2)-(2.5) and (2.7) (compare the argument following (2.11 (2.11)). Thus, by Theorem 2.1 there exists a $\mu(x) < \infty$ such that

$$\lim_{n\to\infty} \frac{1}{nM} T(\underline{0},nMx) = \mu(x) \quad \text{w.p.1 and in } L^1. \tag{3.10}$$

It is easy to see that (3.10) defines $\mu(x)$ uniquely, for if $x \in \mathcal{V}_M \cap \mathcal{V}_N$, then

$$\lim_{n\to\infty} \frac{1}{nM} T(\underline{0},nMx) = \lim_{n\to\infty} \frac{1}{nMN} T(\underline{0},nMNx) = \lim_{n\to\infty} \frac{1}{nN} T(\underline{0},nNx).$$

Similarly one sees that $\mu(\cdot)$ is homogeneous on

$$\mathcal{V} := \bigcup_{M\geq 1} \mathcal{V}_M ,$$

in the sense that for rational $r \geq 0$

$$\mu(rx) = r\mu(x).$$

Next we show that μ is continuous on $\mathcal{V}$. First note that if $x \in \mathcal{V}_M$ and $y \in \mathcal{V}_N$, then $x-y \in \mathcal{V}_{MN}$ and

$$\mu(y)-\mu(x) = \lim_{n\to\infty} \frac{1}{nMN}(T(\underline{0},nMNy)-T(\underline{0},nMNx)).$$

Since, clearly,

$$T(\underline{0},nMNy) \leq T(\underline{0},nMNx)+T(nMNv,nMNy),$$

and $T(nMNx, nMNy)$ has the same distribution as $T(\underline{0}, nMN(y-x))$ it follows that

$$\mu(y)-\mu(x) \leq \lim_{n\to\infty} \frac{1}{nMN} T(nMNx, nMNy) \text{ in probability } = \mu(y-x).$$

It is also obvious that

(3.11) $$\mu((\pm x(\sigma(1)), \pm x(\sigma(2)), \ldots, \pm x(\sigma(d))) = \mu((x(1), x(2), \ldots, x(d)))$$

for any permutation $(\sigma(1), \ldots, \sigma(d))$ of $(1, \ldots, d)$, since our whole model is invariant under permutations or reflections of the coordinates. Thus,

(3.12) $$\mu((x(1), \ldots, x(d))) \leq \sum_{i=1}^{d} \mu((0, \ldots, 0, x(i), 0, \ldots, 0))$$

$$= \sum_{i=1}^{d} |x(i)|\mu(1,0,\ldots,0) = \mu \sum_{i=1}^{d} |x(i)|$$

and

(3.13) $$|\mu(y)-\mu(x)| \leq \mu(y-x) \leq \mu \sum_{i=1}^{d} |x(i)-y(i)| .$$

We can therefore extend μ by continuity to any z by defining

$$\mu(z) = \lim_{\substack{x\to z \\ x\in \mathcal{V}}} \mu(x).$$

The resulting $\mu(\cdot)$ is Lipschitz continuous on compact sets - in fact satisfies (3.11)-(3.13) - and is homogeneous, that is

(3.14) $$\mu(0) = 0 \quad \text{and} \quad \mu(\lambda x) = \lambda\mu(x), \quad \lambda > 0.$$

The last property of $\mu(\cdot)$ which we need is

(3.15) If $\mu(x) = 0$ for some $x \neq 0$, then $\mu(x) = 0$ for all x.

To prove (3.15) assume $\mu((x(1), \ldots, x(d))) = 0$ for some x with $x(1) \neq 0$. Then, by (3.11) and (3.13)

$$2|x(1)|\mu = \mu((2x(1),0,\dots,0)) \le \mu((x(1),\dots,x(d)))$$
$$+\mu((x(1),-x(2),\dots,-x(d))) = 0.$$

Then by (3.12) $\mu(\cdot) \equiv 0$.

Now we can prove (1.9) when $\mu > 0$ with

$$B_0 := \{x:\mu(x) \le 1\} .$$

This set is compact if $\mu > 0$, because in this case by (3.15)

(3.16) $\mu(x)$ is bounded away from 0 on $\{|x| = 1\}$,

and μ is homogeneous. B_0 is also convex by (3.13) and (3.14). B_0 is invariant under permutations and reflections of the coordinates (since $\mu(\cdot)$ has these symmetry properties). Finally, we show that given $\varepsilon > 0$ one has w.p.1

(3.17) $-\varepsilon|y| + \mu(y) \le T(\underline{0},y) \le \mu(y) + \varepsilon|y|$ for all sufficiently large y.

This relation does not depend on $\mu > 0$. To prove (3.17) consider a realization of the $t(e)$ for which the limit relation in (3.10) holds for all $x \in \mathcal{U}$ and the event in Lemma 3.6 occurs for all ε of the form m^{-1}, $m \ge 2$. The set of realizations with these properties has probability one. Finally, for any realization with these properties (3.17) holds. Indeed, If (3.17) would fail, it would fail for some $0 < \varepsilon < \frac{1}{2}$ along some sequence y_n, with $|y_n| \to \infty$ and $|y_n|^{-1}y_n \to z$ for some z with $|z| = 1$. We would then be able to find $m > (K_4+2)\varepsilon^{-1}$, an M, and $z' \in \mathcal{U}_M$ such that for all large n

$$|y_n - \lfloor\frac{|y_n|}{M}\rfloor Mz'| \le \frac{1}{m}|y_n|, \quad |\mu(\frac{y_n}{|y_n|}) - \mu(z')| \le \frac{1}{m}$$

and

(3.18) $$|T(\underline{0},y_n) - \mu(y_n)| > \varepsilon|y_n|$$

for n large enough. Then, for large n

$$|T(\underline{0},y_n) - T(\underline{0},\lfloor\frac{|y_n|}{M}\rfloor Mz')| \le T(y_n,\lfloor\frac{|y_n|}{M}\rfloor Mz') \le K_4\frac{|y_n|}{m},$$

$$\left|\frac{1}{|y_n|}T(\underline{0},\lfloor\frac{|y_n|}{M}\rfloor Mz')-\mu(z')\right| \le \frac{1}{m} ,$$

and consequently

$$\left|\frac{1}{|y_n|}T(\underline{0},y_n)-\mu(\frac{y_n}{|y_n|})\right| \le (K_4+1)\frac{1}{m}+\left|\mu(z')-\mu(\frac{y_n}{|y_n|})\right| \le (K_4+2)\frac{1}{m} .$$

Since we chose m such that $(K_4+2)m^{-1} < \varepsilon$, we arrived at a contradiction to (3.18). Thus (3.17) holds w.p.1.

If $\mu > 0$, then (1.9) easily follows from (3.14), (3.16) and (3.17). It is even more obvious from (3.17) that (1.10) holds when $\mu = 0$.

The fact that B_0 has a nonempty interior is immediate from $B_0 = \{x:\mu(x) \le 1\}$ and (3.13), (3.14).

There remains (1.11). This can be proved by the argument following (2.19). To be specific

$$T(\underline{0},v) \ge \min\{t(e):e \text{ incident to } v\} . \tag{3.19}$$

If we denote the right hand side of (3.19) by $Y(v)$, then the family of random variables $\{Y(v):v(i) \text{ is even for } 1 \le i \le d\}$ is independent, and each $Y(v)$ has the distribution of $\min\{t_1,\dots,t_{2d}\}$. Moreover, if (1.8) fails, then for each $\lambda \ge 1$

$$\begin{aligned}
&\sum_{\text{all } v(i) \text{ even}} P\{Y(v) \ge \lambda|v|\} \\
&= \sum_{k=0}^{\infty}\{\#\text{ of } v \text{ with all } v(i) \text{ even and } |v| = 2k\} \\
&\qquad \cdot P\{\min\{t_1,\dots,t_{2d}\} \ge 2\lambda k\} \\
&\ge K_3 \sum_{k=0}^{\infty}(k+1)^{d-1}P\{\min\{t_1,\dots,t_{2d}\} \ge 2\lambda k\} \\
&\ge \frac{K_4}{d}\, E\{\min\{t_1^d,\dots,t_{2d}^d\}\} = \infty .
\end{aligned}$$

Thus it follows from the Borel-Cantelli lemma that for each λ $Y(v) \ge \lambda|v|$ infinitely often w.p.1, and this, together with (3.19) yields (1.11). ///

The proof of Theorem 3.1 is almost a complete copy of the above proof with $\hat{T}$, $\hat{B}$ and $\hat{\mu}$ substituted for T, B and μ, respectively. Only towards the end of the proof a replacement for (3.6) is needed. This is provided by the next lemma.

(3.20) LEMMA. Without any assumptions on F there exists a K_4 such that for all $0 < \varepsilon \leq d^{-1}$

$$P\{\hat{T}(u,v) \leq K_4 \ \varepsilon|u| \ \text{ for all } \ u,v \in \mathbb{Z}^d \ \text{ with } \ |u-v| \leq \varepsilon|u| \ \text{ and } \ |u| \ \text{ sufficiently large}\} = 1.$$

Proof: We shall prove that for some K_5 and K_6 and all $\lambda \geq 2d$

$$(3.21) \qquad P\{\hat{T}(u,v) \geq K_5(|u-v|+\lambda)\} \leq K_6(|u-v|+\lambda)^{-2d} .$$

With this replacement for 3.5, the proof of Lemma 3.6 goes through if one everywhere replaces T by $\hat{T}$. The proof of (3.21) is based on (2.28)-(2.30). Without loss of generality take $u = \underline{0}$ and write n for $|v|$. Let $r = (u_0, e_1, \dots, e_n, u_n)$ be the path on $\mathbb{Z}^d$ from $\underline{0} = u_0$ to $v = u_n$ which consists of d pieces parallel to the successive coordinate axes, i.e., r goes from 0 to $(v(1),0,\dots,0)$, from there to $(v(1),v(2),0,\dots,0)$, etc., and finally from $(v(1),\dots,v(d-1),0)$ to $v = (v(1),\dots,v(d))$. Consider the sets $\tilde{S}(u_i)$. $S(\underline{0}) \subset \tilde{S}(\underline{0}) = \tilde{S}(u_0)$, $S(v) \subset \tilde{S}(v) = S(u_n)$, (see (2.27)), and each $\tilde{S}(u_i)$ is $\mathbb{Z}^d$-connected and $\tilde{S}(u_{i+1})$ has a vertex in common with $\tilde{S}(u_i)$ (by (2.29) and (2.30), respectively). Thus, there exists a path s from $\underline{0}$ to v which only contains edges in $\bigcup_{i=0}^{n} \tilde{S}(u_i)$. Consequently

$$(3.22) \qquad \hat{T}(\underline{0},v) \leq T(s) \leq \sum_{i=0}^{n} \Delta(u_i)$$

(see the definition of $\Delta(u)$ after (2.27)).

The difficulty now is that the $\Delta(u_i)$ are not independent, since the random sets $\tilde{S}(u_i)$ can intersect. To circumvent this difficulty we resort to a truncation argument. Let v and $\lambda \geq 2d$ be fixed and choose K_7 so large that

$$(3.23) \qquad \frac{1}{4} K_7 \log 3 \geq 2d+2.$$

Next, set $\nu = K_7 \log(n+\lambda)$ and replace $n(u)$ by

$\hat{n}(u)$:= minimal k for which there exists a vertex $z \in D_k(u)$ and a white $\mathbb{Z}^d$-path $(z_0, e_1, z_1, \dots, e_m, z_m)$ from some z_0 adjacent to z to some z_m in the boundary of $D_\nu(u)\}$,

in case such a $k < \nu$ exists. If there is no k with the above property, then we set $\hat{n}(u) = \nu$. Therefore, $\hat{n}(u) \leq \nu$ and $\hat{n}(u)$ depends only on the $t(e)$ for edges $e \subset D_\nu(u)$. We also replace $S(u)$ by

$$\partial_{ext} \mathcal{C}(D_{\hat{n}(u)}(u), b). \tag{3.24}$$

If the boundary in (3.24) lies in $D_\nu(u)$ we define

$\hat{S}(u) := \{v: \exists$ white $\mathbb{Z}^d$-path from v to some point of $\partial_{ext}\mathcal{C}(D_{\hat{n}(u)}(u), b)$ and v has the property that every $\mathbb{Z}^d$-path from v to ∞ must intersect $\partial_{ext}\mathcal{C}(D_{\hat{n}(u)}(u), b)\}$,

$\hat{\Delta}(u)$:= sum of $t(e)$ over all edges e of $\mathbb{Z}^d$ which have at least one endpoint in $\hat{S}(u)$.

If the boundary in (3.24) does not lie in $D_\nu(u)$, then we take $\hat{\Delta}(u) = 0$. With this definition $\hat{\Delta}(u)$ depends only on edges in $D_{\nu+1}(u)$, since one easily sees that $\hat{S}(U) \subset D_\nu(u)$, whenever the boundary in (3.24) is contained in $D_\nu(u)$. In order to replace $\Delta(u_i)$ by $\hat{\Delta}(u_i)$ in (3.22) we need an estimate for $P\{\hat{\Delta}(u) \neq \Delta(u)\}$. First we show that

$$P\{\hat{n}(u) \neq n(u) \quad \text{or} \quad n(u) > \tfrac{\nu}{2}\} \leq K_1(\nu+1)^d 3^{-\frac{\nu}{2}}. \tag{3.25}$$

To prove (3.25) we first observe that $\hat{n}(u) \leq n(u)$, for if $z \in D_k(v)$ with $k < \nu$ has an infinite white cluster, then there is a white path $(z_0, e_1, \dots, e_m, z_m)$ from a neighbor of z to $\partial D_\nu(u)$. Furthermore, if $\hat{n}(u) < n(u) \leq \frac{\nu}{2}$ then there exists a z in the boundary of $D_{\hat{n}(u)}(u)$

with a neighbor z_0 which is connected by a white $\mathbb{Z}^d$-path to $\partial D_\nu(u)$, but $C(z,w)$ finite. Let

$$T = \partial_{ext} C(z,w).$$

$C(z,w)$ is $\mathbb{Z}^d$-connected and a fortiori $\mathcal{L}$-connected, so that T is $\mathbb{Z}^d$-connected (by (2.23)) and as in the proof of the second statement of (2.24), for each $v \in T$ there exists a black v' with $\max_i |v(i)-v'(i)| \leq 1$. T separates $C(z,w)$ - and hence also z as well as z_m - from ∞. For at least one i, $|z(i)-z_m(i)| \geq \frac{\nu}{2}$, since $z \in D_{\hat{n}(u)}(u)$ with $\hat{n}(u) \leq \frac{\nu}{2}$ and $z_m \in \partial D_\nu(u)$. For the sake of argument, let $z(1) \leq \frac{\nu}{2} < \nu \leq z_m(1)$. Then T contains points $(z(1)-\ell_1, z(2), \ldots, z(d))$ and $(z_m(1)+\ell_2, z_m(2), \ldots, z_m(\ell))$ with $\ell_1, \ell_2 > 0$. Consequently

$$\{\hat{n}(u) < n(u) \leq \tfrac{\nu}{2}\} \subset \bigcup_{z \in D_{\nu/2}} \bigcup_{\ell > 0} P\{\exists\, \mathbb{Z}^d\text{-path}$$

$$(u_0, e_1, \ldots, e_{\lfloor \nu/2 \rfloor + \ell + 1}, u_{\lfloor \nu/2 \rfloor + \ell + 1}) \text{ without double points,}$$

$$\text{with } u_0 = (z(1)-\ell, z(2), \ldots, z(d)) \text{ and such that for each}$$

$$j\ \exists\, u_j' \text{ with } \max_i |u_j(i) - u_j'(i)| \leq 1 \text{ and } u_j' \text{ black}\}.$$

As in the proof of (2.38) and (2.40) we therefore have

$$P\{\hat{n}(u) < n(u) \leq \tfrac{\nu}{2}\} \leq (\nu+1)^d \sum_{\ell=1}^{\infty} \{(2d-1)(3^d \pi_0)^{7^{-d}}\}^{\frac{\nu}{2}+\ell}$$

$$\leq (\nu+1)^d\, 3^{-\frac{\nu}{2}} .$$

Together with (2.24) this implies (3.25).

It is now easy to prove that

$$(3.26) \qquad P\{\hat{\Delta}(u) \neq \Delta(u)\} \leq K_2 (\nu+1)^d\, 3^{-\frac{\nu}{4}} .$$

Indeed, if $\hat{n}(u) = n(u)$, then the boundary in (3.24) equals $S(u)$, and hence $\hat{S}(u) = \tilde{S}(u)$, $\hat{\Delta}(u) = \Delta(u)$, unless $S(u)$ is not contained in $D_\nu(u)$. However, the latter event can occur only if the diameter of $S(u) \geq \nu$, so that (3.26) follows from (3.25) and (2.24).

We finally come to the proof of (3.21). By (3.22) and (3.26)

$$(3.27) \qquad P\{\hat{T}(0,v) \geq K_5(|v|+\lambda)\} \leq K_2(n+1)(\nu+1)^d \, 3^{-\frac{\nu}{4}}$$

$$+ P\{\sum_{i=0}^{n} \hat{\Delta}(u_i) \geq K_5(|v|+\lambda)\}.$$

Now $\hat{\Delta}(u_i)$ and $\hat{\Delta}(u_j)$ are independent, whenever $D_{\nu+1}(u_i)$ and $D_{\nu+1}(u_j)$ are disjoint, and this is the case whenever $|i-j| > 2d(\nu+1)$. Thus, in the decomposition

$$\sum_{i=0}^{n} \hat{\Delta}(u_i) = \sum_{j=0}^{2d(\nu+1)} \sum_{\substack{i \equiv j (\mathrm{mod}\ 2d(\nu+1)+1) \\ i \leq n}} \hat{\Delta}(u_i)$$

each of the inner sums in the right hand side is a sum of at most $(2d(\nu+1)+1)^{-1}n+1$ independent random variables, each with the distribution of $\hat{\Delta}(0)$. Now take $K_5 = 2E\hat{\Delta}(0)$. $K_5 < \infty$ by virtue of $\hat{\Delta}(0) \leq D_{d(0)}$ and (2.24) (cf. proof of (2.28)). Therefore, for $p = 2d+1$

$$(3.28) \qquad P\{\sum_{i=0}^{n} \hat{\Delta}(u_i) \geq K_5(|v|+\lambda)\}$$

$$\leq (2d(\nu+1)+1) P\{\sum_{\substack{i \equiv 0 (\mathrm{mod}\ 2d(\nu+1)+1) \\ i \leq n}} [\hat{\Delta}(u_i) - E\hat{\Delta}(u_i)]$$

$$\geq \frac{1}{2} K_5 \frac{(n+\lambda)}{2d(\nu+1)+1}\}$$

$$\leq K_8 \, (\nu+1)\left(\frac{\nu+1}{n+\lambda}\right)^{2p} E\{|\sum_{\substack{i \equiv 0 (\mathrm{mod}\ 2d(\nu+1)+1) \\ i \leq n}} [\hat{\Delta}(u_i) - E\hat{\Delta}(u_i))]|^{2p}\}$$

$$\leq K_9 \, \frac{(\nu+1)^{2p+1}}{(n+\lambda)^{2p}} \left(\frac{n}{\nu+1}\right)^p E|\hat{\Delta}(0) - E\hat{\Delta}(0)|^{2p} \quad \text{(e.g. by [29], Theorem 3)}$$

$$\leq K_{10} \, \frac{(\nu+1)^{p+1}}{(n+\lambda)^p} \quad (K_{12} < \infty \text{ as in } (2.28)).$$

Since $\nu = K_7 \log(n+\lambda)$ and $n = |v|$, (3.21) follows from (3.23), (3.27) and (3.28) for $p = 2d+1$. ///

We close this section with the

Proof of (1.14). The convergence of $n^{-1}a_{0,n}$ to $\hat{\mu}$ in probability is immediate from (2.26). It also follows from (2.26) that

(3.29) $$\limsup \frac{1}{n}\hat{b}_{0,n} \le \limsup \frac{1}{n}\hat{a}_{0n} = \hat{\mu} \quad \text{w.p.1},$$

where, of course

$$\hat{b}_{0,n} = \inf\{\hat{T}(\underline{0},(n,k_2,\ldots,k_d)) : k_2,\ldots,k_d \in \mathbb{Z}\}.$$

We show that Theorem 3.1 implies

(3.30) $$\liminf \frac{1}{n}\hat{b}_{0,n} \ge \hat{\mu} \quad \text{w.p.1}.$$

If $\hat{\mu} = 0$ there is nothing to prove. Assume then that $\hat{\mu} > 0$ and consider a realization of the $t(e)$ for which

(3.31) $$\liminf \frac{1}{n}\hat{b}_{0,n} = \hat{\mu}-2\delta \quad \text{for some } \delta > 0.$$

There then exists a sequence $n_1 < n_3 < \ldots$ and vertices $z_1, z_2, \ldots$ with $z_i(1)$ = first coordinate of $z_k = n_k$, and $\hat{T}(0,z_k) \le \hat{b}_{0,n_k} + \delta$. Thus, by (3.31) $z_k \in B(n_k(\hat{\mu}-\delta))$ for large k. By (3.2) this implies that $\hat{B}_0$ must contain some point x with

$$x(1) \ge \limsup \frac{z_k(1)}{n_k(\hat{\mu}-\delta)} = \frac{1}{\hat{\mu}-\delta}.$$

But then, by symmetry and convexity, $\hat{B}_0$ contains the point

$$\tfrac{1}{2}(x(1),\ldots,x(d)) + \tfrac{1}{2}(x(1),-x(2),\ldots,-x(d)) = (x(1),0,\ldots,0).$$

But then $\hat{B}(t)$ contains $(\lfloor(1-\varepsilon)tx(1)\rfloor,0,\ldots,0)$, for large t, and therefore

$$\liminf \frac{\hat{a}_{0,n}}{n} \le \liminf_{t\to\infty} \frac{\hat{a}_{0,\lfloor(1-\varepsilon)tx(1)\rfloor}}{(1-\varepsilon)tx(1)} \le \liminf_{t\to\infty} \frac{t}{(1-\varepsilon)tx(1)}$$

$$\le \frac{(\hat{\mu}-\delta)}{1-\varepsilon}.$$

This would hold for all $\varepsilon > 0$, in contradiction to (2.26). Thus (3.30) holds and

$$\lim_{n\to\infty} \frac{1}{n} \hat{b}_{0,n} = \hat{\mu} \quad \text{w.p.1.} \tag{3.32}$$

Finally, as in (2.32) one has w.p.1 for all large n

$$b_{0,n} \geq \hat{b}_{0,n} \tag{3.33}$$

and

$$b_{0,n} \leq \Delta(0) + T(0,S(0)) + \hat{a}_{0,n} + \Delta((n,0,\ldots,0)). \tag{3.34}$$

Note that we do not need a term $T((n,0,\ldots,0)), S((n,0,\ldots,0)))$ in the right hand side of (3.34), because $S((n,0,\ldots,0))$ itself already intersects the hyperplane $\{x:x(1) = n\}$. The second part of (1.14) is immediate from (3.32)-(3.34), (2.26) and (2.28).

4. Some inequalities. The widely used Harris inequality (also called FKG inequality in more general setups) states that increasing functions of independent random variables are positively correlated. Our principal aim in this section is to prove an inequality which goes in the opposite direction from Harris' inequality.

We need some definitions to formulate the inequalities. As before our probability space is $\Omega = [0,\infty)^{\mathcal{E}}$, where $\mathcal{E}$ is the edge set of $\mathbb{Z}^d$. For σ-field in Ω we take the obvious σ-field $\mathcal{B} :=$ σ-field generated by the family of coordinate functions $\omega \to \omega(e)$, $e \in \mathcal{E}$.

(4.1) DEFINITION. A random variable X (respectively an event A) is called a cylinder random variable (respectively, cylinder event) if $X(\omega)$ (respectively $I_A(\omega)$) depends on finitely many $\omega(e)$ only.

(4.2) DEFINITION. We say that the cylinder random variable X, or the cylinder event A, is based on $\{e_1,\dots,e_k\}$ if $X(\omega)$, respectively, $I_A(\omega)$, depends on $\omega(e_1),\dots,\omega(e_k)$ only. In this case we write

$$\mathrm{base}(X) \subset \{e_1,\dots,e_k\}, \text{ respectively } \mathrm{base}(A) \subset \{e_1,\dots,e_k\} .$$

(Note that if $X(\omega) =$ constant, then $\mathrm{base}(X) \subset \phi$.)

(4.3) DEFINITION. A random variable X, or event A, is called increasing (decreasing) if $X(\omega)$, respectively $I_A(\omega)$, is an increasing[1)] (decreasing) function of each $\omega(e)$, $e \in \mathcal{E}$.

In the remainder of this section P is a product probability measure on $(\Omega,\mathcal{B})$, so that the $\omega(e)$, $e \in \mathcal{E}$, are independent under P. The simplest form of Harris' inequality is the following:

(4.4) If X_1, X_2 are positive[2)] cylinder random variables which are both increasing or both decreasing then

$$E\, X_1X_2 \geq EX_1EX_2 .$$

Specialization to indicator functions yields

---

1) We call $f(\cdot)$ increasing if $f(x_1) \geq f(x_2)$ for $x_1 \geq x_2$. If $f(x_1) > f(x_2)$ for $x_1 > x_2$ then we call f strictly increasing.

2) In our terminology positive means ≥ 0 and strictly positive means > 0.

(4.5) $$P\{A_1A_2\} \geq P\{A_1\}P\{A_2\}$$

whenever A_1 and A_2 are both increasing or decreasing cylinder events. More generally one has

(4.6) $$P\{A_1 \cap A_2 \cap \dots \cap A_n\} \geq \prod_1^n P\{A_i\}$$

when either all A_i are increasing cylinder events or all A_i are decreasing cylinder events. The intuitive reason behind (4.5) and (4.6) is that the occurrence of an increasing event A_1 makes high values of $\omega(e)$ more likely, and this can only help the occurrence of another increasing event A_2. Thus $P\{A_2|A_1\} \geq P\{A_2\}$ for increasing A_1, A_2, and this is equivalent to (4.5). Assume now that we prevent this positive influence of A_1 on A_2 by insisting that A_1 and A_2 "use different edges" or "occur disjointly" (see (4.7) for the formal definition). Then we can expect the inequality in (4.5) to be reversed because the occurrence of A_1 now merely forces A_2 to occur by means of edges not yet used for A_1. Theorem (4.11), which is a basically a special case of results of [3] states that this is indeed the case.

(4.7) DEFINITION. Let $A_1,\dots,A_n$ be any events. Then

$A_1 \circ \dots \circ A_n = \{\omega: \exists$ disjoint sets of edges $\mathcal{E}_1,\dots,\mathcal{E}_n$ and cylinders $C_1,\dots,C_n$ such that $\text{base}(C_i) \subset \mathcal{E}_i$ and such that $C_i \subset A_i$ and $\omega \in \bigcap_{i=1}^n C_i\}$.

We say that $A_1,\dots,A_n$ occur disjointly if and only if $A_1 \circ \dots \circ A_n$ occurs. ///

It will be part of the proof of the next theorem that $A_1 \circ \dots \circ A_n$ belongs to the completion of $\mathcal{B}$ with respect to P.

(4.8) THEOREM. Let $\{A(k,i):k,i \geq 1\}$ be a family of increasing events, or a family of decreasing events, and let $\{A'(k,i):k,i \geq 1\}$ be a family of events on some probability space $(\Omega',\mathcal{B}',P')$ with the following two properties:

(4.9) For each fixed i the joint distribution[1] of $\{A(k,i):k \geq 1\}$ under P is the same as the joint distribution of $\{A'(k,i):k \geq 1\}$ under P'.

(4.10) The families $\{A'(k,i):k \geq 1\}$, $i = 1,2,\ldots$ are independent under P'.

Then, for arbitrary $n(k) < \infty$ one has

(4.11)
$$P\{\bigcup_{k\geq 1} (A(k,1)\circ A(k,2)\circ\ldots\circ A(k,n(k))\}$$
$$\leq P'\{\bigcup_{k\geq 1} \bigcap_{i=1}^{n(k)} A'(k,i)\}.$$

(4.12) REMARK. For fixed i, the events $A'(1,i)$, $A'(2,i),\ldots$ may of course be dependent. ///

A good illustration of the use of this theorem is the following application to first-passage percolation, which is at the basis of the estimates of the next section. Identify $t(e) = t(e,\omega)$ with $\omega(e)$ and define $T(r)$ and $T(F_0,F_1)$ as in (1.4) and (1.18), respectively. Let $V(\ell,i)$, $\ell \geq 1$, $1 \leq i \leq n(\ell)$ and V, W be sets of vertices of $\mathbb{Z}^d$. Then

(4.13) $P\{\exists$ path r without double points from V to W which for some ℓ passes successively through $V(\ell,1),\ldots,V(\ell,n(\ell))$ and has $T(r) < x\}$
$$\leq \sum_{\ell\geq 1} P\{\sum_{i=0}^{n(\ell)} T'(\ell,i) < x\}, \qquad x > 0,$$

where all $T'(\ell,i)$ are independent and[2]

---

1) By the joint distribution of a family of events $\{A(k)\}$, we mean the joint distribution of $\{I_{A(k)}\}$.

2) $X' \overset{d}{=} X$ means that X' and X have the same distribution

$T'(\ell,i) \overset{d}{=} T(V(\ell,i),V(\ell,i+1))$, $0 \le i \le n(\ell)$, with

$V(\ell,0) = V$, $V(\ell,n(\ell)+1) = W$.

One obtains (4.13) by writing the left hand side of (4.13) as

$$(4.14) \qquad P\{\bigcup_{\ell \ge 1} \bigcup_{m \ge 1} (A(m,\ell,1)\circ \ldots \circ A(m,\ell,n(d))\},$$

where

$$A(m,\ell,i) = \{T(V(\ell,i),V(\ell,i+1)) < x(m,\ell,i),$$

and the union over m (for fixed ℓ) runs over all choices of $x(m,\ell,i)$ as rational numbers which satisfy

$$x(m,\ell,1) + \ldots + x(m,\ell,n(\ell)) < x.$$

The representation (4.14) follows from the fact that any path r from V to W through $V(\ell,1),\ldots,V(\ell,n(\ell))$ can be decomposed into edge disjoint pieces r_i from $V(\ell,i)$ to $V(\ell,i+1)$, $0 \le i \le n(\ell)$ with $T(r) = \sum T(r_i)$. The inequality in (4.13) follows immediately from (4.14) and (4.11)(applied for each fixed ℓ to the union over m in (4.14)).

In the actual application in the proof of Prop. 5.23 there will be some further restrictions on the pieces r_i, which diminishes the $A(m,\ell,i)$, but the idea will remain the same. We shall also ignore the $T'(\ell,i)$ for some of the i. Thus we replace the right hand side of (4.13) by the larger expression

$$\sum_{\ell \ge 1} P\{\textstyle\sum_\ell^* T'(\ell,i) < x\} ,$$

where $\sum_\ell^*$ is a subsum of $\sum_{i=0}^{n(\ell)}$. This can be done because the T' are positive.

Proof of Theorem 4.8. Order $\mathcal{E}$ in an arbitrary way as $\{e(1),e(2),\ldots\}$. We shall first approximate

$$(4.15) \qquad \bigcup_{k \ge 1} (A(k,1)\circ \ldots \circ A(k,n(k))$$

by a set of the form

$$(4.16) \qquad \Gamma = \bigcup_{k=1}^{M} \bigcup_{\wp} \bigcap_{i=1}^{N} C(k,i,\wp),$$

where $\wp$ runs through the finite set of partitions of $\{e(1),\dots,e(L)\}$ into $(N+1)$ sets $\mathcal{E}_1,\dots,\mathcal{E}_{N+1}$, and $C(k,i,\wp)$ is a cylinder with base $\subset \mathcal{E}_i$. This approximation will also serve to prove the measurability of (4.15). We then append to each edge $e(j)$, $1 \le j \le L$, N new edges $e(j,r)$, $1 \le r \le N$, with the associated values $\omega(e(j,r))$ all equal to $\omega(e(j))$. Γ is replaced at the same time by a set $\tilde{\Gamma}$ defined in terms of the $\omega(e(j,r))$ in such a way that Γ and $\tilde{\Gamma}$ have the same probability. It turns out that the right hand side of (4.11) is bounded below by the probability of $\tilde{\Gamma}$ when all $e(j,r)$ are taken as independent. The proof of (4.11) rests on showing, for one j at a time, that making the $\omega(e(j))$, $\omega(e(j,r))$, $r = 1,\dots,N$ into an i.i.d. family (rather than having them equal w.p.1) increases the probability of $\tilde{\Gamma}$.

We turn to the details. Clearly (4.15) equals

$$\lim_{M\to\infty} \bigcup_{k=1}^{M} (A(k,1)\circ\dots\circ A(k,n(k)))$$

and

$$\bigcup_{k\ge 1} \bigcap_{i=1}^{n(k)} A'(k,i) = \lim_{M\to\infty} \bigcup_{k=1}^{M} \bigcap_{i=1}^{n(k)} A'(k,i).$$

It therefore suffices to prove (4.11) when k runs only over the finite set $\{1,\dots,M\}$. Next (4.15) can be written as the increasing limit as $L \to \infty$ of the union over $k = 1,\dots,M$ of the sets

$$\Gamma(k,L) := \{\omega: \exists \text{ disjoint sets of edges } \mathcal{E}_1,\dots,\mathcal{E}_{n(k)} \text{ in } \{1,\dots,L\} \text{ and cylinders } C_1,\dots,C_{n(k)} \text{ such that } C_i \subset A(k,i) \text{ and } \omega \in \bigcap_{i=1}^{n(k)} C_i\}$$

(compare (4.7)). Since k is now restricted to $\{1,\dots,M\}$ $n(k)$ takes only M values, so that for some N we may assume $n(k) \equiv N$ (take $A(k,i) = \Omega$ for $n(k) < i \le N$). If $n(k) \equiv N$ then choosing disjoint subsets $\mathcal{E}_1,\dots,\mathcal{E}_N$ of $\{1,\dots,L\}$ is the same as choosing a partition $\wp$ of $\{1,\dots,L\}$ into $(N+1)$ sets and taking $\mathcal{E}_i$ as the i-th set in this partition. With this interpretation of $\mathcal{E}_i$ we can write $\Gamma(k,L)$ as

$$\bigcup_{\wp} \{\omega : \exists \text{ cylinders } C_1,\dots,C_N \text{ such that } \operatorname{base}(C_i) \subset \mathcal{E}_i, \; C_i \subset A(k,i) \text{ and } \omega \in \bigcap_1^N C_i\} .$$

Finally, for given L, N and $\wp$ we take the C_i above as large as possible. I.e., we define

$$C(k,i,\wp) = \text{union of all cylinders with base} \subset \mathcal{E}_i \text{ and contained in } A(k,i).$$

Then $C(k,i,\wp)$ is again a cylinder in $A(k,i)$ with base contained in $\mathcal{E}_i$ and

$$\Gamma(k,L) = \bigcup_{\wp} \bigcap_{i=1}^{N} C(k,i,\wp) . \tag{4.17}$$

Thus we see that we can write (4.15) as an increasing limit ($M \to \infty$, $L \to \infty$) of sets of the form (4.16). This is the desired approximation.

To show that all sets in this proof belong to $\mathcal{B}^P$, the completion of $\mathcal{B}$ with respect to P, we merely have to show that each $C(k,i,\wp)$ belongs to $\mathcal{B}^P$. For this purpose note that

$$C(k,i,\wp) = \{\omega : \text{if } \bar{\omega}(e(\ell)) = \omega(e(\ell)) \text{ for all } \ell \in \mathcal{E}_i, \text{ then } \bar{\omega} \in A(k,i)\} \tag{4.18}$$

and consequently

$$\begin{aligned} \Omega \setminus C(k,i,\wp) &= \{\omega : \exists\, \bar{\omega} \notin A(k,i) \text{ with } \bar{\omega}(e(\ell)) = \omega(e(\ell)) \text{ for all } \ell \in \mathcal{E}_i\} \\ &= \Pi(k,i) \times [0,\infty)^{\mathcal{E} \setminus \mathcal{E}_i}, \end{aligned}$$

where $\Pi(k,i)$ is the image of $\Omega \setminus A(k,i)$ under the projection $\omega \to \{\omega(e(\ell)) : \ell \in \mathcal{E}_i\}$. The image under such a projection is an analytic set in $[0,\infty)^{\mathcal{E}_i}$ by Theorems III.9 and III.13 of [10]. In particular $\Pi(k,i)$ is universally measurable (see [10], Sect. III.33a) so that

$\Omega \setminus C(k,i,\mathcal{P})$ and $C(k,i,\mathcal{P})$ belong to $\mathcal{B}^P$.

We now introduce the additional edges $e(j,r)$. Specifically, we take N copies $\Omega(1),\ldots,\Omega(N)$ of Ω. We index the coordinates of a point in $\Omega(r)$ by $\mathcal{E}(r) = \{e(j,r): j \geq 1\}$. Thus $\mathcal{E}(r)$ is simply some abstract set which is a copy of $\mathcal{E}$. The coordinate of index $e(j,r)$ of a point ω in $\Omega(r)$ will be denoted by $\omega(e(j,r))$ and $\phi(r)$ will be the one-to-one map from Ω onto $\Omega(r)$ which takes the point $\omega = \{\omega(e(j))\} \in \Omega$ to the point η_r of $\Omega(r)$ with $\eta_r(e(j,r)) = \omega(e(j))$. Finally we introduce certain probability measures P_s, $0 \leq s \leq L$ on $\Omega' := \Omega \times \Omega(1) \times \ldots \times \Omega(N)$. In words, P_s will be the probability measure which assigns probability one to

$$\{\omega(e(j)) = \omega(e(j,r)),\ 1 \leq r \leq N\} \quad \text{when} \quad 1 \leq j \leq s, \tag{4.19}$$

while for each $j > s$ the random variables $\omega(e(j)), \omega(e(j,1)),\ldots,$ $\omega(e(j,N))$ are i.i.d. and independent of all $\omega(e(j'))$, $\omega(e(j',r))$ with $j' \neq j$, $1 \leq r \leq N$. More formally, let B be a Borel set of Ω, and $B'(r)$, $B''(r)$ Borel cylinders of $\Omega(r)$ such that

$$\text{base}(B'(r)) \subset \{e(j,r): 1 \leq j \leq s\},$$

$$\text{base}(B''(r)) \subset \{e(j,r): j > s\}.$$

Then P_s is a complete measure which satisfies

$$P_s\{(\omega,\omega_1,\ldots,\omega_N): \omega \in B, \omega_r \in B'(r) \cap B''(r),\ 1 \leq r \leq N\} \tag{4.20}$$

$$= P\{\omega \in B \cap \bigcap_{r=1}^{N} \phi^{-1}(r)B'(r)\} \cdot \prod_{r=1}^{N} P\{\omega \in \phi^{-1}(r)\, B''(r)\}.$$

We next define the set $\tilde{\Gamma}$ and reduce the problem to showing

$$P_{s+1}\{\tilde{\Gamma}\} \leq P_s\{\tilde{\Gamma}\}, \qquad 0 \leq s < L. \tag{4.21}$$

$\tilde{\Gamma}$ is defined as a subset of $\Omega \times \Omega(1) \times \ldots \times \Omega(N)$ in the following way:

$$\tilde{C}(k,i,\mathcal{P}) := \{(\omega,\omega_1,\ldots,\omega_N): \omega_i \in \phi(i)C(k,i,\mathcal{P})\},$$

$$\tilde{\Gamma}(k,L) := \bigcup_{\mathcal{P}} \bigcap_{i=1}^{N} \tilde{C}(k,i,\mathcal{P}),$$

$$(4.22) \qquad \tilde{\Gamma} := \bigcup_{k=1}^{M} \tilde{\Gamma}(k,L)$$

(compare (4.16), (4.17)). Now for $1 \le j \le L$ (by (4.19) and (4.20))

$$P_L\{(\omega,\omega_1,\ldots,\omega_N):\omega(e(j)) = \omega_r(e(j,r)),\ 1 \le r \le N\} = 1.$$

Since $C(k,i,\mathcal{P})$ is a cylinder with base contained in $\mathcal{E}_i \subset \{1,\ldots,L\}$ this implies that the sets

$$\{(\omega,\omega_1,\ldots,\omega_N):\omega_i \in \phi(i)C(k,i,\mathcal{P})\} \quad \text{and}$$

$$\{(\omega,\omega_1,\ldots,\omega_N):\omega \in C(k,i,\mathcal{P})\}$$

differ only by a P_L-null set. Therefore

$$P_L\{\tilde{\Gamma}\} = P_L\{\bigcup_k \bigcup_{\mathcal{P}} \bigcap_{i=1}^{N} \{(\omega,\omega_1,\ldots,\omega_N):\omega_i \in \phi(i)C(k,i,\mathcal{P})\}\}$$

$$= P_L\{\bigcup_k \bigcup_{\mathcal{P}} \bigcap_{i=1}^{N} \{(\omega,\omega_1,\ldots,\omega_N):\omega \in C(k,i,\mathcal{P})\}\} = P\{\Gamma\}.$$

On the other hand, if we define $\Omega' = \Omega \times \Omega(1) \times \ldots \times \Omega(N)$, and $\mathcal{B}'$ the completion of its standard σ-field with respect to P_0, and

$$A'(k,i) = \phi(i)A(k,i),$$

then (4.9) and (4.10) are satisfied for $P' = P_0$. This is obvious for (4.10) since for each i $A'(k,i)$ is determined by conditions on $\{\omega(e(j,i)):j \ge 1\}$ only, and these families are independent for different i under P_0. Similarly (4.9) follows from the fact that $\{\omega(e(j,i)):j \ge 1\}$ has the same distribution under P_0 as $\{\omega(e(j)):j \ge 1\}$ under P_0 (see (4.20)). Lastly

$$\phi(i)C(k,i,\mathcal{P}) \subset A'(k,i),$$

since $C(k,i,\mathcal{P}) \subset A(k,i)$ by (4.18). Thus

$$\tilde{\Gamma} \subset \bigcup_{k=1}^{M} \bigcap_{i=1}^{N} A'(k,i) \subset \bigcup_{k=1}^{M} \bigcap_{i=1}^{n(k)} A'(k,i)$$

and

$$P_0(\tilde{\Gamma}) \le P_0\{\bigcup_{k=1}^{M} \bigcap_{i=1}^{n(k)} A'(k,i)\} = P'\{\bigcup_{k=1}^{M} \bigcap_{i=1}^{n(k)} A'(k,i)\}.$$

Thus, if we can prove (4.21) we shall have

$$P\{\Gamma\} = P_L\{\tilde{\Gamma}\} \le P_0\{\tilde{\Gamma}\} \le P'\{\bigcup_{k=1}^{M} \bigcap_{i=1}^{n(k)} A'(k,i)\} .$$

This is precisely (4.11) when k runs only over $\{1,\dots,M\}$, and, as we saw before, the general case of (4.11) follows by taking the limit $M \to \infty$. (Note that the right hand side of (4.11) is the same for all P', A' which satisfy (4.9) and (4.10) so that it suffices to deal with the special choice made above.)

The proof of (4.11) has been reduced to (4.21), which we now prove. We restrict ourselves to the case of decreasing $A(k,i)$, the case of increasing $A(k,i)$ being similar. (4.21) will be proved by conditioning on all the variables

$$\{\omega(e(j)),\omega(e(j,r)) : 1 \le r \le N,\ j \ge 1 \quad \text{but} \quad j \ne s+1\}.$$

Let us write Δ_s for this collection of variables. Accordingly $\Delta_s = \bar{\Delta}_s$ will be an abbreviation for

$$\omega(e(j)) = \bar{\omega}(e(j)),\omega(e(j,r)) = \bar{\omega}(e(j,r)),\ 1 \le r \le N,\ j \ge 1,\ j \ne s+1, \tag{4.23}$$

for some specific values $\bar{\omega}$. Also for $\tau = (\tau_0,\tau_1,\dots,\tau_N) \in [0,\infty)^{N+1}$ $(\tau,\bar{\Delta}_s)$ will denote the point ω of Ω' for which (4.23) holds and

$$\omega(e(s+1)) = \tau_0,\ \omega(e(s+1,r)) = \tau_r,\ 1 \le r \le N.$$

The section $\tilde{\Gamma}(\bar{\Delta}_s)$ of $\tilde{\Gamma}$ corresponding to $\bar{\Delta}_s$ is

$$\tilde{\Gamma}(\bar{\Delta}_s) = \{\tau \in [0,\infty)^{N+1} : (\tau,\bar{\Delta}_s) \in \tilde{\Gamma}\} .$$

Now, $(\omega(e(s+1)),\omega(e(s+1,r)),\ 1 \le r \le N))$ is independent of Δ_s, under P_s as well as under P_{s+1} so that for $t = s$ or $s+1$ and fixed $\bar{\Delta}_s$

$$P_t\{\tilde{\Gamma}|\Delta_s = \bar{\Delta}_s\} = P_t\{(\omega(e(s+1)),\omega(e(s+1),r),1 \le r \le N) \in \tilde{\Gamma}(\bar{\Delta}_s)\}. \tag{4.24}$$

The only difference between the right hand sides of (4.24) for $t = s$ and for $t = s+1$ is that for $t = s$, the variables $\omega(e(s+1))$, $\omega(e(s+1),r)$, $1 \le r \le N$ are i.i.d., while for $t = s+1$ they are all equal to $\omega(e(s+1))$ w.p.1.(4.21), will now follow from the stronger relation

$$(4.25) \qquad P_s\{(\omega(e(s+1)),\omega(e(s+1),r), 1 \le r \le N) \in \tilde{\Gamma}(\bar{\Delta}_s)\}$$

$$\ge P_{s+1}\{(\omega(e)s+1)),\omega(e(s+1),r), 1 \le r \le N) \in \tilde{\Gamma}(\bar{\Delta}_s)\}.$$

which will be easy, once we realize that $\tilde{\Gamma}(\bar{\Delta}_s)$ is of a special form, which we shall derive next.

By (4.22)

$$\tilde{\Gamma}(\bar{\Delta}_s) = \bigcup_{k=1}^{M} \bigcup_{\rho} \bigcap_{i=1}^{N} \{\tau:(\tau,\bar{\Delta}_s) \in \tilde{C}(k,i,\rho)\}.$$

Now $C(k,i,\rho)$ is a cylinder with base contained in $\mathcal{E}_i$. Therefore, if $(s+1) \notin \mathcal{E}_i$, then the value of τ does not influence whether $(\tau,\bar{\Delta}_s) \in \tilde{C}(k,i,\rho)$ or not. For such i the set $\{\tau:(\tau,\bar{\Delta}_s) \in \tilde{C}(k,i,\rho)\}$ is either empty or all of $[0,\infty)^{N+1}$. We claim that for $(s+1) \in \mathcal{E}_i$, the last set is of the form

$$(4.26) \qquad \{\tau:(\tau,\bar{\Delta}_s) \in \tilde{C}(k,i,\rho)\} = \{\tau:\tau_i \in I_i\}$$

for some interval $I_i = I_i(k,\rho,\bar{\Delta}_s) = [0,\lambda]$ or $[0,\lambda)$ with a $0 \le \lambda \le \infty$. To see this note that $C(k,i,\rho)$, and hence $\tilde{C}(k,i,\rho)$ is decreasing whenever $A(k,i)$ is decreasing, and also note that $\tilde{C}(k,i,\rho)$ is defined by restrictions on the $\omega(e(j,i))$, $j \in \mathcal{E}_i$, only. Thus, for fixed $\bar{\Delta}_s$ $(\tau,\bar{\Delta}_s) \in \tilde{C}(k,i,\rho)$ is equivalent to the requirement that τ_i lie in some decreasing subset of $[0,\infty)$, i.e., (4.26) holds. The final observation to obtain the required form of $\tilde{\Gamma}(\bar{\Delta}_s)$ is that for each ρ the different $\mathcal{E}_i$ are disjoint. (This is the only place where the disjointness used in the definition of disjoint occurrence, (4.7), comes in.) Thus, for each ρ there is at most one i for which $(s+1) \in \mathcal{E}_i$. If such an i exists we denote it by $i(\rho)$; if no such i exists we can take $i(\rho)$ arbitrary in $[1,N]$. In all cases we can write

$$\bigcap_{i=1}^{N} \{\tau:(\tau,\bar{\Delta}_s) \in \tilde{C}(k,i,\rho)\} = \{\tau:\tau_{i(\rho)} \in J(k,\rho)\}$$

for some interval $J(k,\rho) = [0,\lambda(k,\rho)]$ or $[0,\lambda(k,\rho))$ with a suitable $\lambda(k,\rho) \in [0,\infty]$. (Note that this interval may be empty if $\lambda = 0$ or all of $[0,\infty)$; also $\lambda(k,\rho)$ may depend on $\overline{\Delta}_s$, even though the notation does not indicate this.) Therefore, finally

$$\tilde{\Gamma}(\overline{\Delta}_s) = \bigcup_{k=1}^{M} \bigcup_{\rho} \{\tau : \tau_{i(\rho)} \in J(k,\rho)\} .$$

(4.25), and hence (4.21), is now easy. On the one hand, by (4.24) and the lines following it

$$\begin{aligned} P_{s+1}\{\tilde{\Gamma} | \Delta_s = \overline{\Delta}_s\} &= P_{s+1}\{\bigcup_{k=1}^{M} \bigcup_{\rho}\{\omega(e(s+1),i(\rho)) \in J(k,\rho)\}\} \\ &= P\{\bigcup_{k=1}^{M} \bigcup_{\rho} \{\omega(e(s+1)) \in J(k,\rho)\}\} \\ &= P\{\omega(e(s+1)) \in \bigcup_{k=1}^{M} \bigcup_{\rho} J(k,\rho)\} \\ &= P\{\omega(e(s+1) \in J(k_0,\rho_0)\} \end{aligned}$$

for some k_0, ρ_0. The last equality follows from the special form of the $J(k,\rho)$. Any finite union of intervals $[0,\lambda]$ or $[0,\lambda)$ equals the largest one of these intervals. On the other hand

$$\begin{aligned} P_s\{\tilde{\Gamma} | \Delta_s = \overline{\Delta}_s\} &= P_s\{\bigcup_{k=1}^{M} \bigcup_{\rho}\{\omega(e(s+1),i(\rho)) \in J(k,\rho)\} \\ &\geq P_s\{\omega(e(s+1),i(\rho_0)) \in J(k_0,\rho_0)\} \\ &= P\{\omega(e(s+1)) \in J(k_0,\rho_0)\}. \end{aligned}$$ ///

(4.27) REMARK. The method of proof which consists of "splitting each edge into a number of copies" which are taken independent under P_0 is essentially due to Campanino and Russo [5].

NOTE ADDED AFTER COMPLETION OF TYPESCRIPT. It has come to our attention that Theorem 4.8 is essentially contained in L. Rüschendorf, Comparison of percolation probabilities, J. Appl. Prob. 19 (1982) 864-868 and in C. McDiarmid, General first-passage percolation, Adv. Appl. Prob. 15 (1983) 149-161.

5. The rate of convergence to the time constant. In this section we return to first-passage percolation on $\mathbb{Z}^d$. In partial answer to questions raised by Kingman ([25], Sect. 3.2) and Smythe and Wierman ([31], Sect. 10.2) we prove results about the rate at which

(5.1) $$P\{\frac{1}{n}\theta_{0,n} - \mu > \varepsilon(n)n\} \to 0 \quad \text{and}$$

$$P\{\frac{1}{n}\theta_{on} - \mu < \varepsilon(n)n\} \to 0$$

as $n \to \infty$, when either $\varepsilon(n) = \varepsilon > 0$ or $\varepsilon(n) \to 0$, $\varepsilon(n) > 0$, at a suitable rate. μ is the time constant; recall that we do not distinguish between μ and $\hat{\mu}$ (see Sect. 2 and the Remark following Theorem 2.26). $\theta_{0,n}$ stands for any of the passage times $a_{0,n}$, $b_{0,n}$, $s_{0,n}$ or $t_{0,n}$. We make no moment assumptions beyond those stated explicitly in the theorems. Thus Theorem 5.1 holds without any moment assumptions whatsoever. It is likely that the conditions of Theorem 5.16 can be weakened, but even if the $t(e)$ are bounded, the results of this Theorem seem so far from the true state of affairs that it does not pay to weaken the hypothesis.

Much of this section can be found in a weaker form in [17]. Parts of Theorem 5.2 also appear already in [18] and [26], Sect. 3.2.

(5.2) THEOREM.
(a) For all $\varepsilon > 0$ there exist constants $A_1 = A_1(\varepsilon,F,d)$ $B_1 = B_1(\varepsilon,F,d) > 0$ such that

(5.3) $$P\{\theta_{0,n} < n(\mu-\varepsilon)\} \leq A_1 e^{-B_1 n}, \quad n \geq 0,$$

$$\text{for } \theta = a, b, s \text{ or } t.$$

(b) If $\theta = a$, s or t and $\varepsilon > 0$ (and if we allow $B_1 = \infty$), then (5.3) can be sharpened to

(5.4) $$\lim_{n\to\infty} -\frac{1}{n} \log P\{\theta_{0,n} < n(\mu-\varepsilon)\} = B_1$$

and

(5.5) $$P\{\theta_{0,n} < n(\mu-\varepsilon)\} \leq e^{-B_1 n}.$$
with the same B_1 for $\theta = a,s$ or t.

(c) Set $B_1(\varepsilon,F,d) = 0$ for $\varepsilon < 0$, and $\beta = \sup\{x: F(\mu-x) > 0\}$, where F is the distribution function of the $t(e)$. Set $B_1(0,F,d) = 0$ if $\beta > 0$ and $B_1(0,F,d) = \infty$ if $\beta = 0$. Then (5.4) and (5.5) hold for all ε and $\theta = a,s$ or t and the extended function $B_1(\ ,F,d)$ has the following properties:

$$0 < B_1(\varepsilon,F,d) < \infty \quad \text{for} \quad 0 < \varepsilon < \beta,$$

$$B_1(\varepsilon) = 0 \quad \text{for} \quad \varepsilon < 0, \quad B_1(\varepsilon) = \infty \quad \text{for} \quad \varepsilon \geq \beta,$$

$\varepsilon \to B_1(\varepsilon,F,d)$ is convex and continuous on

$(-\infty,\beta)$ and strictly increasing on $[0,\beta)$.

REMARK. Kingman [26], p.217 seems to say that (5.4) and (5.5) hold with $B_1 > 0$ for any positive superconvolutive sequence $\theta_{0,n}$ (such as $a_{0,n}$, $s_{0,n}$ or $t_{0,n}$). However, Kingman does not give a detailed proof for the strict positivity of B_1, and it is precisely this part of the proof which requires the most effort here. It is therefore of interest to point out that there exist sequences of positive random variables X_n, whose distribution functions F_n satisfy all hypotheses of Theorem 3.4 in [26], and are such that $n^{-1}X_n \to \gamma$ w.p.1 for some constant $0 < \gamma < \infty$, but for which $P\{X_n \leq n(\gamma-\varepsilon)\}$ does not decrease exponentially. One such example is if X_n is the volume of the Wiener Sausage, i.e., X_n = volume of $\bigcup_{0 \leq t \leq n} \{B(t) + A\}$ with $B(t)$ = 3 - dimensional Brownian motion and A the unit ball in $\mathbb{R}^3$ (see [26], Sect. 2.4 and [11]. ///

An important role is played by the critical probability $p_T = p_T(d)$ of Bernoulli percolation. This is defined as follows.

(5.6) $$W = \{v \in \mathbb{Z}^d: \exists \text{ path } r \text{ from } \underline{0} \text{ to } v \text{ with } T(r) = 0\}.$$

If one calls an edge e open if and only if $t(e) = 0$, then W is the so called open cluster of the origin. It consists of all vertices connected to the origin by an open path (i.e., a path all of whose edges are open). Let

$$\#W = \text{number of vertices in } W.$$

Finally, denote by P_p the product measure on $[0,\infty)^{\mathcal{E}}$, under which all $t(e)$ are i.i.d. with

$$P_p\{t(e) = 0\} = 1 - P_p\{t(e) = 1\} = p.$$

Then with E_p denoting expectation with respect to P_p,

(5.7) $$p_T = p_T(d) := \sup\{p: \ E_p\{\#W\} < \infty\}.$$

(Note that $\#W = \infty$ is possible and

$$E_p\{\#W\} \geq P_p\{\#W = \infty\} \cdot \infty$$

so that

$$p_T \leq p_H := \inf\{p: \ P_p\{\#W = \infty\} > 0\}.)$$

(5.8) PROPOSITION. If

$$F(0) = P\{t(e) = 0\} < p_T ,$$

then there exist constants $0 < C,D,E < \infty$, depending on d and F only, such that

$$P\{\exists \text{ selfavoiding path } r \text{ from the origin which contains at least } n \text{ edges but has } T(r) < C\,n\} \leq D\,e^{-En}.$$

(5.9) THEOREM.

(a) Assume that

(5.10) $$\int_0^\infty e^{\gamma x}\, dF(x) < \infty$$

For some $\gamma > 0$. Then for each $\varepsilon > 0$ there exist constants $A_2 = A_2(\varepsilon,F,d)$ and $B_2 = B_2(\varepsilon,F,d) > 0$ such that for $\theta = a,b,s$ or t

(5.11) $$P\{\theta_{0,n} > n(\mu+\varepsilon)\} \leq A_2 e^{-B_2 n}, \quad n \geq 0.$$

(b) If (5.10) holds for all $\gamma > 0$, then for all $\varepsilon > 0$ and $\theta = a,b,s$ or t and $d \geq 2$

(5.12) $$\lim -\frac{1}{n} \log P\{\theta_{0,n} > n(\mu+\varepsilon)\} = \infty$$

(so that the left hand side of (5.11) decreases faster than exponentially).
(c) If $t(e)$ is bounded w.p.1, then for all $\varepsilon > 0$ there exist constants $A_3 = A_3(\varepsilon,F,d)$ and $B_3 = B_3(\varepsilon,F,d) > 0$ such that for $\theta = a,s,b$ or t

(5.13) $$P\{\theta_{0,n} > n(\mu+\varepsilon)\} \leq A_3 \exp(-B_3 n^d).$$

(5.14) REMARK. The estimate

$$\theta_{0,n} \geq \min\{t(e) : e \text{ incident to } \underline{0}\}$$

shows that (5.10) for some $\gamma > 0$ is also necessary for (5.12).
(5.15) REMARK. The asymmetry between (5.4) on the one hand, and (5.12), (5.13) on the other hand make it appear unlikely that a central limit theorem holds for $n^{-1/2}(\theta_{0,n} - n\mu)$. There may, however, be a central limit theorem for $n^{-1/2}(\theta_{0,n} - E\theta_{0,n})$. (Compare Sect 10.2 of [31].) To approach this question we need estimates for $E\theta_{0,n} - n\mu$ and for (5.1) when $\varepsilon(n) \to 0$. A modest beginning to this is made in the next theorem. However, we suspect that the estimates given in this theorem are much worse than the true convergence rates.

(5.16) THEOREM. Assume that (5.10) holds for some $\gamma > 0$. Then there exist constants $0 < C_i = C_i(\varepsilon,F,d)$ such that the B_1 of Theorem 5.2(a) satisfies

(5.17) $\quad B_1(\varepsilon,d,F) \geq \exp(-C_1\varepsilon^{-p})$, $\varepsilon > 0$, with $p = 9d + 3$.

Moreover, for θ = a,s, or t.

(5.18) $$\mu \leq \frac{1}{n}E\theta_{0,n} \leq \mu + C_2(\log n)^{-1/p} ,$$

(5.19) $$\sigma^2(\frac{\theta_{0,n}}{n}) \leq C_3(\log n)^{-2/p} .$$

Finally, for θ = a,b,s or t

(5.20) $$\limsup\, (\log n)^{1/p}|\frac{1}{n}\theta_{0,n} - \mu| < \infty \quad \text{w.p.1.}$$

(5.21) REMARK. Trivially (5.3) holds with $B_1 = \infty$ for $\varepsilon \geq \beta$ for the β of Theorem (5.2 c). Thus (5.17) is only interesting if $0 < \varepsilon < \beta$ (and a fortiori $\mu > 0$ since $\beta \leq \mu$). $\beta = 0 < \mu$ means, by definition) $P\{t(e) \geq \mu > 0\} = 1$. It follows from (6.6) below that this implies $P\{t(e) = \mu\} = 1$, and thus $P\{\theta_{0,n} = n\mu\} = 1$ for θ = a,b,s or t when $\beta = 0 < \mu$. For the case $\mu = 0$ some improvements of (5.18) and (5.20) are given in Sect. 7. ///

The most important step in the proofs is the estimate (5.23) for the distribution of $\theta_{0,n}$ in terms of sums of independent passage times across large strips (these strips are still small with respect to n, though). Prop. 5.23 is obtained by specializing (4.13). The right hand side of (5.23) is then reduced further to estimates for ordinary cylinder passage times by lemma 5.29 , especially part (a). The choice of (the large) N and M is dictated by a tradeoff between the errors caused by the two terms in the right hand side of the estimate in 5.29 (a). The most delicate illustration of this is the proof of (5.17). (see (5.53) and the succeeding estimates).

Not suprisingly we need some definitions and preliminary results to prove these theorems. Recall that H_k denotes the hyperplane $\{x(1) = k\} \subset \mathbb{R}^d$. For a path r we shall write

(5.22) $$H_k < r < H_\ell$$

if all points of r , except possibly its initial and end point, lie strictly between H_k and H_ℓ (i.e. in the set $\{x \in \mathbb{R}^\ell: k < x(1) < \ell\}$. For $M \geq 1$ we shall need the following cylinder passage times, which are variants of the $s_{0,n}$ (see (2.15); $s_{0,n} = s_{0,n}^1$ in the new notation)

$$s_{0,n}^N := \inf \{T(r): \ r \text{ a path from some point } (0,m_2,\ldots, m_d) \text{ with } 0 \leq m_i < N,\ 2 \leq i \leq d, \text{ to } H_n \text{ for which } H_0 < r < H_n\},$$

and

$$\hat{s}_{0,n}^N := \inf \{T(r): \ r \text{ a path from some point } (0,m_2,\ldots,m_d) \text{ with } 0 \leq m_i < N,\ 2 \leq i \leq d, \text{ to } H_n \text{ such that } r \text{, with the exception of its endpoints, is contained in } (0,n) \times (-4n, 4n) \times \ldots \times (-4n,4n)\}.$$

It is clear from these definitions that

$$s_{0,n}^N \leq \hat{s}_{0,n}^N .$$

(5.23) PROPOSITION. Let $X_i(M,N)$, $i \geq 1$, (respectively $\hat{X}_i(M,N)$, $i \geq 1$) be independent random variables, each with the distribution of $s_{0,M}^N$ (respectively $\hat{s}_{0,M}^N$). Then for any $n \geq M \geq N \geq 1$, $x \geq 0$ and $\theta = a,b,s$ or t one has

$$\begin{aligned} &P\{\theta_{0,n} < x\} \\ &\leq \sum_{Q+1 \geq n/(M+N)} \{2d(15\tfrac{M}{N})^d\}^Q \, P\{\hat{X}_1(M,N)+\ldots+\hat{X}_Q(M,N) < x\} \\ &\leq \sum_{Q+1 \geq n/(M+N)} \{2d(15\tfrac{M}{N})^d\}^Q \, P\{X_1(M,N)+\ldots+X_Q(M,N) < x\} \end{aligned}$$

<u>Proof:</u> Since $s_{0,M}^N \leq \hat{s}_{0,M}^N$ it suffices to prove the first inequality. We shall use the following notation: for vertices $v_j = (v_j(1),\ldots, v_j(d))$, $j = 1,2$, of $\mathbb{Z}^d$ we define the distance $\delta(v_1,v_2)$ by

$$\delta(v_1,v_2) = \max_{1 \leq i \leq d} |v_1(i) - v_2(i)|.$$

Now assume $r = (v_0, e_1, \dots, e_p, v_p)$ is a path without double points from the origin to H_n. We shall choose a (random) subset $\{a_0, a_1, \dots, a_Q\}$ of the vertices in r and then apply (4.13) with the $V(\ell, i)$ suitable pieces of hyperplanes near the a_i. We choose the a_i as follows. $a_0 = v_0 = 0$. Assume $a_0, a_1, \dots, a_q$ have already been chosen such that $a_i = v_{\tau(i)}$ with $\tau(0) = 0 < \tau(1) < \dots < \tau(q)$. Then take

$$\tau(q+1) = \min \{t > \tau(q): \delta(a_q, v_t) = M+N\}, a_{q+1} = v_{\tau(q+1)} ,$$

provided such that a t exists; if no such t exists, i.e., if $\delta(v_t, a_q) < M+N$ for all $\tau(q) < t \le p$, then we stop and take $Q = q$. Clearly

$$a_{q+1}(1) \le a_q(1) + M+N$$

and if a_Q is the last vertex of r selected as one of the a's , then

$$n = v_n(1) < a_Q(1) + M+N \le a_0(1) + (Q+1)(M+N) = (Q+1)(M+N) \tag{5.24}$$

so that we must select at least $n/(M+N)$ vertices a_i. In the sequel we shall not use any further the fact that r is a path from $\underline{0}$ to H_n. We only needed this to derive the above estimate $Q+1 \ge n/(M+N)$

Next we define for each $0 \le q \le Q-1$ a triple $\Lambda(q) = (\nu(q), \eta(q), V(q))$, where $\nu(i)$ is one of the integers $1,2,\dots,d$, $\eta(q) = +1$ or -1 , and $V(q)$ is a set of the form

$$\{x \in \mathbb{Z}^d: x(\nu(q)) = \lambda(\nu(q))N , \quad \lambda(j)N \le x(j) < (\lambda(j)+1)N , j \ne \nu(q)\} \tag{5.25}$$

for some integers $\lambda(j) = \lambda(j,q)$. These are chosen as follows. By definition $\delta(a_{q+1}, a_q) = (M+N)$ so that we can choose $\nu(q) \in \{1,\dots,d\}$ such that

$$a_{q+1}(\nu(q)) - a_q(\nu(q)) = \pm(M+N).$$

We take $\eta(q) = +1$ (respectively -1) if this equality holds with the + sign (respectively - sign) in the right hand side. Next, note that between $a_q = v_{\tau(q)}$ and $a_{q+1} = v_{\tau(q+1)}$ the value of $v_t(\nu(q))$ goes from $a_q(\nu(q))$ to $a_q(\nu(q)) + \eta(q)(M+N)$, so that we can define

$$\rho(q) = \max \{t \in [\tau_q, \tau_{q+1}) : \quad v_t(\nu(q)) = N(\lfloor a_q(\nu(q))/N \rfloor + \tfrac{1}{2} (\eta(q) + 1))\},$$

$$\sigma(q) = \min \{t \in (\rho(q), \tau_{q+1}): \quad v_t(\nu(q)) \quad N(\lfloor a_q(\nu(q))/N \rfloor + \tfrac{1}{2}(\eta(q) + 1)) + \eta(q)M\} ,$$

and these times satisfy

$$\tau(q) \leq \rho(q) < \sigma(q) \leq \tau(q+1). \tag{5.26}$$

The piece

$$r(q) := (v_{\rho(q)} \ , \ e_{\rho(q)+1}, \ldots, v_{\sigma(q)})$$

of r lies strictly between the hyperplanes

$$\mathcal{H}'(q) := \{x: x(\nu(q)) = N(\lfloor a_q(\nu(q))/N \rfloor + \tfrac{1}{2}(\eta(q) + 1)\}$$

and

$$\mathcal{H}''(q) := \{x: x(\nu(q)) = N(\lfloor a_q(\nu(q))/N \rfloor + \tfrac{1}{2}(\eta(q) + 1) + \eta(q)M\} ,$$

with the exception of its initial and endpoint, which lie in $\mathcal{H}'(q)$ and $\mathcal{H}''(q)$, respectively. Finally for $V(q)$ we take the unique set of the form (5.25) which contains the vertex $b_q := v_{\rho(q)}$ (see Fig. 5.1). Note that $V(q) \subset \mathcal{H}'(q)$.

By virtue of (5.26) and the fact that $\delta(v_t, a_q) \leq M+N$ for $\tau(q) \leq t \leq \tau(q+1)$ the pieces $r(q)$ have the following properties. For $q_1 \neq q_2$ $r(q_1)$ and $r(q_2)$ are edge disjoint. The piece $r(q)$ connects $V(q)$ to $\mathcal{H}''(q)$, and except for its endpoints lies strictly between $\mathcal{H}'(q)$

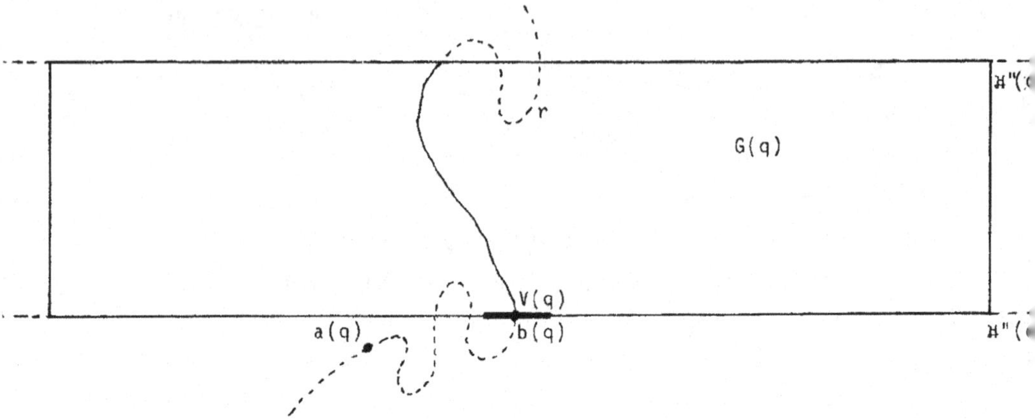

Fig. 5.1 Illustration of b_q, $V(q)$, $r(q)$, $G(q)$ for $d=2$. In this example $\nu(q) = 2$, $\eta(q) = +1$. $r(q)$ is the solidly drawn piece of r, and $V(q)$ the baldly drawn interval in $\mathcal{H}'(q)$. $G(q)$ is the large rectangle between $\mathcal{H}'(q)$ and $\mathcal{H}''(q)$.

and $\mathcal{H}''(q)$. Finally, $r(q)$ is contained in the box $G(q)$ whose j-th side is

$$[\lambda(j,q)N - 2(M+N) \, , \, \lambda(j,q)M + 2(M+N)] \quad \text{if} \quad j \neq \nu(q), \text{ and}$$

$$[N(\lfloor a_q(\nu(q))/N\rfloor + \tfrac{1}{2}(\eta(q) + 1)) \, ,$$
$$N\lfloor a_q(\nu(q))/N\rfloor + \tfrac{1}{2}(\eta(q) + 1)) + \eta(q)M] \quad \text{if} \quad j=\nu(q)$$

Since

$$T(r) \geq \sum_{q=0}^{Q-1} T(r(q)),$$

we obtain

(5.27) $$\{\theta_{0,n} < x\} \subset \bigcup_{Q+1 \geq n/(M+N)} \; \bigcup_{\Lambda(q),\, q<Q} \{ \exists$$

a path r which contains edge disjoint pieces $r(q)$, $0 \leq q \leq Q-1$, which cross the boxes $G(q)$

in the $\nu(q)$-direction from $V(q)$ to $\aleph''(q)$ and satisfy $\sum_{0}^{Q-1} T(r(q)) < x\}$.

Given $\Lambda(q)$, and hence $V(q)$, $\aleph''(q)$ and the box $G(q)$, the minimal passage time of any path which connects $V(q)$ to $\aleph''(q)$ in $G(q)$ is stochastically larger than $\hat{s}^N_{0,M}$ (since $2(N+M) \leq 4M$). For fixed $\Lambda(0)$, $\ldots,\Lambda(Q-1)$, we therefore have, just as in (4.13), that the probability of the event between braces in the right hand side of (5.27) is at most

$$P\{\hat{X}_1(M,N) +\ldots+ \hat{X}_Q(M,N) < x\}. \tag{5.28}$$

To obtain the proposition from (5.27) and (5.28) it remains to count how many choices there are for $\Lambda(0),\ldots,\Lambda(Q-1)$. Assume that $\Lambda(q)$ has already been chosen. As we already observed, (5.26) and the definition of $\Lambda(q+1)$ imply

$$\delta(b_q,a_q) < M+N,\ \delta(a_{q+1},a_q) = M+N \text{ , and}$$
$$\text{similarly } \delta(b_{q+1},a_{q+1}) < M+N.$$

Thus, b_{q+1} must lie within distance $3(M+N) - 2$ of some point of $V(q)$. Therefore, given $V(q)$ there are at most $(N + 6M + 6N)^d$ choices for b_{q+1}, and at most

$$(6\frac{M}{N} + 9)^d$$

choices for $\lambda(1),\ldots,\lambda(d)$ such that

$$\lambda(j)M \leq b_{q+1}(j) < (\lambda(j) + 1)M \quad \text{for \underline{all}} \quad 1 \leq j \leq d.$$

Moreover, there are at most $2d$ choices for $(\nu(q+1), \eta(q+1))$. therefore, given $\Lambda(q)$ there are at most

$$2d(6\frac{M}{N} + 9)^d \leq 2d(15\frac{M}{N})^d$$

choices for $\Lambda(q+1)$,and in total at most

$$\{2d(15\tfrac{M}{N})^d\}^Q$$

choices for $\Lambda(0),\ldots,\Lambda(Q-1)$. the proposition now follows from (5.27) and (5.28).

(5.29) LEMMA.

(a) Let $X_i(M,N)$ and $\hat{X}_i(M,N)$ be as in Prop. 5.23, and let $t_1, t_2, \ldots$ be independent random variables, each with the distribution of $t(e)$. Then for $Q \geq 1$, $M \geq N$,

$$\begin{aligned} &P\{X_1(M,N) + \ldots + X_Q(M,N) < x\} \\ &\leq P\{X_1(M,1) + \ldots + X_Q(M,1) < x+y\} \\ &\quad + N^{Q(d-1)} P\{t_1 + \ldots + t_{Q\lfloor dN/2 \rfloor} \geq y\}, \end{aligned}$$

and the same inequality holds when $X_i(M,N)$ is replaced on both sides by $\hat{X}_i(M,N)$.

(b) Let

(5.31) $$b_{0,n}^N = \inf\{T(r): \ r \text{ a path from some point } (0,m_2,\ldots,m_d) \text{ with } 0 \leq m_i < N, \ 2 \leq i \leq d, \text{ to } H_n\}.$$

Then for any $\varepsilon > 0$

(5.32) $$\begin{aligned} &\lim_{M\to\infty} \max_{N<M} P\{s_{0,M}^N \leq M(\mu-\varepsilon)\} \\ &\leq \lim_{M\to\infty} \max_{N\leq M} P\{b_{0,M}^N \leq M(\mu-\varepsilon)\} = 0 \end{aligned}$$

Proof: (a) We only give the proof for $Q=1$ and for X. The case of general Q and/or $\hat{X}$ is essentially the same. For any vertex v in the hyperplane $H_0 = \{x(1) = 0\}$ consider the cylinder passage time

$$s_M(v) := \inf\{T(r): \ r \text{ a path from } v \text{ to } H_M \text{ with } H_0 < r < H_M\}.$$

Then $s_{0,M} = s_M(\underline{0})$. Note that any r which starts at $v \in H_0$ and satisfies $H_0 < r < H_M$ must have as its first edge the edge between v and $v' = v + (1,0,\ldots,0)$, and in particular passes through v'. Take $v_0 = (\lfloor N/2 \rfloor, \lfloor N/2 \rfloor, \ldots, \lfloor N/2 \rfloor)$ and for each $v \in \{0\} \times [0,N)^{d-1}$ let $\rho(v)$ be a path which goes from v_0 via one edge to v_0', and then from v_0' to v' via a path in $H_1 = \{x(1) = 1\}$ of

$$\sum_{j=2}^{d} |v_0(j) - v(j)|$$

steps. If r is a path from v to H_M with $H_0 < r < H_M$, denote by r' the piece of r from v' to H_M. Then $\rho(v)$ followed by r' is a path from v_0 (via v') to H_M, and with the exception of its end-points this path lies strictly between H_0 and H_M. The passage time of this path is $T(\rho(v)) + T(r')$ so that

$$s_M(v_0) \leq T(\rho(v)) + T(r') \leq T(\rho(v)) + T(r).$$

If we take the inf over r this yields

$$s_M(v_0) \leq T(\rho(v)) + s_M(v). \tag{5.33}$$

If we choose $v \in \{0\} \times [0,N)^{d-1}$ such that

$$s_M(v) = s_{0,M}^N = \min\{s_M(u) : u \in \{0\} \times [0,N)^{d-1}\},$$

then we find

$$s_M(v_0) \leq s_{0,M}^N + \max\{T(\rho(v)) : v \in \{0\} \times [0,N)^{d-1}\}.$$

Thus

$$P\{s_{0,M}^N < x\} \leq P\{s_M(v_0) < x+y\} + \sum_{v \in \{0\} \times [0,N)^{d-1}} P\{T(\rho(v)) \geq y\}.$$

Part (a) follows since $X_i(M,N)$ and $X_i(M,1)$ have the distribution of $s_{0,M}^N$ and $s_{0,M}$, respectively, and each $\rho(v)$ contains at most $dN/2$ edges.

(b) The inequality between the two probabilities in this part is clear. To prove that both probabilities have limit zero, it suffices to show that for some $\eta=\eta(\varepsilon) > 0$

(5.34) $$\lim_{M\to\infty} \max_{N \le 2\eta M} P\{b_{0,M}^{N} \le M(\mu-\varepsilon)\} = 0 .$$

Indeed, if we set for v with $v(1) = 0$

$$b_M(v) = \inf \{T(r): \ r \text{ a path from } v \text{ to } H_M\} ,$$

then for $N \le M$

$$b_{0,M}^{N} \ge b_{0,M}^{M} = \min\{b_M(v) : 0 \le v_i < M\}$$

$$\ge \min_{0\le \underline{k}\eta \le 2} \min\{b_M(v): \ \delta(v,\lfloor \underline{k}\eta M\rfloor) < \eta M\} .$$

Here $\underline{k}\eta$ stands for the vector $(0,k_2\eta,\dots,k_d\eta)$, and $\lfloor \underline{k}\eta M\rfloor$ for $(0,\lfloor k_2\eta M\rfloor,\dots,\lfloor k_d\eta M\rfloor)$, while $0 \le \underline{k}\eta \le 2$ means $0 \le k_i\eta \le 2$ for $2 \le i < \ell$. The minimum over $\underline{k}$ only runs over a fixed number of terms (once η is fixed) and each

$$\min \{b_M(v) : \delta(v,\lfloor \underline{k}\eta M\rfloor) < \eta M\}$$

has the same distribution as $b_{0,M}^{2\eta M}$. Thus, if (5.34) holds for any $\eta > 0$ then it holds for $\eta = 1$.

To prove (5.34) for suitable η we follow the idea of part (a). We now wish to find an analogue of (5.33) which compares $b_M(v_0)$ with $b_M(v)$ (v_0 as before). For $\rho(v)$ we now use a path from v_0 to a point near v, of passage time close to $\hat{T}(v_0,v)$. We remind the reader of the sets $\tilde{S}(u)$ from the proof of Theorem 2.26. The properties of the $\tilde{S}(u)$ which we need are firstly (2.28) and (2.29). Also the fact that any path r on $\mathbb{Z}^d$ from u to ∞ must intersect $\tilde{S}(u)$ (since it must intersect $S(u)$ by (2.23) and $S(u) \subset \tilde{S}(u)$ by (2.27)). Finally

(5.35) $$\tilde{S}(u) \subset D_{d(u)}(u) = \{w : \ \delta(w,u) \le d(u)\}$$

for some random $d(u) =$ diameter $S(u)$ (see proof of (2.26)) which satisfies (see Lemma (2.24))

$$(5.36) \qquad P\{d(u) > k\} \leq K_2(k+3)^{d-1} 3^{-\frac{k}{4}} .$$

Now assume

$$(5.37) \qquad \{d(u) < M \text{ for all } v \in \{0\} \times [0,N)^{d-1}\}$$

occurs. Then if r is any path from $v \in \{0\} \times [0,N)^{d-1}$ to H_M , r
must intersect $\tilde{S}(v)$ (since its endpoint lies outside the box $D_{\tilde{d}(v)}(v)$
which contains $S(v)$. Let u_1 be an intersection of r with $\tilde{S}(v)$
and let r' be the piece of r from u_1 to H_M. By definition of
$\hat{T}(v_0,v)$

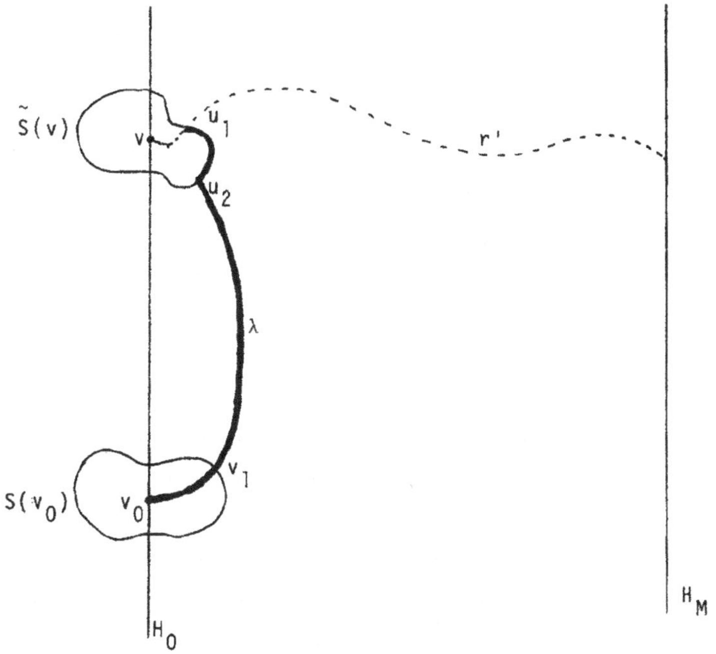

Fig. 5.2. $\rho(u)$ is the boldly drawn path .

(see (2.25)) there exist a path λ from some vertex $v_1 \in S(v_0)$ to
some vertex u_2 of $S(v) \subset \tilde{S}(v)$ with $T(\lambda) \leq \hat{T}(v_0,v) + 1$. Now con-

struct the path $\rho(v)$ from v_0 to u_1 by first connecting v_0 to v_1 by a path in $D_{d(v_0)}(v_0)$, then following λ from v_1 to u_2, and finally by connecting u_2 to u_1 by a path with all its edges in $\tilde{S}(v)$. The first piece of ρ from v_0 to v_1 can be chosen inside $D_{d(v_0)}(v_0)$ because $v_1 \in \tilde{S}(v_0)$ and by virtue of (5.35). This piece has a passage time at most equal to

$$Z := \sum_{e \in D_{d(v_0)}(v_0)} t(e).$$

The last piece of $\rho(v)$ from u_2 to u_1 can be chosen with all its vertices in $\tilde{S}(v)$ by virtue of (2.29). Its passage time is at most $\Delta(v)$. Since $\rho(v)$ followed by r' is a path from v_0 to H_M we see that

$$b_M(v_0) \leq T(\rho(v)) + T(r').$$
$$\leq Z + \hat{T}(v_0, v) + 1 + \Delta(v) + T(r).$$

By taking the inf over all r from v to H_M we obtain, whenever (5.37) occurs,

$$b_M(v_0) \leq Z + \hat{T}(v_0, v) + 1 + \Delta(v) + b_M(v) .$$

This is the desired analogue of (5.33).

The rest of the proof is still quite similar to (a). Firstly we choose $v \in \{0\} \times [0,N)^{d-1}$ such that $b_M(v) = b^N_{0,M}$. Then we obtain

$$b_M(v_0) \leq \max \{(Z + \Delta(v) + 1) : \delta(v, v_0) \leq N\}$$
$$+ \max \{\hat{T}(v_0, v) : v \in \{0\} \times [0,N)^{d-1}\} + b^N_{0,M} .$$

Secondly, we observe that the probability that (5.37) occurs tends to 1 as $M \to \infty$ (by (5.36) and $N \leq M$). Thirdly, as $M \to \infty$

$$\frac{1}{M} \max \{(Z + \Delta(v) + 1) : \delta(v, v_0) \leq M\} \to 0 \text{ in probability,}$$

because the distribution of Z is independent of M, and (2.28). Finally, by (1.14)

$$P\{b_M(v_0) \leq M(\mu - \tfrac{1}{2}\varepsilon)\} \to 0 \ , \ M \to \infty.$$

Thus, to obtain (5.34) it suffices to show that for a suitable $\eta=\eta(\varepsilon)>0$ and $N \leq 2\eta M$

$$P\{\max \{\hat{T}(v_0,v): \ v \in \{0\} \times [0,N)^{d-1}\} > \tfrac{\varepsilon}{4}M\} \to 0.$$

This, however, is contained in Theorem 3.1, because

$$\max \{\hat{T}(v_0,v): \ v \in \{0\} \times [0,N)^{d-1}\}$$

has the same distribution as

$$\max \{\hat{T}(\underline{0},v): \ v \in \{0\} \times [-N/2,N/2)^{d-1}\}$$

and for $N \leq 2\eta M$ the event

$$\max \{\hat{T}(0,v): \ v \in \{0\} \times [-M/2,M/2)^{d-1}\} > \tfrac{\varepsilon}{4}M$$

is contained in the event

$$\hat{B}(\tfrac{\varepsilon}{4}M) \quad \text{does not contain} \quad [-\eta M,\eta M)^d \ .$$

The probability of the last event goes to zero as $M \to \infty$ when η is sufficiently small by (3.2) or (3.3).

Proof of Theorem 5.2. Part (a) is almost immediate from Prop. 5.23 and (5.32). For $\mu = 0$ there is nothing to prove since $P\{\Theta_{0,n} < 0\} = 0$. For $\mu > 0$, take $x = n(\mu-\varepsilon)$ and

$$N = \min \left(M, \left\lfloor\frac{M\varepsilon}{4\mu}\right\rfloor\right)$$

in (5.23), and use that for $Q + 1 \geq n/(M+N)$

$$(5.38)\qquad P\{X_1(M,N)+\ldots+X_Q(M,N)<n(\mu-\varepsilon)\}$$
$$\le \inf_{\gamma\ge 0}\{e^{\gamma n(\mu-\varepsilon)}(E\{e^{-\gamma X_1(M,N)}\})^Q\}$$
$$\le \inf_{\gamma\ge 0}\{e^{\gamma(M+N)(\mu-\varepsilon)}[e^{\gamma(M+N)(\mu-\varepsilon)}(e^{-\gamma M(\mu-\frac{\varepsilon}{2})}$$
$$+P\{X_1(M,N)<M(\mu-\tfrac{\varepsilon}{2})\})]^Q\}$$

Now use (5.29)(b) to choose $M = M(\varepsilon,F,d)$ and $\gamma = \gamma(\varepsilon,F,d)$ such that

$$e^{\gamma(M+N)(\mu-\varepsilon)}(e^{-\gamma M(\mu-\frac{\varepsilon}{2})}+P\{X_1(M,N)<M(\mu-\tfrac{\varepsilon}{2})\})$$
$$\le e^{-\gamma M\varepsilon/4}+e^{2\gamma M\mu}P\{s_{0,M}^N<M(\mu-\tfrac{\varepsilon}{2})\}$$
$$\le (60d)^{-d}(\max(2,\tfrac{8\mu}{\varepsilon}))^{-d}\le (60d\tfrac{M}{N})^{-d}\ .$$

For such M and γ Prop. 5.23 and (5.38) yield

$$P\{\theta_{0,n}<n(\mu-\varepsilon)$$
$$\le e^{\gamma(M+N)\mu}\sum_{Q+1\ge n/(M+N)}2^{-dQ}\le e^{2\gamma M\mu}2^{d+1-dn/2M}\ .$$

(b) and (c). For these parts the proof of Cor. 3.3 in [17] can be copied verbatim; most of these statements are also proven in [18] and [26], Sect. 3.2. Only (5.5) is not explicitly proven in [17], but it follows immediately from the fact that

$$x_n := P\{\theta_{0,n}<n(\mu-\varepsilon)\}$$

satisfies $x_{n+m}\ge x_n x_m$ for $\theta = a,s$ or t (see [17], Proof of Cor. 3.3). It is well known (see [26], 3.2.8) that this implies

$$(x_n)^{1/n}\le \lim_{k\to\infty}(x_k)^{1/k}\ .$$

<u>Proof of Prop. 5.8.</u> This is a greatly simplified version of the proof of Theorem 5.2, which is similar to the proof of Prop. 1 in [22]. If

$r = (\underline{0}, e_1, \ldots, e_n, v_n)$ is a selfavoiding path starting at $\underline{0}$ we define the times $\tau(i)$ by $\tau(0) = 0$, $a_i = v_{\tau(i)}$,

$$\tau(q+1) = \min \{\tau > \tau(q) : \delta(a_q, v_\tau) = M\} \quad ,$$

which is essentially the same as in the proof of (5.23); M will be chosen below. We continue till we arrive at a Q such that $\delta(a_Q, v_\tau) < M$ for all $\tau > \tau(Q)$. We claim that

$$Q + 1 \geq (2M+1)^{-d}\, n \ .$$

This is so because r is selfavoiding, and consequently cannot have more than $(2M+1)^d$ vertices in a set $\{v: \ \delta(a,v) \leq M\}$ for fixed a. This implies

$$\tau(q+1) - \tau(q) \leq (2M+1)^d, \ n - \tau(Q) \leq (2M+1)^d \ .$$

The lower bound for Q is immediate from this. We now decompose the event $\{T(r) < Cn\}$ according to the values of $a_1, \ldots, a_Q$. We obtain that the probability in Prop. 5.8 is bounded above by

$$\sum_{Q+1 \geq n(2M+1)^{-d}} \ \sum_{a_1, \ldots, a_Q} P\{r \text{ passes successively through } a_1, \ldots, a_Q \text{ and } \sum_0^{Q-1} T(a_i, a_{i+1}) < Cn\}.$$

By virtue of (4.13) this expression is bounded by

$$\sum_{Q+1 \geq n(2M+1)^{-d}} \ \sum_{a_1, \ldots, a_Q} P\{\sum_0^{Q-1} T'(a_i, a_{i+1}) < Cn\} \ ,$$

where the $T'(a_i, a_{i+1})$ are independent copies of the $T(a_i, a_{i+1})$, $0 \leq i \leq Q-1$, and the sum over the a_i is restricted to $\delta(a_i, a_{i+1}) = M$, $0 \leq i \leq Q-1$ $(a_0 = \underline{0})$. The standard methods now yield

$$\sum_{a_1, \ldots, a_Q} P\{\sum_0^{Q-1} T'(a_i, a_{i+1}) < Cn\} \leq e^{\xi Cn} \sum_{a_1, \ldots, a_Q} \prod_0^{Q-1} E e^{-\xi T(a_i, a_{i+1})}$$

$$= e^{\xi Cn} \Big[\sum_{\delta(\underline{0}, a) = M} E e^{-\xi T(\underline{0}, a)} \Big]^Q$$

(sum successively over $a_Q, a_{Q-1}, \dots, a_1$). Finally we choose M such that

$$\sum_{\delta(\underline{0},a)=M} P\{T(\underline{0},a) = 0\} \le \frac{1}{2} .$$

This is possible, because $F(0) < p_T$ means that

$$E\{\#W\} = \sum_v P\{v \in W\} = \sum_v P\{T(\underline{0},v) = 0\} < \infty .$$

Once M has been fixed in this way we can choose ξ so large and $C > 0$ so small that for $n \le (2M+1)^d(Q+1)$

$$e^{\xi Cn}\Big[\sum_{\delta(\underline{0},a)=M} Ee^{-\xi T(\underline{0},a)}\Big]^Q \le e^{\xi Cn}\left(\frac{3}{4}\right)^Q \le e^{\xi C(2M+1)^d}\left(\frac{7}{8}\right)^Q .$$

Proof of Theorem 5.9. Part (a) is the same as Theorem 3.2 and 3.4b of [17]. We repeat a good part of the proof since it is needed for part (b). It suffices to consider $\theta = t$ only, since $t_{0,n} \ge_1 a_{0,n}$, $s_{0,n} \ge b_{0,n}$. Let $\varepsilon > 0$ be given. Since $n^{-1}t_{0,n} \to \mu$ in L^1 we can choose ν such that

$$Et_{0,\nu} \le \nu\left(\mu + \frac{\varepsilon}{5}\right).$$

Part (a) can be proved by using the $t_{0,\nu}$. However, for parts (b) and (c) we need an extra truncation, and we introduce it now. We define

$$\begin{aligned} t_{m,n}(k) = \inf\{T(r):& r \text{ a path from } (m,0,\dots,0) \text{ to } (n,0,\dots,0) \\ & \text{such that, with the exception of its endpoints,} \\ & r \text{ is contained in } (m,n)\times(-k,k)\times\dots\times(-k,k)\}. \end{aligned}$$

For fixed ν

$$\lim_{k\to\infty} E\, t_{0,\nu}(k) = E\, t_{0,\nu}$$

so that we can choose a k for which

$$E\, t_{0,\nu}(k) \le \nu\left(\mu + \frac{2\varepsilon}{5}\right).$$

Now

$$t_{0,\pi\nu}(k) \le \sum_{j=0}^{\pi-1} t_{j\nu,(j+1)\nu}(k)$$

and the random variables $t_{j\nu,(j+1)\nu}(k)$, $j \ge 0$, are independent, all have the same distribution as $t_{0,\nu}(k)$, and hence expectation $\le \nu(\mu + \frac{2\varepsilon}{5})$, and for the γ for which (5.10) holds one also has

$$E \exp \gamma t_{j\nu,(j+1)\nu}(k) = E \exp \gamma t_{0,\nu}(k) < \infty,$$

since

(5.39) $$t_{a,b}(k) \le \sum_{j=0}^{b-1} t(e_j),$$

where e_j is the edge between $(j,0,\dots,0)$ and $((j+1),0,\dots,0))$. Standard methods now show that

$$P\{t_{0,\pi\nu}(k) \ge \pi\nu(\mu + \frac{3\varepsilon}{5})\} \le A_4 e^{-B_4\pi}, \quad \pi \ge 1,$$

for some $0 < A_4, B_4 < \infty$. If $n = \pi\nu + \tau$, with $0 \le \tau < \nu$ use

$$t_{0,n}(k) \le t_{0,\pi\nu}(k) + t_{\pi\nu,n}(k)$$

and

$$P\{t_{\pi\nu,n}(k) \ge \frac{\varepsilon}{5}n\} \le e^{-\frac{1}{5}\gamma\varepsilon n} E e^{\gamma t_{\pi\nu,n}(k)} \le e^{-\frac{1}{5}\gamma\varepsilon n}\{\int e^{\gamma x} dF(x)\}^{\tau}$$

(again by (5.39)). It follows that in all cases

(5.40) $$P\{t_{0,n}(k) \ge n(\mu + \frac{4}{5}\varepsilon)\} \le A_2 e^{-B_2 n},$$

for suitable $0 < A_2, B_2 < \infty$ (see [17], proof of Theorem 3.2 for more details). This completes the proof of part (a).

To obtain an estimate for $P\{t_{0,n} > n(\mu+\varepsilon)\}$ which decreases faster than exponentially we consider a number of independent "parallel" paths each of which has a passage time with the same distribution as $t_{0,n}(k)$. Specifically, if $v(1) = 0$, let

$$t_n(v;k) = \inf\{T(r): r \text{ a path from } v \text{ to } (n,v(2),\dots,v(d)) \text{ such that with the exception of its endpoints } r \text{ is contained in } (0,n)\times(v(2)-k,v(2)+k)\times\dots\times(v(d)-k,v(d)+k))\}.$$

Of course $t_{0,n}(k) = t_n(\underline{0};k)$. Clearly all $t_n(v,k)$ with $v(1) = 0$ have the same distribution as $t_{0,n}(k)$. Moreover, the family $\{t_n(v;k): v(1) = 0,\ v(2),\dots,v(d) \in 2k\mathbb{Z}\}$ are independent, since $t_n(0,2j_2k,\dots,2j_dk)$ is an infimum over paths in the "tube" $(0,n)\times(2j_2k-k,2j_2k+k)\times\dots\times(2j_dk-k,2j_dk+k)$, and these tubes are disjoint for different $\underline{j} := (0,j_2,\dots,j_d)$. To construct a cylinder path from 0 to $(n,0,\dots,0)$ which makes use of $t_n(\underline{j}k;k)$ we first connect the origin to the point $(1,2j_2k\dots,2j_dk)$ by going along the edge from $\underline{0}$ to $(1,0,\dots,0)$, and then connect $(1,0,\dots,0)$ to $(1,2j_2k,\dots,2j_dk)$ in the hyperplane $H_1 = \{x(1) = 1\}$ by a path of at most $2(j_2+\dots+j_d)k$ steps. Call the resulting path from $\underline{0}$ to $(1,2j_2k,\dots,2j_dk)$ $\rho(\underline{j})$. Construct a similar path $\tau(\underline{j})$ from $(n-1,2j_2k,\dots,2j_dk)$ to $(n,0,\dots,0)$. Any path r from $2\underline{j}k$ to $2\underline{j}k+(n,0,\dots,0)$ for which $H_0 < r < H_n$ passes through $(1,2j_2k,\dots,2j_dk)$ and $(n-1,2j_2k,\dots,2j_dk)$. If r' denotes the piece of r between these points, then $\rho(\underline{j})$, r' and $\tau(\underline{j})$ together make up a cylinder path from 0 to $(n,0,\dots,0)$ (see Fig. 5.3), so that

$$t_{0,n} \le T(\rho(\underline{j})) + T(r) + T(\tau(\underline{j})) \,.$$

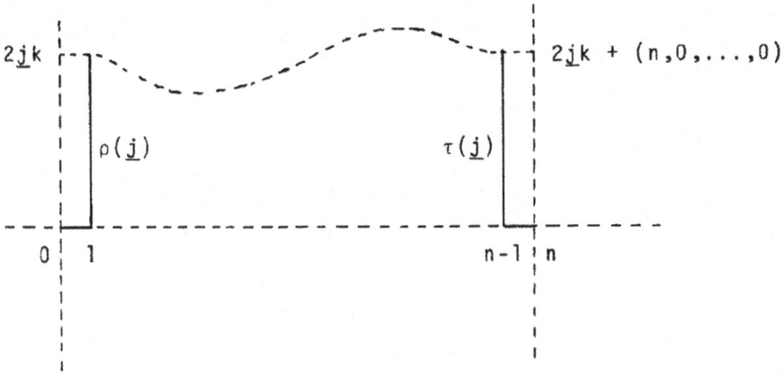

Fig. 5.3. $\rho(\underline{j})$ and $\tau(\underline{j})$ are the solidly drawn paths.

Taking the inf over all r which are contained in $(0,n)\times(2j_2k-k,$ $2j_2k+k)\times\ldots\times(2j_dk-k,2j_dk+k)$ (except for their endpoints) yields

$$t_{0,n} \le T(\rho(\underline{j})) + T(\tau(\underline{j})) + t_n(2\underline{j}k;k).$$

Consequently, for any p, if $\{t_i\}$ is a family of independent random variables, each with the distribution of $t(e)$, and k such that (5.40) holds, then

$$P\{t_{0,n} > n(\mu+\varepsilon)\} \le P\{\max\{T(\rho(\underline{j})) + T(\tau(\underline{j})) : |\underline{j}| \le p\} \ge \tfrac{1}{5}\, n\varepsilon\}$$

$$+ \prod_{|\underline{j}|\le p} P\{t_n(2\underline{j}k;k) \ge n(\mu + \tfrac{4\varepsilon}{5})\}$$

$$\le (2p+1)^{d-1} P\{t_1+\ldots+t_{2pk(d-1)+2} \ge \tfrac{1}{5}\, n\varepsilon\}$$

$$+ [A_2 \exp - B_2 n]^{(2p+1)^{d-1}} .$$

Here $|\underline{j}|$ stands for $\sum_2^d |j_\ell|$ and in the second inequality we used that there are $(2p+1)^{d-1}$ vectors $\underline{j} = (0,j_2,\ldots,j_d)$ with $|\underline{j}| \le p$, the fact that $T(\rho(\underline{j}))$ and $T(\tau(\underline{j}))$ together contain at most $2+2(d-1)pk$ edges (all of which are distinct if $n > 2$), and finally (5.40).

It is now easy to complete the proof of the theorem by observing that for any $\xi \ge 0$

$$P\{t_1+\ldots+t_{2pk(d-1)+2} \ge \tfrac{1}{5}\, n\varepsilon\} \le e^{-\frac{1}{5}\xi n\varepsilon}\{Ee^{\xi t(e)}\}^{2pk(d-1)+2} .$$

(5.12) follows from the above estimates by letting $p \to \infty$ and $\xi \to \infty$ with n in such a way that

$$(2pk(d-1)+2)\log Ee^{\xi t(e)} + (d-1)\log(2p+1) \le \tfrac{1}{10}\,\xi n\varepsilon .$$

Finally, if $t(e)$ is bounded, say $t(e) \le \Gamma$, then

$$P\{t_1+\ldots+t_{2pk(d-1)+2} \ge \tfrac{1}{5}\, n\varepsilon\} = 0$$

as soon as

$$2pk(d-1)+2 < \frac{\varepsilon}{5\Gamma} n .$$

(5.13) is now immediate.

(5.41) LEMMA. For $m < n$ there exists a random variable $s'_{m,n}$ with the same distribution as $s_{0,n-m}$, but independent of $s_{0,m}$ and such that

(5.42) $$s_{0,n} \le s_{0,m} + s'_{m,n} .$$

Consequently, if $Et^2(e) < \infty$,

(5.43) $$\frac{1}{n} Es_{0,n} \ge \frac{1}{2n} Es_{0,2n} \ge \inf_k \frac{1}{k} Es_{0,k} = \mu, \ n \ge 1,$$

and for $K \ge 0$, $n \ge 1$

(5.44) $$\sum_{\ell=1}^{K} \sigma^2\left(\frac{s_{0,2^\ell}}{n2^\ell}\right) \le \sigma^2\left(\frac{s_{0,n}}{n}\right) + 2\left(E\left(\frac{s_{0,n}}{n}\right)\right)^2 - 2\left(E\left(\frac{s_{0,n2^K}}{n2^K}\right)\right)^2$$
$$\le \sigma^2\left(\frac{s_{0,n}}{n}\right) + 2\left(E\left(\frac{s_{0,n}}{n}\right)\right)^2 - 2\mu^2 .$$

Proof: These results are easy and well known. (5.42) says that $\{s_{m,n}\}$ is superconvolutive (compare [18]) and is (4.2.7) of [19] or (5.12) of [31]. (5.43) is now immediate (see also Theorem 4.25 of [19] or Lemma 5.2 of [31]). Finally, (5.44) is obtained by summing (2.21) of [31] and using (5.43) (see also (2.22)-(2.24) of [31]).

(5.45) COROLLARY. Let

$$C_4 = (4 \log 2)Et^2(e).$$

Then for every $m \ge 10$ there exists an M in $[m,m^2]$ such that

(5.46) $$\sigma^2\left(\frac{s_{0,M}}{M}\right) \le \frac{1}{\log M} C_4 .$$

Proof: By (5.44) with $n = 1$

$$\sum_{\ell=1}^{\infty} \sigma^2\left(\frac{s_{0,2^\ell}}{2^\ell}\right) \le \sigma^2(s_{0,1}) + 2(Es_{0,1})^2 \le 2Es_{0,1}^2 \le 2Et^2(e).$$

Consequently

$$\min\{\sigma^2(\frac{s_{0,2^\ell}}{2^\ell}) : m \le 2^\ell \le m^2\}$$

$$\le (\lfloor\frac{2 \log m}{\log 2}\rfloor - \lfloor\frac{\log m}{\log 2}\rfloor)^{-1} \sum_{m\le 2^\ell \le m^2} \sigma^2(\frac{s_{0,2^\ell}}{2^\ell})$$

$$\le 2(\lfloor\frac{2 \log m}{\log 2}\rfloor - \lfloor\frac{\log m}{\log 2}\rfloor)^{-1} E t^2(e),$$

which implies the Corollary.

Proof of (5.17). We restrict ourselves to $0 < \varepsilon < \beta \le \mu$; this is permissible as remarked in (5.21). We shall use $B(\varepsilon)$ as abbreviation for $B_1(\varepsilon,F,d)$. $C_i = C_i(F,d)$ will denote various constants which satisfy $0 < C_i < \infty$ and which depend on d and F only. Also, $\varepsilon_0 = \varepsilon_0(F,d) \le 1$ will be a small strictly positive constant. Its value actually depends on d and C_4-C_6 only (see (5.59) and the lines following (5.61), (5.66) and (5.67)).

The main step consists of an inductive argument of the following form. Assume that $0 < \varepsilon_0 < \beta$ has already been chosen. We can choose D such that

$$B(\varepsilon_0) \ge \exp(-D\varepsilon_0^{-p}) \quad \text{and} \quad D \ge 1. \tag{5.47}$$

Assume next that we have some $0 < \varepsilon_k \le \varepsilon_0$ for which

$$B(\varepsilon_k) \ge \exp(-D\varepsilon_k^{-p}).$$

If

$$B(\varepsilon_k) > \exp(-\varepsilon_k^{-p}), \tag{5.48}$$

then define

$$\eta_k = \inf\{\eta : \eta \le \varepsilon_k \text{ and } B(\varepsilon) > \exp(-\varepsilon^{-p}) \text{ for } \eta \le \varepsilon \le \varepsilon_k\}.$$

If $\eta_k \le 0$, then

$$B(\varepsilon) > \exp(-\varepsilon^{p}) \quad \text{for} \quad 0 < \varepsilon \le \varepsilon_k$$

and we can stop our inductive procedure. However, if $\eta_k > 0$, then

$$B(\eta_k) = \exp(-\eta_k^{-p})$$

(since $B(\cdot)$ is continuous on $(0,\beta)$, by Theorem 5.2(c) and we define

(5.49) $$\varepsilon_{k+1} = \frac{1}{2}\eta_k .$$

If (5.48) fails, then we take $\eta_k = \varepsilon_k$ and again define ε_{k+1} by (5.49). We then have, in any case that for $\eta_k \le \rho \le \varepsilon_k$

(5.50) $$B(\rho) \ge \exp(-D\rho^{-p}).$$

The fundamental part of the proof consists in showing that (5.50) also holds for $\rho = \varepsilon_{k+1}$. Once this is done, (5.50) holds on all the intervals $[\eta_k, \varepsilon_k]$, while for

$$\varepsilon_{k+1} = \frac{1}{2}\eta_k < \rho < \eta_k$$

one has by the monotonicity of B

$$B(\rho) \ge B(\varepsilon_{k+1}) \ge \exp(-D\varepsilon_{k+1}^{-p}) \ge \exp(-D2^p\rho^{-p}),$$

so that (5.17) follows with $C_1 = D2^p$ for $0 \le \varepsilon < \varepsilon$. By increasing C_1 we can make it valid on the whole interval $0 \le \varepsilon < \beta$.

We now turn to the proof of (5.50) for $\rho = \varepsilon_{k+1}$. From now on D is chosen in such a way that (5.47) holds, and by virtue of the above construction we may assume that

(5.51) $$\exp(-D\eta_k^{-p}) \le B(\eta_k) \le \exp(-\eta_k^{-p}) .$$

The starting point of the proof is Prop. 5.23, which with $x = n(\mu-\varepsilon_{k+1})$ yields

(5.52) $$P\{\theta_{0,n} < n(\mu-\varepsilon_{k+1})$$

$$\le \sum_{Q+1 \ge n/(M+N)} \left(30d\,\frac{M}{N}\right)^{dQ} P\{X_1(M,N)+\ldots+X_Q(M,N) < n(\mu-\varepsilon_{k+1})\}.$$

First we replace $X_i(M,N)$ by $X_i(M,1)$ with the help of Lemma 5.29(a). We have for $Q+1 \geq n/(M+N)$

$$(5.53) \qquad P\{X_1(M,N)+\ldots+X_Q(M,N) < n(\mu-\varepsilon_{k+1})$$

$$\leq P\{X_1(M,N)+\ldots+X_Q(M,N) < (Q+1)(M+N)(\mu-\varepsilon_{k+1})$$

$$\leq P\{X_1(M,1)+\ldots+X_Q(M,1) < QM(\mu-\tfrac{1}{2}\varepsilon_{k+1})$$

$$+ N^{Q(d-1)}P\{t_1+\ldots+t_{Q\lfloor dN/2\rfloor} \geq \tfrac{1}{2}QM\varepsilon_{k+1}-QN\mu-2M\mu\} \ .$$

Now we choose $C_5 > 0$ such that

$$(5.54) \qquad (\exp \gamma(\mu-\frac{1}{2C_5}))\cdot(Ee^{\gamma t(e)})^{d/2} = \frac{1}{4}$$

(with $\gamma > 0$ such that (5.10) holds), and

$$(5.55) \qquad M \in [(B(n_k))^{-2},(B(n_k))^{-4}]$$

such that

$$(5.56) \qquad \sigma^2(\frac{S_{0,M}}{M}) \leq C_4 \frac{1}{\log M} \ ,$$

(this is possible by (5.46)), and finally

$$(5.57) \qquad N = \lfloor C_5\varepsilon_{k+1}M\rfloor = \lfloor \tfrac{1}{2}C_5\eta_k M\rfloor \ .$$

Then the last term in the right hand side of (5.53) is at most

$$(5.58) \qquad N^{Q(d-1)}e^{\gamma QN\mu+2\gamma M\mu}e^{-\frac{1}{2}\gamma QM\varepsilon_{k+1}}(Ee^{\gamma t(e)})^{Q\lfloor dN/2\rfloor}$$

$$\leq e^{2\gamma M\mu}\{N^{(d-1)/N}e^{\gamma(\mu-\frac{1}{2}N^{-1}M\varepsilon_{k+1})}(Ee^{\gamma t(e)})^{d/2}\}^{QN}$$

$$\leq e^{2\gamma M\mu}\{\tfrac{1}{4}N^{(d-1)/N}\}^{QN} \quad \text{(by (5.57) and (5.54))}.$$

Now note that for $\varepsilon_0 \leq 1$

$$N = \lfloor C_5 \varepsilon_{k+1} M \rfloor \geq \frac{1}{2} C_5 \eta_k (B(\eta_k))^{-2} - 1 \quad \text{(by (5.49) and (5.55))}$$

$$\geq \frac{1}{2} C_5 \eta_k \exp(2\eta_k^{-p}) - 1 \quad \text{(by (5.51)} \geq \frac{1}{2} C_5 \varepsilon_0 \exp(2\varepsilon_0^{-p}) - 1.$$

Thus, if we choose ε_0 so small that

$$N_0 := \frac{1}{2} C_5 \varepsilon_0 \exp(2\varepsilon_0^{-p}) - 1$$

satisfies

$$N_0^{(d-1)/N_0} \leq 4/e \quad \text{and} \quad N_0 \geq e, \tag{5.59}$$

then the last member of (5.53) is at most

$$e^{2\gamma M\mu} \, e^{-QN} ,$$

and

$$\sum_{Q+1 \geq n/(M+N)} \left(\frac{30dM}{N}\right)^{dQ} N^{Q(d-1)} \tag{5.60}$$

$$\cdot P\{t_1 + \ldots + t_{Q \lfloor dN/2 \rfloor} \geq \frac{1}{2} QM\varepsilon_{k+1} - QN\mu - 2M\mu\}$$

$$\leq e^{2\gamma M\mu} \sum_{Q+1 \geq n/(M+N)} \left\{\left(\frac{120d}{C_5 \eta_k}\right)^d e^{-N}\right\}^Q .$$

Note that (by (5.55) and (5.51))

$$\left(\frac{120d}{C_5 \eta_k}\right)^d e^{-N} \leq \left(\frac{120d}{C_5}\right)^d \eta_k^{-d} \exp(1 - \tfrac{1}{2} C_5 \eta_k M) \tag{5.61}$$

$$\leq C_6 \eta_k^{-d} \exp(-\tfrac{1}{2} C_5 \eta_k \exp(2\eta_k^{-p}))$$

$$\leq C_6 \varepsilon_0^{-d} \exp(-\tfrac{1}{2} C_5 \varepsilon_0 \exp(2\varepsilon_0^{-p})) .$$

The last inequality requires ε_0 to be sufficiently small. If necessary we may reduce ε_0 still further so that the last member of (5.61) does not exceed e^{-1}. We may also assume $C_5 \varepsilon_0 \leq 1$ so that $N \leq M$. In this case (5.60) is bounded by

$$e^{2\gamma M\mu} \sum_{(Q+1) \geq n/(M+N)} e^{-Q} \leq e^{2+2\gamma M\mu} e^{-n/2M} . \tag{5.62}$$

Next we estimate

$$(5.63)\quad P\{X_1(M,1)+\ldots+X_Q(M,1) < QM(\mu-\tfrac{1}{2}\varepsilon_{k+1})$$

$$\leq \{\inf_{\delta\geq 0} e^{\delta M(\mu-\frac{1}{2}\varepsilon_{k+1})} Ee^{-\delta X_1(M,1)}\}^Q .$$

Since $X_1(M,1)$ has the same distribution as $s_{0,M}$ we have

$$Ee^{-\delta X_1(M,1)} \leq e^{-\delta M(\mu-\frac{1}{4}\varepsilon_{k+1})} + e^{-\delta M(\mu-\eta_k)} P\{M(\mu-\eta_k)$$

$$\leq s_{0,M} < M(\mu-\tfrac{1}{4}\varepsilon_{k+1})\} + P\{s_{0,M} < M(\mu-\eta_k)\}$$

$$= I + II + III, \text{ say.}$$

Since $Es_{0,M} \geq M\mu$ (by (5.43)) we have by Chebyshev's inequality and (5.56), (5.55) and (5.51)

$$II \leq e^{-\delta M(\mu-\eta_k)} \frac{16\sigma^2(s_0,M)}{\varepsilon_{k+1}^2 M^2}$$

$$\leq e^{-\delta M(\mu-\eta_k)} \frac{64C_4}{\eta_k^2 \log M} \leq e^{-\delta M(\mu-\eta_k)} 32C_4\eta_k^{p-2} .$$

Also, by (5.5), (5.55) and (5.51)

$$III \leq \exp(-MB(\eta_k)) \leq \exp(-(B(\eta_k))^{-1}) \leq \exp(-\exp(\eta_k^{-p})).$$

Finally we choose $\delta > 0$ such that

$$\frac{1}{4}\delta M\varepsilon_{k+1} = \frac{1}{8}\delta M\eta_k = -\log \frac{1}{3e}\left(\frac{N}{30dM}\right)^d \sim -\log \frac{1}{3e}\left(\frac{C_5\eta_k}{60d}\right)^d .$$

We then obtain

$$(5.64)\quad e^{\delta M(\mu-\frac{1}{2}\varepsilon_{k+1})}(I) = e^{-\frac{1}{4}\delta M\varepsilon_{k+1}} = \frac{1}{3e}\left(\frac{N}{30dM}\right)^d .$$

Also

$$(5.65)\qquad e^{\delta M(\mu - \frac{1}{2}\varepsilon_{k+1})}(II) \le e^{\delta M \eta_k} 32C_4\eta_k^{p-2}$$

$$= (3e)^8\left(\frac{30dM}{N}\right)^{8d} 32C_4\eta_k^{p-2} \le \frac{1}{3e}\left(\frac{N}{30dM}\right)^d ,$$

provided

$$\eta_k^{p-2}\left(\frac{M}{N}\right)^{9d} \le (32C_4)^{-1}(3e)^{-9}(30d)^{-9d} .$$

Since

$$(5.66)\qquad \eta_k^{p-2}\left(\frac{M}{N}\right)^{9d} \le \eta_k^{p-2}\left(\frac{4}{C_5\eta_k}\right)^{9d} = \eta_k\left(\frac{4}{C_5}\right)^{9d} \le \varepsilon_0\left(\frac{4}{C_5}\right)^{9d} ,$$

we may assume that ε_0 is chosen so small that (5.65) holds. Finally, by similar estimates

$$(5.67)\qquad e^{\delta M(\mu - \frac{1}{2}\varepsilon_{k+1})}(III) \le e^{\delta M\mu}\exp(-\exp(\eta_k^{-p}))$$

$$\le \left\{3e\left(\frac{30dM}{N}\right)^d\right\}^{8\mu\eta_k^{-1}} \exp(-\exp(\eta_k^{-p})) \le \frac{1}{3e}\left(\frac{N}{30dM}\right)^d ,$$

if ε_0 is chosen small enough. When the estimates (5.52), (5.53), (5.58)-(5.67) are combined we obtain

$$P\{\theta_{0,n} \le n(\mu-\varepsilon_{k+1})\} \le e^{2+2\gamma M\mu - n/2M}$$

$$+ \sum_{Q+1\ge n/(M+N)} \left(\frac{30dM}{N}\right)^{dQ}\left\{e^{\delta M(\mu - \frac{1}{2}\varepsilon_{k+1})}(I + II + III)\right\}^Q$$

$$\le e^{2+2\gamma M\mu}\, e^{-n/2M} + \sum_{Q+1\ge n/(M+N)} e^{-Q}$$

$$\le e^2(e^{2\gamma M\mu}+1)e^{-n/2M} .$$

Therefore,

$$B(\varepsilon_{k+1}) = \lim -\frac{1}{n} \log P\{\theta_{0,n} \leq n(\mu-\varepsilon_{k+1})\}$$

$$\geq \frac{1}{2M} \geq \frac{1}{2}(B(\eta_k))^4 \quad \text{(by (5.55))}$$

$$\geq \frac{1}{2} \exp(-4D\eta_k^{-p}) \text{ (by (5.51))} \geq \exp(-D\varepsilon_{k+1}^{-p}) \text{ (by (5.49),}$$

$$D \geq 1,\ \eta_k \leq \varepsilon_0 \leq 1 \text{ and } p \geq 3).$$

This is just (5.50) for $\rho = \varepsilon_{k+1}$. The proof of (5.17) is complete.

We prepare for the proof of (5.18)-(5.20) with a lemma whose proof is very similar to the proof of (5.17). We therefore shall only sketc the required estimates. We remind the reader of the definition of $\hat{s}_{0,n}^N$ before (5.23). To save writing indices we write $\hat{s}_{0,n}$ for $\hat{s}_{0,n}^1 = \inf\{T(r): r$ a path from $\underline{0}$ to H_n, such that r, with the exception of its endpoints is contained in $(0,n) \times (-4n,4n)^{d-1}\}$.

(5.68) LEMMA. Fix $D \geq 1$ so large that $2C_1D^{-p} \leq 1$ and such that for all $M \geq 2$

(5.69) $$e^{-\gamma(D-2\mu-1)} \leq \{Ee^{\gamma t(e)}\}^{-d} \cdot \{60dM\}^{-d(\log M)^{1/p}/M},$$

where $\gamma > 0$ is fixed so that (5.10) holds. Then for every sufficiently large m there exists an M in $[m,m^2]$ such that

(5.70) $$E\hat{s}_{0,M} \leq M\mu + 4D(\log M)^{-1/p},$$

and

(5.71) $$\sigma^2\left(\frac{\hat{s}_{0,M}}{M}\right) \leq \frac{1}{\log M} C_4 .$$

Proof: Observe first that for every $m \geq 10$ there exists an M in $[m,m^2]$ for which (5.71) holds. The proof is identical to that of (5.46). Indeed (5.46) is based entirely on the fact that $\{s_{0,n}\}$ satisfies equation (5.42) with $n = 2m$. The relation is true, because if r' and r'' are paths from 0 to H_m with $H_0 < r', r'' < H_m$, then the path r, obtained by combining r' with the translate of r'' by $v:=$ endpoint of r', is a path from 0 to

H_{2m} with $H_0 < r < H_{2m}$. In the same way one shows that $\{\hat{s}_{0,n}\}$ satisfies (5.42) with $n = 2m$. Indeed, if r', r'' above are in addition contained in $(0,n) \times (-4n,4n)^{d-1}$ (with the exception of their endpoints), then r is contained in $(0,2n) \times (-8n,8n)^{d-1}$, except for its endpoints. Thus (5.42) holds with s replaced by $\hat{s}$ and (5.71) follows.

Now assume that M satisfies (5.71), but that

$$(5.72) \qquad 4\lambda_M := E\left(\frac{\hat{s}_{0,M}}{M}\right) - \mu \geq 4D(\log M)^{-1/p} .$$

We shall show that this implies for M large and suitable $E_i = E_i(M) \in (0,\infty)$

$$(5.73) \qquad P\{s_{0,n} \leq n(\mu+\lambda_M)\} \leq E_1 \exp -nE_2, \; n \geq 0.$$

This, however, is impossible because of Theorem 2.18. Thus also (5.72) cannot be true and the lemma will follow.

By the above we only have to prove that (5.71) plus (5.72) imply (5.73) and we do this now. By (5.23) and (5.29a)

$$P\{s_{0,n} \leq n(\mu+\lambda_M)\} \leq \sum_{Q+1 \geq n/(M+N)} \left(\frac{30dM}{N}\right)^{dQ}$$

$$[P\{\hat{X}_1(M,1)+\ldots+\hat{X}_Q(M,1) \leq QM(\mu+2\lambda_M)\}$$

$$+N^{Q(d-1)}P\{t_1+\ldots+t_{Q\lfloor dN/2\rfloor} \geq QM\lambda_M-(QN+2M)(\mu+\lambda_M)\}].$$

We take $N = \lfloor M(\log M)^{-1/p}\rfloor + 1$ and estimate (assuming that (5.72) holds and that M and Q are large)

$$P\{t_1+\ldots+t_{Q\lfloor dN/2\rfloor} \geq QM\lambda_M-(QN+2M)(\mu+\lambda_M)$$

$$\geq QM(D-2\mu-1)(\log M)^{-1/p}-2M(\mu+1)\}$$

as in (5.58)-(5.62). Taking (5.69) into account this last probability is at most

$$e^{2\gamma M(\mu+1)}[\exp(-\gamma M(D-2\mu+1)(\log M)^{-1/p})\{Ee^{\gamma t(e)}\}dM(\log M)^{-1/p}]^Q$$

$$\leq e^{2\gamma M(\mu+1)}(\frac{1}{60dM})^{dQ} .$$

Also, analogously to (5.63) we have

$$P\{\hat{X}_1(M,1)+\ldots+\hat{X}_Q(M,1) \leq QM(\mu+2\lambda_M)\}$$

$$\leq \inf_{\delta\geq 0}\{e^{\delta M(\mu+2\lambda_M)}Ee^{-\delta\hat{s}_{0,M}}\}^Q$$

and

$$E\exp(-\delta\hat{s}_{0,M}) \leq \exp-\delta M(\mu+3\lambda_M)+\exp-\delta M(\mu-\lambda_M)\cdot P\{\hat{s}_{0,M} \leq M(\mu+3\lambda_M)\}$$

$$+P\{\hat{s}_{0,M} \leq M(\mu-\lambda_M)\}.$$

Again by Chebyshev's inequality and (5.71), (5.72)

$$P\{\hat{s}_{0,M} \leq M(\mu+3\lambda_M) = E\hat{s}_{0,M}-M\lambda_M\} \leq \frac{\sigma^2(\hat{s}_{0,M})}{M^2\lambda_M^2} \leq \frac{C_4}{\lambda_M^2\log M} ,$$

while, by virtue of (5.5), (5.17) and the obvious inequality $s_{0,M} \leq \hat{s}_{0,M}$, one has

$$P\{\hat{s}_{0,M} \leq M(\mu-\lambda_M)\} \leq P\{s_{0,M} \leq M(\mu-\lambda_M) \leq \exp-MB(\lambda_M)$$

$$\leq \exp(-M\exp(-C_1\lambda_M^{-p})) \leq \exp(-M^{1/2})$$

(recall (5.72) and $2C_1D^{-p} \leq 1$). Finally we take δ such that

$$\exp(\delta M\lambda_M) = (\lambda_M^2\log M)^{1/4} .$$

We leave it to the reader to imitate (5.64)-(5.67) and to deduce (5.73) from these estimates.

Proof of (5.18) and (5.19). $n^{-1}Es_{0,n} \geq \mu$ by (5.43), while $n^{-1}Ea_{0,n} \geq n^{-1}Et_{0,n} \geq \mu$ follows from $t_{0,n+m} \leq t_{0,m} + t_{m,n}$ (see [31], p.78

and (2.10)). This proves the first inequality of (5.18).

The second inequality in (5.18) only needs to be proved for $\theta = t$, because $a_{0,n}, s_{0,n} \le t_{0,n}$. To prove the inequality for $\theta = t$ we construct cylinder paths from $\underline{0}$ to $(n,0,\dots,0)$ with $n = \pi M + \tau$, $0 < \tau \le M$, and $n^{1/4} \le M \le n^{1/2}$ such that (5.70) and (5.71) hold. We do this by combining a number of cylinder paths between[1] H_{jM} and $H_{(j+1)M}$, $0 \le j < \pi$, and a last piece from $H_{\pi M}$ to $(n,0,\dots,0)$. The idea of this construction goes back to Hammersley and Welsh ([19], pp.83,84). To carry out this construction begin with taking $v_0 = \underline{0}$. If $v_j \in H_{jM}$ has been found let r_j be a path from v_j to $H_{(j+1)M}$ such that, with the exception of its endpoints, r_j is contained in

$$(jM,(j+1)M) \times (v_j(2)-4M, v_j(2)+4M) \times \dots \times (v_j(d)-4M, v_j(d)+4M)$$

and such that $T(r_j)$ is minimal among all such paths. Then $T(r_j)$ has the same distribution as $\hat{s}_{0,M}$. Denote the endpoint of r_j in $H_{(j+1)M}$ by v_{j+1}. We do this for $j = 0,\dots,\pi-1$. Finally, we connect v_π to $(n,0,\dots,0)$ by a path r_π which goes from v_π to $(\pi M,0,\dots,0)$ in the hyperplane $H_{\pi M}$, and from there to $(n,0,\dots,0)$ along the first coordinate axis (see Fig. 5.4). We choose r_π such that it contains

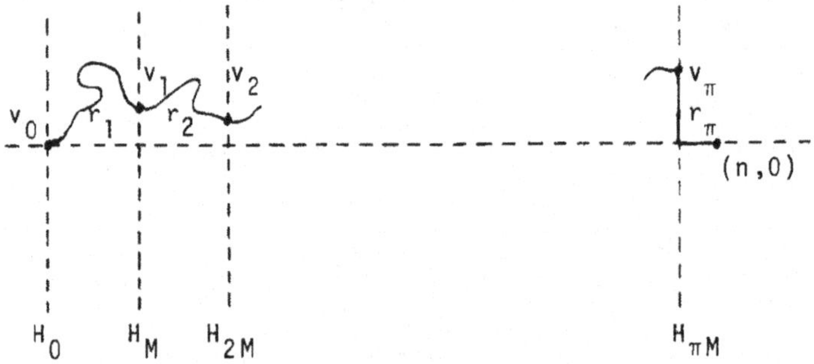

Fig. 5.4.

$$\rho := \tau + \sum_{i=2}^{d} v_\pi(i)$$

edges. The path r obtained by concatenating $r_0,\dots,r_\pi$ is a cylinder path from $\underline{0}$ to $(n,0,\dots,0)$ so that

[1] Recall that $H_\ell = \{x \in \mathbb{R}^d : x(1) = \ell\}$.

$$t_{0,n} \leq T(r) \leq \sum_{j=0}^{\pi} T(r_j).$$

Moreover, the $T(r_j)$, $0 \leq j \leq \pi-1$ are i.i.d. with the distribution of $\hat{s}_{0,M}$, while conditional on v_π and $T(r_0),\ldots,T(r_{\pi-1})$, the distribution of $T(r_\pi)$ is that of $t_1+\ldots+t_\rho$ (with $t_1, t_2, \ldots$ independent and each with the distribution of $t(e)$).

We now use the above construction of a cylinder path to estimate the distribution of $t_{0,n}$. Since $t_{0,n} \leq T(r) \leq \sum T(r_j)$ we have by virtue of (5.70)

$$(5.74) \qquad P\{t_{0,n} \geq n\mu+(4D+2)n(\log M)^{-1/p}\}$$

$$\leq P\{\sum_{j=0}^{\pi-1} T(r_j) \geq \pi E\hat{s}_{0,M}+\pi M(\log M)^{-1/p}\}+P\{\rho \geq M\pi^{3/4}\}$$

$$+P\{t_1+\ldots+t_{\lfloor M\pi^{3/4}\rfloor} \geq \pi M(\log M)^{-1/p}\} .$$

By Chebyshev's inequality and (5.71) the first probability in the right hand side of (5.74) is at most

$$\frac{1}{\pi M^2}(\log M)^{\frac{2}{p}} \quad \sigma^2(\hat{s}_{0,M}) \leq C_4 \frac{1}{\pi}(\log M)^{\frac{2}{p}-1}$$

As for the second term, $\iota \leq M$ while

$$\sum_{i=2}^{d} v_\pi(i) = \sum_{j=0}^{\pi-1} \sum_{i=2}^{d}(v_{j+1}(i)-v_j(i))$$

is the sum of the π i.i.d. random variables $\sum_i(v_{j+1}(i)-v_j(i))$, which we can choose in such a way that they have zero expectation (see [17], proof of $B_s(\varepsilon) = B_t(\varepsilon)$ in Sect. 3). Therefore (since $\lfloor v_{j+1}(i)-v_j(i)\rfloor$ $\leq 4M$ by the definition of the $\hat{s}_{0,M}$)

$$P\{\rho \geq M\pi^{3/4}\} \leq \frac{\pi}{M^2(\pi^{3/4}-1)^2}\,\sigma^2\{\sum_{i=2}^{d}(v_1(i)-v_0(i))\}$$

$$\leq \frac{4\pi}{M^2\pi^{3/2}}(d-1)^2(4M)^2 \leq \frac{64d^2}{\pi^{1/2}} .$$

Finally, since $\pi \geq n/(2M) \geq \frac{1}{2} n^{1/2} \geq \frac{1}{2} M$ we have for large n

$$P\{t_1+\ldots+t_{\lfloor M\pi^{3/4}\rfloor} \geq \pi M(\log M)^{-1/p}\}$$

$$\leq M\pi^{3/4}\sigma^2(t(e))\{\pi M(\log M)^{-1/p}-\pi^{3/4}MEt(e)\}^{-2}$$

$$\leq \frac{4}{M\pi^{5/4}}(\log M)^{2/p}\sigma^2(t(e)) .$$

In total, for n large and $n^{1/4} \leq M \leq n^{1/2}$, and consequently $\pi \geq n/(2M) \geq \frac{1}{2} n^{1/2} \geq \frac{1}{2} M$, we find

$$(5.75) \quad P\{t_{0,n} \geq n\mu+(4D+2)n(\log M)^{-1/p}\} \leq C_7\pi^{-1/2} \leq 2C_7 n^{-1/4}$$

Since for all large n we can choose $M \in [n^{1/4},n^{1/2}]$ such that (5.70) and (5.71) hold, by Lemma (5.68), and then choose π, τ such that $n = \pi M+\tau$, $0 < \tau \leq M$ we have proved (5.75) for all large n.

Finally we prove (5.18). Take $C_8 = 4^{1/p}(4D+2)$ so that $(4D+2)(\log M)^{-1/p} \leq C_8(\log n)^{-1/p}$. Then, using (5.39) and (5.75) we have

$$Et_{0,n} \leq n\mu+C_8n(\log n)^{-1/p}+E\{t_{0,n};t_{0,n} > n\mu+C_8n(\log n)^{-1/p}\}$$

$$\leq n(\mu+C_8(\log n)^{-1/p})$$

$$+(E\{\sum_{j=0}^{n-1} t(e)\}^2)^{1/2}(P\{t_{0,n} > n(\mu+C_8(\log n)^{-1/p})\}^{1/2}$$

$$\leq n(\mu+C_8(\log n)^{-1/p})+n(Et^2(e))^{1/2}(2C_7 n^{-1/4})^{1/2}$$

$$\leq n(\mu+(C_8+1)(\log n)^{-1/p}) \text{ (for large } n).$$

This completes the proof of (5.18).

For (5.19) we take $C_9 = C_2+C_8+(2C_1)^{1/p}$

$$\sigma^2(\theta_{0,n}) \le C_9^2 n^2(\log n)^{-2/p}$$

$$+E\{(\theta_{0,n}-E\theta_{0,n})^2; |\theta_{0,n}-E\theta_{0,n}| > C_9 n(\log n)^{-1/p}\}$$

$$\le C_9^2 n^2(\log n)^{-2/p}$$

$$+(E\{\theta_{0,n}^4\})^{1/2}(P\{\theta_{0,n} \ge n\mu+C_8 n(\log n)^{-1/p}\})^{1/2}$$

$$+(E\theta_{0,n})^2 P\{\theta_{0,n} \le n\mu+(C_2-C_9)n(\log n)^{-1/p}\} .$$

Since (for $\theta = a$, s or t)

$$\theta_{0,n} \le t_{0,n} \le \sum_{0}^{n-1} t(e_j) \quad \text{(see (5.39))},$$

the second term in the right hand side is bounded by

$$n^2(E\{t^4(e)\}\,2C_7)^{1/2} n^{-1/8},$$

while the third term is bounded by

$$n^2(Et(e))^2 \exp -nB_1((2C_1)^{1/p}(\log n)^{-1/p}) ,$$

$$\le n^2(Et(e))^2 \exp(-n \exp(-C_1(2C_1)^{-1}\log n)$$

$$= n^2(E(t(e))^2 \exp(-n^{1/2})$$

by virtue of (5.5) and (5.17).

<u>Proof of (5.20)</u>. It is immediate from the Borel-Cantelli lemma, (5.5) and (5.17) that for $\theta = a,s$ or t and any $C' \ge (2C_1)^{1/p}$

(5.76) $\qquad \theta_{0,n} \ge n(\mu - C'(\log n)^{-1/p})$ eventually w.p.1.

Clearly (5.76) also holds for $\theta = b$ if $\mu = 0$. To prove (5.76) for $\theta = b$ when $\mu > 0$, denote for any path r from $\underline{0}$ to H_n, the last vertex of r in H_0 by w_r. Then in the notation of the proof of (5.29a).

$$(5.77)\quad P\{b_{0,n} < x\} \le P\{\exists \text{ path } r \text{ starting at } \underline{0} \text{ with } T(r) < x \text{ and } \max_{2\le i\le d} |w_r(i)| \ge 2n\} + \sum_{|w(i)|<2n} P\{s_n(w) < x\}.$$

Now any path r with $w_r(i) \ge 2n$ must have crossed the hyperplane $\{x:x(i) = 2n\}$ so that

$$P\{\exists \text{ path } r \text{ starting at } \underline{0} \text{ with } T(r) < x \text{ and } w_r(i) \ge 2n\} \le P\{b_{0,2n} < x\}.$$

Also $s_n(w)$ has the same distribution as $s_{0,n}$ so that with $x = n(\mu - C'(\log n)^{-1/p})$ (5.77) yields

$$P\{b_{0,n} < n(\mu - C'(\log n)^{-1/p})\} \le 2dP\{b_{0,2n} < 2n(\tfrac{1}{2}\mu)\}$$
$$+(4n)^{d-1}P\{s_{0,n} < n(\mu - C'(\log n)^{-1/p})\}.$$

By (5.3), (5.5) and (5.17) we obtain that (5.76) still holds for $\theta = b$ if $C' \ge (2C_1)^{1/p}$. This proves one half of (5.20).

To prove the other half of (5.20) it suffices to show for suitable C'' that

$$(5.78)\quad t_{0,n} \le n(\mu + C''((\log n)^{-1/p}) \text{ eventually w.p.1}$$

(since $a_{0,n}, b_{0,n}, s_{0,n} \le t_{0,n}$). But it follows from (5.75) that (5.78) holds with $C'' = C_8$ if we restrict n to the sequence $\{k^5\}_{k\ge 1}$. In addition, analogously to (5.39)

$$\max_{k^5 \le n \le (k+1)^5} t_{0,n} - t_{0,k^5} \le \sum_{j=k^5}^{(k+1)^5-1} t(e_j),$$

and Chebyshev's inequality implies that

$$\sum_k P\{\sum_{j=k^5}^{(k+1)^5-1} t(e_j) > k^5(\log k)^{-1/p}\} < \infty .$$

Thus (5.78) and (5.20) follow.

(5.79) REMARK. If we combine (5.18), (5.19) and (5.44) we obtain

$$\sum_{\ell=1}^{K} \sigma^2\left(\frac{s_{0,n2^\ell}}{n2^\ell}\right) \leq C_{10}(\log n)^{-1/p} .$$

As in Cor. 5.45 this shows that for each m there exists an $M \varepsilon [m,m^2]$ for which

$$\sigma^2\left(\frac{s_{0,M}}{M}\right) \leq C_1 \ (\log M)^{-1-1/p} .$$

This can then be used to improve the estimates in Theorem 5.16 somewhat. However, this only seems to reduce the value of p, but does not allow us to replace $(\log n)^{-1/p}$ by $n^{-\varepsilon}$ for some $\varepsilon > 0$ in (5.18)-(5.20).

6. Properties of the time constant μ and the limit set B_0.

We established as part of Theorem 3.1 that B_0, the limit set of $t^{-1}\hat{B}(t)$ is either all of $\mathbb{R}^d$, if $\mu = 0$, or is compact and convex, if $\mu > 0$. It is of interest to give a simple criterion in terms of F, the distribution of t(e), to distinguish between these two cases. As the next theorem shows this depends only on the atom of F at the origin. (See (5.7) for p_T.)

(6.1) THEOREM. $\mu = \mu(F,d) = 0$ if and only if $F(0) \geq p_T(d)$.

(6.2) REMARK. $p_T(2) = \frac{1}{2}$ ([23], pp. 54-57) but p_T is unknown in all higher dimensions. $p_T(d)$ is clearly decreasing in d, and probably even strictly decreasing. It is known that $p_T(d) < \frac{1}{2}$ for $d \geq 3$ ([23], pp. 271, 272) and that

$$(6.3) \qquad p_T(d) \geq \frac{1}{2d-1} \quad \text{and} \quad \limsup_{d\to\infty} d\, p_T(d) \leq 1 .$$

(The first inequality in (6.3) follows from a simple Peierls estimate; the expected number of selfavoiding open paths of n steps is at most $2d\,(2d-1)^{n-1}\,(F(0))^n$. This tends to zero if $F(0) < (2d-1)^{-1}$. The second inequality in (6.3) is implied by [8], Theorem 1.) ///

Some simple but very crude bounds for μ are given by the next theorem.

(6.4) THEOREM. Let $\lambda(F) := \inf\{x : F(x) > 0\}$ be the left endpoint of supp(F). If F is not concentrated in one point, then

$$(6.5) \qquad \mu < \int x\, dF(x) = E\, t(e).$$

If in addition $\lambda(F) > 0$, then

$$(6.6) \qquad \lambda(F) < \mu$$

(6.7) REMARK. Trivially, if F is concentrated on one point then $\mu = \lambda(F) = E\,t(e)$. Better bounds for μ then (6.5) and (6.6) have been derived for $\mu(F,2)$ for various distribution functions F (see [31], Sect. 7.2, [30], [21] and [1]. E.g. if F is the exponential distribution function (1.19), then these sources state

$$.29853 \leq \mu(F,2) \leq .458.$$

If F_p is the Bernoulli distribution with mass at 0 and 1 only and $F_p(0) = p$, $F_p(\{1\}) = 1-p$, then, since $p_T(2) = \frac{1}{2}$, $\mu(F_p,2) = 0$ if and only if $p \geq \frac{1}{2}$. For $p < \frac{1}{2}$ there exist constants $0 < C_i, \delta_i < \infty$

such that

$$C_1\left(\tfrac{1}{2} - p\right)^{\delta_1} \le \mu(F_p,2) \le C_2\left(\tfrac{1}{2} - p\right)^{\delta_2}, \quad p < \tfrac{1}{2}. \tag{6.8}$$

The bounds in (6.8) were first pointed out to us by J.T. Cox (private communication). The left hand inequality (with p and $1-p$ interchanged) is derived in [23], Theorem 11.1 ; the right hand inequality will be proven later. In general the information about μ is scanty and the bounds unsatisfactory. ///

The only other general property of μ known to us is the following continuity property.

(6.9) THEOREM. If $F_n \Rightarrow F$, then $\mu(F_n,d) \to \mu(F,d)$, i.e. μ is continuous under weak convergence.

Even less information is available for $\hat{B}_0$ than for μ. We already saw (see proof of (1.14) at the end of Sect. 3) that the symmetry and convexity of $\hat{B}_0$ imply that the points of $\hat{B}_0$ on the coordinate axes have maximal distance to $\underline{0}$ among the points of B_0. Thus

$$\hat{B}_0 \subset \left[-\frac{1}{\mu}, \frac{1}{\mu}\right]^d . \tag{6.10}$$

Also, by convexity, $\hat{B}_0$ constains the convex hull of the points $(0, 0, \ldots, \pm\mu, 0, \ldots, 0)$ on the coordinate axes (the diamond in Fig. 6.1).

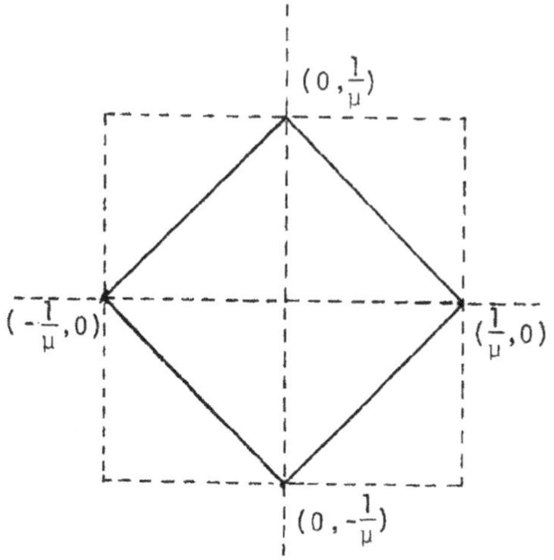

Fig. 6.1. $\hat{B}_0$ lies between the solidly drawn diamond and the dashed square.

Monte-Carlo experiments ([15], [28]) at first gave the impression that $\hat{B}_0$ is a Euclidean disc when $d = 2$ and F the exponential distribution (1.19). Few believe that $\hat{B}_0$ is really a disc in this case, because for high dimensions and exponential F, $\hat{B}_0$ is not a ball (see Cor. 8.4). The only non-trivial exact knowledge about the shape of $\hat{B}_0$ seems to be remarkable result of Durret and Liggett [14], which shows that $\hat{B}_0$ can have flat edges if the atom of F at the left endpoint of its support, $F(\lambda(F))$, is large enough, and when $\lambda_F > 0$. Note that for any vertex v

$$(6.11) \qquad T(\underline{0},v) \geq \lambda(F)\ |v| = \lambda(F) \sum_{1}^{d} |v_i|,$$

because any path from $\underline{0}$ to v contains at least $|v|$ edges, and, by definition, the passage time of each edge is at least λ . Consequently $\mu \geq \lambda$ and

$$(6.12) \qquad \hat{B}_0 \subset \{x \in \mathbb{R}^d : |x| \leq 1/\lambda\}$$

Durrett and Liggett show that part of the boundary of $\hat{B}_0$ can actually be as far out as allowed by (6.12), i.e. lie on $\{x : |x| = 1/\lambda\}$. They only discuss the two dimensional case, but the argument should carry over to higher dimensions. To state their result precisely we must bring in the critical probability of oriented percolation on $\mathbb{Z}^d$. If e is an edge of $\mathbb{Z}^d$ between the neighboring vertices v and w with $w_i = v_i + 1$, $w_j = v_j$, $j \neq i$, for some i, then e is directed from v to w. (Thus all edges are directed upwards). A <u>directed path</u> is a path $r = (v_0, e_1, \ldots, e_n, v_n)$ on $\mathbb{Z}^d$ such that e_i is the edge between v_i and v_{i+1} and e_i is directed from v_i to v_{i+1} $0 \leq i < n$. Now take each edge, independently of all others, open (closed) with probability $p(q: = 1-p)$ and denote the corresponding probability measure on the configurations of open and closed edges by P_p. In analogy with the undirected case (see (5.6) and the lines following it) define

$$\hat{W} = \{v \in (\mathbb{Z}_+)^d : \exists \text{ open directed path from } \underline{0} \text{ to } v\}$$

$$\hat{p}_T = \hat{p}_T(d) = \sup\{p: E_p\{\#\hat{W}\} < \infty\},$$

$$\hat{\theta}_p = \hat{\theta}_p(d) = P_p\{\#\hat{W} = \infty\}$$

and

$$\hat{p}_H = \hat{p}_H(d) = \inf\{p: \hat{\theta}_p > 0\}.$$

(6.13) THEOREM. For first-passage percolation on $\mathbb{Z}^2$, if $\lambda(F) > 0$ and $F(\lambda(F)) > \hat{p}_H(2)$ then there exists an $\alpha = \alpha(F) > 0$ such that

$$\{x: |x| = 1/\lambda,\ |x(1) - x(2)| \le \alpha(F)/\lambda(F)\} \subset \hat{B}_0$$

(see Fig. 6.2).

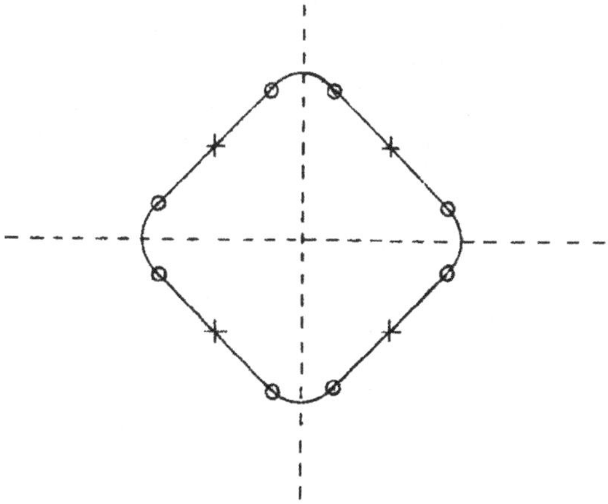

Fig. 6.2. Illustration of a $\hat{B}_0$ with straight edges. The point marked by + are $(\pm(2\lambda)^{-1}, \pm(2\lambda)^{-1})$ and the points marked by o are $(\pm(1+\alpha)(2\lambda)^{-1}, \pm(1-\alpha)(2\lambda)^{-1})$ and $(\pm(1-\alpha)(2\lambda)^{-1}, \pm(1+\alpha)(2\lambda)^{-1})$. The fact that $\hat{B}_0$ is not a full diamond is implied by (6.6) and (6.15).

(6.14) REMARK. Durrett's estimates ([13], (3.12); see also [14], Theorem 8) show that

$$\alpha(F) \ge C\{F(\lambda(F)) - \hat{p}_H(2)\}$$

for a suitable $C > 0$. ///

In the opposite direction of Theorem 6.13 one has the following result (in any dimension)

(6.15) PROPOSITION. If $\lambda(F) > 0$ and F is not concentrated on $\{\lambda\}$, then $\hat{B}_0$ is strictly contained in the "diamond"

$$\{x \in \mathbb{R}^d : |x| = 1/\lambda\} . \tag{6.16}$$

If in addition $F(\lambda(F)) < \hat{p}_T(d)$, then $\hat{B}_0$ lies entirely in the interior of the diamond (6.16).

(6.17) REMARK. For $d = 2$ $\hat{p}_H(2) = \hat{p}_T(2)$ so that Theorems 6.13 and (6.15) show that this common value is the critical value for $F(\lambda(F))$ for $\hat{B}_0$ to contain a point on the boundary of the diamond (6.16).

(6.18) REMARK. Theorem 6.9 implies that $\hat{B}_0$ is in a certain sense continuous in F. Indeed the proof of Theorem 3.1 together with (6.9) shows that for any line L through $\underline{0}$ the endpoints of the segment $L \cap \hat{B}_0$ are continuous in F, and this continuity is even uniform in L.

Proof of Theorem 6.1. If $F(0) < p_T$ then it is immediate from (5.8) that $b_{0,n} \geq Cn$ eventually w.p.1, for the strictly positive C of Prop. 5.8. Thus $\mu = \lim n^{-1} b_{0,n} \geq C$.

In the other direction, assume $F(0) \geq p_T$ but $\mu > 0$. We shall show that this contradicts (5.3). Indeed, let W be as in (5.6). Then, clearly W can contain a vertex v with $v(1) \geq n$ only if there exists a path r from the origin to $H_n = \{x : x(1) = n\}$ with $T(r) = 0$ In particular $b_{0,n} = 0$ in this case. By (5.3) this has probability at most $A_1 \exp(-n B_1(\mu/2))$, with $B_1(\mu/2) > 0$. Since all coordinates play the same role we have

$$P\{W \text{ contains any vertex } v \text{ with some } |v(i)| > n\} \leq A_1 e^{-nB_1(\mu/2)} .$$

and

$$E\{\#W\} = \sum_{k=1}^{\infty} P\{\#W \geq k\}$$

$$\leq A_1 \sum_{k=1}^{\infty} \exp\{-\tfrac{1}{2} B_1(\mu/2)(k^{1/d} - 1)\} < \infty$$

But W depends only on which edges e have $t(e) = 0$ and which edges have $t(e) > 0$ and $E\{\#W\}$ equals $E_{F(0)}\{\#W\}$ in the notation introduced in the beginning of Sect. 5. Thus $E_{F(0)}\{\#W\} < \infty$. By definition of p_T this means $F(0) \leq p_T$. In fact one can conclude $F(0) < p_T$, by means of [23], Corollary 5.1, so that we have derived at contradiction. (The conclusion $\mu = 0$ for the boundary case $F(0) = p_T$ can also be obtained by means of the continuity theorem 6.9.)

Proof of Theorem 6.4: The inequality (6.5) is Theorem 4.1.9 of [19]. We already explained with (6.11) that $b_{0,n} \geq n\lambda$ so that $\mu \geq \lambda$. To show that this inequality is strict unless F is a one-point distribution, we use a crude version of Janson [21]. First we estimate the number of paths on $\mathbb{Z}^d$ from $\underline{0}$ to $H_n = \{x(1) = n\}$ which contain exactly k edges, $k \geq n$, by

$$(6.19) \qquad (2d)^k \; P\{U_1 + \ldots + U_k = n\},$$

where the U_i are independent random variables with

$$P\{U_i = \pm 1\} = \frac{1}{2d}, \quad P\{U_i = 0\} = 1 - \frac{1}{d}.$$

(6.19) holds because of the 2d possibilities for each of the k steps of the path 2d-2 leave the first coordinate unchanged, while there is one possibility each to increase or decrease the first coordinate by one. In turn (6.19) is bounded above by

$$(6.20) \qquad (2d)^k \, e^{-\xi n} \left(1 - \frac{1}{d} + \frac{1}{d}\cosh\xi\right)^k \leq e^{-\xi n} \left(2d-1 + e^{\xi}\right)^k,$$

for any $\xi \geq 0$. Therefore, if $t_1, t_2, \ldots$ are independent random variables, each with the same distribution as $t(e)$, then for $\rho \geq 0$, $\xi \geq 0$

$$(6.21) \qquad P\{b_{0,n} \leq nx\} < \sum_{k \geq n} \{ \text{\# of paths from } \underline{0} \text{ to } H_n \text{ of } k \text{ steps}\}. \; P\{t_1 + \ldots + t_k \leq nx\}$$

$$\leq e^{\rho n x} \; e^{-\xi n} \sum_{k \geq n} [(2d-1 + e^{\xi}) \; E \, e^{-\rho t(e)}]^k .$$

Now

$$Ee^{-\rho t(e)} = e^{-\rho\lambda} F(\lambda) + \int_{(\lambda,\infty)} e^{-\rho y} \, dF(y)$$

$$= e^{-\rho\lambda} \{1 - \int_{(\lambda,\infty)} (1-e^{\rho(\lambda-y)}) dF(y)\} \leq e^{-\rho\lambda} .$$

Since $\lambda > 0$ we can for each $\xi > 0$ choose ρ such that

$$(6.22) \qquad (2d-1 + e^{\xi}) \; e^{-\rho\lambda} \leq \frac{1}{2} .$$

For such ξ, ρ the last expression in (6.21) is at most

$$(6.23) \qquad 2[e^{\rho(x-\lambda)}\{1-\int_{(\lambda,\infty)}(1-e^{-\rho(y-\lambda)})dF(y)\}(1+(2d-1)e^{-\xi})]^n .$$

If $F(\lambda) < 1$, then it is easy to see that we can choose ξ and ρ large such that (6.22) holds and the expression between square brackets in (6.23) is at most

$$e^{\rho(x-\lambda)} \frac{1}{2}(1 + F(\lambda)) .$$

This is < 1 for $x - \lambda$ sufficiently small but strictly positive. Thus for some $x > \lambda$ $P\{b_{0,n} \le nx\} \to 0$ exponentially in n, whence

$$\mu = \lim \frac{1}{n} b_{0,n} \ge x > \lambda \qquad \text{w.p.1.}$$

<u>Sketch of proof of (6.8)</u>. As pointed out we only have to prove the second inequality of (6.8) here. For this we use the so called "reach process".

$$y_t := \sup\{n : s_{0,n} \le t\} .$$

It is known ([31], Theorem 6.3) that

$$\lim_{t\to\infty} \frac{y_t}{t} = \frac{1}{\mu} . \qquad \text{w.p.1.}$$

On the other hand it is not hard to see that in the Bernoulli case

$$(6.24) \qquad \lim_{t\to\infty} \frac{y_t}{t} \ge 1 + E\, y_0 .$$

In fact, if $y_0 = a$, then there exists a path r_1 from $\underline{0}$ to H_a of zero passage time. Let the endpoint of r_1 be (a,b). The edge between (a,b) and $(a+1,b)$ must have passage time 1, by definition of y_0 Moreover,

$$y_t \ge 1 + y_0 + y'_{t-1} ,$$

where y'_{t-1} is the reach process starting from $(a+1,b)$. Specifically

$$y'_{t-1} = \sup\{n : \exists \text{ path } r \text{ from } (a+1,b) \text{ to } H_n \text{ with } H_{a+1} < r < H_n \text{ and } T(r) \le t-1\} - (a+1).$$

Since the event $\{y_0 = a\}$ is independent of the edges to the right of H_{a+1}, y_{t-1} is also independent of y_0 and has the same distribution as y_{t-1}. Continuing in this way one obtains that y_t is bounded by the sum of t independent random variables, each with the distribution of $1 + y_0$. This proves (6.24).

In view of (6.24) it suffices to show that in the Bernoulli case

$$E_p y_0 = \sum_{n=1}^{\infty} P_p\{y_0 \geq n\} = \sum_{n=1}^{\infty} P_p\{s_{0,n} = 0\}$$
$$\geq C_2^{-1} \left(\tfrac{1}{2} - p\right)^{-\delta_2}, \quad p < \tfrac{1}{2},$$

where the subscript p is used to indicate that $F(0) = p$. Now let $C(n)$ be the event

(6.25) $$C(n) = \{\exists \text{ path } r \text{ from } \underline{0} \text{ to } \{n\} \times [-n,n] \text{ such that } r \text{ minus its endpoints lies in } (0,n) \times [-n,n] \text{ and such that } T(r) = 0\}.$$

Then clearly $C(n) \subset \{s_{0,n} = 0\} = \{y_0 \geq n\}$ and by Lemma 4.1 of [23] (compare also (8.108) in [23])

$$P_p\{C(n)\} \geq (2p)^{4n(n+1)} P_{\frac{1}{2}}\{C(n)\}, \; p \leq \tfrac{1}{2}.$$

Finally, the proof of Lemma 8.4 in [23] shows that for some $C_3, \alpha > 0$

(6.26) $$P_{\frac{1}{2}}\{C(n)\} \geq C_3 n^{\alpha - 1}.$$

Putting these estimates together we have

$$E_p y_0 \geq \sum_{n < (\frac{1}{2}-p)^{-1/2}} (2p)^{4n(n+1)} P_{\frac{1}{2}}\{C(n)\}$$
$$\geq C_4 \left(\tfrac{1}{2}-p\right)^{-\alpha/2}. \qquad ///$$

REMARK. It is also possible to obtain (6.26) by combining the method of proof of Cor. 3.15 in [3] with (8.7) in [23].

Proof of Theorem 6.9. The proof is essentially the same as in [9], with the sets $\tilde{S}(u)$ from the proof of Theorem 2.26 taking the place of the sets $\Delta(u)$ of [7]. As in [9], Lemma 1 one first shows that

$$\limsup \mu(F_n,d) \le \mu(F,d)$$

if $F_n \Rightarrow F$. One is therefore done when $\mu(F,d) = 0$, and in the remainder one may assume $F(0) < p_T$ (this time by Theorem 6.1). Thus also (5.8) applies; this takes the place of (2.17) in [9]. Finally, we mention what takes the place of the circuits $C(e)$ on pp. 814-818 of [9]. For an edge e with endpoints v', v'' the role of $C(e)$ will be taken by $S(e) := \partial_{ext}\, \mathcal{C}_j(\{v',v''\},b)$, the exterior boundary of the black cluster of v' and v'' (see Sect. 2, after (2.21) for the definitions). As in [9] there exists a constant β such that, for given e, there exist at most β^n possibilities for an $\mathcal{L}$-connected cluster $\mathcal{C}$ containing $\{v',v''\}$ and having exactly n vertices (see [23 (5.22)). Also, if z is the maximal degree of any vertex of $\mathcal{L}$, then $\partial_{ext}\,\mathcal{C}$ contains at most z times as many vertices as $\mathcal{C}$. One then obtains analogously to (2.13) of [9] that for distinct edges $e_1,\dots,e_k$ and fixed shells $\bar{S}_1,\dots,\bar{S}_k$ with disjoint interiors

$$(6.27)\qquad P\{S(e_i) = \bar{S}_i,\ 1 \le i \le k\} \le \xi^{\sum_{i=1}^{k} (|\bar{S}_i|-2z-1)} ,$$

where $|\bar{S}_i|$ is the total number of vertices in $\bar{S}_i$, and ξ can be made as small as desired, by choosing M in (2.21) large. With (6.27) taking the place of (2.13) of [9], one can now copy the proof of Lemma 2 and Theorem 3 of [9], with only obvious changes, to obtain Theorem 6.9 (see the next remark, though).

REMARK. In the reduction of Theorem 3 to Lemma 2 in [9], an appeal is made to [6] to first prove 6.9 in the case where supp(F) and supp(F_n) are all contained in a fixed finite interval [0,B]. One can avoid going through all of [6], by observing that we are only interested in the case with $F(0) < p_T$ so that (5.8) applies. Therefore, when comparing the distribution of $\hat{b}_{0,n}$ for F and F_k one may restrict oneself to paths of length $\le An$ for some $A > 0$ (cf [9], pp. 815, 816). We can also choose the $t(e)$ with distribution F_k and F simultaneously as $F_k^{-1}(U(e))$ and $F^{-1}(U(e))$, respectively, for independent uniform variables on [0,1], $U(e)$, $e \in \mathbb{Z}^d$. In this situation, if $a^k_{0,n}$ and $a_{0,n}$ denote the point to point passage times according to $F_k^{-1}(U(e))$ and $F^{-1}(U(e))$, respectively, then

$$P\{|a_{0,n} - a^k_{0,n}| > \varepsilon n\} \le P\{a_{0,n} \text{ or } a^k_{0,n} \text{ is achieved on a path of length } > An\} + P\{\exists \text{ path } r \text{ from the origin of length} \le An \text{ with } \sum_{e\in r} |F_k^{-1}(U(e)) - F^{-1}(U(e))| > \varepsilon n\}$$

The first term in the right hand side goes to zero with n, uniformly in k, by virtue of (5.8) (when A is sufficiently large). The second term in the right hand side is at most

$$(2d)^{An} P\{ \sum_{i=1}^{An} |F_k^{-1}(U_i) - F^{-1}(U_i)| > \varepsilon n\} ,$$

where $U_1, U_2, \ldots$ are independent uniform random variables on [0,1]. This expression goes to zero as $k \to \infty$, uniformly in n, when $F_k^{-1}(U_i)$, $F^{-1}(U_i) \in [0,B]$ for some fixed $B < \infty$. From this one quickly obtains 6.9 in the case when $\mathrm{supp}(F_k)$ and $\mathrm{supp}(F)$ are all contained in some fixed $[0,B]$, $B < \infty$. ///

<u>The proof of Theorem 6.13</u> is essentially the same as that of Theorem 9 of [14]. The fact that [14] takes F to be a translated geometric distribution plays no role. Also, the arguments for site first-passage percolation of [14] carry over to the model used in these notes. The idea behind the proof is to consider the so called edge process

$$R_n = \max\{x : (x,n-x) \text{ is connected to some point } (y,-y) \text{ with } y \le 0 \text{ by a directed path } r \text{ with } T(r) = n\lambda\}.$$

Note that a point $(x,n+1-x)$ on the line $x(1) + x(2) = n+1$ is connected to $(-y,y)$ by a directed path r with $T(r) = (n+1)\lambda$ if and only if at least one of the two points $(x-1, n+1-x)$ and $(x,n-x)$ on the line $x(1) + x(2) = n$ is connected to $(-y,y)$ by a directed path r' with $T(r') = n\lambda$, and the edge between this point and $(x,n-x)$ has passage time λ. Now define for $m < n$

$$X_{m,n} = -R_m + \sup\{x : (x,n-x) \text{ is connected to some point } (y,m-y) \text{ with } y \le R_m \text{ by a directed path } r \text{ with } T(r) = n-m\}.$$

Then if we set $R_0 = 0$ one has $X_{0,n} = R_n$, and by means of the observations above one easily verifies that $X_{m,n}$ satisfies (2.2)-(2.5) and (2.7) (see also [27]). Therefore

$$\lim \frac{1}{n} R_n = \alpha' \quad \text{exists} \quad \text{w.p.1} .$$

Similarly

$$\lim \frac{1}{n} L_n = 1 - \alpha' \qquad \text{w.p.1} ,$$

where

$$L_n = \min\{x: (x,n-x) \text{ is connected to some point } (y,-y) \text{ with } y \geq 0 \text{ by a directed path } r \text{ with } T(r) = n\lambda\}.$$

If $\alpha' > \frac{1}{2}$, then for large n the points

$$(L_n, n-L_n) \quad \text{and} \quad (R_n, n-R_n) \tag{6.28}$$

on the line $x(1) + x(2) = n$ will be connected to the line $x(1) + x(2) = 0$ by a path of passage time $n\lambda$. This is responsible for the straight edge from $((1-\alpha')/\lambda, \alpha'/\lambda)$ to $(\alpha'/\lambda, (1-\alpha')/\lambda)$ in the boundary of $\hat{B}_0$. To carry out the details, one must show that with positive probability the points (6.28) - or points close to them - are not only connected to the line $x(1) + x(2) = 0$, but even to the origin itself, with paths of passage time $n\lambda$. Moreover, one must show that $\alpha' > \frac{1}{2}$ for $F(\lambda) > \hat{p}_H$. The reader is referred to [13], Sect. 3 or [14] for details.

Proof of Proposition 6.15. $\hat{B}_0$ must be contained in the diamond (6.16) by virtue of (6.11). $\hat{B}_0$ cannot equal (6.16), by virtue of (6.6) and (6.10). This proves the first part of the proposition.

The second statement of the proposition is also given for $d = 2$ and F a (translated) geometric distribution in Theorem 15 of [14], but the proof if [14] needs some minor modifications. We base our proof on (4.13). Before applying (4.13) we show that if $F(\lambda) < \hat{p}_T$, then we can choose N such that

$$\sum_{v \in L_N} P\{T(\underline{0},v) \leq N\lambda\} \leq (2^8 N)^{-1}, \tag{6.29}$$

where L_k denotes the line $x(1) + x(2) = k$. To obtain (6.29) we call an edge e open (closed) if $t(e) = \lambda$ $(t(e) > \lambda)$. Then (6.11) implies that for a $v \in L_N$ $T(\underline{0},v) \leq N\lambda$ is possible only if there exists an open path of N edges from $\underline{0}$ to v. Such a path must necessarily be a directed path and v must belong to $\hat{W}$ (see definitions preceding (6.13)). Thus, the left hand side of (6.29) is at most equal to

$$E_p\{\#(W \cap L_N)\}.$$

Clearly this is less than $(2^8 N)^{-1}$ for some N if $E_p\{\#W\} = \sum_N E_p\{\#W \cap L_N\} < \infty$, that is, for $p < \hat{p}_T$. Thus (6.29) holds.

Now apply (4.13) with the following choices: $V = \{\underline{0}\}$, $W = L_{kN}$, $x = kN\lambda(1+\varepsilon)$ and $V(\ell,i)$ a point on L_{iN} such that as ℓ runs from 1 to ∞, $V_{\ell i}$ runs through all of L_{iN}. Finally we take $n(\ell) = k$

for all ℓ. Now any path from $\underline{0}$ to L_{kN} must successively pass through some point of $L_N, L_{2N}, \ldots, L_{kN}$, so that the left hand side of (4.13) is equal to $P\{T(\underline{0}, L_{kN}) < kN\lambda(1+\varepsilon)\}$ for these choices. The right hand side of (4.13) can be written as

$$(6.30) \qquad \sum_{v(1),\ldots,v(k)} P\{\sum_{i=1}^{k} T'(v(i-1),v(i)) < kN\lambda(1+\varepsilon)\},$$

where $v(0) = \underline{0}$, and $v(1),\ldots,v(k)$ run independently through $L_N,\ldots,L_{kN}$, respectively, and $T'(v(i-1),v(i))$ are independent random variables with the distribution of $T(v(i-1,v(i))$. (6.30) can be viewed as an expectation for a branching random walk. This brings us back to the method of estimation of [14]. Rather than explain this connection in detail we explicitly estimate (6.30) by

$$(6.31) \qquad e^{\xi kN\lambda(1+\varepsilon)} \sum_{v(1),\ldots,v(k)} \prod_{i=1}^{k} E\{e^{-\gamma T(v(i-1),v(i))}\}$$

$$= \left[\sum_{v\varepsilon L_N} e^{\xi N\lambda(1+\varepsilon)} E\{e^{-\xi T(\underline{0},v)}\}\right]^k, \quad \xi \geq 0$$

(by induction on k). If we take into account that (by (6.11) $T(\underline{0},V) \geq \lambda|v| \geq \lambda N$ for $v \varepsilon L_N$, we see that the expectation in square brackets in (6.31) is at most

$$(6.32) \qquad e^{\xi N\lambda(1+\varepsilon)}\left\{\sum_{v\varepsilon L_N,|v|<2N} e^{-\xi N\lambda} P\{T(\underline{0},v \leq N\lambda(1+2\varepsilon)\}\right.$$

$$\left. + 4Ne^{-\xi N\lambda(1+2\varepsilon)} + 2\sum_{k=0}^{\infty} e^{-\xi\lambda(2N+k)}\right\}$$

$$\leq e^{\xi N\lambda\varepsilon} \sum_{v\varepsilon L_N,|v|<2N} P\{T(\underline{0},v) \leq N\lambda(1+2\varepsilon)\}$$

$$+ 4Ne^{-\xi N\lambda\varepsilon} + 2(1-e^{-\xi\lambda})^{-1} e^{-\xi N\lambda(1-\varepsilon)}$$

Finally, by virtue of (6.29) we can choose $0 < \varepsilon \leq 1/N$ such that

$$\sum_{v\varepsilon L_N,|v|<2N} P\{T(\underline{0},v) \leq N\lambda(1+2\varepsilon)\} \leq (2^7 N)^{-1}$$

and then $\xi > 0$ such that (6.32) becomes strictly less than 1. By (4.13) we then obtain that for such an $\varepsilon > 0$, $P\{T(\underline{0},L_{kN} < kN\lambda(1+\varepsilon)\}$

decreases exponentially in k, so that $\hat{B}_0$ is contained in $\{x: x(1) + x(2) \leq (\lambda(1+\varepsilon))^{-1}\}$. Symmetry then shows $\hat{B}_0 \subset \{x: |x| \leq (\lambda(1+\varepsilon))^{-1}\}$ and the last statement of Prop. 6.15 is proven.

7. The rate of convergence to the time constant revisited: the case $\mu = 0$ (and $d = 2$). In this brief section we discuss the rate at which $n^{-1}\theta_{0,n}$ converges to 0 when the time constant μ is 0. Since $\theta_{0,n} \geq 0$ Theorem 5.2 with $B_1 = \infty$ becomes trivial. The upper bounds for $\theta_{0,n}$ in Theorem 5.9 and 5.16 remain valid of course, but, as we shall see, for $d = 2$ we can do considerably better. Actually Theorem 7.1 is formulated for all d, and even Theorem 7.2 remains valid for all d (for the reason given in the beginning of the proof of (7.1b)). But when $d \geq 3$ Theorems 7.1 and 7.2 are incomplete. By Theorem 6.1 $\mu = 0$ if and only if $F(0) \geq p_T(d)$ so that we want results for this whole regime. For $d = 2$ we know (cf [23], pp. 54,55)

$$p_T(2) = p_H(2) = \frac{1}{2} ,$$

so that Theorems 7.1 and 7.2 cover all cases with $\mu = 0$ when $d = 2$. Unfortunately, for $d \geq 3$ and $p_T(d) \leq F(0) \leq \frac{1}{2}$ we have no improvements over (5.18) and (5.20) (except for (7.1a)).

(7.1) THEOREM (a) In any dimension, if $F(0) > p_H(d)$, then $b_{0,n}$ is bounded w.p.1

(b) If any dimension, if $F(0) > \frac{1}{2}$, then $b_{0,n}$ and $s_{0,n}$ are bounded w.p.1 and the families $\{a_{0,n}\}_{n \geq 0}$ and $\{t_{0,n}\}_{n \geq 0}$ are tight.

(7.2) THEOREM Let $d = 2$ and $F(0) = \frac{1}{2}$.

(a) There exists an $\alpha \in (0,1)$ such that if

$$E\{t(e)\} = \int x dF(x) < \infty , \tag{7.3}$$

then the family $\{n^{\alpha-1}\theta_{0,n}\}_{n\geq 1}$ is tight for $\theta = a,b,s$ or t and

$$\limsup_{n\to\infty} n^{\alpha-1} (\log n)^{-1} b_{0,n} \leq \limsup_{n\to\infty} n^{\alpha-1} (\log n)^{-1} s_{0,n} < \infty \text{ w.p.1.} \tag{7.4}$$

(b) Let α be as in (a) and $\beta = \max\{2, \frac{4\alpha}{1-\alpha}\}$ and assume that

$$E\{t^{\beta}(e)\} = \int t^{\beta} dF(x) < \infty . \tag{7.5}$$

(Note that (7.5) is stronger than (7.3).) Then also

$$\limsup_{n\to\infty} n^{\alpha-1} (\log n)^{-1} a_{0,n} \leq \limsup_{n\to\infty} n^{\alpha-1} (\log n)^{-1} t_{0,n} < \infty \text{ w.p.1.} \tag{7.6}$$

Proof of Theorem 7.1. Most of this proof rests on the following simple idea. Call an edge e open (closed) if $t(e) = 0$. Then any path in an open cluster has zero passage time. If $F(0) > p_H(d)$, then there exist infinite open clusters, so that one can travel arbitrarily far in zero time. We next give some more details

(a) If $F(0) > p_H(d)$ then (by definition) there is a strictly positive probability that $\mathbb{Z}^d$ has an infinite connected subgraph all of whose edges are open. We call the maximal open subgraphs open clusters. Let $W(v)$ be the open cluster which contains the vertex v. $W(v)$ may consists of v only, if none of the edges incident to v are open. However,

$$P\{\#W(v) = \infty\} > 0 , \tag{7.7}$$

and this probability is independent of v. We claim that an infinite cluster must necessarily be unbounded on the left or right, i.e. for each $n < \infty$

$$P\{\#W(v) = \infty \text{ , but } W(v) \text{ is contained in } [-n,n] \times \mathbb{R}^{d-1}\} = 0. \tag{7.8}$$

Before proving (7.8) we show how it implies (a). From (7.8) and symmetry it follows that

$$P\{W(v) \text{ contains vertices in } [n,\infty) \times \mathbb{R}^{d-1} \text{ for each } n \geq 0\} \geq \tfrac{1}{2}P\{\#W(v) = \infty\} > 0$$

It then follows from the ergodic theorem, applied to the translation invariant process $\{I[\ W(v) \text{ contains vertices in } [n,\infty) \times \mathbb{R}^{d-1} \text{ for each } n \geq 0]\}_{v\in\mathbb{Z}^d}$, that w.p.1 there exists a $W(v)$ which intersects each H_n, $n \geq 0$ (compare [20], Lemmas 3.1 and 5.1) Now let r be a path which connects $\underline{0}$ to such a $W(v)$. We can then for each $n \geq 0$ continue r by a path inside $W(v)$ to H_n. By the initial remarks of this proof this implies $b_{0,n} \leq T(r)$ for al $n \geq 0$.

We next prove (7.8). Assume $\#W(v) = \infty$ but $W(v) \subset [-n,n] \times \mathbb{R}^{d-1}$. Then there exists a maximal $k \in [-n,n]$ such that $\#W(v) \cap H_k = \infty$ and consequently a finite ℓ such that

$$\#W(v) \cap H_k = \infty \text{ but } W(v) \subset [-n,k] \times \mathbb{R}^{d-1} \cup [-\ell,\ell]^d. \tag{7.9}$$

It therefore suffices to show that for each n,k,ℓ the probability of (7.9) is zero. Let $v_1, v_2, \ldots$ be the vertices of $W(v) \cap H_k$ outside $[-\ell,+\ell]^d$. If (7.9) occurs, all the edges between v_i and $v_i + (1,0,\ldots 0)$, $i=1,2\ldots$ have to be closed. However, these edges are independent of all edges in $[-n,k] \times \mathbb{R}^{d-1} \cup [-\ell,\ell]^d$. Therefore, given any infinite sequence of v_i in H_k, and given that these v_i are connected to $\underline{v}$ by an open path in $[-n,k] \times \mathbb{R}^{d-1} \cup [-\ell,\ell]^d$, the conditional probability of all the edges between a v_i and $v_i + (1,0,\ldots,0)$ being closed is zero. This implies that (7.9) has probability zero, and (7.8) follows.

(b) It suffices to prove this part for $d=2$ only, since restriction to paths r in the two-dimensional set $\{x: x(3)=x(4)=\ldots=x(d)=0\}$ can only raise $\theta_{0,n}$ for $\theta = a,b,s$ or t. For the remainder of this proof take $d=2$, $F(0) > \frac{1}{2}$. Then essentially the same argument as in part (a) can be used to show that $s_{0,n}$ is bounded. We merely have to use the known result that if $d=2$, $F(0) > \frac{1}{2}$, then for all v

(7.10) $P\{v$ is connected to all H_n with $n > v(1)$ by open paths inside $[v(1),\infty) \times \mathbb{R}^{d-1}\} > 0$

(see [31], pp30-35). Again by the ergodic theorem we can choose a v with $v(1) = 1$ with the property between braces in (7.10). The path r and the connecting paths inside $W(v)$ to H_n can therefore be chosen such that all their points, except the initial point $\underline{0}$, lie in the open right half plane. Thus also $s_{0,n}$ is bounded w.p.1 if $d=2$, $F(0) > \frac{1}{2}$.

To prove the tightness of $\{a_{0,n}\}$ and $\{t_{0,n}\}$ it suffices to consider $\{t_{0,n}\}$, since $0 \le a_{0,n} \le t_{0,n}$. To do this note that if r_1 is a path from $\underline{0}$ to the half line $\{n\} \times [0,\infty)$ and r_2 is a path from $(n,0)$ to $\{0\} \times [0,\infty)$ such that $H_0 < r_1, r_2 < H_n$ (see Fig. 7.1 and see (5.22) for notation), then r_1 and r_2 must intersect. We can then combine a piece of r_1 and a piece of r_2 to construct a path r from $\underline{0}$ to $(n,0)$ such that $H_0 < r < H_n$. Clearly $T(r) \le T(r_1) + T(r_2)$ in this construction. Now let $\mathcal{E}_1(\mathcal{E}_2)$ denote the event that there exists a path r_1 from $\underline{0}$ to $\{n\} \times [0,\infty)$ (r_2 from $(n,0)$) such that $H_0 < r_1 < H_n$ and $T(r_1) \le L$ ($H_0 < r_2 < H_n$ and $T(r_2) \le L$). Then from the above,

$$P\{t_{0,n} \le 2L\} \ge P\{\mathcal{E}_1 \cap \mathcal{E}_2\}.$$

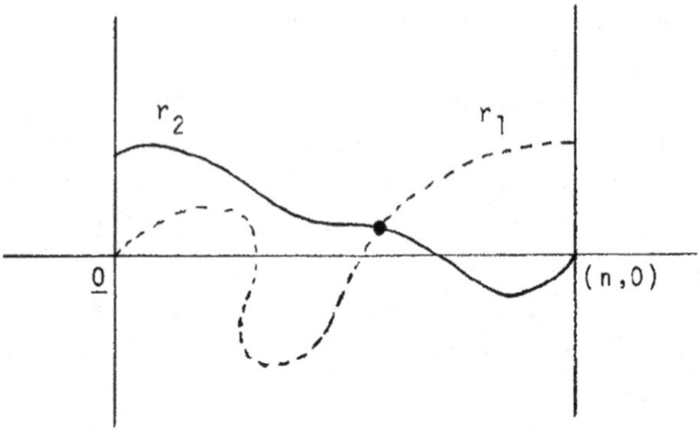

Fig. 7.1 r_1 is dashed, r_2 the solid path. r goes along r_1 from $\underline{0}$ to the intersection of r_1 and r_2 at the solid circle, and continues along r_2 to $(n,0)$.

Both $\mathcal{E}_1$ and $\mathcal{E}_2$ are decreasing events, and by symmetry they have the same probability. Therefore, by Harris' inequality (4.5)

$$P\{\mathcal{E}_1 \cap \mathcal{E}_2\} \geq P\{\mathcal{E}_1\}P\{\mathcal{E}_2\} = (P\{\mathcal{E}_1\})^2.$$

Finally, let $\mathcal{E}_1^-$ be the event that there exists a path r^- from $\underline{0}$ to $\{n\} \times (-\infty,0]$ with $H_0 < r^- < H_n$ and $T(r^-) \leq L$. Then $\mathcal{E}_1 \cup \mathcal{E}_1^- = \{s_{0,n} \leq L\}$ since a path from $\underline{0}$ to H_n must end on $\{n\} \times [0,\infty)$ or on $\{n\} \times (-\infty,0]$. Thus, again using Harris' inequality and symmetry we have

$$1 - P\{s_{0,n} \leq L\} = P\{(\mathcal{E}_1 \cup \mathcal{E}_1^-)^c\} \geq P\{\mathcal{E}_1^c\}\, P\{(\mathcal{E}_1^-)^c\} = (1 - P\{\mathcal{E}_1\})^2.$$

Combining all the inequalities we find

$$P\{t_{0,n} \leq 2L\} \geq [1 - (1 - P\{s_{0,n} \leq L\})^{1/2}]^2 . \tag{7.11}$$

Since the $s_{0,n}$ are bounded w.p.1, the required tightness follows by letting L go to ∞.

Proof of Theorem 7.2(a). We begin the argument exactly as in the proof of (6.8). Let $y_0 = \max\{n: s_{0,n} = 0\}$. Then, $y_0 = a_1$ means that

(7.12) for some $b_1 \in \mathbb{Z}$ there exists a path r_0 from $\underline{0}$ to (a_1,b_1) such that $T(r_0) = 0$, $H_0 < r_0 < H_{a_1}$, but there does not exist a similar path to any point of H_{a_1+1}.

The event in (7.12) depends only on edges which (except for their endpoint) lie in $(-\infty,a_1+1) \times \mathbb{R}^{d-1}$, and is therefore independent of all edges in $[a_1+1,\infty) \times \mathbb{R}^{d-1}$. Moreover, the conditional distribution of the edge f_1 between (a_1,b_1) and (a_1+1,b_1), given that (7.12) occurs for that b_1 is given by

$$(7.13) \qquad P\{t(f_1) \le x \mid (7.12) \text{ occurs}\} = P\{t(e) \le x \mid t(e) > 0\}$$
$$= \frac{F(x) - F(0)}{1 - F(0)} .$$

This follows easily if one observes that the only information about $t(f_1)$ contained in (7.12) is that $t(f_1) > 0$.

Now for given (a_i,b_i), $i \ge 1$, we find a_{i+1} such that there exists a b_{i+1} and a path r_i from (a_i+1,b_i) to (a_{i+1},b_{i+1}) with

$$T(r_i) = 0, \qquad H_{a_i+1} < r < H_{a_{i+1}}.$$

We chose a_{i+1} maximal among all integers $> a_i$ with the above property. We denote the edge between (a_i,b_i) and (a_i+1,b_i) by f_i. As before, the choice of (a_i,b_i), $i \le k$, and of (r_j,f_j), $0 \le j \le k-1$ only depends on the edges in $(-\infty,a_k+1] \times \mathbb{R}^{d-1}$ and the only information about $t(f_k)$ contained in the above choice is that $t(f_k) > 0$. Therefore, conditionally on (a_i,b_i), $0 \le i \le k$, and $r_0,f_1,r_1,\dots,f_{k-1},r_{k-1}$, the variables $t(f_k)$ and r_k are independent, with the distribution (7.13) and the distribution of r_0, respectively. We successively choose the a_{i+1}, b_{i+1}, r_i, f_{i+1} for all i.

For each k $r_0, f_1, r_2, f_2, \ldots, r_{k-1}$ can be combined to a cylinder path from $\underline{0}$ to H_{a_k} with passage time $\sum_{i=1}^{k-1} t(f_i)$, because the r_i hav zero passage time. Therefore

$$s_{0,a_k} \le \sum_{i=1}^{k-1} t(f_i).$$

By construction the $t(f_i)$ are i.i.d with distribution given in (7.13) so that (by virtue of (7.3))

$$\frac{1}{k} \sum_{i=1}^{k-1} t(f_i) \to \frac{1}{1-F(0)} \int x dF(x) < \infty \quad \text{w.p.1.} \tag{7.14}$$

Also, by construction the $\{a_{i+1} - a_i\}_{i>1}$ are i.i.d, each with the distribution of $y_0 + 1$. With $C(n)$ as in (6.25) this implies

$$P\{a_{i+1} - a_i \ge n\} \ge P\{s_{0,n} = 0\} \ge P\{C(n)\}.$$

Therefore, by (6.26),

$$\begin{aligned} P\{a_k \ge n\} &\ge P\{\, a_{i+1} - a_i \ge n \quad \text{for some} \quad i \in [1,k-1]\} \\ &\ge 1 - (1 - C_3 n^{\alpha-1})^{k-1} \ge 1 - \exp(-\tfrac{1}{2} C_3 k n^{\alpha-1}). \end{aligned} \tag{7.15}$$

Finally, the fact that $s_{0,n}$ is increasing in n shows that

$$\begin{aligned} P\{s_{0,n} \le x\} &\ge P\{s_{0,a_k} \le x \quad \text{and} \quad a_k \ge n\} \\ &\ge P\{\sum_{i=1}^{k-1} t(f_i) \le x\} - P\{a_k < n\} \\ &\ge P\{\sum_{i=1}^{k-1} t(f_i) \le x\} - \exp(-\tfrac{1}{2} C_3 k n^{\alpha-1}). \end{aligned} \tag{7.16}$$

By taking $k = Ln^{1-\alpha}$ and $x = L^2 n^{1-\alpha}$ and letting L go to ∞, the tightness of $\{n^{\alpha-1} s_{0,n}\}_{n\ge1}$ follows. This implies the tightness of $\{n^{\alpha-1} \theta_{0,n}\}_{n\ge1}$ for $\theta = t$ by means of (7.11), and then for $\theta = a$

or b by means of $b_{0,n} \leq a_{0,n} \leq t_{0,n}$.

To prove (7.4) note that the argument which proves (7.16) also shows that

$$\{s_{0,n} > Cn^{1-\alpha} \log n\} \subset \{\sum_{i \leq Dn^{1-\alpha}\log n} t(f_i)$$

$$> Cn^{1-\alpha} \log n\} \cup \{a_{\lfloor Dn^{1-\alpha}\log n\rfloor} < n\}.$$

If we take

$$D\,(1 - F(0))^{-1} \quad \int x dF(x) < C,$$

then the first event in the right hand side occurs w.p.1 only for finitely many n (by (7.14)). Moreover, by (7.15)

$$\sum_n P\{a_{\lfloor Dn^{1-\alpha}\log n\rfloor} < n\} \leq \sum_n \exp(-\tfrac{1}{4}DC_3 \log n) < \infty$$

for $DC_3 > 4$.(7.4) follows easily by an application of the Borel-Cantelli lemma.

(b) We only indicate the <u>proof of (7.6)</u> under (7.5). Set

$$\nu(n) = (n \log n)^{\frac{1}{\alpha}},$$

$$k(n) = \lfloor D(\nu(n))^{1-\alpha} \log n\rfloor = \lfloor Dn^{\frac{1-\alpha}{\alpha}} (\log n)^{\frac{1}{\alpha}}\rfloor,$$

$$x(n) = C\,(\nu(n))^{1-\alpha} \log n\;.$$

Now use (7.11) and (7.16) with n replaced by $\nu(n)$ and $k=k(n)$, $x=x(n)$ to show that for suitable C and D

$$\sum_n P\{t_{0,\nu(n)} > 2x(n) = 2C(\nu(n))^{1-\alpha} \log n\} < \infty\;.$$

Thus

$$\limsup (\nu(n))^{\alpha-1} (\log \nu(n))^{-1} t_{0,\nu(n)} < \infty \quad \text{w.p.1}$$

Then one "interpolates" between $t_{0,\nu(n)}$ and $t_{0,\nu(n+1)}$ by using

$$\max\{t_{0,k} - t_{0,\nu(n)} : \nu(n) \le k \le \nu(n+1)\}$$

$$\le \sum_{\nu(n) \le k \le \nu(n+1)} t(e_i) = O(\nu(n+1) - \nu(n))$$

$$= O(\nu(n)^{1-\alpha} \log \nu(n)) \quad \text{w.p.1} \quad ,$$

which follows from (5.39) and the Borel-Cantelli lemma.

8. Asymptotics for large dimensions. In this section we investigate the behavior of $\mu(F,d)$ and $B_0(F,d)$ as $d \to \infty$. We find that μ is usually of the order $\frac{1}{d}\log d$, so that the intersection of the boundary of B_0 with the coordinate axes has a distance of the order $d/\log d$ from the origin. On the other hand, the distance to the origin of the intersection of the boundary of B_0 with the diagonal $x(1) = \ldots = x(d)$ is only of order $\sqrt{d}$. As a corollary we obtain that for a large class of distributions F, $B_0(F,d)$ is not an Euclidean ball in high dimension.

The class of distribution functions to be considered in this section is described in the following definition, in which a is taken > 0.

(8.1) DEFINITION. $\mathcal{C}(a)$ is the class of distribution functions with $F(0) = 0$ which have a density f on some interval $[0,b]$, $b > 0$, which satisfies

$$f(x) = a(1+o(\frac{1}{|\log x|})), \quad x \downarrow 0. \qquad ///$$

The precise results are stated next.

(8.2) THEOREM. There exists a universal constant $0 < C_1 < \infty$ such that for all $F \in \mathcal{C}(a)$ with $a > 0$ and $\int x dF(x) < \infty$ one has for large d

$$C_1 \frac{\log d}{ad} \le \mu(F,d) \le 11 \frac{\log d}{ad} .$$

(8.3) THEOREM. Assume that $F \in \mathcal{C}(a)$ and that F has a finite first moment. Let μ^* be the time constant in the direction of the diagonal, i.e.,

$$\mu^* = \mu^*(F,d) := \lim_{n\to\infty} \frac{1}{n} \inf\{T(r): r \text{ a path from } \underline{0} \text{ to the hyperplane } \{x: x(1)+\ldots+x(d) = n\}\} \quad \text{w.p.1.}$$

Then

$$\frac{1}{6ea} \le \liminf_{d\to\infty} d\mu^*(F,d) \le \limsup d\mu^*(F,d) \le \frac{1}{2a} .$$

(8.4) COROLLARY. If $F \in \mathcal{C}(a)$ has a finite first moment then $B_0(F,d)$ is not a Euclidean ball for sufficiently large d.

(8.5) REMARK. The values of d for which the results in (8.2) and (8.4) become valid may still depend on F. One can of course obtain explicit estimates by restricting F further. We have not pursued this except for the exponential distribution of (1.19). For this F one obtains that $B_0(F,d)$ is not a Euclidean ball as soon as $d \geq 10^6$.

(8.6) REMARK. Theorems 8.2 and 8.3 remain valid if one orients all edges in the positive direction and restricts oneself to directed paths in $\{x : x(i) \geq 0$ for all $i\}$. In fact for this case Cox and Durrett [8] prove that $d\mu^*(F,d)$ has a limit as $d \to \infty$ for F the exponential distribution (1.19). ///

Theorem 8.3 is essentially copied from [8]. The lower bound in (8.2) is proven very much as (6.6). One bounds by brute force the expected number of paths from $\underline{0}$ to H_n with a passage time at most $C_1 n \frac{\log d}{ad}$. For sufficiently small C_1 this expectation decreases exponentially. The upper bound in (8.2) requires a more elaborate proof. Basically we estimate the first and second moment of the number of directed paths from $\underline{0}$ to H_1 with a passage time at most $C_2(ad)^{-1} \log d$. We then obtain more or less by a Chebyshev argument that with high probability there is at least one such path. In turn this leads to the estimate

$$(8.7) \qquad E\tilde{s}_{0,1} \leq 11 \frac{\log d}{ad} \quad \text{for large } d,$$

where

$$\tilde{s}_{0,1} = \inf\{T(r) : r \text{ a path from } \underline{0} \text{ to } H_1 \text{ such that } r, \text{ except for its final point, is contained in } [0,1) \times \mathbb{R}^{d-1}\}.$$

The upper bound in (8.2) follows easily from (8.7).

We now derive the estimates for Theorem 8.2 in a series of lemmas. Throughout we assume that $F \in \mathcal{C}(a)$ for some fixed $a > 0$ and that F has a finite first moment. S_k will denote the sum of k independent random variables, each with the distribution F.

(8.8) LEMMA. Uniformly in $1 \leq n \leq 2 \log \frac{1}{x}$

$$n!(ax)^{-n} P\{S_n \leq x\} \to 1 \quad \text{as} \quad x \downarrow 0.$$

<u>Proof</u>: If F has density f on $[0,b]$, then for $x \leq b$

$$P\{S_n \le x\} = \int \cdots \int_{x_1+\ldots+x_n \le x} f(x_1)\ldots f(x_n)\,dx_1\ldots dx_n$$

$$= a^n\left(1+o\left(\frac{1}{|\log x|}\right)\right)^n \int \cdots \int_{x_1+\ldots+x_n \le x} dx_1\ldots dx_n$$

$$= (\text{set } y_i = x_1+\ldots+x_i)\, a^n(1+o(1)) \int_0^x dy_n \int_0^{y_n} dy_{n-1} \cdots \int_0^{y_2} dy_1$$

$$= \frac{(ax)^n}{n!}(1+o(1)). \qquad ///$$

Next we introduce a class of "increasing" paths from $\underline{0}$ to H_1. Set $\nu = \lfloor \frac{3}{4} \log d \rfloor$ and let $\mathcal{P}$ be the class of paths $r = (v_0 = \underline{0}, e_1, v_1, \ldots, e_\nu, v_\nu)$ of ν steps such that $v_{j+1} - v_j$ equals some coordinate vector $\xi_i = (\delta_{i,1}, \ldots, \delta_{i,d})$ with $2 \le i \le \frac{d}{2}+1$ when $0 \le j < \nu-1$, while $v_\nu - v_{\nu-1} = (1,0,\ldots,0)$. Thus, each path in $\mathcal{P}$ moves around in H_0 for $\nu-1$ steps, each one of these steps being a positive one in one of the directions $2,\ldots,\lfloor\frac{d}{2}\rfloor+1$. The last step is in the positive first coordinate direction and takes the path from H_0 to H_1. Finally, we define

$$N = \text{number of paths } r \text{ in } \mathcal{P} \text{ with } T(r) \le \frac{3\nu}{ad}.$$

(8.9) LEMMA.

$$EN \sim \frac{1}{d}\sqrt{\frac{2}{\pi\nu}}\left(\frac{3}{2}e\right)^{\lfloor\frac{3}{4}\log d\rfloor} \to \infty, \quad d \to \infty.$$

Proof. Since each of the first $\nu-1$ steps of any path in $\mathcal{P}$ can be chosen in $\lfloor d/2 \rfloor$ ways $\#\mathcal{P} = \lfloor d/2 \rfloor^{\nu-1}$. Therefore

$$EN = (\#\mathcal{P})\, P\{S_\nu \le \frac{3\nu}{ad}\} = \lfloor\frac{d}{2}\rfloor^{\nu-1} P\{S_\nu \le \frac{3\nu}{ad}\}$$

$$\sim \lfloor\frac{d}{2}\rfloor^{\nu-1} \frac{1}{\nu!}\left(\frac{3\nu}{d}\right)^\nu \sim \frac{1}{d}\sqrt{\frac{2}{\pi\nu}}\left(\frac{3}{2}e\right)^{\lfloor\frac{3}{4}\log d\rfloor}$$

by virtue of Lemma (8.8) and Stirling's formula. ///

Unfortunately it is not as simple to estimate $E\{N^2\}$. For this we define a random walk on $(\mathbb{Z}_+)^{\lfloor d/2 \rfloor}$. Let $X_1', X_1'', X_2', X_2'', \ldots$ be independent random vectors, each with the distribution

$$P\{X'_j = \xi_i\} = P\{X''_j = \xi_i\} = (\lfloor d/2 \rfloor)^{-1}, \ 2 \le i \le \frac{d}{2}+1, \ j = 1,2,\ldots$$

and let

$$Y'_n = \sum_{j=1}^{n} X'_j, \quad Y''_n = \sum_{1}^{n} X''_j .$$

An important quantity is the probability that $Y'_{\cdot}$ and $Y''_{\cdot}$ over coincide for two successive times:

$$\rho = \rho(\lfloor d/2 \rfloor) := P\{\exists\, n \ge 0 \text{ such that } Y'_n = Y''_n \text{ and } Y'_{n+1} = Y''_{n+1}\}.$$

It is known (cf. [8]) that

(8.10) $$\rho(\lfloor d/2 \rfloor) = \frac{2}{d} + O(\frac{1}{d^2}), \ d \to \infty.$$

(8.11) LEMMA. For large d

$$E\{N^2\} \le 4(EN)^2 .$$

Proof: Let r' and r'' stand for generic paths in $\mathcal{P}$. Sums over r' or r'' without restrictions run over all of $\mathcal{P}$. Then

(8.12) $$E\{N^2\} = \sum_{r'} \sum_{r''} P\{T(r') \le \frac{3\nu}{ad}, T(r'') \le \frac{3\nu}{ad}\}$$

$$\le \sum_{r'} \sum_{r''} P\{T(r') \le \frac{3\nu}{ad}, \sum{}^{*} t(e) \le \frac{3\nu}{ad}\} ,$$

where $\sum^*$ runs over the edges of r'' which do not belong to r'. For given r', r'' the latter edges have passage times which are independent of the passage times of the edges of r'. Therefore, if r' and r'' have exactly k edges in common, then the probability in the last member of (8.12) equals

$$P\{S_\nu \le \frac{3\nu}{ad}\} \cdot P\{S_{\nu-k} \le \frac{3\nu}{ad}\} .$$

Thus

(8.13) $$E\{N^2\} \le \sum_{k=0}^{\nu} P\{S_\nu \le \frac{3\nu}{ad}\} P\{S_{\nu-k} \le \frac{3\nu}{ad}\} \cdot (\text{number of pairs } r', r''$$
$$\text{with exactly } k \text{ edges in common}).$$

To estimate the number of pairs r', r'' with k edges in common we must distinguish two cases. This is due to the fact that the last step of each path in $\mathcal{P}$ is in the positive first coordinate direction. If r' and r'' have the same one but last vertex, then their last edges automatically coincide. Let $r' = (v_0' = \underline{0}, e_1', \dots, e_\nu', v_\nu')$ and $r'' = (v_0'' = \underline{0}, e_1'', \dots, e_\nu'', e_\nu'')$. If r' and r'' have k edges in common, then there exist indices $0 \le \lambda(1) < \lambda(2) \dots < \lambda(k)$ such that

(8.14) $$v'_{\lambda(i)} = v''_{\lambda(i)} \text{ and } v'_{\lambda(i)+1} = v''_{\lambda(i)+1}, \; 1 \le i \le k.$$

The first case to consider is when $\lambda(k) \le \nu-3$. We simply have

(8.15) $$(\# \text{ of pairs } r', r'' \text{ for which there exist } \lambda(1),\dots,\lambda(k) \le \nu-3 \text{ satisfying } (8.14)) \le \lfloor\tfrac{d}{2}\rfloor^{2\nu-2} P\{\text{there exist } 0 \le \lambda(1) < \dots < \lambda(k) \le \nu-3 \text{ such that } Y'_{\lambda(i)} = Y''_{\lambda(i)} \text{ and } Y'_{\lambda(i)+1} = Y''_{\lambda(i)+1}\} \le \lfloor\tfrac{d}{2}\rfloor^{2\nu-2}(\rho(\lfloor d/2\rfloor))^k.$$

The second case to consider is when the k-th edge which r' and r'' have in common is $e_\nu' = e_\nu''$. This means $\lambda(k) = \nu-1$. $\lambda(k) = \nu-2$ is not possible, for then $v'_{\nu-1} = v''_{\nu-1}$ and $e'_\nu = e''_\nu$ is a $(k+1)$-th edge in common to r' and r''. In this second case we shall decompose the collection of pairs r', r'' further according to the number of edges they have in common at the end. If r', r'' have k edges in common and $\lambda(k) = \nu-1$, define j as the largest index $\le k$ for which $\lambda(k-\ell+1) = \nu-\ell$, $1 \le \ell \le j$. Then $e'_{\nu-\ell+1} = e''_{\nu-\ell+1}$, $1 \le \ell \le j$ and, provided $k < \nu$, $\lambda(k-j+1) \ge \lambda(k-j)+2$ (with $\lambda(0) = 0$). Thus for $k < \nu$

(8.16) $$(\# \text{ of pairs } r', r'' \text{ for which there exist } \lambda(1),\dots,\lambda(k) \text{ satisfying } (8.14) \text{ and } \lambda(k) = \nu-1)$$
$$\le \lfloor\tfrac{d}{2}\rfloor^{2\nu-2} \sum_{j=1}^{k} P\{\text{there exist } 0 \le \lambda(1) < \dots < \lambda(k-j) \text{ and } \lambda(k-j+1) \ge \lambda(k-j)+2 \text{ such that } Y'_{\lambda(i)} = Y''_{\lambda(i)} \text{ and } Y'_{\lambda(i)+1} = Y''_{\lambda(i)+1}, \; 1 \le i \le k-j, \; Y'_{\lambda(k-j)+1} \ne Y''_{\lambda(k-j)+1} \text{ and such that } Y'_\sigma = Y''_\sigma \text{ for } \lambda(k-j+1) \le \sigma \le \lambda(k-j+1)+j-1\}$$
$$\le \lfloor\tfrac{d}{2}\rfloor^{2\nu-2} \sum_{j=1}^{k} P\{\text{there exist } 0 \le \lambda(1) < \dots < \lambda(k-j) \text{ such that } Y'_{\lambda(i)} = Y''_{\lambda(i)} \text{ and } Y'_{\lambda(i)+1} = Y''_{\lambda(i)+1}, 1 \le i \le k-j\}$$
$$\cdot P\{Y_1' \ne Y_1'' \text{ but } \exists\, \lambda \ge 2 \text{ such that } Y'_\lambda = Y''_\lambda\}$$

$$\cdot P\{Y'_\sigma = Y''_\sigma \text{ for } 0 \le \sigma \le j-1\}$$

$$= \lfloor \tfrac{d}{2} \rfloor^{2\nu-2} \sum_{j=1}^{k} (\rho(\lfloor d/2 \rfloor)^{k-j} P\{Y'_1 \ne Y''_1 \text{ but } \exists\ \lambda \ge 2$$

$$\text{such that } Y'_\lambda - Y''_\lambda\} \cdot \lfloor \tfrac{d}{2} \rfloor^{-j+1} .$$

To estimate the last probability we observe that Cox and Durrett [8], formula (2.2), have shown

$$(8.17) \qquad P\{Y'_1 \ne Y''_1 \text{ but } \exists\ \lambda \ge 2 \text{ such that } Y'_\lambda = Y''_\lambda\}$$
$$= \lfloor \tfrac{d}{2} \rfloor \{\rho(\lfloor \tfrac{d}{2} \rfloor) - \lfloor \tfrac{d}{2} \rfloor^{-1}\} \le \lfloor \tfrac{d}{2} \rfloor^{-2} + 0(\tfrac{1}{d^3}).$$

Finally, if $k = \nu$, then necessarily $r' = r''$ and the number of such pairs is $\lfloor d/2 \rfloor^{\nu-1}$. Combining this with (8.13) and (8.15)-(8.17) we obtain

$$(8.18) \qquad E\{N^2\} = \lfloor \tfrac{d}{2} \rfloor^{\nu-1} P\{S_\nu \le \tfrac{3\nu}{ad}\} \cdot (1 + \lfloor \tfrac{d}{2} \rfloor^{\nu-1} \sum_{k=0}^{\nu-1} P\{S_{\nu-k} \le \tfrac{3\nu}{ad}\}$$

$$\cdot [(\rho(\lfloor \tfrac{d}{2} \rfloor))^k + \sum_{j=1}^{k} (\rho(\lfloor \tfrac{d}{2} \rfloor))^{k-j} (\tfrac{4}{d^2} + 0(\tfrac{1}{d^3})) \lfloor \tfrac{d}{2} \rfloor^{-j+1}]).$$

Since $\rho(\lfloor \frac{d}{2} \rfloor) \sim 2/d + 0(d^{-2})$ (cf. (8.10)) we may replace the factor in square brackets on the right hand side of (8.18) by $(2/d)^k(1+o(1))$ with $o(1)$ tending to zero as $d \to \infty$, uniformly in $k \le \nu$. Moreover, as in Lemma 8.9

$$\lfloor \tfrac{d}{2} \rfloor^{\nu-1} P\{S_\nu \le \tfrac{3\nu}{ad}\} = EN.$$

Finally, by Lemma 8.8

$$P\{S_{\nu-k} \le \tfrac{3\nu}{ad}\} = P\{S_\nu \le \tfrac{3\nu}{ad}\} \frac{\nu!}{(\nu-k)!} (\tfrac{d}{3\nu})^k \cdot (1+o(1)).$$

Substituting these estimates into (8.18) gives

$$(8.19) \qquad E\{N^2\} \le EN + (EN)^2 (1+o(1)) \sum_{k=0}^{\nu-1} \frac{\nu!}{(\nu-k)!} (\tfrac{2}{3\nu})^k$$

$$\le EN + 3.5(EN)^2 \le 4(EN)^2$$

for large d (recall that $EN \to \infty$ as $d \to \infty$).

Proof of the upper bound in Theorem 8.2. From Lemma 8.11 and Schwarz' inequality we obtain

$$P\{N \neq 0\} \geq \frac{(E\{N\})^2}{E\{N^2\}} \geq \frac{1}{4}$$

for large d. Taking into account the definition of N and $\mathbb{P}$ we see that this means

(8.20) $P\{\exists$ path r from $\underline{0}$ to H_1 which, with the exception of its final point, is contained in $([0,1) \times \mathbb{R}^{d-1})$ $\cap$ span of $\{\xi_1,\dots,\xi_{\lfloor d/2\rfloor+1}\}$, and has $T(r) \leq \frac{3\nu}{ad}\} \geq \frac{1}{4}$.

Let $\mathcal{S}$ be the subspace spanned by $\xi_1,\dots,\xi_{\lfloor d/2\rfloor+1}$ and let $\mathcal{E}_i$ be the event that there exists a path r from ξ_i to H_1 which, with the exception of its final point, is contained in $[0,1) \times \mathbb{R}^{d-1} \cap (\xi_i+\mathcal{S})$, and has $T(r) \leq \frac{3\nu}{ad}$. By translation invariance and (8.20) $P\{\mathcal{E}_i\} \geq 1/4$. for $i > 1$. Now let f_i be the edge between $\underline{0}$ and ξ_i, and $\mathcal{F}_i$ the event $\{t(f_i) \leq \frac{9 \log d}{ad}\} \cap \mathcal{E}_i$. If $\mathcal{F}_i$ occurs for any $i > 1$, then $\tilde{s}_{0,1}$ (which is defined after (8.7)) is at most

$$t(f_i) + \frac{3\nu}{ad} \leq \frac{9 \log d}{ad} + \frac{3}{ad}\cdot\frac{3}{4} \log d = \frac{42 \log d}{4ad} .$$

Also, since $F \in \mathcal{C}(a)$

$$P\{\mathcal{F}_i\} \geq \frac{1}{4} P\{t(f_i) \leq \frac{9 \log d}{ad}\} \sim \frac{9}{4} \frac{\log d}{d} .$$

Note further that the events $\mathcal{F}_i$, $\lfloor d/2\rfloor+2 \leq i \leq d$, are independent, because for $i \neq j$, $\xi_i,\xi_j \notin \mathcal{S}$ a path with first edge f_i and whose other edges are contained in $\xi_i+\mathcal{S}$ cannot have a point in common with another path with first edge f_j and other edges in $\xi_j+\mathcal{S}$. Finally, none of the events $\mathcal{F}_i, \lfloor d/2\rfloor+2 \leq i \leq d$ involve the edge f_1 which goes directly from $\underline{0}$ to H_1, so that

$$\tilde{s}_{0,1} \leq \frac{41 \log d}{ad} + I\{\bigcap_{\lfloor\frac{d}{2}\rfloor+2\leq i\leq d} \mathcal{F}_i^c\} t(f_1),$$

and consequently, for large d

$$E\tilde{s}_{0,1} \leq \frac{42 \log d}{4ad} + P\{\text{none of the } \mathcal{F}_i, \lfloor\tfrac{d}{2}\rfloor+2 \leq i \leq d, \text{occur}\} \int x \, dF(x)$$

$$\leq \frac{42 \log d}{4ad} + \{1 - \frac{9 \log d}{4d}(1+o(1))\}^{\frac{d}{2}-2} \int x \, dF(x) \leq \frac{11 \log d}{ad} .$$

This proves (8.7).

It is easy to obtain the upper bound in Theorem 8.2 from (8.7) by the same argument as used in [19], Theorem 4.2.5 or [31], Lemma 5.2. If r is a path from $\underline{0}$ to H_{n-1} such that $H_0 < r < H_{n-1}$ with end-point v, then we can construct a path r' from $\underline{0}$ to H_n such that $H_0 < r' < H_n$, by continuing with a path r'' from v to H_n, as long as r'' (with the exception of its final point) is contained in $[n-1,n) \times \mathbb{R}^{d-1}$. The infimum of $T(r'')$ over all such r'' is independent of r' and has the same distribution as $\tilde{s}_{0,1}$. Consequently, for $n \geq 1$

$$Es_{0,n} \leq Es_{0,n-1} + E\tilde{s}_{0,1} \leq \cdots \leq Es_{0,1} + (n-1)E\tilde{s}_{0,1} .$$

Thus

$$\mu = \lim \frac{1}{n} Es_{0,n} \leq E\tilde{s}_{0,1}$$

and the proof is complete.

Proof of the lower bound in Theorem 8.2. Clearly $P\{b_{0,n} \leq C_1 n(ad)^{-1}\log d\}$ is no bigger than the sum over all paths r from $\underline{0}$ to H_n of $P\{T(r) \leq C_1 n(ad)^{-1}\log d\}$. It therefore suffices to show that the latter sum tends to 0 exponentially fast in n for sufficiently small $C_1 > 0$. As in the proof of (6.6) we decompose the sum according to the number of edges k in r. The number of paths from $\underline{0}$ to H_n of k steps is at most

$$(2d)^k e^{-\rho n}\left(1 - \frac{1}{d} + \frac{1}{d}\cosh \rho\right)^k$$

for any $\rho \geq 0$ (cf. (6.20)). Of course there are no such paths at all for $k < n$ and in any case the number of such paths is also bounded by $(2d)^k$. Next, let $b > 0$ be so small that the density f of $F \in \mathcal{C}(a)$ satisfies $f(y) \leq 2a$ for $0 \leq y \leq b$. Also, let $0 \leq x \leq b$ be so small that $z^{-1}\exp(-b/z) \leq a$ for all $z \leq x$. Then for any path r with $k \geq n$ steps one has with $\gamma = k/(nx)$

$$\begin{aligned}(8.21)\qquad P\{T(r) \leq nx\} &= P\{S_k \leq nx\} \leq e^{\gamma nx} E e^{-\gamma S_k} \\ &\leq e^{\gamma nx}\left\{\int_0^b e^{-\gamma y} f(y)\,dy + e^{-\gamma b}(1-F(b))\right\}^k \\ &\leq e^{\gamma nx}\left\{2a\int_0^b e^{-\gamma y}\,dy + e^{-\gamma b}(1-2ab)\right\}^k \leq \left\{e^{\gamma nx/k}\,\frac{3a}{\gamma}\right\}^k = \left(\frac{3eanx}{k}\right)^k .\end{aligned}$$

In view of these observations we have for large d and any choice of $\rho_k \geq 0$

$$P\{b_{0,n} \leq C_1 n(ad)^{-1}\log d\}$$

$$\leq \sum_{n\leq k\leq 10eC_1 n\log d} \exp\{-\rho_k n + \frac{k}{d}(\cosh \rho_k - 1)\}\left(\frac{6enC_1\log d}{k}\right)^k$$

$$+ \sum_{k>10eC_1 n \log d} \left(\frac{6enC_1\log d}{k}\right)^k .$$

Now choose $\rho_k = \log \frac{2dn}{k}$, and then replace k by yn. Then the first sum in the right hand side is seen to be bounded by

$$\sum_{k=n}^{10eC_1 n\log d} \left(\frac{ke}{2dn}\right)^n\left(\frac{6enC_1\log d}{k}\right)^k \leq 10eC_1 n \log d$$

$$\cdot \left[\left(\frac{e}{2d}\right)\max\{y^{1-y}(6eC_1\log d)^y : 1 \leq y \leq 10eC_1\log d\}\right]^n .$$

A little calculus shows that for sufficiently small C_1 and large d the expression between square brackets is less than one, and Theorem 8.2 follows.

<u>Proof of Theorem 8.3</u>. The lower bound in 8.3 can be proved essentially as in [8], p.153, or as the lower bound in 8.2 above. Indeed, the probability that there is any path r of n steps or more with a passage time $\leq Cn(ad)^{-1}$ is for large d at most

$$\sum_{k=n}^{\infty}(2d)^k\left(\frac{3enC}{dk}\right)^k \qquad (\text{cf. } (8.21)).$$

This sum goes to zero exponentially in n whenever $C < (6e)^{-1}$.

As for the upper bound for μ^*, it was shown in [8], p.161 that

$$\mu^*(F,d) \leq \frac{\int_0^{t(d)} x\,dF(x)}{F(t(d))}, \tag{8.22}$$

for any $t(d)$ for which $F(t(d)) > \hat{p}_H(d)$, where $\hat{p}_H(d)$ is the critical probability for oriented percolation introduced before (6.13). Moreover $d\hat{p}_H(d) \to 1$ as $d \to \infty$. Since

$$F(t) = \int_0^t f(x)dx \sim at, \; t \downarrow 0, \text{ for } F \in C(a),$$

it follows that we can take $t(d) \sim (ad)^{-1}$ and the upper bound in (8.3) is an easy consequence of (8.22)

Proof of Corollary 8.4. As observed after (1.14) the boundary of B_0 intersects the coordinate axes at distance $(\mu(d))^{-1}$ from the origin. This distance is of order $d/\log d$ under the hypotheses of Theorem 8.2. On the other hand, the same proof as used for (1.14) in Sect. 3 shows that the boundary of B_0 intersects the diagonal $x(1) = x(2) = \ldots = x(d)$ in the point $((d\mu^*(d))^{-1}, (d\mu^*(d))^{-1}, \ldots, (d\mu^*(d))^{-1})$, which is at distance $d^{-1/2}(\mu^*(d))^{-1}$ from the origin. Under the hypotheses of Theorem 8.3 this is only of order $d^{1/2} = o(d/\log d)$. Thus for large d B_0 is not a Euclidean ball.

9. Maximal flows and other open problems

(9.1) The principal new problem which we want to discuss here concerns the maximal flow discussed in Sect. 1. For $\underline{n} \in (\mathbb{Z}_+)^{d-1}$ we define

(9.2) $$\phi_{\underline{n},m} = \text{maximal flow in (the restriction of } \mathbb{Z}^d \text{ to)}$$
$$[0,n_1] \times \ldots \times [0,n_{d-1}] \times [0,m] \text{ between}$$
$$F_0 := [0,n_1] \times \ldots \times [0,n_{d-1}] \times \{0\} \text{ and}$$
$$F_m := [0,n_1] \times \ldots \times [0,n_{d-1}] \times \{m\}.$$

This flow is, of course, restricted by the capacities $\{t(e)\}$ of the edges e in $\mathbb{Z}^d$, and is therefore random. The question is under what conditions on the $t(e)$ and m as a function of n does

(9.3) $$\frac{1}{|n_1 \cdots n_{d-1}|} \phi_{\underline{n},m}$$

have a limit as $n_1, \ldots, n_{d-1} \to \infty$, and in what sense does convergence take place? For $d = 2$ this question is largely answered in [17] by the use of the max-flow min-cut theorem. Write $B_{\underline{n},m}$ for the box $[0,n_1] \times \ldots \times [0,n_{d-1}] \times [0,m]$. Define a <u>cut</u> in $B_{\underline{n},m}$ to be a minimal set of edges in $B_{\underline{n},m}$ which separates F_0 from F_m (cf. (9.2) for F_0, F_m). Assign to a set E of edges the <u>value</u>

$$V(E) := \sum_{e \in E} t(e).$$

Then the max-flow min-cut theorem states that

(9.4) $$\phi_{\underline{n},m} = \min\{V(E): E \text{ a cut in } B_{\underline{n},m}\} .$$

We can find the asymptotic behavior of $\phi_{\underline{n},m}$ for $d = 2$ by means of (9.4), because in dimension two it is possible to give a simple description of the cuts in $B_{\underline{n},m}$. Indeed, let $\mathcal{L}$ be the graph $\mathbb{Z}^2 + (\frac{1}{2},\frac{1}{2})$. Its vertices are at $(i+\frac{1}{2}, j+\frac{1}{2})$, $i,j \in \mathbb{Z}$ and $(i_1+\frac{1}{2}, j_1+\frac{1}{2})$ and $(i_2+\frac{1}{2}, j_2+\frac{1}{2})$ are connected by an edge in $\mathcal{L}$ if and only if $|i_1-i_2| + |j_1-j_2| = 1$. One easily sees (see Fig. 9.1) that each edge e of $\mathbb{Z}^2$ intersects a unique edge e^* of $\mathcal{L}$ and vice versa. $\mathcal{L}$ is the so-called dual graph of $\mathbb{Z}^2$ ([33], especially proof of Theorem 29). Call a pair of edges e and e^* of $\mathbb{Z}^2$ and $\mathcal{L}$, respectively, <u>associated</u> if they intersect; in this case define

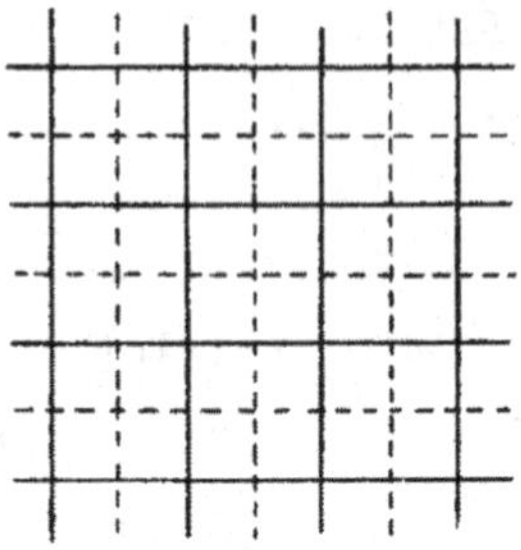

Fig. 9.1 $\mathbb{Z}^2$ and $\mathcal{L}$. The edges of $\mathbb{Z}^2$ are drawn as solid segments and those of $\mathcal{L}$ as dashed segments.

$t(e^*) = t(e)$. We also define for any collection E^* of edges of $\mathcal{L}$, its value

$$V(E^*) = \sum_{e^* \in E^*} t(e^*).$$

In dimension two we have the following simple criterion for a set of edges $E \subset B_{n,m} = [0,n] \times [0,m]$ to be a cut. E is a cut if and only if the collection of edges e^* associated to the edges of E is the set of edges of a self-avoiding path r^* on $\mathcal{L}^*$ from $\{-\frac{1}{2}\} \times [\frac{1}{2}, m - \frac{1}{2}]$ to $\{n + \frac{1}{2}\} \times [\frac{1}{2}, m - \frac{1}{2}]$ inside $[-\frac{1}{2}, n + \frac{1}{2}] \times [\frac{1}{2}, m - \frac{1}{2}]$ (cf. [31], Theorem 2.1, [34], Theorem 4; see also Fig. 9.2). In view of (9.4) this shows that for $d = 2$

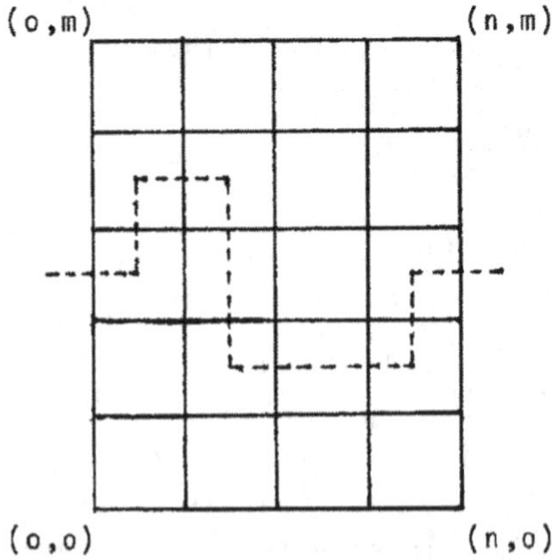

Fig. 9.2 The dashed path is a cut for $B_{n,m}$.

$$\phi_{n,m} = \min\{\sum_1^{\nu} t(e_i^*) : (v_0^*, e_1^*, \ldots, e_\nu^*, v_\nu^*) \text{ is a path on } \mathcal{L} \text{ from}$$
$$\{-\tfrac{1}{2}\} \times [\tfrac{1}{2}, m-\tfrac{1}{2}] \text{ to } \{n+\tfrac{1}{2}\} \times [\tfrac{1}{2}, m-\tfrac{1}{2}] \text{ inside}$$
$$[-\tfrac{1}{2}, m+\tfrac{1}{2}] \times [\tfrac{1}{2}, m-\tfrac{1}{2}]\}.$$

Since $\mathcal{L}$ is isomorphic to $\mathbb{Z}^2$ we see that $\phi_{n,m}$ has a distribution closely related to that of $s^m_{0,n+1}$ (see the lines following (5.22) for definition). It is proved in [17] that

$$\lim \frac{1}{n} \phi_{n,m} = \mu \quad \text{w.p.1},$$

whenever $m \to \infty$ with n in such a way that $n^{-1} \log m(n) = 0$.

If $d \geq 3$ the problem becomes far more difficult because we do not have a good description of the cuts. Consider $d = 3$. Following Aizenman et al. [2], it seems natural to associate with each edge e of $\mathbb{Z}^3$ a so-called plaquette, namely the unique unit square with corners in $\mathbb{Z}^3 + (\frac{1}{2}, \frac{1}{2}, \frac{1}{2})$ which is intersected by e (and perpendicular to e). Then each plaquette π^* is associated with a unique edge e and vice versa, and we can set $v(\pi^*) = t(e)$ if π^* is associated with e. Also for any collection E^* of plaquettes we can set

$$V(E^*) = \sum_{\pi^* \varepsilon E^*} v(\pi^*),$$

and analogously to the above we have

$$\text{(9.5)} \quad \phi_{\underline{n},m} = \min\{V(E^*) : E^* \text{ a set of plaquettes in } [-\tfrac{1}{2}, n_1+\tfrac{1}{2}]$$
$$\times [-\tfrac{1}{2}, n_2+\tfrac{1}{2}] \times [\tfrac{1}{2}, m-\tfrac{1}{2}] \text{ such that the associated}$$
$$\text{edges form a cut in } B_{\underline{n},m}\}.$$

However, since we do not really know what class of E^* figure in the min in (9.5) it is difficult to use (9.5). Nevertheless it is shown in [24] that (9.3) has a limit $\nu > 0$ for $d = 3$ and under a number of conditions. In [24] also various analogues of (1.13) are considered. It is not entirely obvious what the analogues of $a_{0,n}, b_{0,n}, t_{0,n}$ and $s_{0,n}$ should be, but following [2], Sect. 1.(ii), we take these to be the infimum of the values of all sets of plaquettes which separate the bottom $[0,n_1] \times [0,n_2] \times \{-N\}$ from the top $[0,n_1] \times [0,n_2] \times \{N\}$ in $[0,n_1] \times [0,n_2] \times [-N,N]$ for arbitrarily large N <u>and</u> whose boundary is partially or totally specified. Here the boundary of a set E^* of plaquettes is the collection of edges of $\mathbb{Z}^3 + (\frac{1}{2},\frac{1}{2},\frac{1}{2})$ which belong to an odd number of plaquettes in E^*. For example in an analogue $\tau(\underline{n})$ of $t_{0,n}$ we specify this boundary to consist of the four segments

$$[-\tfrac{1}{2}, n_1 + \tfrac{1}{2}] \times \{y\} \times \{\tfrac{1}{2}\}, \; y = -\tfrac{1}{2} \text{ or } n_2 + \tfrac{1}{2},$$

$$\{x\} \times [-\tfrac{1}{2}, n_2 + \tfrac{1}{2}] \times \{\tfrac{1}{2}\}, \; x = -\tfrac{1}{2} \text{ or } n_1 + \tfrac{1}{2}.$$

We show that $(n_1 n_2)^{-1}\tau(\underline{n})$ has the same limit as $n^{-2}\phi_{(n,n),n}$.

The proofs in [24] are based on a multiparameter subadditive ergodic theorem of Ackoglu and Krengel. The main difficulty is to find a method to combine a cut in $B_{\underline{n},m}$ with a cut in $[n_1+1, 2n_1+1] \times [0,n_2] \times [0,n]$ to make a cut in $[0, 2n_1+1] \times [0,n_2] \times [0,m]$.

The results of [24] are unsatisfactory in so far as they apply only for small $F(0)$. Our main reason for studying the flows $\phi_{\underline{n},m}$ was to derive lower bounds for the resistance in $B_{\underline{n},m}$ between F_0 and F_m, when $t(e)$ represents the conductivity of the edge e, as done for $d = 2$ in [17], Sect. 6 and [23], Ch.11. In order to be useful for resistance estimates we would like to prove

$$\lim \frac{1}{n^2} \phi_{(n,n),n} > 0$$

or at least

$$\liminf \frac{1}{n^2} \phi_{(n,n),n} > 0$$

whenever $F(0) < 1 - p_H(3)$. Of course generalizations to dimensions > 3 are also desired.

We list a number of <u>further open problems</u>. Several of these have been mentioned before or appear in the lists of Hammersley and Welsh [19]), Smythe and Wierman [31], Ch.10, and Kingman, [25], Sect. 3.

(9.6) Find the value of the time constant $\mu(F,d)$, and by extension, find the shape $B_0(F,d)$. This seems to be an (almost?) impossible problem, so that even good estimates of μ would be useful.

(9.7) The next major problem seems to us the question of limit theorems for $\gamma_n^{-1}(\theta_{0,n} - \delta_n)$ for suitable δ_n and $\gamma_n \to \infty$. It is not at all clear that such limit theorems exist. Sect. 5 only gives weak indications in this direction. At least one should try to improve the estimates in (5.17)-(5.20). We suspect that $(\log n)^{1/p}$ can be replaced by n^q for some $q > 0$ in (5.13) - (5.20).

As we indicated in Remark 5.15, it is probably better for a limit theorem to center at $\delta_n = E\theta_{0,n}$ than at $\delta_n = n\mu$.

(9.8) This is the so-called "height-problem" (the next two problems are related). It discusses the deviations from the first coordinate axis of paths from $\underline{0}$ to $(n,0,\ldots,0)$ or to H_n which have a passage time close to $\theta_{0,n}$. For simplicity consider the case of $d = 2$, and for any path r from $\underline{0}$ define

$$M(r) := \max\{x(2):(x(1),x(2)) \text{ is a point on } r\},$$

and if $\theta = a$

$$h_n^\theta = \max\{M(r):r \text{ a path from } \underline{0} \text{ to } (n,0,\ldots,0) \text{ with } T(r) = \theta_{0,n}\}.$$

For $\theta = b$ $(n,0,\ldots,0)$ should be replaced by H_n in the last definition, and similar definitions can be given for $\theta = t$ or s. Question: Does

$$\frac{1}{n} h_n^\theta \to 0 \quad \text{w.p.1 or in probability?}$$

In other words, is the maximal height of the paths of minimal passage time from $\underline{0}$ to $(n,0,\ldots,0)$ or H_n of smaller order than n?

The following variant of the problem may be interesting. Set for $\varepsilon > 0$

$$h_n^a(\varepsilon) = \max\{M(r):r \text{ a path from } \underline{0} \text{ to } (n,0,\ldots,0) \text{ with } T(r) \le n(\mu+\varepsilon)\}.$$

(One may replace $n(\mu+\varepsilon)$ by $a_{0,n} + n\varepsilon$ in this definition.) The question now is whether

(9.9) $$\lim_{\varepsilon \downarrow 0} \limsup_{n\to\infty} \frac{1}{n} h_n^a(\varepsilon) = 0$$

in some sense. We prove the following lemma. For simplicity we restrict ourselves to the case with $Et(e) < \infty$.

(9.10) LEMMA. Assume that $Et(e) < \infty$. Then (9.9) holds if and only if $\mu > 0$ and $\hat{B}_0$ does not contain any segment $\{\frac{1}{\mu}\} \times [-\delta,+\delta]$ with $\delta > 0$.

(9.11) REMARK. Since B_0 contains the point $(\frac{1}{\mu},0)$ and is contained in $[\frac{1}{\mu},\frac{1}{\mu}] \times [-\frac{1}{\mu},\frac{1}{\mu}]$ (cf. (6.10)) the condition in the lemma says that B_0 does not have a flat edge near its rightmost point.

Proof of Lemma 9.10 If $\mu > 0$ and B_0 contains the interval $\{\mu\} \times [-\delta,+\delta]$, then by Theorem 3.1 $\hat{B}(t)$ eventually contains the point $((1-\varepsilon)t/\mu,(1-\varepsilon)t\delta)$ so that

$$\limsup \frac{1}{n}T(\underline{0},(n,n\delta\mu)) \leq \frac{\mu}{1-\varepsilon} \quad \text{w.p.1.}$$

By symmetry

$$P\{\frac{1}{n}T((n,n\delta\mu),(2n,0)) \leq \frac{\mu}{1-\varepsilon}+\varepsilon\} \to 1.$$

Now choose ε so small that

$$\frac{2\mu}{1-\varepsilon}+\varepsilon < 2\mu+4\varepsilon.$$

Since the concatenation of any path r' from $\underline{0}$ to $(n,n\delta\mu)$ and a path r'' from $(n,n\delta\mu)$ to $(2n,0)$ gives a path from $\underline{0}$ to $(2n,0)$ we obtain

$$P\{h^a_{2n}(4\varepsilon) \geq \delta\mu n\} \geq P\{T(\underline{0},(n,n\delta\mu)) \leq \frac{\mu}{1-\varepsilon}n \text{ and }$$
$$T((n,n\delta\mu),(2n,0)) \leq (\frac{\mu}{1-\varepsilon}+\varepsilon)n\} \to 1 .$$

Thus, if B_0 contains an interval $\{\mu\} \times [-\delta,\delta]$ with $\delta > 0$ then (9.9) fails in any reasonable sense. Similar arguments show that (9.9) does not hold for $\mu = 0$.

Conversely, assume $\mu > 0$ an that for some $\delta > 0$

$$P\{\lim_{\varepsilon \downarrow 0} \limsup \frac{1}{n} h^a_n(\varepsilon) \geq \delta\} \geq \delta. \tag{9.12}$$

Even though it is not necessary for our argument, we point out that the event in the left hand side of (9.12) is independent of any finite number of edges, and hence a constant w.p.1, by Kolmogorov's zero-one law. Thus, if (9.12) holds, then we may replace the right hand side by 1.

Under the assumption (9.12) we can find with positive probability a sequence of paths r_ℓ from $\underline{0}$ to $(n_\ell,0)$, such that

(9.13) $\quad T(r_\ell) \le n_\ell(\mu + \frac{1}{\ell})$ and r_ℓ contains a point (k_ℓ, m_ℓ)

$$\text{with } \frac{m_\ell}{n_\ell} \ge \frac{\delta}{2} .$$

By virtue of (1.13) and Theorem 3.1 we may in addition assume

(9.14) $\quad \frac{1}{n_\ell}T(r_\ell) \to \mu$ and any accumulation point of $\frac{1}{n_\ell \mu}(k_\ell, m_\ell)$

belongs to B_0.

Since B_0 is contained in $[-\frac{1}{\mu}, \frac{1}{\mu}] \times \mathbb{R}$ (9.14) implies in particular that

$$\limsup \frac{k_\ell}{n_\ell} \le 1.$$

If this lim sup equals 1, then, by virtue of (9.13) and (9.14) B_0 contains a point $(\frac{1}{\mu}, y)$ with $y \ge \delta/(2\mu)$ and we are done, since B_0 is symmetric with respect to the first coordinate axis, and convex. Assume therefore that

(9.15) $$\limsup \frac{k_\ell}{n_\ell} = a < 1.$$

Denote by r'_ℓ (r''_ℓ) the piece of r from $\underline{0}$ to (k_ℓ, m_ℓ) (from (k_ℓ, m_ℓ) to $(n_\ell, 0)$). Now note that for any $\eta > 0$

(9.16) $$P\{T(r''_\ell) \le (n_\ell - k_\ell)(\mu - \eta) \text{ and } k_\ell \le n_\ell \tfrac{1}{2}(1+a)$$

$$\le \sum_{k=-\infty}^{n_\ell(1+a)/2} P\{\exists \text{ path } r'' \text{ from } (n_\ell, 0) \text{ to } H_k \text{ with}$$

$$T(r'') \le (n_\ell - k)(\mu - \eta)\}$$

$$\le \sum_{m=n_\ell(1-a)/2}^{\infty} P\{b_{0,m} \le m(\mu - \eta)\} = O(e^{-B_1(\eta) n_\ell (1-a)/2}),$$

by virtue of (5.3). Thus we may assume that for all $\eta > 0$

(9.17) $$T(r'_\ell) = T(r_\ell) - T(r''_\ell) \le (\mu+\eta)n_\ell - (\mu-\eta)(n_\ell - k_\ell)$$

$$= (\mu - \eta)k_\ell + 2\eta n_\ell \quad \text{for large } \ell.$$

Note that the estimates in (9.16) also show that we may assume $\liminf k_\ell / n_\ell \ge 0$, since

$$\limsup \frac{1}{n_\ell} T(r''_\ell) \le \limsup \frac{1}{n_\ell} T(r_\ell) = \mu .$$

Actually $\liminf k_\ell/n_\ell = 0$ also has probability zero, since $m_\ell \ge \frac{\delta}{2} n_\ell$ implies

$$P\{T(r'_\ell) \le \frac{\delta}{4} \mu n_\ell\} \le P\{b_{0,\frac{\delta}{2}n_\ell} \le \frac{\delta}{4} \mu n_\ell\} = O(\exp(-n_\ell B_1(\frac{\delta\mu}{4}))),$$

so that

$$\liminf \frac{1}{n_\ell} T(r'_\ell) \ge \frac{\delta}{4} \mu$$

and (by (9.16))

$$\limsup \frac{1}{n_\ell} T(r_\ell) \ge \frac{\delta}{4}\mu + \limsup \frac{1}{n_\ell} T(r''_\ell) \ge (\frac{\delta}{4} + 1)\mu \quad \text{a.e. on}$$

$$\{\liminf n_\ell^{-1} k_\ell = 0\} .$$

Thus, we may assume

(9.18) $$0 < \liminf \frac{1}{n_\ell} k_\ell \le a.$$

Again by (1.13) and (9.17) we must now have

(9.19) $$\lim \frac{1}{k_\ell} T(r'_\ell) = \mu .$$

Finally, by Theorem 3.1 again we may assume that B_0 contains any accumulation point of

$$\{\frac{1}{T(r'_\ell)} (k_\ell, m_\ell)\}_{\ell \ge 1}$$

since $(k_\ell, m_\ell) \in \hat{B}(T(r'_\ell))$. In view of (9.18) and (9.19) this shows that B_0 contains a point $(\frac{1}{\mu}, y)$ with

$$y \ge \limsup \frac{m_\ell}{T(r'_\ell)} = \limsup \frac{1}{\mu k_\ell} m_\ell \ge \frac{\delta}{2a\mu} > 0.$$

Thus, (9.9) implies that B_0 contains an interval $\{\frac{1}{\mu}\} \times [-\eta, \eta]$ for some $\eta > 0$. ///

(9.20) Hammersley and Welsh, [19], Sect. 8.3 point out that it would be useful to estimate $E\tilde{h}_n$, where (compare (5.22) for notation)

$\tilde{h}_n = \min\{\sum_2^d |k(i)|: k$ such that there exists a path r from $\underline{0}$

to $(n,k(2),\dots,k(d))$ with $H_0 < r < H_n$ and $T(r) = s_{0,n}\}$.

Hammersley and Welsh show that

(9.21) $$Et_{0,2n} \geq 2Et_{0,n} - (2E\tilde{h}_{n-1}+2)Et(e)$$

and hence by iteration

$$\frac{1}{n}Et_{0,n} \leq \frac{1}{2n}Et_{0,2n} + \frac{1}{n}(E\tilde{h}_{n-1}+1)Et(e)$$

$$\leq \frac{1}{4n}Et_{0,4n} + \{\frac{1}{2n}(E\tilde{h}_{2n-1}+1) + \frac{1}{n}(E\tilde{h}_{n-1}+1)\}Et(e)$$

$$\leq \dots \leq \mu + Et(e)\sum_{\ell=0}^{\infty}(n2^{\ell})^{-1}(E\tilde{h}_{n2^{\ell}-1}+1).$$

Thus a good estimate for $E\tilde{h}_n$ would lead to an estimate for

$$0 \leq Es_{0,n}-\mu \leq Et_{0,n}-\mu \; .$$

By the use of (2.21) in [31] (compare also Lemma 5.41) this might lead to good estimates for $\sigma^2(s_{0,n})$ and then by the methods of Sect. 5 to improved estimates for the speed of convergence of $n^{-1}\theta_{0,n}$ to μ.

Note that one could also derive similar estimates by means of the following quantities

$$m_n(r) = \max\{\sum_2^d |k(i)|: (n+1,k(2),\dots,k(d)) \text{ or } (n-1,k(2),\dots,k(d)) \text{ a point of } r\}$$

$$\underline{h}_{2n}^t(\varepsilon) = \min\{m_n(r): r \text{ a path from } \underline{0} \text{ to } (2n,0,\dots,0) \text{ with } H_0 < r < H_{2n} \text{ and } T(r) \leq t_{0,2n}+\varepsilon n\} \; .$$

$h_{2n}^t(\varepsilon)$ is something like the height in the middle of a path from $\underline{0}$ to $(2n,0,\dots,0)$ with passage time close to $t_{0,2n}$. An argument similar to that of [19] for (9.21) gives in the case when $t(e)$ is <u>bounded</u> by Γ w.p.1

$$Et_{0,2n} + \varepsilon_n n \geq 2Et_{0,n} - (2E\underline{h}_{2n}^t(\varepsilon_n)+2)\Gamma$$

and consequently for $\sum \varepsilon_{n2^\ell} < \infty$,

$$\frac{1}{n} E t_{0,n} \le \mu + \sum_{\ell=0}^{\infty} \frac{1}{n2^\ell}\{E\underline{h}^t_{n2^{\ell+1}}(\varepsilon_{n2^\ell})+1\}\Gamma + \frac{1}{2}\sum_{\ell=0}^{\infty} \varepsilon_{n2^\ell} .$$

(9.22) $\underline{h}^t_{2n}$ measures height near the midpoint of paths from $\underline{0}$ to $(2n,0,\dots,0)$. If we shift these paths we could consider the height in H_0 of paths from $(-n,0,\dots,0)$ to $(n,0,\dots,0)$ and study its behavior for large n. This lead H. Furstenberg to ask the author the following question: Do there exist doubly infinite geodesics through $\underline{0}$? A doubly infinite geodesic is a "doubly infinite path" $(\dots e_{-2},v_{-2}, e_{-1},v_{-1},e_0,v_0,e_1,v_1,\dots)$ such that for any $k < \ell$

$$\sum_{k+1}^{\ell} t(e_i) = T(v_k,v_\ell).$$

Of course one-sided geodesics starting at $\underline{0}$ can be found by taking a sequence of paths r_n from $\underline{0}$ to $(\underline{n},0,\dots,0)$ such that $T(r_n) - a_{0,n} \to 0$ and then to use a selection method to find $r = (\underline{0},e_1,v_1, e_2,\dots)$ such that each initial piece $(\underline{0},e_1,v_1,\dots,e_k,v_k)$ belongs to r_n for infinitely many n. It is also obvious that there exist doubly infinite geodesics when $F(0)$ is so large that there exists an infinite cluster of edges e with $t(e) = 0$.

(9.23) A <u>route for</u> $\theta_{0,n}$ is defined as a path r satisfying the proper restrictions in the definition of $\theta_{0,n}$ and with $T(r) = \theta_{0,n}$ (see Sect. 1 or (2.11), (2.12) for $a_{0,n}$, $b_{0,n}$, and (2.14), (2.15) for $t_{0,n}$ and $s_{0,n}$). Thus a route for $\theta_{0,n}$ exists if and only if the inf in the definition of $\theta_{0,n}$ is actually a minimum. It was shown in various stages (cf. [31], Sect. 4.3) that routes exist for each $\theta_{0,n}$ in dimension 2. It is easy to see from (5.9) that routes exist for each $\theta_{0,n}$ in dimension d if $F(0) < p_T(d)$. Also, for sufficiently large $F(0)$ one can imitate the proof of (2.26) to construct routes whose edges have zero passage time with the exception of the edges in a shell around the initial and final point. Still, the question remains whether routes for $\theta_{0,n}$ always exist. Most likely it will be hardest to decide this when $F(0) = p_T(d)$ (see also next problem).

(9.24) Hammersley and Welsh [19], Sect. 8.2 already raised the question how long routes are (if they exist). Thus, let us define

$$N_n^\theta = \inf\{\text{number of edges in } r : r \text{ a route for } \theta_{0,n}\}.$$

(Some modification will be needed if routes do not always exist.) One expects that $\lim n^{-1}N_n^\theta$ exists w.p.1 if $F(0) \neq p_T$. Results in this direction can be found in [31], Ch.8, [22] and [37]. Zhang and Zhang [37] proved that this limit does indeed exist for $d = 2$, $F(0) > \frac{1}{2}$. For $F(0) = p_T(d)$ the situation is much less clear. Even in dimension 2, the behavior of N_n^θ when $F(0) = \frac{1}{2}$ seems unknown. Probably $n^{-1}N_n^\theta \to \infty$ in this case.

(9.25) Which of the results of first-passage percolation generalize to situations in which one uses different distributions for $t(e)$ for edges in different directions, or if one drops the assumption of the independence of all $t(e)$. The existence of $\lim n^{-1}a_{0,n}$ for instance requires only stationarity, and Y. Derriennic (private communication) observed that the proofs of Theorem 1.7 and (1.13) go through for bounded $t(e)$ under the assumption of stationarity under all the shifts only. He raised the question what the minimal conditions are for the existence of $\lim n^{-1}b_{0,n}$ w.p.1.

Derriennic also raised the following question. Let all $t(e)$ be independent, but with distribution function F_i for those e which are parallel to the i-th coordinate axis, $1 \leq i \leq d$. Under what condition can the time constant in direction x ($\mu(x)$ of the proof of Theorem 1.7 in Sect. 3) be zero for some x and not for other x. It is easy to give such an example for $d = 3$. E.g. take F_3 concentrated on $[1,\infty)$. Then certainly $\mu((0,0,1)) \geq 1$ since it takes at least one unit of time to move a step upwards. However, if $F_1 = F_2$ with $F_1(0) = F_2(0) > \frac{1}{2}$, then if one restricts oneself to paths in $x(3) = 0$, the time constant for first-passage percolation in the plane $x(3) = 0$ is zero, by Theorem 6.1. A fortiori, in the original three dimensional problem $\mu((x(1),x(2),0)) = 0$ for any $x(1)$, $x(2)$. It appears that this phenomenon cannot occur in dimension 2, i.e., that for $d = 2$ either $\mu(x) \equiv 0$ or $\mu(x) > 0$ for all $x \neq 0$ (except in trivial degenerate cases).

REFERENCES

[1] R. Ahlberg and S. Janson (1984). Upper bounds for the connectivity constant, preprint, Uppsala Univ.

[2] M. Aizenman, J.T. Chayes, L. Chayes, J. Fröhlich and L. Russo (1983). On a sharp transition from area law to perimeter law in a system of random surfaces, Comm. Math. Phys. 92, 19-69.

[3] J. van den Berg and H. Kesten (1985). Inequalities with applications to percolation and reliability, to appear in J. Appl. Prob.

[4] L. Breiman (1968). Probability, Addison-Wesley Publ. Co.

[5] M. Campanino and L. Russo (1985). An upper bound on the critical percolation probability for the three-dimensional cubic lattice, to appear in Ann. Probab.

[6] J.T. Cox (1980). The time constant of first-passage percolation on the square lattice, Adv. Appl. Prob. 12, 864-879.

[7] J.T. Cox and R. Durrett (1981). Some limit theorems for percolation processes with necessary and sufficient conditions, Ann. Probab. 9, 583-603.

[8] J.T. Cox and R. Durrett (1983). Oriented percolation in dimension $d \geq 4$: bounds and asymptotic formulas, Math. Proc. Comb. Phil. Soc. 93, 151-162.

[9] J.T. Cox and H. Kesten (1981). On the continuity of the time constant of first-passage percolation, J. Appl. Prob. 18, 809-819.

[10] C. Dellacherie and P.A. Meyer (1975). Probabilités et potentiel, Ch. I-IV, Hermann.

[11] M. Donsker and S.R.S. Varadhan (1975). Asymptotics of the Wiener sausage, Comm. Pure Appl. Math. 28, 525-565.

[12] J. Dugundji (1966). Topology, Allyn and Bacon, Inc.

[13] R. Durrett (1984). Oriented percolation in two dimensions, Ann. Probab. 12, 999-1040.

[14] R. Durrett and T.M. Liggett (1981). The shape of the limit set in Richardson's growth model, Ann. Probab. 9, 186-193.

[15] M. Eden (1961). A two dimensional growth process, Proc. Fourth Berkeley Symp. Math. Stat. Prob., Vol IV, 223-239, Univ. of Cal. Press.

[16] D.R. Fulkerson (1975). Flow networks and combinational operations research pp. 139-171 in Studies in graph theory, D.R. Fulkerson ed., Math. Assoc. of America.

[17] G. Grimmett and H. Kesten (1984). First-passage percolation, network flows and electrical resistances, Z. Wahrsch. verw. Geb. 66, 335-366.

[18] J.M. Hammersley (1974). Postulates for subadditive processes, Ann. Probab. 2, 652-680.

[19] J.M. Hammersley and D.J.A. Welsh (1965). First-passage perco-colation, subadditive processes, stochastic networks and gen-ralized renewal theory, pp. 61-110 in Bernoulli, Bayes, Laplace Anniversary Volume, J. Neyman and L.M. LeCam eds. Springer-Verlag.

[20] T.E. Harris (1960). A lower bound for the critical probability in a certain percolation process, Proc. Cambr. Phil. Soc. 56, 13-20.

[21] S. Janson (1981). An upper bound for the velocity of first-passage percolation, J. Appl. Prob. 18, 256-262.

[22] H. Kesten (1980). On the time constant and path length of first-passage percolation, Adv. Appl. Prob. 12, 848-863.

[23] H. Kesten (1982). Percolation theory for mathematicians, Birkhäuser-Boston.

[24] H. Kesten (1984). Surfaces with minimal random weights and maximal flows: a higher dimensional version of first-passage percolation, preprint.

[25] J.F.C Kingman (1973). Subadditive ergodic theory, Ann. Probab. 1, 883-909.

[26] J.F.C. Kingman (1975). Subadditive processes, pp. 168-223 in Lecture Notes in Mathematics, Vol. 539, Springer-Verlag.

[27] T.M. Liggett (1985). An improved subadditive ergodic theorem, Ann. Probab. 13.

[28] D. Richardson (1973). Random growth in a tesselation, Proc. Cambr. Phil. Soc. 74, 515-528.

[29] H.P. Rosenthal (1970). On the subspaces of $L^p (p > 2)$ spanned by sequences of independent random variables, Isr. J. Math. 8, 273-303.

[30] R.T. Smythe (1980). Percolation models in two and three dimensions, pp. 504-511 in Biological Growth and Spread, Lecture Notes in Biomath., Vol 38, Springer-Verlag.

[31] R.T. Smythe and J.C. Wierman (1978). First-passage percolation on the square lattice, Lecture Notes in Mathematics, Vol. 671, Springer-Verlag.

[32] E.H. Spanier (1966). Algebraic topology, McGraw-Hill.

[33] H. Whitney (1932). Non-separable planar graphs, Trans. Amer. Math. Soc. 34, 339-362.

[34] H. Whitney (1933). Planar graphs, Fund. Math. 21, 73-84.

[35] J.C. Wierman (1980). Weak moment conditions for time coordinates in first-passage percolation models, J. Appl. Prob. 17, 968-978.

[36] J.C. Wierman and W. Reh (1978). On conjectures in first-passage percolation theory, Ann. Probab. 6, 388-397.

[37] Y. Zhang and Y.C. Zhang (1984). A limit theorem for N_{0n}/n in first-passage percolation, Ann. Probab. 12, 1068-1076.

INDEX

AN INTRODUCTION TO STOCHASTIC PARTIAL DIFFERENTIAL EQUATIONS

John B. WALSH

The general problem is this. Suppose one is given a physical system governed by a partial differential equation. Suppose that the system is then perturbed randomly, perhaps by some sort of a white noise. How does it evolve in time? Think for example of a guitar carelessly left outdoors. If u(x,t) is the position of one of the strings at the point x and time t, then in calm air u(x,t) would satisfy the wave equation $u_{tt} = u_{xx}$. However, if a sandstorm should blow up, the string would be bombarded by a succession of sand grains. Let $\dot{W}_{xt}$ represent the intensity of the bombardment at the point x and time t. The number of grains hitting the string at a given point and time will be largely independent of the number hitting at another point and time, so that, after subtracting a mean intensity, $\dot{W}$ may be approximated by a white noise, and the final equation is

$$u_{tt}(x,t) = u_{xx}(x,t) + \dot{W}(x,t),$$

where $\dot{W}$ is a white noise in both time and space, or, in other words, a two-parameter white noise.

One peculiarity of this equation - not surprising in view of the behavior of ordinary stochastic differential equations - is that none of the partial derivatives in it exist. However, one may rewrite it as an integral equation, and then show that in this form there is a solution which is a continuous, though non-differentiable, function.

In higher dimensions - with a drumhead, say, rather than a string - even this fails: the solution turns out to be a distribution, not a function. This is one of the technical barriers in the subject: one must deal with distribution-valued solutions, and this has generated a number of approaches, most involving a fairly extensive use of functional analysis.

Our aim is to study a certain number of such stochastic partial differential equations, to see how they arise, to see how their solutions behave, and to examine some techniques of solution. We shall concentrate more on parabolic equations than on hyperbolic or elliptic, and on equations in which the perturbation comes from something akin to white noise.

In particular, one class we shall study in detail arises from systems of branching diffusions. These lead to linear parabolic equations whose solutions are generalized Ornstein-Uhlenbeck processes, and include those studied by Ito, Holley and Stoock, Dawson, and others. Another related class of equations comes from certain neurophysiological models.

Our point of view is more real-variable oriented than the usual theory, and, we hope, slightly more intuitive. We regard white noise $\dot{W}$ as a measure on Euclidean space, $W(dx, dt)$, and construct stochastic integrals of the form $\int f(x,t)dW$ directly, following Ito's original construction. This is a two-parameter integral, but it is a particularly simple one, known in two-parameter theory as a "weakly-adapted integral". We generalize it to include integrals with respect to martingale measures, and solve the equations in terms of these integrals.

We will need a certain amount of machinery: nuclear spaces, some elementary Sobolev space theory, and weak convergence of stochastic processes with values in Schwartz space. We develop this as we need it.

For instance, we treat SPDE's in one space dimension in Chapter 3, as soon as we have developed the integral, but solutions in higher dimensions are generally Schwartz distributions, so we develop some elementary distribution theory in Chapter 4 before treating higher dimensional equations in Chapter 5. In the same way, we treat weak convergence of $\underline{\underline{S}}'$-valued processes in Chapter 6 before treating the limits of infinite particle systems and the Brownian density process in Chapter 8.

After comparing the small part of the subject we can cover with the much larger mass we can't, we had a momentary desire to re-title our notes: "An Introduction to an Introduction to Stochastic Partial Differential Equations";

which means that the introduction to the notes, which you are now reading, would be the introduction to "An Introduction ... ", but no. It is not good to begin with an infinite regression. Let's just keep in mind that this is an introduction, not a survey. While we will forego much of the recent work on the subject, what we do cover is mathematically interesting and, who knows? Perhaps even physically useful.

CHAPTER ONE

WHITE NOISE AND THE BROWNIAN SHEET

Let $(E,\underline{\underline{E}},\nu)$ be a σ-finite measure space. A white noise based on ν is a random set function W on the sets $A \in \underline{\underline{E}}$ of finite ν-measure such that

(i) $W(A)$ is a $N(0,\nu(A))$ random variable;

(ii) if $A \cap B = \phi$, then $W(A)$ and $W(B)$ are independent and

$W(A \cap B) = W(A) + W(B)$.

In most cases, E will be a Euclidean space and ν will be Lebesgue measure. To see that such a process exists, think of it as a Gaussian process indexed by the sets of $\underline{\underline{E}}$: $\{W(A),\ A \in \underline{\underline{E}},\ \nu(A) < \infty\}$. From (i) and (ii) this must be a mean-zero Gaussian process with covariance function C given by

$$C(A,B) = E\{W(A)\ W(B)\} = \nu(A \cap B).$$

By a general theorem on Gaussian processes, if C is positive definite, there exists a Gaussian process with mean zero and covariance function C. Now let $A_1,\ldots,A_n$ be in $\underline{\underline{E}}$ and let $a_1,\ldots,a_n$ be real numbers.

$$\sum_{i,j} a_i a_j\ C(A_i,A_j) = \sum_{i,j} a_i a_j \int I_{A_i}(x)\ I_{A_j}(x)dx$$

$$= \int (\sum_i a_i\ I_{A_i}(x))^2 dx \geq 0.$$

Thus C is a positive definite, so that there exists a probability space $(\Omega,\underline{\underline{F}},P)$ and a mean zero Gaussian process $\{W(A)\}$ on $(\Omega,\underline{\underline{F}},P)$ such that W satisfies (i) and (ii) above.

There are other ways of defining white noise. In case $E = R$ and ν = Lebesgue measure, it is often described informally as the "derivative of Brownian motion". Such a description is possible in higher dimensions too, but it involves the Brownian sheet rather than Brownian motion.

Let us specialize to the case $E = \mathbf{R}^n_+ = \{(t_1,\ldots,t_n): t_i \geq 0,\ i=1,\ldots,n\}$ and ν = Lebesgue measure. If $t = (t_1,\ldots,t_n) \in \mathbf{R}^n_+$, let $(0,t] = (0,t_1] \times \cdots \times (0,t_n]$. The Brownian sheet on $\mathbf{R}^n_+$ is the process $\{W_t,\ t \in \mathbf{R}^n_+\}$ defined by $W_t = W\{(0,t]\}$. This is a mean-zero Gaussian process. If $s = (s_1,\ldots,s_n)$ and $t = (t_1,\ldots,t_n)$, its covariance function is

(1.1) $$E\{W_s W_t\} = (s_1 \wedge t_1)\cdots(s_n \wedge t_n)$$

If we regard W(A) as a measure, W_t is its distribution function. Notice that we can recover the white noise in $\mathbf{R}_+^n$ from W_t, for if $\mathbf{R}$ is a rectangle, W(R) is given by the usual formula (if n = 2 and $0 \le u \le s$, $0 \le v \le t$, $W((u,v),(s,t)] = W_{st} - W_{sv} - W_{ut} - W_{uv}$). If A is a finite union of rectangles, W(A) can be computed by additivity, and a general Borel set A of finite measure can be approximated by finite unions of rectangles A_n in such a way that

$$E\{(W(A) - W(A_n))^2\} = \nu(A - A_n) + \nu(A_n - A) \to 0 .$$

Interestingly, the Brownian sheet was first introduced by a statistician, J. Kitagawa, in 1951 in order to do analysis of variance in continuous time. To get an idea of what this process looks like, let's consider its behavior along some curves in $\mathbf{R}_+^2$, in the case n = 2, ν = Lebesgue measure.

1). W vanishes on the axes. If $s = s_o > 0$ is fixed, $\{W_{s_o t}, t \ge 0\}$ is a Brownian motion, for it is a mean-zero Gaussian process with covariance function $C(t,t') = s_o(t \wedge t')$.

2). Along the hyperbola st = 1, let

$$X_t = W_{e^t, e^{-t}}.$$

Then $\{X_t, -\infty<t<\infty\}$ is an Ornstein-Uhlenbeck process, i.e. a strictly stationary Gaussian process with mean zero, variance 1, and covariance function

$$C(s,t) = E\{W_{e^s,e^{-s}} W_{e^t,e^{-t}}\} = e^{-|s-t|}.$$

3). Along the diagonal the process $M_t = W_{tt}$ is a martingale, and even a process of independent increments, although it is not a Brownian motion, for these increments are not stationary. The same is true if we consider W along <u>increasing</u> paths in $\mathbf{R}_+^2$.

4). Just as in one parameter, there are scaling, inversion, and translation transformations which take one Brownian sheet into another.

<u>Scaling:</u> $$A_{st} = \frac{1}{ab} W_{a^2 s, b^2 t}.$$

<u>Inversion:</u> $$C_{st} = st\, W_{\frac{1}{s}\frac{1}{t}}\ ; \quad D_{st} = s\, W_{\frac{1}{s} t}.$$

<u>Translation by (s_o,t_o):</u> $$E_{st} = W_{s_o+s,t_o+t} - W_{s_o+s,t_o} - W_{s_o,t_o+t} + W_{s_o t_o}.$$

Then A, C, D, and E are Brownian sheets, and moreover, E is independent of $F^*_{s_o t_o} = \sigma\{W_{uv} : u \leq s_o \text{ or } v \leq t_o\}$.

The easiest way to see this in the case of A, C and D is to notice that they are all mean zero Gaussian processes with the right covariance function. In the case of E, we can go back to white noise, and notice that $E_{st} = W((s_o,s] \times (t_o,t])$. The result then follows immediately from the properties (i) and (ii).

5). Another interesting transformation is this: let $U_{st} = e^{-s-t} W_{e^{2s},e^{2t}}$. Then $\{U_{st}, -\infty<s, t<\infty\}$ is an <u>Ornstein-Uhlenbeck sheet</u>. This is a stationary Gaussian process on R^2 with covariance function $E\{U_{st}U_{uv}\} = e^{-|u-s|-|t-v|}$. If we look at U along any line, we get a one-parameter Ornstein-Uhlenbeck process. That is, if $V_s = U_{s,a+bs}$, then $\{V_s, -\infty<s<\infty\}$ is an Ornstein-Uhlenbeck process.

SAMPLE FUNCTION PROPERTIES

The Brownian sheet has continuous paths, but we would not expect them to be differentiable - indeed, nowhere-differentiable processes such as Brownian motion can be embedded in the sheet, as we have just seen. We will see just how continuous they are. This will give us an excuse to derive several beautiful and useful inequalities, beginning with an elegant result of Garsia, Rodemich, and Rumsey.

Let $\Psi(x)$ and $p(x)$ be positive continuous functions on $(-\infty,\infty)$ such that both Ψ and p are symmetric about 0, p(x) is increasing for $x > 0$ and $p(0) = 0$, and Ψ is convex with $\lim_{x\to\infty} \Psi(x) = \infty$. If R is a cube in R^n, let e(R) be the length of its edge and $|R|$ its volume. Let R_1 be the unit cube.

<u>THEOREM 1.1</u>. If f is a measurable function on R_1 such that

$$\int_{R_1}\int_{R_1} \Psi\left(\frac{f(y)-f(x)}{p(|y-x|/\sqrt{n})}\right)dx\,dy = B < \infty, \tag{1.2}$$

then there is a set K of measure zero such that if $x,y \in R_1 - K$

$$|f(y) - f(x)| \leq 8 \int_0^{|y-x|} \Psi^{-1}\left(\frac{B}{u^{2n}}\right)dp(u). \tag{1.3}$$

If f is continuous, (1.3) holds for all x and y.

PROOF. If $Q \subset R_1$ is a rectangle and $x,y \in Q$, then $|y-x| \leq \sqrt{n}\, e(Q)$. Ψ is increasing, so that (3.2) implies

(1.4) $$\int_Q\int_Q \Psi\left(\frac{f(y)-f(x)}{p(e(Q))}\right) dxdy \leq B.$$

Let $Q_0 \supset Q_1 \supset \cdots$ be a sequence of subcubes of R_1 such that

$$p(e(Q_j)) = \frac{1}{2}\, p(e(Q_{j-1})).$$

For any cube Q, let $f_Q = \frac{1}{|Q|}\int_Q f(x)dx$.

Since Ψ is convex

$$\Psi\left(\frac{f_{Q_j}-f_{Q_{j-1}}}{p(e(Q_{j-1}))}\right) \leq \frac{1}{|Q_{j-1}|}\int_{Q_{j-1}} \Psi\left(\frac{f_{Q_j}-f(x)}{p(e(Q_{j-1}))}\right) dx$$

$$\leq \frac{1}{|Q_{j-1}||Q_j|}\int_{Q_{j-1}}\int_{Q_j} \Psi\left(\frac{f(y)-f(x)}{p(e(Q_{j-1}))}\right) dx\, dy$$

By (1.4) this is

$$\leq \frac{B}{|Q_{j-1}||Q_j|}\,.$$

If Ψ^{-1} is the (positive) inverse of Ψ

(1.5) $$|f_{Q_j}-f_{Q_{j-1}}| \leq p(e(Q_{j-1}))\, \Psi^{-1}\left(\frac{B}{|Q_j||Q_{j-1}|}\right).$$

Now $p(e(Q_{j-1})) = 4|(p(e(Q_{j+1})) - p(e(Q_j))|$, so this is

$$= 4\Psi^{-1}\left(\frac{B}{|Q_j||Q_{j-1}|}\right)|p(e(Q_{j+1})) - p(e(Q_j))|.$$

Ψ^{-1} increases so if $e(Q_{j+1}) \leq u \leq e(Q_j)$, then $|Q_{j-1}||Q_j| \geq u^{2n}$ and

$$\Psi^{-1}\left(\frac{B}{|Q_{j-1}||Q_j|}\right) \leq \Psi^{-1}\left(\frac{B}{u^{2n}}\right).$$

Set $v_j = e(Q_j)$. Then from (1.5)

(1.6) $$|f_{Q_j}-f_{Q_{j-1}}| \leq 4\int_{v_j}^{v_{j-1}} \Psi^{-1}\left(\frac{B}{u^{2n}}\right) dp(u).$$

Sum this over j:

(1.7) $$\limsup |f_{Q_j}-f_{Q_0}| \leq 4\int_0^{v_0} \Psi^{-1}\left(\frac{B}{u^{2n}}\right) dp(t).$$

By the Vitali theorem, if x is not in some null set K, then $f_{Q_j} \to f(x)$ for any sequence Q_j of cubes decreasing to $\{x\}$. If x and y are in $R_1 - K$, and if Q_0 is the smallest cube containing both, then, since $v_0 \leq |y-x|$,

$$|f(x) - f_{Q_0}| \leq 4\int_0^{|y-x|} \Psi^{-1}\left(\frac{B}{u^{2n}}\right) dp(u).$$

he same inequality holds for y, proving the theorem.

This is purely a real-variable result - f is deterministic - but it can often be used to estimate the modulus of continuity of a stochastic process. One usually computes $E\{B\}$ to show $B < \infty$ a.s. Everything hinges on the choice of Ψ and p. f we let Ψ be a power of x, we get a venerable result of Kolmogorov.

COROLLARY 1.2 (Kolmogorov). Let $\{X_t, t\epsilon R_1\}$ be a real-valued stochastic process. uppose there are constants $k > 1$, $K > 0$ and $\varepsilon > 0$ such that for all $s,t \in R_1$

$$E\{|X_t - X_s|^k\} \leq K|t-s|^{n+\varepsilon} .$$

hen

(i) X has a continuous version;

(ii) there exist constants C and γ, depending only on n, k, and ε, and a random variable Y such that with probability one, for all $s, t \in R_1$

$$|X_t - X_s| \leq Y\, |t-s|^{\varepsilon/k}\left(\log \frac{\gamma}{|t-s|}\right)^{2/k}$$

and

$$E\{Y^k\} \leq CK;$$

(iii) if $E\{|X_t|^k\} < \infty$ for some t, then

$$E\{\sup_{t\epsilon R_1} |X_t|^k\} < \infty.$$

PROOF. We will apply Theorem 1.1 to the paths of X. We will use s and t instead of x and y, and the function f(x) is replaced by the sample path $X_t(\omega)$ for a fixed ω. Choose $\Psi(x) = |x|^k$ and $p(x) = |x|^{\frac{2n+\varepsilon}{k}}\left(\log \frac{\gamma}{|x|}\right)^{2/k}$. If $\gamma = \sqrt{n}\, e^{\frac{k}{n}}$, p will be increasing on $(0, \sqrt{n})$. Notice that the quantity B in (1.2) is now random, for it depends on ω. Let us take its expectation. By Fubini's Theorem

$$E\{B\} = n^{n+\frac{\varepsilon}{2}} \int_{R_1}\int_{R_1} \frac{E\{|X_t - X_s|^k\}}{|t-s|^{2n+\varepsilon}\log^2\left(\frac{\gamma}{|t-s|}\right)}\, ds\, dt$$

$$\leq n^{n+\frac{\varepsilon}{2}} K \int_{R_1}\int_{R_1} \frac{ds\, dt}{|t-s|^n \log^2(\gamma|t-s|^{-1})}$$

If t is fixed, the integral over R_1 with respect to s is dominated by the integral over the ball of radius $\sqrt{n}$ centered at t, since the ball contains R_1, whose diameter is $\sqrt{n}$. Let σ_n be the area of the unit sphere in R^n. Then integrate in polar coordinates:

$$\leq n^{n+\frac{\varepsilon}{2}} \sigma_n K \int_0^{\sqrt{n}} \frac{r^{n-1}dr}{r^n(\log \frac{\gamma}{2})^2}$$

$$= n^{n+1+\frac{\varepsilon}{2}} \frac{\sigma_n}{k} K.$$

If we integrate by parts twice in (1.3), we get

$$|X_t - X_s| \leq 8B^{1/k}[\,|t-s|^{\frac{\varepsilon}{k}}(\log \gamma|t-s|^{-1})^{\frac{2}{b}}(1 + \frac{2}{\varepsilon} n)]$$

$$+ \frac{4n}{k\varepsilon} \int_0^{|t-s|} \log^{\frac{2}{k}-1}(\frac{\gamma}{u})u^{\frac{\varepsilon}{k}-1} du.$$

The integral is dominated by $|t-s|^{\varepsilon/k}(\log \gamma|t-s|^{-1})^{2/k}$ for small enough values of $|t-s|$ - and for all $|t-s| \leq \sqrt{n}$ if $k \geq n$ - so that for a suitable constant A, we have

$$\leq 8AB^{1/k} \; |t-s|^{\varepsilon/k}(\log \gamma|t-s|^{-1})^{2/k}.$$

Then we take $Y = 8AB^{1/k}$, proving (i) and (ii).

To see (iii), just note that if s, t $\in R_1$, then $|t-s| \leq \sqrt{n}$ so that

$$\sup_t |X_t| \leq |X_{t_0}| + Y\, n^{\varepsilon/2k}(\log \frac{\gamma}{\sqrt{n}})^{2/k}.$$

Since X_{t_0} and Y are in L^k, so is $\sup_t |X_t|$.

Q.E.D.

The great flexibility in the choice of Ψ and p is useful, but it has its disadvantages: one always suspects he could have gotten a better result had he chosen them more cleverly. For example, if we take p(x) to be $|x|^{\frac{2n+\varepsilon}{k}} (\log\frac{\gamma}{|x|})^{1/k} (\log \log \frac{\gamma}{|x|})^{2/k}$, we can improve the modulus of continuity of X to $Y|t-s|^{\varepsilon/k}(\log \gamma|t-s|^{-1})^{1/k}(\log \log \gamma|t-s|^{-1})^{2/k}$, and so on.

If we apply Theorem 1.1 to Gaussian processes, we get the following result.

<u>COROLLARY 1.3.</u> Let $\{X_t, t\in R_1\}$ be a mean zero Gaussian process, and set

$$p(u) = \max_{|s-t|\leq|u|\sqrt{n}} E\{|X_t-X_s|^2\}^{1/2}.$$

If $\int_0^1 (\log \frac{1}{u})^{1/2} dp(u) < \infty$, then X has a continuous version whose modulus of continuity $\Delta(\delta)$ satisfies

$$\Delta(\delta) \le C \int_0^\delta (\log \frac{1}{u})^{1/2} dp(u) + Yp(\delta) \tag{1.8}$$

where C is a universal constant and Y is a random variable such that $E\{\exp(Y^2/256)\} \le \sqrt{2}$.

PROOF. Let $\Psi(x) = e^{x^2/4}$. Note that $U_{st} \overset{\text{def}}{=} \dfrac{X_t - X_s}{p(|t-s|/\sqrt{n})}$ is Gaussian with mean zero and variance $\sigma^2(s,t) \le 1$. We can use the Gaussian density to calculate directly that

$$E\{B\} = \int_{R_1}\int_{R_1} E\{\exp(\tfrac{1}{4} U_{st})\} ds\, dt \le \sqrt{2}.$$

Thus $B < \infty$ a.s. Now $\Psi^{-1}(u) = \sqrt{4 \log \frac{1}{u}}$ so Theorem 1.1 implies that if $|s-t| \le \delta$, and if s,t are not in some exceptional null-set,

$$|X_t - X_s| \le 16 |\log B|^{1/2} p(\delta) + 16\sqrt{2n} \int_0^\delta |\log \frac{1}{u}|^{1/2} dp(u). \tag{1.9}$$

It now follows easily that X has a continuous version and that $\Delta(\delta)$ is bounded by the right hand side of (1.9), proving (1.8), with $C = 16\sqrt{2n}$ and $Y = 16|\log B|^{1/2}$. Q.E.D.

This theorem usually gives the right order of magnitude for $\Delta(\delta)$, but it does not always give the best constants.

To apply this to the Brownian sheet, note that if $s = (s_1,\dots,s_n)$, $t = (t_1,\dots,t_n)$, then $E\{(W_t - W_s)^2\} \le \sum_{i=1}^n |t_i - s_i| \le \sqrt{n}\,|t-s|$. Thus $p(u) = \sqrt{nu}$. If $\rho(\delta) = 16\sqrt{2} \int_0^\delta \left|\frac{\log \frac{1}{u}}{2u}\right|^{1/2} du$ then Corollary 1.3 gives

PROPOSITION 1.4. W has a continuous version with modulus of continuity

$$\Delta(\delta) \le n\rho(\delta) + Y\sqrt{\delta} \tag{1.10}$$

where Y is a random variable with $E\{e^{Y^2/16}\} < \infty$. Moreover, with probability one, for all $t \in R_1$ simultaneously:

(1.11) $$\limsup_{|h|\to 0} \frac{W_{t+h} - W_h}{\sqrt{2|h| \log 1/|h|}} \leq 16\sqrt{2n}$$

Here (1.11) follows from (1.10) on noticing that $\frac{\rho(|h|)}{\sqrt{2h \log 1/|h|}} \to 16\sqrt{2}$ as $h \to 0$. The constants are not best possible. Orey and Pruitt have shown that the right-hand side of (1.11) is $\sqrt{n}$.

This gives the modulus of continuity of W_t. There is also a law of the iterated logarithm.

<u>THEOREM 1.5.</u> (i) $$\limsup_{s,t\to\infty} \frac{W_{st}}{\sqrt{4st \log \log st}} = 1 \quad \text{a.s.}$$

(ii) $$\limsup_{s,t\to 0} \frac{W_{st}}{\sqrt{4st \log \log \frac{1}{st}}} = 1 \quad \text{a.s..}$$

We will not prove this except to remark that (ii) is a direct consequence of (i) and the fact that $st\, W_{\frac{1}{s}\frac{1}{t}}$ is a Brownian sheet.

SOME REMARKS ON THE MARKOV PROPERTY

In order to keep our notation simple, let us consider only the case $n = 2$, so that the Brownian sheet becomes a two-parameter process W_{st}. We first would like to examine the analogue of the strong Markov property of Brownian motion: that Brownian motion restarts after each stopping time. We don't have stopping times in our set-up, but we can define stopping points. Let $(\Omega, \underline{\underline{F}}, P)$ be a probability space. Recall that $\underline{\underline{F}}_t^* = \sigma\{W_s : s_i \leq t_i$ for at least one $i=1,2.\}$ A random variable $T = (T_1, T_2)$ with values in R_+^2 is a <u>weak stopping point</u> if the set

$$\{T_1 < t_1, T_2 < t_2\} \in \underline{\underline{F}}_t^* .$$

The main example we have in mind is this: let $\underline{\underline{F}}_{t_1}^1 = \sigma\{W_{s_1 s_2} : s_1 \leq t_1\}$. If T_2 is a stopping time relative to the filtration $(\underline{\underline{F}}_{t_2}^2)$ and if $S_1 \geq 0$ is measurable relative to $\underline{\underline{F}}_{T_2}^2$, then (S_1, T_2) is a weak stopping point.

For a weak stopping point T, set

$$\underline{\underline{F}}_T^* = \{A \in \underline{\underline{F}} : A \cap \{T_1 < t_1, T_2 < t_2\} \in \underline{\underline{F}}_t^*\}.$$

This is clearly a σ-field . Set

$$W_t^T = W((T,T+t]), \quad t \in \mathbb{R}_+^2 ,$$

where the mass of the (random) rectangle (T,T+t] is computed from W_s by the usual formula.

THEOREM 1.6. Let T be a finite weak stopping point. Then the process $\{W_t^T, t \in \mathbb{R}_+^n\}$ is a Brownian sheet, independent of $\underline{\underline{F}}_T^*$.

PROOF. We approximate T from above as follows:

Write $T = (T^1,\dots,T^n)$ and define $T_m = (T_m^1,\dots,T_m^n)$ by $T_m^i = j\,2^{-m}$ if $(j-1)2^{-m} \leq T^i < j\,2^{-m}$. Let $\{r_i\}$ be any enumeration of the lattice points $(j_1 2^{-m},\dots,j_n 2^{-m})$, and note that $\{T_m = r_i\} \in \underline{\underline{F}}_{r_i}^*$ for all i. Now for each t, $W(T,T+t] = \lim W(T_m,T_m+t]$, by continuity of W. For any set $A \in \underline{\underline{F}}_T^* \subset \underline{\underline{F}}_{T_m}^*$, and any Borel set B

$$P\{W(T_m,T_m+t] \in B;\ A\}$$
$$= \sum_i P\left\{W(r_i,r_i+t] \in B;\ A \cap \{T_m = r_i\}\right\}.$$

But $A \cap \{T_m = r_i\} \in \underline{\underline{F}}_{r_i}^*$ so, by the independence property of white noise this is

$$= \sum_i P\{W(r_i,r_i+t] \in B\}\, P\left\{A \cap \{T_m = r_i\}\right\}$$
$$= P\{W_t \in B\}P\{A\}.$$

Thus, for each m, $\{W(T_m,T_m+t]\}$ is a Brownian sheet, independent of $\underline{\underline{F}}_T^*$. The same is therefore true of $\{W(T,T+t]\}$ in the limit. Q.E.D.

Notice that this is merely a random version of the translation property given at the beginning of this chapter.

A second, quite different type of Markov property is this. For any set $D \subset \mathbb{R}_+^2$ let $\underline{\underline{G}}_D = \sigma\{W_t,\ t\in D\}$, and let $\underline{\underline{G}}_D^* = \bigcap_{\varepsilon>0} \underline{\underline{G}}_{D^\varepsilon}$ where D^ε is an open ε-neighborhood of D. We say W satisfies Lévy's Markov property for D if $\underline{\underline{G}}_D$ and $\underline{\underline{G}}_{D^c}$ are conditionally independent given $\underline{\underline{G}}_{\partial D}^*$, where ∂D is the boundary of D. We say W satisfies Lévy's sharp Markov property if $\underline{\underline{G}}_D$ and $\underline{\underline{G}}_{D^c}$ are conditionally independent given $\underline{\underline{G}}_{\partial D}$.

It is rather easy to show W satisfies Lévy's sharp Markov property relative to a rectangle; this follows from the independence property of white noise. With slightly more work, one can show this also holds for finite unions of rectangles. It is more surprising to learn that it does not hold for all sets D. Indeed, consider the following example.

<u>EXAMPLE</u>. Let D be the triangle with corners at (0,0), (1,0) and (0,1). Since W_t vanishes on the axes, $\underline{\underline{G}}_{\partial D} = \{W_{s,1-s},\ 0 \le s \le 1\}$. Let us notice that W(D) is measurable with respect to both $\underline{\underline{G}}_D$ and $\underline{\underline{G}}_{D^c}$.

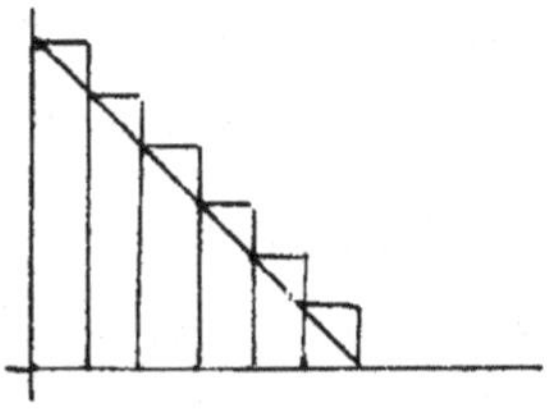

Call the above union of rectangles D_n. The mass of each rectangle is given in terms of the W_{t_j}, where the t_j are the corners of the rectangles, and hence is $\underline{\underline{G}}_{D^c}$-measurable. Since $W(D_n) \to W(D)$, $W(D) \in \underline{\underline{G}}_{D^c}$. A similar argument shows $W(D) \in \underline{\underline{G}}_D$. (Moreover if V_ε is an ε-neighborhood of D, $W(D_n) \in \underline{\underline{G}}_{V^\varepsilon}$ for large n, hence $W(D) \in \underline{\underline{G}}^*_{\partial D}$ too). On the other hand, we claim W(D) is <u>not</u> measurable with respect to the sharp boundary field, and Lévy's sharp Markov property does not hold. We need only show $\hat{D} \stackrel{\text{def}}{=} E\{W(D) | \underline{\underline{G}}_{\partial D}\}$ is not equal to W(D).

Since all the random variables are Gaussian, $\hat{D}$ will be a linear combination of the $W_{s,1-s}$, determined by

$$E\{(W(D)-\hat{D})\, W_{s,1-s}\} = 0, \quad \text{all } 0 \le s \le 1.$$

Let

$$\hat{D} = 2 \int_0^1 W_{u,1-u}\, du.$$

Then
$$E\{(W(D)-\hat{D})W_{s,1-s}\} = s(1-s) - 2\int_0^s (1-s)u\, du - 2\int_0^1 s(1-u)\, du$$

$$= 0.$$

But $\hat{D} \neq W(D)$. Indeed, $E\{\hat{D}\, W(D)\} = 2\int_0^1 u(1-u)du = \frac{1}{3}$, while $E\{W(D)^2\} = |D| = \frac{1}{2}$.

Thus W does not satisfy Lévy's sharp Markov property relative to D. It is ot hard to see that it satisfies Lévy's Markov property, i.e. that the germ field $^{*}_{\partial D}$ is a splitting field. In fact this is always the case: W_t satisfies Lévy's arkov property relative to all Borel sets D, but we will not prove it here.

THE PROPAGATION OF SINGULARITIES

The Brownian sheet is far from rotation invariant, even far away from the xes. Any unusual event has a tendency to propagate parallel to the axes. Let us ook at an example of this.

We need two facts about ordinary Brownian motion. Let $\{B_t, t \geq 0\}$ be a tandard Brownian motion.

1.) For any fixed t, $\limsup\limits_{h \downarrow 0} \dfrac{B_{t+h} - B_t}{\sqrt{2h \log \log 1/h}} = 1$ a.s.

2.) For a.e. ω, there exist uncountably many t for which

$$\limsup_{h \downarrow 0} \frac{B_{t+h}(\omega) - B_t(\omega)}{\sqrt{2h \log \log 1/h}} = \infty .$$

he first fact is well-known, and the second a consequence of the fact that the exact odulus of continuity of Brownian motion is $\sqrt{2h \log 1/h}$, not $\sqrt{2h \log \log 1/h}$.

Indeed, Lévy showed that if $d(h) = (2h \log 1/h)^{1/2}$, then for any $\varepsilon > 0$ and $< b$, there exist $a < s < t < b$ for which $|B_t - B_s| > (1-\varepsilon)\, d(t-s)$. Thus we can hoose $s_1 < t_1$ in (a,b) such that $|B_{t_1} - B_{s_1}| > \frac{1}{2} d(t_1 - s_1)$. Having chosen $s_1, \ldots, s_n$ nd $t_1, \ldots, t_n$, choose s_n' such that $s_n' \in (s_n, t_n)$, $s_n' \leq s_n + 2^{-n}$ and such that $|B_{t_n} - B_s| > \frac{n}{n+1} d(t_n - s)$ for all $s \in (s_n, s_n')$, which we can do by continuity. Next, hoose $s_{n+1} < t_{n+1}$ in (s_n, s_n') for which $|B_{t_{n+1}} - B_{s_{n+1}}| > \frac{n+1}{n+2} d(t_{n+1} - s_{n+1})$. Now let $_0 \in \bigcap\limits_n [s_n, t_n]$. If $h_n = t_n - s_0$, $|B_{s_0 + h_n} - B_{s_0}| > \frac{n}{n+1} d(h_n)$. this shows that there is dense set of points s for which

$$\limsup_{h \downarrow 0} \frac{|B_{s+h} - B_s|}{\sqrt{2h \log 1/h}} = 1,$$

hich is more than we claimed.

One can see there are uncountably many such s by modifying the construction slightly. In the induction step, just break each interval (s_n, s_n') into three parts, throw away the middle third, and operate with each of the two remaining parts as above. See Orey and Taylor's article for a detailed study of these singular points.

PROPOSITION 1.7. Fix s_0. Then, with probability one,

$$\limsup_{h \downarrow 0} \frac{W_{s_0+h,t} - W_{s_0 t}}{\sqrt{2h \log \log 1/h}} = \sqrt{t}, \tag{1.12}$$

simultaneously for all $t > 0$.

PROOF. (1.12) holds a.s. for each fixed t by the law of the iterated logarithm, hence it holds for a.e. t by Fubini. We must show it holds for all t. Set

$$L_t = \limsup_{h \downarrow 0} \frac{W_{s_0+h,t} - W_{s_0 t}}{\sqrt{2h \log \log 1/h}} .$$

It is easy to see that L_t is well-measurable relative to the fields $\underline{\underline{F}}_t^2 = \underline{\underline{F}}_{(0,\infty)\times(0,t]}$ for it is a measurable function of $W_{\cdot,t}$ which, being continuous and adapted to $(\underline{\underline{F}}_t^2)$, is itself well-measurable. By Meyer's section theorem, if $P\{\exists t \ni L_t \neq \sqrt{t}\} > 0$, there exists a finite stopping time T (relative to the $\underline{\underline{F}}_t^2$) such that $P\{L_T \neq \sqrt{T}\} > 0$.

But now let $B_{st} = W_{s,T+t} - W_{sT}$. B is again a Brownian sheet (apply Theorem 1.5 to the weak stopping point (0,T)) so that if $\delta > 0$ we have

$$|L_{T+\delta} - L_T| \leq \limsup \frac{|B_{s_0+h,\delta} - B_{s_0,\delta}|}{\sqrt{2h \log \log 1/h}} = \sqrt{\delta} .$$

It follows that if $L_T(\omega) \neq \sqrt{T}$, then $L_{T+\delta} \neq \sqrt{T+\delta}$ for small enough δ, i.e. $L_t \neq \sqrt{t}$ for a set of positive Lebesgue measure, a contradiction.

PROPOSITION 1.8. Fix $t_0 > 0$ and let $S \geq 0$ be a random variable which is $\underline{\underline{F}}_{t_0}^2$ measurable. Suppose that

$$\limsup_{h \downarrow 0} \frac{W_{S+h,t} - W_{S,t}}{\sqrt{2h \log \log 1/h}} = \infty \quad \text{a.s.} \tag{1.13}$$

or $t = t_0$. Then (1.13) also holds for all $t \geq t_0$. If S is $\sigma\{W_{st_0}, s \geq 0\}$ - measurable, then (1.13) holds for all $t \geq 0$.

PROOF. We can assume without loss of generality that $t_0 = 1$. Set $B_{st} = W((S,1),(S+s, 1+t)]$. Note that (S,1) is a weak stopping point, so B_{st} is a Brownian sheet. By Proposition 1.7, it satisfies the log log law for all $t > 0$. Thus, if $t' = 1 + t$

$$\limsup_{h \downarrow 0} \frac{W_{S+h,t'} - W_{S,t'}}{\sqrt{2h \log \log 1/h}} \geq \limsup_{h \downarrow 0} \frac{W_{S+h,1} - W_{S,1}}{\sqrt{2h \log \log 1/h}}$$

$$- \limsup_{h \downarrow 0} \frac{B_{ht} - B_{0t}}{\sqrt{2h \log \log 1/h}}$$

$$= \infty - \sqrt{t} = \infty.$$

This proves (1.13) for all $t \geq 1$. Suppose $S \in \sigma\{W_{s1}, s \geq 0\}$. To see that (1.13) follows for $t < 1$ as well, set $\hat{W}_{st} = tW_{s\frac{1}{t}}$. Then $\hat{W}_{st}$ is a Brownian sheet, and $\hat{W}_{sl} = W_{sl}$ for all s. Clearly $S \in \sigma\{\hat{W}_{s1}, s \geq 0\}$. Thus $\hat{W}$ satisfies (1.13) for all $t > 1$, which implies that W satisfies (1.13) at $\frac{1}{t}$ for all $t > 1$. Q.E.D.

REMARKS. If we call a point at which the law of the iterated logarithm fails a singular point, the above proposition tells us that such singularities propagate vertically. By symmetry, there are singularities of the same type propagating horizontally. One can visualize these propagating singularities as wrinkles in the sheet.

THE BROWNIAN SHEET AND THE VIBRATING STRING

It is time to connect the Brownian sheet with our main topic, stochastic partial differential equations: the Brownian sheet gives the solution to a vibrating string problem.

Let us first modify the sheet as follows. Let D be the half plane $\{(s,t) : s + t \geq 0\}$ and put $\hat{R}_{st} = D \cap (-\infty,s] \times (-\infty,t]$. If W is a white noise, define $\hat{W}_{st} = W(\hat{R}_{st})$.

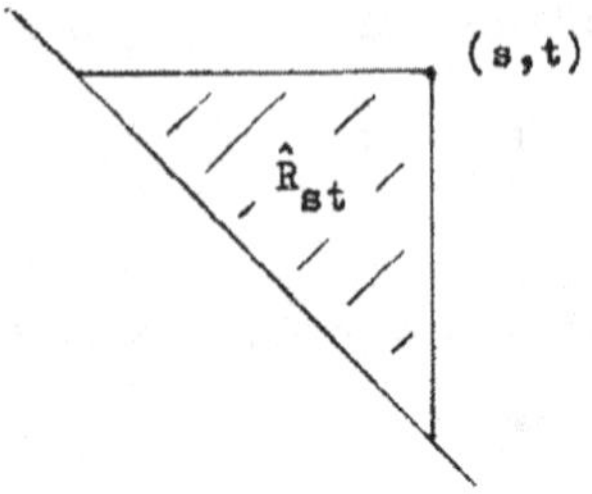

Then $\hat{W}$ is not a Brownian sheet: instead of vanishing on the coordinate axes, it vanishes on $\{s + t = 0\}$. However, it is easily seen that $W_{st} \overset{\text{def}}{=} \hat{W}_{st} - \hat{W}_{so} - \hat{W}_{ot}$, $s,t \geq 0$ is a Brownian sheet, and that the processes $s \to \hat{W}_{so}$ and $t \to \hat{W}_{ot}$ are independent continuous processes of independent increments. We can use this to read off many of the properties of $\hat{W}$ from those of W. In particular, the singularities of $\hat{W}$ propagate exactly like those of W.

Now let us put the sheet back in the closet for the moment and let us consider a vibrating string driven by white noise. One can imagine a guitar left outdoors during a sandstorm. The grains of sand hit the strings continually but irregularly. The number of grains hitting a portion dx of the string during a time ds will be essentially independent of those hitting a different portion dy during a time dt. Let W(dx,dt) be the (random) measure of the number hitting in (dx,dt), centered by subtracting the mean. Then W will be essentially a white noise, and we expect the position V(t,x) of the string to satisfy the inhomogeneous wave equation driven by a white noise. In order to avoid worrying about boundary conditions, we will assume that the string is infinite, and that it is initially at rest. Thus V should satisfy

$$(1.14)\quad \begin{cases} \dfrac{\partial^2 V}{\partial t^2}(x,t) = \dfrac{\partial^2 V}{\partial x^2}(x,t) + W(dx,dt), & t > 0,\ -\infty < x < \infty \\ V(x,0) = \dfrac{\partial V}{\partial t}(x,0) = 0, & -\infty < x < \infty. \end{cases}$$

Putting aside questions of the existence and uniqueness of solutions to (1.14), let us recall how to solve it when the driving term is a smooth function. If f(t,x) is smooth and bounded, then the solution to

$$\begin{cases} \dfrac{\partial^2 V}{\partial t^2} = \dfrac{\partial^2 V}{\partial x^2} + f \\ V(x,0) = \dfrac{\partial V}{\partial t}(x,0) = 0 \end{cases}$$

is given by

(1.15) $$V(x,t) = \frac{1}{2}\int_0^t \int_{x+s-t}^{x+t-s} f(y,s)\,dy\,ds,$$

which can be checked by differentiating. Now let us rotate coordinates by 45°. Let $u = (s-y)/\sqrt{2}$, and $v = (s+y)/\sqrt{2}$, and set $\hat{V}(u,v) = V(y,s)$, $\hat{f}(u,v) = f(y,s)$. Then (1.15) implies that

$$\hat{V}(u,v) = \frac{1}{2}\int_0^v \int_{-v}^{u} \hat{f}(u',v')\,du'\,dv',$$

or

(1.16) $$\hat{V}(u,v) = \frac{1}{2}\int_{\hat{R}_{uv}}\int \hat{f}(u',v')\,du'\,dv'.$$

By a slight act of faith, we see that the solution of (1.14) should be given by (1.15), with f dy ds replaced by W(dy,ds), or, in the form (1.16), that

$$V(u,v) = \frac{1}{2}\int_{D \cap R_{uv}}\int dW$$

or, finally, that

$$\hat{V}(u,v) = \frac{1}{2}\hat{W}_{uv}, \quad u,v \geq 0,$$

where $\hat{W}$ is the modified Brownian sheet defined above.

We can conclude that the shape of the vibrating string at time t is just the cross-section of the sheet $\frac{1}{2}\hat{W}$ along the -45° line $u + v = \sqrt{2t}$.

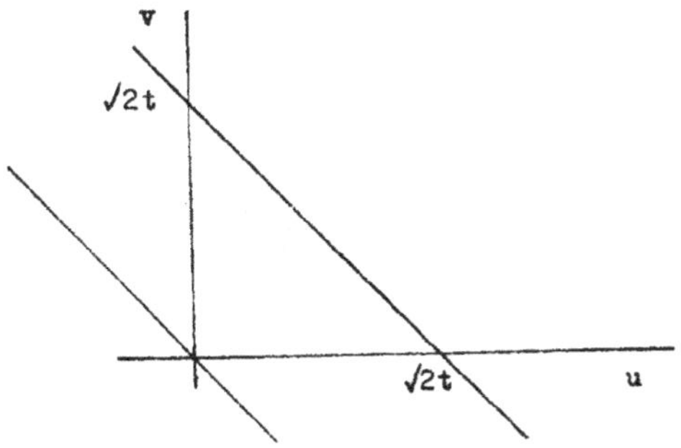

This gives us a complete representation of the solution of (1.14), and it also gives us an interpretation of the propagating singularities. Like W, the sheet $\hat{W}$ has singularities propagating parallel to the axes of the uv-plane. In the xt-plane these propagate along the lines $x = c + t$ and $x = c - t$ respectively. Thus these propagating singularities correspond to travelling waves which move along the string with velocity one, the speed of propagation in the equation (1.14). In general, the tendency of unusual events to propagate parallel to the axes of the Brownian sheet can be understood as the propagation of waves in our vibrating string.

It also explains the rather puzzling failure of the Brownian sheet to satisfy the sharp Markov property. In fact, the initial conditions for the vibrating string involve not only the position V, but also the velocity $\frac{\partial V}{\partial t}$, and in order to calculate the velocity, one must know V in some neighborhood. This is exactly why we needed the germ field $\underline{\underline{G}}^*_{\partial D}$ for the Markov property in the above example. A more delicate analysis of the Markov property would show that the minimal splitting field is in fact made up of the values of V and its derivative on the boundary.

__Exercise 1.1.__ For each fixed x, show that there exists a standard Brownian motion $\{B_s, s \geq 0\}$ such that $V(x,t) = \frac{1}{2} B(t^2)$, all $t \geq 0$. Show that with probability one, $\lim\limits_{t \downarrow 0} \frac{1}{t} V(x,t)$ does not exist, while $\lim\limits_{t \downarrow 0} \int_0^t \frac{1}{s} V(x,s)ds = 0$.
Discuss the initial condition $\frac{\partial V}{\partial t}(x,0) = 0$.

WHITE NOISE AS A DISTRIBUTION

One thinks of white noise on $\mathbf{R}$ as the derivative of Brownian motion. In two or more parameters, white noise can be thought of as the derivative of the Brownian sheet, and this can be made rigorous.

The Brownian sheet W_{st} is nowhere-differentiable in the ordinary sense, but its derivatives will exist in the sense of Schwartz distributions. Thus define

$$\dot{W}_{st} = \frac{\partial^2 W_{st}}{\partial s\, \partial t},$$

that is, if $\phi(s,t)$ is a test function with compact support in $\mathbf{R}^2_+$, $\dot{W}$ is the distribution

$$\dot{W}(\phi) = \int_{\mathbf{R}^2_+}\int W_{uv} \frac{\partial^2(\phi)}{\partial u \partial v}(u,v)\, dudv.$$ If we may anticipate

the introduction of stochastic integrals, let us note that this is almost surely

$$= \iint \phi \, dW.$$

Formally, if $\phi(u,v) = I_{\{0 \leq u \leq s,\ 0 \leq v \leq b\}}$ then

$$\dot{W}(\phi) = \int_0^s \int_0^t \frac{\partial^2 W}{\partial u \partial v} \, du \, dv = W_{st} = \iint \phi \, dW,$$

but it takes some work to make it rigorous. We leave it as an exercise.

If we regard the "measure" W as a distribution, then certainly $W(\phi) = \iint \phi \, dW$. In other words $\dot{W}(\phi) = W(\phi)$ so that $\dot{W}$ and W are the same distribution.

Note that in R^n, $\dot{W}$ would be the nth mixed partial:

$$\dot{W} = \frac{\partial^n}{\partial t_1 \cdots \partial t_n} W_{t_1 \cdots t_n} .$$

CHAPTER TWO

MARTINGALE MEASURES

We will develop a theory of integration with respect to martingale measures. We think of them as white noises, but we treat them differently. Instead of considering set functions on R^{d+1} with all coordinates treated symmetrically, we break off one coordinate to play the role of "time" and think of the remaining coordinates as "space".

Let us begin with some remarks on random set functions and vector-valued measures. Let $(E,\underline{\underline{E}})$ be a Lusin space, i.e. a measurable space homeomorphic to a Borel subset of the line. (This includes all Euclidean spaces and, more generally, all Polish spaces.)

Suppose $U(A,\omega)$ is a function defined on $\underline{\underline{A}} \times \Omega$, where $\underline{\underline{A}} \subset \underline{\underline{E}}$ is an algebra, and such that $E\{U(A)^2\} < \infty$, $A \in \underline{\underline{A}}$. Suppose that U is finitely additive: if $A \cap B = \phi$, $A, B \in \underline{\underline{A}}$, then $U(A \cup B) = U(A) + U(B)$ a.s.

In most interesting cases U will not be countably additive if we consider it as a real-valued set function. However, it may become countably additive if we consider it as a set function with values in, say, $L^2(\Omega,\underline{\underline{F}},P)$. This is the case, for instance, with white noise. Let $\|U(A)\|_2 = E\{U^2(A)\}^{1/2}$ be the L^2-norm of $U(A)$.

We say U is <u>σ-finite</u> if there exists an increasing sequence $(E_n) \subset \underline{\underline{E}}$ whose union is E, such that for all n

(i) $\underline{\underline{E}}_n \subset \underline{\underline{A}}$, where $\underline{\underline{E}}_n = \underline{\underline{E}}|_{E_n}$;

(ii) $\sup\{\|U(A)\|_2 : A \in \underline{\underline{E}}_n\} < \infty$.

Define a set function μ by

$$\mu(A) = \|U(A)\|_2^2.$$

A σ-finite additive set function U is countably additive on $\underline{\underline{E}}_n$ (as an L^2-valued set function) iff

$$A_j \in \underline{\underline{E}}_n,\ \forall n,\ A_j \downarrow \phi \Rightarrow \lim_{j\to\infty} \mu(A_j) = 0 \tag{2.1}$$

If U is countably additive on $\underline{\underline{E}}_n$, $\forall n$, we can make a trivial further extension: if $A \in \underline{\underline{E}}$, set $U(A) = \lim_{n\to\infty} U(A \cap E_n)$ if the limit exists in L^2, and let U(A) be undefined otherwise. This leaves U unchanged on each $\underline{\underline{E}}_n$, but may change its value on some sets $A \in \underline{\underline{E}}$ which are not in any $\underline{\underline{E}}_n$. We will assume below that all our countably additive set functions have been extended in this way. We will say that such a U is a <u>σ-finite L^2-valued measure</u>.

<u>DEFINITION</u>. Let $(\underline{\underline{F}}_t)$ be a right continuous filtration. A process $\{M_t(A), \underline{\underline{F}}_t, t \geq 0, A \in \underline{\underline{A}}\}$ is a <u>martingale measure</u> if

(i) $M_o(A) = 0$;

(ii) if $t > 0$, M_t is a σ-finite L^2-valued measure;

(iii) $\{M_t(A), \underline{\underline{F}}_t, t \geq 0\}$ is a martingale.

<u>Exercise 2.1</u>. Let $\nu_t(A) = \sup\{E\{M_t(B)^2\}: B \subset A, B \in \underline{\underline{E}}\}$. Show that $t \to \nu_t(A)$ is increasing. Conclude that for each T, the same family (E_n) works for all M_t, $t \leq T$.

It is not necessary to verify the countable additivity for all t; one t will do, as the following exercise shows.

<u>Exercise 2.2</u>. If N is a σ-finite L^2-valued measure and $(\underline{\underline{F}}_t)$ a filtration, show that

$$M_t(A) = E\{N(A)|\underline{\underline{F}}_t\} - E\{N(A)|\underline{\underline{F}}_0\}$$

is a martingale measure.

Note: One commonly gets such an M_t by first defining it for a small class of sets and then constructing the L^2-valued measure from these. This is, in fact, exactly what one does when constructing a stochastic integral, although the fact that the result is a vector-valued measure is usually not emphasized. In the interest of a speedy development, we will assume that the L^2-measure has already been constructed. Thus we know how to integrate over dx for fixed t - this is the Bochner integral - and over dt for fixed sets A - this is the Ito integral. The problem facing us now is to integrate over dx and dt at the same time.

There are two rather different classes of martingale measures which have been popular, orthogonal martingale measures and martingale measures with a nuclear covariance.

DEFINITION. A martingale measure M is orthogonal if, for any two disjoint sets A and B in $\underline{\underline{A}}$, the martingales $\{M_t(A),\ t \geq 0\}$ and $\{M_t(B),\ t \geq 0\}$ are orthogonal.

Equivalently, M is orthogonal if the product $M_t(A)M_t(B)$ is a martingale for any two disjoint sets A and B. This is in turn equivalent to having $\langle M(A), M(B)\rangle_t$, the predictable process of bounded variation, vanish.

DEFINITION. A martingale measure M has nuclear covariance if there exists a finite measure η on $(E,\underline{\underline{E}})$ and a complete ortho-normal system (ϕ_k) in $L^2(E,\underline{\underline{E}},\eta)$ such that $\eta(A) = 0 \Rightarrow \mu(A) = 0$ for all $A \in \underline{\underline{E}}$ and

$$\sum_k E\{M_t(\phi_k)^2\} < \infty,$$

where $M_t(\phi_k) = \int \phi_k(x)M_t(dx)$ is a Bochner integral.

The canonical example of an orthogonal martingale measure is a white noise. If W is a white noise on $E \times \mathbb{R}_+$, let $M_t(A) = W(A \times [0,t])$. This is clearly a martingale measure, and if $A \cap B = \phi$, $M_t(A)$ and $M_t(B)$ are independent, hence orthogonal. Any martingale measure derived from a white noise this way will also be called a white noise.

If $(E,\underline{\underline{E}},\eta)$ is a finite measure space, if $f \in L^2(E,\underline{\underline{E}},\eta)$ and if $\{B_t,\ t \geq 0\}$ is a standard (one-dimensional) Brownian motion, then the measure defined by

$$M_t(A) = B_t \int_A f(x)\eta(dx)$$

has nuclear covariance, since for any CONS (ϕ_k), $\sum_k E\{M_t^2(\phi_k)\} = \|f\|_2^2$. More generally, if $B^1, B^2, \ldots$ are iid standard Brownian motions and if $a_1, a_2, \ldots$ are real numbers such that $\sum a_k^2 < \infty$, then

$$M_t(A) = \sum_k a_k B_t^k \int_A \phi_k(x)\eta(dx)$$

has nuclear covariance.

Note that it is only in exceptional cases, such as when E is a finite set, that a white noise will have nuclear covariance.

WORTHY MEASURES

Unfortunately, it is not possible to construct a stochastic integral with respect to all martingale measures - we will give a counter-example at the end of the chapter - so we will need to add some conditions. These are rather strong, and, though sufficient, are doubtless not necessary. However, they are satisfied for both orthogonal martingale measures and those with a nuclear covariance.

Let M be a σ-finite martingale measure. By restricting ourselves to one of the E_n, if necessary, we can assume that M is finite. We shall also restrict ourselves to a fixed time interval $[0,T]$. The extension to infinite measures and the interval $[0,\infty]$ is routine.

DEFINITION. The covariance functional of M is

$$\bar{Q}_t(A,B) = \langle M(A),\ M(B)\rangle_t .$$

Note that $\bar{Q}_t$ is symmetric in A and B and biadditive: for fixed A, $\bar{Q}_t(A,\cdot)$ and $\bar{Q}_t(\cdot,A)$ are additive set functions. Indeed, if $B \quad C = \phi$,

$$\begin{aligned}\bar{Q}_t(A,\ B \quad C) &= \langle M(A),\ M(B) + M(C)\rangle_t \\ &= \langle M(A),\ M(B)\rangle_t + \langle M(A),\ M(C)\rangle_t \\ &= \bar{Q}_t(A,B) + \bar{Q}_t(A,C).\end{aligned}$$

Moreover, by the general theory,

$$|\bar{Q}_t(A,B)| \le Q_t(A,A)^{1/2}\ Q_t(B,B)^{1/2}.$$

A set $A \times B \times (s,t] \subset E \times E \times R_+$ will be called a rectangle. Define a set function Q on rectangles by

$$Q(A \times B \times (s,t]) = \bar{Q}_t(A,B) - \bar{Q}_s(A,B),$$

and extend Q by additivity to finite disjoint unions of rectangles, i.e. if $A_i \times B_i \times (s_i,t_i]$ are disjoint, $i = 1,\ldots,n$, set

$$(2.2) \qquad Q\Big(\bigcup_{i=1}^{n} A_i \times B_i \times (s_i,t_i]\Big) = \sum_{i=1}^{n} \big(\bar{Q}_{t_i}(A_i,B_i) - \bar{Q}_{s_i}(A_i,B_i)\big).$$

Exercise 2.3. Verify that Q is well-defined, i.e. if

$\Lambda = \bigcup_{i=1}^{n} A_i \times B_i \times (s_i,t_i] = \bigcup_{j=1}^{m} A'_j \times B'_j \times (s'_j,t'_j]$, each representation gives the same value for $Q(\Lambda)$ in (2.5). (Hint: use biadditivity.)

If $a_1,\dots,a_n \in R$ and if $A_1,\dots,A_n \in \underline{\underline{E}}$ are disjoint, then for any $s < t$

$$(2.3) \qquad \sum_{j=1}^{n} \sum_{j=1}^{n} a_i a_j \, Q(A_i \times A_j \times (s,t]) \geq 0,$$

for the sum is

$$= \sum_{i,j} a_i a_j \big(\langle M(A_i), M(A_j)\rangle_t - \langle M(A_i), M(A_j)\rangle_s\big)$$

$$= \Big\langle \sum_i a_i (M_t(A_i) - M_s(A_i)), \sum_i a_i (M_t(A_i) - M_s(A_i))\Big\rangle \geq 0.$$

DEFINITION. A signed measure $K(dx\,dy\,ds)$ on $\underline{\underline{E}} \times \underline{\underline{E}} \times \underline{\underline{B}}$ is positive definite if for each bounded measurable function f for which the integral makes sense,

$$(2.4) \qquad \int_{E\times E\times R_+} f(x,s)f(y,s)K(dxdyds) \geq 0 \quad .$$

For such a positive definite signed measure K, define

$$(f,g)_K = \int_{E\times E\times R_+} f(x,s)g(y,s)K(dxdyds).$$

Note that $(f,f)_K \geq 0$ by (2.4).

Exercise 2.4. Suppose K is symmetric in x and y. Prove Schwartz' and Minkowski's inequalities

$$(f,g)_K \leq (f,f)_K^{1/2}(g,g)_K^{1/2}$$

and

$$(f+g,\ f+g)_K^{1/2} \leq (f,f)_K^{1/2} + (g,g)_K^{1/2}.$$

It is not always possible to extend Q to a measure on $\underline{\underline{E}} \times \underline{\underline{E}} \times \underline{\underline{B}}$, where $\underline{\underline{B}}$ = Borel sets on R_+, as the example at the end of the chapter shows. We are led to the following definition.

DEFINITION. A martingale measure M is worthy if there exists a random σ-finite measure $K(\Lambda,\omega)$, $\Lambda \in \underline{\underline{E}} \times \underline{\underline{E}} \times \underline{\underline{B}}$, $\omega \in \Omega$, such that

(i) K is positive definite and symmetric in x and y;

(ii) for fixed A, B, $\{K(A \times B \times (0,t]),\ t \geq 0\}$ is predictable;

(iii) for all n, $E\{K(E_n \times E_n \times [0,T])\} < \infty$;

(iv) for any rectangle Λ, $|Q(\Lambda)| \leq K(\Lambda)$.

We call K the dominating measure of M.

The requirement that K be symmetric is no restriction. If not, we simply replace it by $K(dx\,dy\,ds) + K(dy\,dx\,ds)$. Apart from this, however it is a strong condition on M. We will show below that it holds for the two important special cases mentioned above: both orthogonal martingale measures and those with nuclear covariance are worthy. In fact, we can state with confidence that we will have no dealings with unworthy measures in these notes.

If M is worthy with covariation Q and dominating measure K, then K + Q is a positive set function. The σ-field $\underline{\underline{E}}$ is separable, so that we can first restrict ourselves to a countable subalgebra of $\underline{\underline{E}} \times \underline{\underline{E}} \times \underline{\underline{B}}$ upon which $Q(\cdot,\omega)$ is finitely additive for a.e. ω. Then K + Q is a positive finitely additive set function dominated by the measure 2K, and hence can be extended to a measure. In particular, for a.e. ω $Q(\cdot,\omega)$ can be extended to a signed measure on $E \times E \times \underline{\underline{B}}$, and the total variation of Q satisfies $|Q|(\Lambda) \leq K(\Lambda)$ for all $\Lambda \in \underline{\underline{E}} \times \underline{\underline{E}} \times \underline{\underline{B}}$. By (2.3), Q will be positive definite.

Orthogonal measures and white noises are easily characterized. Let $\Delta(E) = \{(x,x)\colon x \in E\}$, be the diagonal of E.

PROPOSITION 2.1. A worthy martingale measure is orthogonal iff Q is supported by $\Delta(E) \times R_+$.

PROOF. $Q(A \times B \times [0,t]) = \langle M(A), M(B)\rangle_t$.
If M is orthogonal and $A \cap B = \phi$, this vanishes hence $|Q|[E \times E - \Delta(E)) \times R_+] = 0$, i.e. supp $Q \subset \Delta(E) \times R_+$. Conversely, if this vanishes for all disjoint A and B, M is evidently orthogonal. Q.E.D.

STOCHASTIC INTEGRALS

We are only going to do the L^2-theory here - the bare bones, so to speak. It is possible to extend our integrals further, but since we won't need the extensions in this course, we will leave them to our readers.

Let M be a worthy martingale measure on the Lusin space $(E,\underline{\underline{E}})$, and let Q_M and K_M be its covariation and dominating measures respectively. Our definition of the stochastic integral may look unfamiliar at first, but we are merely following Ito's construction in a different setting.

In the classical case, one constructs the stochastic integral as a process rather than as a random variable. That is, one constructs $\{\int_0^t f\, dB,\ t \geq 0\}$ simultaneously for all t; one can then say that the integral is a martingale, for instance. The analogue of "martingale" in our setting is "martingale measure". Accordingly, we will define our stochastic integral as a martingale measure.

Recall that we are restricting ourselves to a finite time interval [0,T] and to one of the E_n, so that M is finite. As usual, we will first define the integral for elementary functions, then for simple functions, and then for all functions in a certain class by a functional completion argument.

__DEFINITION.__ A function $f(x,s,\omega)$ is __elementary__ if it is of the form

$$f(x,s,\omega) = X(\omega) I_{(a,b]}(s)\, I_A(x), \tag{2.5}$$

where $0 \leq a < t$, X is bounded and $\underline{\underline{F}}_a$-measurable, and $A \in \underline{\underline{E}}$. f is __simple__ if it is a finite sum of elementary functions. We denote the class of simple functions by $\underline{\underline{S}}$.

__DEFINITION.__ The __predictable σ-field__ $\underline{\underline{P}}$ on $\Omega \times E \times R_+$ is the σ-field generated by $\underline{\underline{S}}$. A function is __predictable__ if it is $\underline{\underline{P}}$-measurable.

We define a norm $\| \ \|_M$ on the predictable functions by

$$\|f\|_M = E\{(|f|, |f|)_K\}^{1/2}.$$

Note that we have used the absolute value of f to define $\|f\|_M$, so that

$$(f,f)_Q \le \|f\|_M^2.$$

Let $\underline{\underline{P}}_M$ be the class of all predictable f for which $\|f\|_M < \infty$.

PROPOSITION 2.2. Let $f \in \underline{\underline{P}}_M$ and let $A = \{(x,s): |f(x,s)| \ge \varepsilon\}$. Then

$$E\{K(A \times E \times [0,T])\} \le \frac{1}{\varepsilon} \|f\|_M \, E\{K(E \times E \times [0,T])\}$$

PROOF.
$$\begin{aligned} \varepsilon\, E\{K(A \times E \times [0,T])\} &\le E\{\textstyle\int |f(x,t)| K(dx\,dy\,dt)\} \\ &= E\{(|f|, 1)_K\} \\ &\le E\{(|f|, |f|)_K^{1/2} K(E \times E \times [0,T])\} \\ &\le \|f\|_M \, E\{K(E \times E \times [0,T])\}^{1/2} \end{aligned}$$

where we have used Schwartz' inequality in two forms (see Exercise 2.4).

Q.E.D.

Exercise 2.5. Use Proposition 2.2 to show $\underline{\underline{P}}_M$ is complete, and hence a Banach space.

PROPOSITION 2.3. $\underline{\underline{S}}$ is dense in $\underline{\underline{P}}_M$.

PROOF. If $f \in \underline{\underline{P}}_M$, let $f_N(x,s) = \begin{cases} f(x,s) & \text{if } |f(x,s)| < N \\ 0 & \text{otherwise} \end{cases}$.

Then $\|f-f_N\|_M^2 = E\{\int |f(x,s) - f_N(x,s)|\ |f(y,s) - f_N(y,s)| K(dxdyds)\}$ which goes to zero by monotone convergence. Thus the bounded functions are dense. If f is a bounded step function, i.e. if there exist $0 \le t_0 < t_1 < \ldots < t_n$ such that $t \to f(x,t)$ is constant on each $(t_j, t_{j+1}]$, then f can be uniformly approximated by simple functions. Thus the simple functions are dense in the step functions. It remains to show that the step functions are dense in the bounded functions.

To simplify our notation, let us suppose that $K(E \times E \times ds)$ is absolutely continuous with respect to Lebesgue measure. [We can always make a preliminary time change to assure this.] If $f(x,s,\omega)$ is bounded and predictable, set

$$f_n(x,s,\omega) = 2^{-n} \int_{(k-1)2^{-n}}^{k2^{-n}} f(x,u,\omega)du \quad \text{if } k2^{-n} \le s < (k+1)2^{-n}.$$

Fix ω and x. Then $f_n(x,s,\omega) \to f(x,s,\omega)$ for a.e. s by either the martingale convergence theorem or Lebesgue's differentiation theorem, take your choice. It follows easily that $\|f - f_n\|_M \to 0$. Q.E.D.

We can now construct the integral with a minimum of interruption. If $f(x,s,\omega) = X(\omega)\, I_{(a,b]}(s)\, I_A(x)$ is an elementary function, define a martingale measure $f \cdot M$ by

(2.6) $$f \cdot M_t(B) = X(\omega)(M_{t \wedge b}(A \cap B) - M_{t \wedge a}(A \cap B)).$$

<u>LEMMA 2.4</u>. $f \cdot M$ is a worthy martingale measure. Its covariance and dominating measures $Q_{f \cdot M}$ and $K_{f \cdot M}$ are given by

(2.7) $$Q_{f \cdot M}(dx\, dy\, ds) = f(x,s)\, f(y,s)\, Q_M(dx\, dy\, ds)$$

(2.8) $$K_{f \cdot M}(dx\, dy\, dx) = |f(x,s)f(y,s)| K_M(dx\, dy\, ds).$$

Moreover

(2.9) $$E\{f \cdot M_t(B)^2\} \le \|f\|_M^2 \quad \text{for all } B \in \underline{\underline{E}},\ t \le T.$$

<u>PROOF</u>. $f \cdot M_t(B)$ is adapted since $X \in \underline{\underline{F}}_a$; it is square integrable, and a martingale. $B \to f \cdot M_t(B)$ is countably additive (in L^2), which is clear from (2.6). Moreover

$$f \cdot M_t(B) f \cdot M_t(C) - \int_{B \times C \times [0,t]} f(x,s)f(y,s)Q_M(dx\, dy\, ds)$$

$$= X^2[(M_{t \wedge b}(A \cap B) - M_{t \wedge a}(A \cap B)(M_{t \wedge b}(A \cap C) - M_{t \wedge a}(A \cap C))$$
$$- \langle M(A \cap B), M(A \cap C) \rangle_{t \wedge b} + \langle M(A \cap B), M(A \cap C) \rangle_{t \wedge a}]$$

which is a martingale. This proves (2.7), and (2.8) follows immediately since $K_{f \cdot M}$ is positive and positive definite. (2.9) then follows easily.

Q.E.D.

We now define $f \cdot M$ for $f \in \underline{\underline{S}}$ by linearity.

<u>Exercise 2.6</u>. Show that (2.7)-(2.9) hold for $f \in \underline{\underline{S}}$.

Suppose now that $f \in \underline{\underline{P}}_M$. By Prop. 2.6 there exist $f_n \in \underline{\underline{S}}$ such that

$\|f-f_n\|_M \to 0$. By (2.9), if $A \in \underline{\underline{E}}$ and $t \le T$,

$$E\{(f_m\cdot M_t(A) - f_n\cdot M_t(A))^2\} \le \|f_m - f_n\|_M \to 0$$

as $m, n \to \infty$. It follows that $(f_n\cdot M_t(A))$ is Cauchy in $L^2(\Omega,\underline{\underline{F}},P)$, so that it converges in L^2 to a martingale which we shall call $f\cdot M_t(A)$. The limit is independent of the sequence (f_n).

THEOREM 2.5. If $f \in \underline{\underline{P}}_M$, then $f\cdot M$ is a worthy martingale measure. It is orthogonal if M is. Its covariance and dominating measures respectively are given by

$$Q_{f\cdot M}(dx\ dy\ ds) = f(x,s)f(y,s)\ Q_M(dx\ dy\ ds); \tag{2.10}$$

$$K_{f\cdot M}(dx\ dy\ ds) = |f(x,s)\ f(y,s)|K_M(dx\ dy\ ds). \tag{2.11}$$

Moreover, if $g \in \underline{\underline{P}}_M$ and $A, B \in \underline{\underline{E}}$, then

$$\langle f\cdot M(A),\ g\cdot M(B)\rangle_t = \int_{A\times B\times[0,t]} f(x,s)f(y,s)Q_M(dx\ dy\ ds); \tag{2.12}$$

$$E\{(f\cdot M_t(A))^2\} \le \|f\|_M^2. \tag{2.13}$$

PROOF. $f\cdot M(A)$ is the L^2 limit of the martingales $f_n\cdot M(A)$, and is hence a square-integrable martingale. For each n

$$f_n\cdot M_t(A)\ f_n\cdot M_t(B) - \int_{A\times B\times[0,t]} f_n(x,s)f_n(y,s)\ Q_M(dx\ dy\ ds) \tag{2.14}$$

is a martingale. $f_n\cdot M(A)$ and $f_n\cdot M(B)$ each converge in L^2, hence their product converges in L^1. Moreover

$$E\{|\int_{A\times B\times[0,t]} (f_n(x,s)f_n(y,s) - f(x,s)f(y,s))Q_M(dx\ dy\ ds)|\}$$

$$\le E\{\int_{E\times E\times[0,T]} |f_n(x)|\,|f_n(y)-f(y)|K_M(dx\ dy\ ds)\}$$

$$+ E\{\int_{E\times E\times[0,T]} |f_n(x)-f(x)|\ |f(y)|K_M(dx\ dy\ ds)\}$$

$$\le E\{(|f_n|,\ |f-f_n|)_K + (|f-f_n|,\ |f|)_K\}$$

By Schwartz:

$$\le (\|f_n\|_M + \|f\|_M)\|f_n - f\|_M \to 0\ .$$

Thus the expression (2.14) converges in L^1 to

$$f\cdot M_t(A)\ f\cdot M_t(B) - \int_{A\times B\times[0,t]} f(x,s)f(y,s)\ Q_M(dx\ dy\ ds)$$

which is therefore a martingale. The latter integral, being predictable, must therefore equal $\langle f\cdot M(A),\ f\cdot M(B)\rangle_t$, which verifies (2.10), and (2.11) follows.

This proves (2.12) in case $g = f$, and the general case follows by polarization. (2.13) then follows from (2.11).

To see that $f\cdot M_t$ is a martingale measure, we must check countable additivity. If $A_n \subset E$, $A_n \downarrow \phi$, then

$$E\{f\cdot M_t(A_n)^2\} \le E\{\int_{A_n\times A_n\times[0,t]} |f(x,s)f(y,s)|K(dx\ dy\ ds)\}$$

which goes to zero by monotone convergence.

If M is orthogonal, Q_M sits on $\Delta(E) \times [0,T]$, hence, by (2.10), so does $Q_{f\cdot M}$. By Proposition 2.4, $f\cdot M$ is orthogonal. Q.E.D.

Now that the stochastic integral is defined as a martingale measure, we define the usual stochastic integrals by

$$\int_{A\times[0,t]} f\ dM = f\cdot M_t(A)$$

and

$$\int_{E\times[0,t]} f\ dM = f\cdot M_t(E).$$

while

$$\int f\ dM = \lim_{t\to\infty} f\cdot M_t(E).$$

When it is necessary we will indicate the variables of integration. For instance

$$\int_{A\times[0,t]} f(x,s)M(dx\ ds) \qquad \text{and} \qquad \int_0^t \int_A f(x,s)dM_{xs}$$

both denote $f\cdot M_t(A)$.

It is frequently necessary to change the order of integration in iterated stochastic integrals. Here is a form of stochastic Fubini's theorem which will be useful.

Let $(G, \underline{\underline{G}}, \mu)$ be a finite measure space and let M be a martingale with dominating measure K.

<u>THEOREM 2.6</u>. Let $f(x,s,\omega,\lambda)$, $x \in E$, $s \ge 0$, $\omega \in \Omega$, $\lambda \in G$ be a $\underline{\underline{P}} \times \underline{\underline{G}}$-measurable function. Suppose that

$$(2.15)\qquad E\{\int_{E\times E\times[0,T]\times G} |f(x,s,\omega,\lambda)\ f(y,s,\omega,\lambda)|K(dx\ dy\ ds)\mu(d\lambda)\} < \infty.$$

Then

$$(2.16)\quad \int_G \Big[\int_{E\times[0,t]} f(x,s,\lambda)\, M(dxds)\Big]\mu(d\lambda) = \int_{E\times[0,t]} \Big[\int_G f(x,s,\lambda)\mu(d\lambda)\Big] M(dxds).$$

PROOF. If $f(x,s,\omega,\lambda) = X(\omega)\, I_{(a,b]}(s)\, I_A(x) g(\lambda)$, then both sides of (2.16) equal

$$X\big(M_{t\wedge b}(A) - M_{t\wedge a}(A)\big) \int g(\lambda)\mu(d\lambda).$$

Both sides of (2.16) are additive in f, so this also holds for finite sums of such f. If f is $\underline{\underline{P}} \times \underline{\underline{G}}$ - measurable and satisfies (2.16), we can apply an argument similar to the proof of Proposition 2.6 to show that there exists a sequence (f_n) of such functions such that

$$E\{ \int |f(x,s,\lambda) - f_n(x,s,\lambda)||f(y,s,\lambda) - f_n(y,s,\lambda)| K(dx\, dy\, ds)\, \mu(d\lambda)\}$$
$$= \int \|f(\lambda) - f_n(\lambda)\|_M^2 \, \mu(d\lambda) \to 0 .$$

We see that the integral in brackets on the right hand side of (2.16) is $\underline{\underline{P}}$-measurable, (Fubini) so that the integral makes sense, providing that $\| \int f(\lambda)\, \mu(d\lambda) \|_M < \infty$.

On the left-hand side we can take a subsequence if necessary to have $\|f(\lambda) - f_n(\lambda)\|_M \to 0$ for η - a.e.λ. This implies that for a.e.λ $\int f_n(\lambda) dM \to \int f(\lambda) dM$ in L^2, hence in measure. Using Fubini's theorem again we see that $\int f_n(\omega,\lambda) dM \to \int f(\omega,\lambda) dM$ in $P \times \mu$-measure, hence the latter integral is measurable in the pair (ω,λ). It follows that $\int f(\lambda) dM$ is μ-measurable for fixed ω, so that the integral on the left-hand side of (2.16) makes sense.

We must show that both sides converge as $n \to \infty$. Set $g_n = f - f_n$.

$$\| \int g_n(\lambda)\mu(d\lambda) \|_M = E\{ \int_{E\times E\times[0,T]} \int_G g_n(x,s,\lambda)\mu(d\lambda) K(dxdyds) \int_G g_n(y,s,\lambda')\mu(d\lambda')\}$$
$$= \int_{G\times G} E\{(g_n(\lambda),\, g_n(\lambda'))_K\}\mu(d\lambda)\mu(d\lambda');$$

by Schwartz, this is

$$\le \int_{G\times G} E\{(|g_n(\lambda)|,\, |g_n(\lambda)|)_K\}^{1/2} E\{(|g_n(\lambda')|,\, |g_n(\lambda')|)_K\}^{1/2} \mu(d\lambda)\mu(d\lambda')$$
$$= \Big(\int_G \|g_n(\lambda)\|_M \mu(d\lambda)\Big)^2$$
$$\le \mu(G) \int_G \|f(\lambda) - f_n(\lambda)\|_M^2 d\lambda$$

which tends to zero by (2.15).

This implies that the right-hand side of (2.16) converges. On the left,

$$E\{\int_G (\int g_n(x,s,\lambda)M(dxds))^2\mu(d\lambda)\}$$
$$= \int_G E\{(\int g_n(x,s,\lambda)M(dxds))^2\}\mu(d\lambda)$$
$$\leq \int_G \|g_n(\lambda)\|_M^2 \mu(d\lambda) \to 0.$$

By choosing a subsequence if necessary, we see that for a.e. ω,

$\int_G (\int g_n(x,s,\lambda)M(dxds))^2\mu(d\lambda) \to 0$, hence $\int_G (\int f - f_n)dM)d\mu \to 0$, and the left hand side of (2.16) converges too. Q.E.D.

ORTHOGONAL MEASURES

The remainder of this chapter concerns special properties of measures which are orthogonal or have nuclear covariance. We must certify their worthiness, so that the foregoing integration theory applies.

We should admit here that although we are handling a wide class of martingale measures in this chapter, our main interest is really in orthogonal measures. This is not because the theory is simpler - it is only simpler at the beginning - but because the problems which motivated this study involved white noises and related orthogonal measures.

The theory of integration does simplify, at least initially, if the integrator is orthogonal. For instance, the covariance measure Q sits on the diagonal and is positive, so that Q = K. Instead of having two measures on $E \times E \times R_+$, we need only concern ourselves with a single measure ν on $E \times R_+$ where $\nu(A \times [0,t]) = Q(A \times A \times [0,t])$, and this leads to several rather pleasant consequences which we will detail below.

The proof that an orthogonal measure M is worthy comes down to finding a good version of the increasing process $\langle M(A)\rangle_t$, one which is a measure in A and is right continuous in t.

We will fix our attention on a fixed time interval $0 \leq t \leq T$, and we continue to assume that $E = E_n$, so that M is finite. Define

$$\mu(A) = E\{M_T(A)^2\} = E\{\langle M(A)\rangle_T\}.$$

1° $\langle M(\cdot)\rangle_t$ is an additive set function,

i.e. $A \cap B = \phi \Rightarrow \langle M(A)\rangle_t + \langle M(B)\rangle_t = \langle M(A \cup B)\rangle_t$ a.s.

Indeed, $\langle M(A \cup B)\rangle_t = \langle M(A) + M(B)\rangle_t$

$$= \langle M(A)\rangle_t + \langle M(B)\rangle_t + 2\langle M(A), M(B)\rangle_t,$$

and the last term vanishes since M is orthogonal.

2° $A \subset B \Rightarrow \langle M(A)\rangle_t \le \langle M(B)\rangle_t$

$s \le t \Rightarrow \langle M(A)\rangle_s \le \langle M(B)\rangle_s$

3° μ is a σ-finite measure: it must be σ-finite since M_T is, and additivity follows by taking expectations in 1°.

The increasing process $\langle M(A)\rangle_t$ is finitely additive for each t by 1°, but it is better than that. It is possible to construct a version which is a measure in A for each t.

<u>THEOREM 2.7</u>. Let $\{M_t(A), \underline{\underline{F}}_t, 0 \le t \le T, A \in \underline{\underline{E}}\}$ be an orthogonal martingale measure. Then there exists a family $\{\nu_t(\cdot), 0 \le t \le T\}$ of random σ-finite measures on $(E,\underline{\underline{E}})$ such that

(i) $\{\nu_t, 0 \le t \le T\}$ is predictable;

(ii) for all $A \in \underline{\underline{E}}$, $t \to \nu_t(A)$ is right-continuous and increasing;

(iii) $P\{\nu_t(A) = \langle M(A)\rangle_t\} = 1$ all $t \ge 0$, $A \in \underline{\underline{E}}$.

<u>PROOF</u>. We can reduce this to the case $E \subset \mathbf{R}$, for E is homeomorphic to a Borel set $F \subset \mathbf{R}$. Let $h: E \to F$ be the homeomorphism, and define $M_t(A) = M_t(h^{-1}(A))$, $\mu(A) = \mu(h^{-1}(A))$. If we find a $\bar{\nu}_t$ satisfying the conclusions of the theorem and if $\bar{\nu}_t(\mathbf{R} - F) = 0$, then $\nu_t = \bar{\nu}_t \circ h$ satisfies the theorem. Thus we may assume E is a Borel subset of $\mathbf{R}$.

Since M is σ-finite, there exist $E_n \uparrow E$ for which $\mu(E_n) < \infty$. Then there are compact $K_n \subset E_n$ such that $u(E_n - K_n) < 2^{-n}$. We may also assume $K_n \subset K_{n+1}$ all n. It is then enough to prove the theorem for each K_n. Thus we may assume E is compact in $\mathbf{R}$ and $\mu(E) < \infty$.

Define $F_t(x) = \langle M(-\infty,x]\rangle_t$, $-\infty < x < \infty$.

Then

a) $\quad x \le x' \Rightarrow F_t(x) \le F_t(x')$;

$\quad t \le t' \Rightarrow F_t(x') - F_t(x) \le F_{t'}(x') - F_{t'}(x)$

b) $$E\{\sup_{\substack{t\le T \\ x_1 \le x \le x_2 \\ x \in Q}} |F_t(x) - F_t(x_1)|\} \le \mu((x_1,x_2])$$

Indeed (a) follows from 2°. To see (b), note that for fixed $t \le T$, $F_t(x) - F_t(x_1) \le \langle M(x_1,x_2]\rangle_t \le \langle M(x_1,x_2]\rangle_T$ a.s. by 2°. By right continuity this holds simultaneously for all $t \le T$ and all rational x in $(x_1,x_2]$. But then (b) follows since $E\{\langle M(x_1,x_2]\rangle_T\} = \mu((x_1,x_2])$.

Define $\overline{F}_t(x) = \inf\{F_{t'}(x'): x' > x,\ t' > t,\ x', t' \in Q\}$. This will be the "good" version of F. We claim that $\overline{F}_t$ is the distribution function of a measure.

c) $\quad t_1 \le t_2$ and $x_1 \le x_2 \Rightarrow \overline{F}_{t_1}(x_1) \le \overline{F}_{t_2}(x_2)$;

d) $\quad \overline{F}_t(x)$ is right continuous in the pair (x,t);

e) $\quad$ for fixed x, $P\{\overline{F}_t(x) = F_t(x), \text{ all } t \le T\} = 1$.

Indeed, (c) is clear and (d) and (e) follow from the uniform convergence guaranteed by (b). To see (e), for instance, choose rational t_n and x_n which strictly decrease to t and x respectively. Then

$$F_t(x) < \overline{F}_t(x) \le F_{t_n}(x_n) = F_{t_n}(x) + \left(F_{t_n}(x_n) - F_{t_n}(x)\right).$$

But $F_{t_n}(x) \to F_t(x)$ by right continuity and the term in square brackets tends to zero in probability by (b).

Let ν_t be the distribution on R generated by the distribution function $\overline{F}_t$. Note that ν_t does not charge $R-E$, for E is compact; and if $(a,b] \subset R - E$, $\nu_t(a,b] = \overline{F}_t(b) - \overline{F}_t(a) = F_t(b) - F_t(a) \le F_T(b) - F_T(a)$ by (e). This is true simultaneously for all rational t. Since $\overline{F}$ is right continuous we have a.s.

$$0 \le \sup_{t\le T} \nu_t(a,b] \le F_T(b) - F_T(a)$$

and the latter has expectation $\mu\{(a,b]\} = 0$.

Note that $\{\nu_t: 0 \le t \le T\}$ is predictable, for it is determined by $\overline{F}_t(x)$, $x \in Q$, hence by $F_t(x)$, $x \in Q$ by (e), and the F_t are predictable.

If $t < t'$, $\bar{F}_{t'}(x) - \bar{F}_t(x)$ is a distribution function, so $t \to \nu_t(A)$ is increasing for each A. It is right continuous in t if $A = (0,x]$, some x, or if $A = \mathbb{R}$ (for $E \subset \mathbb{R}$ is compact). Then right continuity for all Borel A follows by a monotone class argument.

To show (iii), note that if $A = (-\infty,x]$,

$$M_t^2(A) - \nu_t(A) \quad \text{is a martingale,} \tag{2.17}$$

for then $\nu_t(A) = F_t(x) = \langle M(A)\rangle_t$. Let $\underline{\underline{G}}$ be the class of A for which (2.17) holds. $\underline{\underline{G}}$ must contain finite unions of intervals of the form (a,b], and it contains $\mathbb{R}$, for $\nu_t(-\infty,x] = \nu_t(\mathbb{R})$ for large x.
It is closed under complementation, for

$$M_t^2(A^c) - \nu_t(A^c) = M_t^2(\mathbb{R}) - \nu_t(\mathbb{R}) - \left(M_t^2(A) - \nu_t(A)\right) - 2\,M_t(A)M_t(A^c)$$

and each of the terms on the right hand side is a martingale if $A \in \underline{\underline{G}}$. $\underline{\underline{G}}$ is also closed under monotone convergence. If $A_n \uparrow A$, for instance, $M_t(A_n)$ converges in L^2 to $M_t(A)$, and $\nu_t(A_n)$ increases to $\nu_t(A)$, hence the martingale $M_t^2(A_n) - \nu_t(A_n)$ converges in L^1 to $M_t^2(A) - \nu_t(A)$. The latter must therefore be a martingale. The case where $A_n \downarrow A$ follows by complementation. Thus $\underline{\underline{G}}$ contains all Borel sets. Q.E.D.

Now $t \to \nu_t(A)$ is increasing, so that we can define a measure ν on $E \times \mathbb{R}_+$ by defining $\nu(A \times (0,t]) = \nu_t(A)$ and extending it to $\underline{\underline{E}} \times \underline{\underline{B}}$, where $\underline{\underline{B}}$ is the class of Borel subsets of $\mathbb{R}_+$. This gives us the following.

<u>COROLLARY 2.8.</u> Let M be an orthogonal martingale measure. Then there exists a random σ-finite measure $\nu(dxds)$ on $E \times \mathbb{R}_+$ such that $\nu_t(A) = \nu(A \times [0,t])$ for all $A \in \underline{\underline{E}}$, $t > 0$.

We can get the covariance measure Q of M directly from ν. Set $\Delta = \Delta(E) \times \mathbb{R}_+$ where $\Delta(E)$ is the diagonal of $E \times E$ and let $\pi\colon \Delta \to E \times \mathbb{R}_+$ be defined by $\pi(x,x,t) = (x,t)$. Then we define Q by

$$Q(\Lambda) = \nu(\pi(\Lambda \cap \Delta)), \quad \Lambda \in \underline{\underline{E}} \times \underline{\underline{E}} \times \underline{\underline{B}}.$$

Then

$$\begin{aligned} Q(A \times B \times [0,t]) &= \nu(A \cap B \times [0,t]) \\ &= \langle M(A \cap B)\rangle_t \\ &= \langle M(A), M(B)\rangle_t, \end{aligned}$$

so Q is indeed the covariance measure of M. Since Q is positive and positive definite, we can set $K = Q$ and we have:

COROLLARY 2.9. An orthogonal martingale measure is worthy.

We noted above that a white noise gives rise to an orthogonal martingale measure. It is easy to characterize white noises among martingale measures.

PROPOSITION 2.10. Let M be an orthogonal martingale measure, and suppose that for each $A \in \underline{\underline{E}}$, $t \to M_t(A)$ is continuous. Then M is a white noise if and only if its covariance measure is deterministic.

PROOF. Rather than use Q, let us use the measure ν of Corollary 2.8, which is equivalent.

If M is a white noise on $E \times R_+$ based on a measure μ, it is easy to see that $\nu = \mu$, so ν is deterministic.

Conversely, if M is orthogonal and if ν is deterministic, then for $B \in \underline{\underline{E}}$, both $M_t(B)$ and $M_t^2(B) - \nu(B \times [0,t])$ are martingales.

To show M is a white noise, we must show it gives disjoint sets independent Gaussian values. One can see it is sufficient to show the following: if $B_1,\ldots,B_n$ are disjoint sets, then $\{M_t(B_1), t \geq 0\},\ldots,$ $\{M_t(B_n), t \geq 0\}$ are independent mean zero Gaussian processes with independent increments. This reduces to the following calculation.

Let

$$N_s = \exp\{i \sum_{j=1}^{n} \lambda_j(M_{t+s}(B_j) - M_s(B_j))\} .$$

By Ito's formula

$$N_s = 1 + \sum_{j=1}^{n} \int_t^{t+s} i\,\lambda_j\, N_u\, dM_u(B_j) - \frac{1}{2} \sum_{j=1}^{n} \lambda_j^2 \int_t^{t+s} N_u \nu(B_j \times du)$$

where we have used the fact that $d < M(B_j), M(B_k)>_t$ vanishes by orthogonality if $j \neq k$. Let $f(x) = E\{N_s | \underline{\underline{F}}_t\}$ and note that

$$f(s) = 1 - \frac{1}{2} \sum_{j=1}^{n} \lambda_j^2 \int_t^{t+s} f(u) d\nu(B_j \times du).$$

This has the unique solution

$$f(s) = \prod_{j=1}^{n} e^{-\frac{1}{2}\lambda_j^2 \nu(B_j \times (s,t])},$$

from which we see that the increments $M_{t+s}(B_j) - M_t(B_j)$ are independent of $\underline{\underline{F}}_t$ and of each other, and are Gaussian with mean zero and variance $\nu(B_j \times (s,t])$. Thus M is a white noise based on ν. Q.E.D.

NUCLEAR COVARIANCE

We will develop some of the particular properties of martingale measures with nuclear covariance. In particular, we will show they are worthy.

Suppose M is a martingale measure on $(E,\underline{\underline{E}})$ with nuclear covariance. Then there is a measure η and a CONS (ϕ_n) in $L^2(E,\underline{\underline{E}},\eta)$ such that

(2.18) $$\sum_n E\{M_T(\phi_n)^2\} < \infty.$$

We continue to assume $E = E_n$ for some n, so that M is finite, not just σ-finite.

PROPOSITION 2.11. For each x there exists a square-integrable martingale $\{M_t(x),\ t \geq 0\}$ such that for a.e. ω, $x \to M_t(x)$ $L^2(E,\ \underline{\underline{E}},\ \eta)$. Moreover

(i) $M_t(A) = \int_A M_t(x)\eta(dx)$, $A \subset E$;

(ii) $Q_M(A \times B \times [0,t]) = \int_{A\times B} \langle M(x), M(y)\rangle_t\ \eta(dx)\eta(dy)$.

Furthermore, there exists a predictable increasing process C_t and a positive predictable function $\sigma(x,t)$ such that $E\{\int_0^T \sigma^2(x,s)dC_s\eta(dx)\} < \infty$ and

(iii) $K_M(A \times B \times [0,t]) = \int_{A\times B\times[0,t]} \sigma(x,s)\sigma(y,s)\eta(dx)\eta(dy)dC_s$.

Finally

(iv) $\sum_n E\{M_t(\phi_n)^2\} = E\{\int_E M_t(x)^2\eta(dx)\}$.

PROOF. We will be rather cavalier in handling null-sets and measurable versions below. We leave it to the reader to supply the details.

The map $\phi \to M_t(\phi)$ is linear, and since $\sum M_t^2(\phi_n) < \infty$ a.s., it is a bounded linear functional on $L^2(E, \underline{\underline{E}}, \eta)$. It thus corresponds to a function which we denote $M_t(x)$, such that for $\phi \in L^2$,

$$M_t(\phi) = \int M_t(x)\phi(x)\eta(dx).$$

In fact

$$M_t(x,\omega) = \sum M_t(\phi_n)\phi_n(x).$$

This series converges in L^2 for a.e. ω hence, taking a subsequence if necessary, it converges in $L^2(\Omega,\underline{\underline{F}},P)$ for η-a.e. x. For each such x, $M_t(x)$ must be a martingale, being the L^2-limit of martingales. By modifying $M_t(x)$ on a set of η-measure zero, we can assume $M_t(x)$ is a martingale for all x.

Now

$$M_t(A)M_t(B) = \int_{A\times B} M_t(x)M_t(y)\ \eta(dx)\eta(dy)$$

so that

$$\langle M(A),\ M(B)\rangle_t = \int_{A\times B} \langle M(x),\ M(y)\rangle_t\ \eta(dx)\eta(dy).$$

This proves (i) and (ii).

$\underline{\underline{E}}$ is separable, so it is generated by a countable sub-algebra $\underline{\underline{A}}$. Let $\underline{\underline{M}}$ be the smallest class of martingales which contains $M_t(A)$, all $A \in \underline{\underline{A}}$, and which is closed under L^2-convergence. Then one can show that there exists an increasing process C_t such that for all $N \in \underline{\underline{M}}$, $d\langle N\rangle_t \ll dC_t$. Consequently, by Motoo's theorem, $\langle N\rangle_t = \int_0^t h(s)dC_s$ for some predictable h. This holds in particular for all the $M_t(\phi_k)$, for these are in $\underline{\underline{M}}$, and hence for the $M_t(x)$.

Furthermore, one can see by polarization that there exists a function $h(x,y,s)$ such that

$$\langle M(x),\ M(y)\rangle_t = \int_0^t h(x,y,s)dC_s.$$

But since $\langle M(x),\ M(y)\rangle \le \langle M(x)\rangle^{1/2}\ \langle M(y)\rangle^{1/2}$, we see that $|h(x,y,s)| \le h(x,x,s)^{1/2}\ h(y,y,s)^{1/2}$. Set $\sigma^2(x,s) = h(x,x,s)$. Then

$$Q_M(A \times B \times [0,t]) \le \int_{E\times E\times[0,t]} \sigma(x,s)\sigma(y,s)\eta(dx)\eta(dy)dC_s$$

which identifies $K(dx\ dy\ ds) = \sigma(x,s)\sigma(y,s)\eta(dx)\eta(dy)dC_s$. This is clearly positive and positive definite. To see that it is finite, write

$$E\{K(E \times E \times [0,T])\} = E\Big\{ \int_0^T [\int_E \sigma(x,s)\eta(dx)]^2 dC_s\Big\}$$

$$\le \eta(E)\ E\Big\{ \int_E \int_0^T \sigma^2(x,s)dC_s\eta(dx)\Big\}$$

$$= \eta(E)\, E\{ \int_E M_t(x)^2\, \eta(dx)\} < \infty.$$

Finally, to see (iv), note that

$$\sum E\{M_t(\phi_n)^2\} = \sum E\{(\int M_t(x)\phi_n(x)\eta(dx))^2\}$$
$$= E\{ \sum_n M_n^2(t)\}$$

where $M_n(t) = \int M_t(x)\phi_n(x)\eta(dx)$. By the Plancharel theorem, this is:

$$= E\{ \int M_t(x)^2\eta(dx)\} .$$

Q.E.D.

REMARK. Note from (iv) of the proposition that the sum in (2.18) does not depend on the particular CONS (ϕ_n).

Exercise 2.7. Suppose M has nuclear covariance on $L^2(E, \underline{\underline{E}}, \eta)$. Let f be predictable and set

$$k^2(x) = E\{ \int_0^T f^2(x,s)\; d < M(x)>_s\}.$$

Show that if $\int k^2(x)\; \eta(dx) < \infty$, then

(i) $\|f\|_M < \infty$ (so $f\cdot M$ is defined);

(ii) $f\cdot M_t(x) \overset{\text{def}}{=} \int_0^t f(x,s)dM_s(x)$ exists as an Ito integral for η-a.e. x;

(iii) $f\cdot M_t(A) = \int_A f\cdot M_t(x)\; \eta(dx)$;

(iv) for any CONS (ϕ_n) in $L^2(E, \underline{\underline{E}}, \eta)$ and $t \le T$,

$$\sum_n E\{f\cdot M_t(\phi_n)^2\} \le \int_E k^2(x)\eta(dx).$$

so that $f\cdot M$ has nuclear covariance.

AN EXAMPLE

We will construct D. Bakry's example of a martingale measure which is not an integrator.

Let U be a random variable, uniformly distributed on [0,1]. Let $s_n = 1 - 1/n$, n = 1,2,... and define a filtration $(\underline{\underline{F}}_t)$ as follows.

$$\underline{\underline{F}}_{s_n} = \sigma\{[2^n U]\} \qquad ([n] = \text{greatest integer} \le n)$$

$$\underline{\underline{F}}_t = \underline{\underline{F}}_{s_n} \quad \text{if} \quad s_n \le t < s_{n+1}$$

and define

$$M_t(A) = P\{U \in A \mid \underline{\underline{F}}_t\}, \quad A \subset [0,1], \quad t \ge 0.$$

If $K = [2^n U]$, let $J_n = [K\,2^{-n}, (K+1)2^{-n})$ and put $H_n = [K2^{-n}, (K+1/2)2^{-n})$ and $L_n = [(K+1/2)2^{-n}, (K+1)2^{-n})$. Then $J_n = H_n \cup L_n$ and all three are $\underline{\underline{F}}_{s_n}$-measurable. Note that $U \in J_n$ for all n.

Then M is a martingale measure and

(i) if $t < 1$, then $M_t(dx) = 2^n I_{J_n}(x)dx$;

(ii) if $t \ge 1$, then $M_t(A) = I_A(U)$.

If $t < 1$ and $s_n \le t < s_{n+1}$, then J_n is $\underline{\underline{F}}_{s_n}$-measurable and the conditional distribution of U given $\underline{\underline{F}}_t = \underline{\underline{F}}_{s_n}$ is uniform on J_n, which implies (i), while if $t \ge 1$, U is $\underline{\underline{F}}_t$ measurable, which implies (ii).

Thus $M_t(\cdot)$ is a (real-valued) measure of total mass one, not just an L^2-measure. However there exist bounded predictable f for which $\int f(x,s)dM$ does not exist.

Set

$$f(x,t) = \begin{cases} 2 & \text{if } x \in H_n \text{ and } s_n < t \le s_{n+1} \\ -2 & \text{if } x \in L_n \text{ and } s_n < t \le s_{n+1} \\ 0 & \text{otherwise} \end{cases}$$

Then f is adapted and $t \to f(x,t)$ is left continuous, so f is predictable. $t \to M_t$ is constant on each $[s_n, s_{n+1})$, and it jumps at each s_n. f is a sum of simple functions so, if $f\cdot M$ exists, we can compute it directly.

$$\int_{[0,1]\times(s_n,s_{n+1}]} f(x,t)dM = 2(M_{s_{n+1}}(H_n) - M_{s_n}(H_n)) - 2(M_{s_{n+1}}(L_n) - M_{s_n}(L_n)) .$$

Now by (i) this is

$$= I_{\{J_{n+1} = H_n\}} - I_{\{J_{n+1} = L_n\}} .$$

Since J_{n+1} is either H_n or L_n, with probability 1/2 each, the above is either 1 or -1, with probability 1/2 each. But now if $\int_{[0,1]\times(0,1)} f\,dM$ exists, it must equal $\sum_n \int_{[0,1]\times(s_n,s_{n+1}]} f\,dM$ which diverges since each term has absolute value 1.

Evidently the dominating measure K does not exist. To see why, let us calculate the covariance measure Q. Let $N_t = M_t(dx)M_t(dy)$. For $t = s_{n+1}$,

$$Q(dx,\ dy,\ \{s_{n+1}\}) = \Delta\langle M(dx),\ M(dy)\rangle_{s_{n+1}} = N_{s_{n+1}} - E\{N_{s_{n+1}} | \underline{\underline{F}}_{s_n}\}$$

$$= \left[4^n I_{\{x,y\ \in\ J_{n+1}\}} - 4^{n+1} P\{x,y \in J_{n+1} | \underline{\underline{F}}_{s_n}\}\right] dxdy.$$

If x and y are both in H_n or both in L_n, $P\{x,y \in J_{n+1} | \underline{\underline{F}}_{s_n}\} = 1/2$ by (i). If one is in H_n and one in L_n, they can't both be J_{n+1}, (which equals H_n or L_n) and the conditional expectation vanishes. Thus it is

$$= 4^n\left[I_{\{x,y\ \in\ J_{n+1}\}} - 2I_{\{x,y\ \in\ H_n\}} - 2I_{\{x,y\ \in\ L_n\}}\right]dxdy.$$

The term in brackets is ±1 on the set $\{x,y \in J_n\}$ and zero off. Thus

$$|Q(dx\ dy \times\{s_{n+1}\})| = 4^n I_{\{x,y\ \in\ J_n\}} dxdy$$

and hence

$$|Q([0,1] \times [0,1] \times \{s_{n+1}\})| = 4^n \int_0^1 \int_0^1 I_{J_n}(x) I_{J_n}(y) dxdy$$

$$= 1$$

Thus, if K exists, it must dominate Q and

$$K([0,1] \times [0,1] \times (0,1)) \geq \sum_n 1 = \infty.$$

CHAPTER THREE

EQUATIONS IN ONE SPACE DIMENSION

We are going to look at stochastic partial differential equations driven by white noise and similar processes. The solutions will be functions of the variables x and t, where t is the time variable and x is the space variable. There turns out to be a big difference between the case where x is one-dimensional and the case where $x \in \mathbf{R}^d$, $d \geq 2$. In the former case the solutions are typically, though not invariably, real-valued functions. They will be non-differentiable, but are usually continuous. On the other hand, in $\mathbf{R}^d$, the solutions are no longer functions, but are only generalized functions.

We will need some knowledge of Schwartz distributions to handle the case $d \geq 2$, but we can treat some examples in one dimension by hand, so to speak. We will do that in this chapter, and give a somewhat more general treatment later, when we treat the case $d \geq 2$.

THE WAVE EQUATION

Let us return to the wave equation of Chapter one:

$$(3.1)\qquad \begin{cases} \dfrac{\partial^2 V}{\partial t^2} = \dfrac{\partial^2 V}{\partial x^2} + \dot{W}\,, & t > 0,\ x \in \mathbf{R}; \\ V(x,0) = 0, & x \in \mathbf{R} \\ \dfrac{\partial V}{\partial t}(x,0) = 0, & x \in \mathbf{R}. \end{cases}$$

White noise is so rough that (3.1) has no solution: any candidate for a solution will not be differentiable. However, we can rewrite it as an integral equation which will be solvable. This is called a weak form of the equation.

We first multiply by a C^∞ function $\phi(x,t)$ of compact support and integrate over $\mathbf{R} \times [0,T]$, where $T > 0$ is fixed. Assume for the sake of argument that $V \in C^{(2)}$.

$$\int_0^T\int_{\mathbf{R}} [V_{tt}(x,t)-V_{xx}(x,t)]\phi(x,t)\,dx\,dt = \int_0^T\int_{\mathbf{R}} \phi(x,t)\dot{W}(x,t)\,dx\,dt.$$

Integrate by parts twice on the left-hand side. Now ϕ is of compact support in x, but it may not vanish at t = 0 and t = T, so we will get some boundary terms:

$$\int_0^T \int_R V(x,t)[\phi_{tt}(x,t)-\phi_{xx}(x,t)]dxdt + \int_R [\phi(x,\cdot)V_t(x,\cdot)\Big|_0^T - \phi_t(x,\cdot)V(x,\cdot)\Big|_0^T]dx$$

$$= \int_0^T \int_R \phi(x,t)\dot{W}(x,t)dxdt.$$

If $\phi(x,T) = \phi_t(x,T) = 0$, the boundary terms will drop out because of the initial conditions. This leads us to the following.

DEFINITION. We say that V is a weak solution of (3.1) providing that V(x,t) is locally integrable and that for all T > 0 and all C^∞ functions $\phi(x,t)$ of compact support for which $\phi(x,T) = \phi_t(x,T) = 0, \forall x$, we have

(3.2)
$$\int_0^T \int_R V(x,t)[\phi_{tt}(x,t) - \phi_{xx}(x,t)]dxdt = \int_0^T \int_R \phi dW.$$

The above argument is a little unsatisfying; it indicates that if V satisfies (3.1) in some sense, it should satisfy (3.2), while it is really the converse we want. We leave it as an exercise to verify that if we replace $\dot{W}$ by a smooth function f in (3.1) and (3.2), and if V satisfies (3.2) and is in $C^{(2)}$, then it does in fact satisfy (3.1).

THEOREM 3.1. There exists a unique continuous solution to (3.2), namely $V(x,t) = \frac{1}{2}\hat{W}(\frac{t-x}{\sqrt{2}}, \frac{t+x}{\sqrt{2}})$, where $\hat{W}$ is the modified Brownian sheet of Chapter One.

PROOF. Uniqueness: if V_1 and V_2 are both continuous and satisfy (3.2), then their difference $U = V_2 - V_1$ satisfies

$$\iint U(x,t)[\phi_{tt}(x,t) - \phi_{xx}(x,t)]dxdt = 0$$

Let f(x,t) be a C^∞ function of compact support in $R \times (0,T)$. Notice that there exists a $\phi \in C^\infty$ with $\phi(x,T) = \phi_t(x,T) = 0$ such that $\phi_{tt} - \phi_{xx} = f$. Indeed, if $C(x,t;\, x_o,t_o)$ is the indicator function of the cone $\{(x,t):\, t<t_o, x_o+t-t_o < x < x_o+t_o-t\}$, then

$$\phi(x_o,t_o) = \int_R \int_0^T f(x,t)C(x,t;\, x_o,t_o)dxdt.$$

Thus

$$\iint U(x,t)\ f(x,t)\ dxdt = 0,$$

so $U = 0$ a.e., hence $U \equiv 0$ a.e.

To show existence, let us rotate coordinates by 45°. Let $u = (t-x)/\sqrt{2}$, $v = (t+x)/\sqrt{2}$ and set $\hat{\phi}(u,v) = \phi(x,t)$ and $\hat{W}(dudv) = W(dxdt)$. Note that $\phi_{tt} - \phi_{xx} = 2\hat{\phi}_{uv}$. Define $\hat{R}(u,v;u_o,v_o) = I_{\{u\leq u_o, v\leq v_o\}}$. The proposed solution is

$$\hat{V}(u,v) = \frac{1}{2} \iint_{\{u'+v'>0\}} \hat{R}(u',v';u,v)\hat{W}(du'dv').$$

Now V satisfies (3.2) iff the following vanishes:

$$(3.3)\quad \iint_{\{u'+v'>0\}} \Big[\iint_{\{u+v>0\}} \frac{1}{2} \hat{R}(u,v;u',v')\hat{W}(dudv)\Big] 2\hat{\phi}_{uv}(u',v')du'dv' - \iint_{\{u+v>0\}} \hat{\phi}(u,v)\hat{W}(dudv).$$

We can interchange the order of integration by the stochastic Fubini's theorem of Chapter Two:

$$= \iint_{\{u+v>0\}} \Big[\int_v^\infty\int_u^\infty \hat{\phi}_{uv}(u',v')du'dv' - \hat{\phi}(u,v)\Big]\hat{W}(dudv).$$

But the term in brackets vanishes identically, for $\hat{\phi}$ has compact support.

QED

The literature of two-parameter processes contains studies of stochastic differential equations of the form

$$(3.4)\qquad d\hat{V}(u,v) = f(\hat{V})d\hat{W}(u,v) + g(\hat{V})dudv$$

where $\hat{V}$ and $\hat{W}$ are two parameter processes, and $d\hat{V}$ and $d\hat{W}$ represent two-dimensional increments, which we would write $\hat{V}(dudv)$ and $\hat{W}(dudv)$. These equations rotate into the non-linear wave equation

$$V_{tt}(x,t) = V_{xx}(x,t) + f(V)\dot{W}(x,t) + g(V)$$

in the region $\{(x,t): t > 0, -t<x<t\}$. One specifies Dirichlet boundary conditions. Because of the special nature of the region, it is enough to give V on the boundary; one does not have to give V_t as well.

We have of course just finished solving the linear case ($f\equiv 1$, $g\equiv 0$). We will not pursue this any further here but we will treat a non-linear parabolic equation in the next section.

AN EXAMPLE ARISING IN NEUROPHYSIOLOGY

Let us look at a particular parabolic SPDE. The general type of equation has many applications, but this particular example came up in connection with a study of neurons. These nerve cells are the building blocks of the nervous system, and operate by a mixture of chemical, biological and electrical properties, but in this particular mathematical oversimplifcation they are regarded as long, thin cylinders, which act much like electrical cables. If such a cylinder extends from 0 to L, and if we only keep track to the x coordinate, we let $V(x,t)$ be the electrical potential at the point x and time t. This potential is governed by a system of non-linear PDE's, called the Hodgkin-Huxley equations, but in certain ranges of values of V, they are well-approximated by the cable equation $V_t = V_{xx} - V$. (The variables have been scaled to make the coefficients all equal to one.)

The surface of the neuron is covered with synapses, thru which it receives impulses of current. If the current arriving at (x,t) is $F(x,t)$, the system will satisfy the inhomogeneous PDE

$$V_t = V_{xx} - V + F.$$

Even if the system is at rest, the odd random impulse will arrive, so that F will have a random component. The different synapses are more-or-less independent, and there are an immense number of them, so that one would expect the impulses to arrive according to a Poisson process. The impulses may be of different amplitudes and even of different signs (impulses can be either "excitatory" or "inhibitory").

We thus expect that F can be written as $F = \overline{F} + \Pi$, where $\overline{F}$ is deterministic and Π is a compound Poisson process, centered so that it has mean zero. Since the equation is linear, its solution will be the sum of the solutions of the PDE $V_t = V_{xx} - V + \overline{F}$, and of the SPDE $V_t = V_{xx} - V + \Pi$. We can study the two separately, and, since the first is familiar, we will concentrate on the latter.

The impulses are generally small, and there are many of them, so that in fact Π is very nearly a white noise $\dot{W}$. This leads us to study the SPDE $V_t = V_{xx} - V + \dot{W}$.

One final remark. The response of the neuron to a current impulse may depend on the local potential, so that instead of $\dot{W}$, we have a term $f(V)\dot{W}$ in the above equation. f is often assumed to have the form $f(V) = V-V_o$, where V_o is a constant.

Let W be a white noise on a probability space $(\Omega, \underline{\underline{F}}, P)$, let $(\underline{\underline{F}}_t)$ be a filtration such that W_t is adapted and such that, if $A \subset [t,\infty) \times R$, $W(A)$ is independent of $\underline{\underline{F}}_t$. Consider

$$(3.5)\qquad \begin{cases} \dfrac{\partial V}{\partial t} = \dfrac{\partial^2 V}{\partial x^2} - V + f(V,t)\dot{W}, & t > 0,\ 0 < x < L; \\ \dfrac{\partial V}{\partial x}(0,t) = \dfrac{\partial V}{\partial x}(L,t) = 0, & t > 0; \\ V(x,0) = V_o(x), \ t > 0. & \end{cases}$$

We assume that V_o is $\underline{\underline{F}}_o$-measurable and that $E\{V_o(x)^2\}$ is bounded, and that f satisfies a uniform Lipschitz condition, so that there exists a constant K such that

$$(3.5a)\qquad \begin{cases} |f(y,t) - f(x,t)| \leq K|y-x|, \\ |f(y,t)| \leq K(1+t)(1+|y|) \end{cases}$$

for all $x,y \in [0,L]$ and $t > 0$.

The homogeneous form of (3.5) is called the <u>cable equation</u>. We have specified reflecting boundaries for the sake of concreteness, but there is no great difficulty in treating other boundary conditions.

The Green's function for the cable equation can be gotten by the method of images. It is given by

$$G_t(x,y) = \frac{e^{-t}}{\sqrt{4\pi t}} \sum_{n=-\infty}^{\infty} \left[\exp\left(-\frac{(y-x-2nL)^2}{4t}\right) + \exp\left(-\frac{(y+x-2nL)^2}{4t}\right)\right].$$

We won't need to use this explicitly. We will just need the following facts, which can be seen directly:

$$(3.6)\qquad \int_0^L G_s(x,y)G_t(y,z)\,dy = G_{s+t}(x,z), \text{ and } G_t(x,y) = G_t(y,x);$$

for each $T > 0$ there is a constant C_T such that

$$(3.7)\qquad G_t(x,y) \leq \frac{C_T}{\sqrt{t}} \exp\left(-\frac{|y-x|^2}{4t} - t\right).$$

Define $G_t(\phi,y) = \int_0^L G_t(x,y)\phi(x)\,dx$ for any function ϕ on $[0,L]$ for which the integral exists. Then $G_t(x,y)$ satisfies the homogeneous cable equation (i.e. it satisfies (3.5) with $f \equiv 0$) except at $t = 0$, and $G_o(\phi,y) = \phi(y)$. After integrating by parts, we have

(3.8) $$G_t(\phi,y) = \phi(y) + \int_0^t G_s(\phi''-\phi;y)ds$$

for all test functions ϕ for which $\phi'(0) = \phi'(L) = 0$.

Once again we pose the problem in a weak form.

Let $\phi \in C^\infty(R)$, with $\phi'(0) = \phi'(L) = 0$. Multiply (3.5) by $\phi(x)$ and integrate over both variables:

$$\int_0^L V(x,t)\phi(x)dx = \int_0^L V_o(x)\phi(x)dx + \int_0^t\int_0^L \left(\frac{\partial^2 V}{\partial x^2} - V\right)(x,s)\phi(x)dsdx$$
$$+ \int_0^t\int_0^L f(V(x,s),s)\phi(x)W(dxds).$$

Integrate by parts over x and use the boundary conditions on V and ϕ to get the following weak form of (3.5):

For each $\phi \in C^\infty(R^n)$ of compact support such that $\phi'(0) = \phi'(L) = 0$,

(3.9) $$\int_0^L (V(x,t)-V_o(x))\phi(x)dx = \int_0^t\int_0^L V(x,s)(\phi''(x)-\phi(x))dxds+\int_0^t\int_0^L f(V(x,s),s)\phi(x)W(dxds)$$

Exercise 3.1. (3.9) can be extended to smooth functions $\psi(x,t)$ of two variables which satisfy $\frac{\partial\psi}{\partial x}(0,t) = \frac{\partial\psi}{\partial x}(L,t) = 0$ for each t. Show that (3.9) implies that

$$\int_0^L \left[V(x,t)\,\psi(x,t) - V_o(x)\psi(x,0)\right]dx$$

(3.10) $$= \int_0^t\int_0^L V(x,s)\left(\frac{\partial^2\psi}{\partial x^2} - \psi + \frac{\partial\psi}{\partial t}\right)(x,s)\,dxds$$
$$+ \int_0^t\int_0^L f(V(x,s),s)\psi(x,s)W(dx\,ds)\ .$$

Exercise 3.2. Show that (3.5) and (3.9) are equivalent if things are smooth, i.e. show that if V_0 and $\dot{W}$ are smooth functions and if $V \in C^{(2)}$, then (3.9) implies (3.5).

THEOREM 3.2. There exists a unique process $V = \{V(x,t),\ t\geq 0,\ 0<x<L\}$ which is L^2-bounded on $[0,L] \times [0,T]$ for any T and which satisfies (3.9) for all $t \geq 0$. If

$V_o(x)$ is bounded in L^p for some $p \geq 2$, then $V(x,t)$ is L^p-bounded on $[0,L] \times [0,T]$ for any T.

PROOF. We will suppress the dependence of f on s, and write $f(x)$ rather than $f(x,s)$.

Uniqueness: a solution of (3.9) must satisfy (3.10), so fix t and let $\psi(y,s) = G_{t-s}(\phi,y)$. Then $\psi(y,t) = \phi(y)$ and, by (3.8), $\psi_{xx} - \psi + \psi_s = 0$. Thus (3.10) becomes

$$\int_0^L V(x,t)\phi(x)dx = \int_0^L V_o(y)G_t(\phi,y)dy + \int_0^t\int_0^L f(V(y,s))G_{t-s}(\phi,y)W(dyds).$$

Let us refine this. $E\{V^2(x,t)\}$ is bounded in $[0,L]$, so for a.e. ω $V^2(x,t)$ will be integrable with respect to x (Fubini's theorem). Let ϕ approach a delta function, e.g. take ϕ of the form $(2\pi n)^{-1/2}\exp(-\frac{(y-x)^2}{2n})$ and let $n \to \infty$. The above equation will tend to

$$(3.11) \qquad V(x,t) = \int_0^L V_o(y)G_t(x,y)dy + \int_0^t\int_0^L f(V(y,s))G_{t-s}(x,y)W(dyds)$$

a.s. for a.e. pair (t,x). (To see this, apply Lebesgue's differentiation theorem to the left hand side, and note that $G_t(\phi,y) \to G_t(x,y)$ on the right.)

If V_1 and V_2 both satisfy (3.11), let $U = V_2 - V_1$, and define $F(x,t) = E\{U^2(x,t)\}$ and $H(t) = \sup_x F(x,t)$, which is finite by hypothesis. Then from (3.11)

$$F(x,t) = \int_0^t\int_0^L E\{(f(V_2(y,s)) - f(V_1(y,s))^2\}G_{t-s}^2(x,y)dy\,ds$$

$$\leq K^2 \int_0^t\int_0^L F(y,s)G_{t-s}^2(x,y)dy\,ds$$

by (3.5a). Thus

$$H(t) \leq K^2 \int_0^t H(s) \int_0^L G_{t-s}^2(x,y)dy\,ds$$

$$\leq K^2 C\int_0^t H(s) \frac{ds}{\sqrt{t-s}}$$

by (3.7). Iterate this:

$$\leq (K^2 C)^2 \int_0^t\int_0^s H(u) \frac{du\,ds}{\sqrt{(s-u)(t-s)}}.$$

Interchange the order in the integral, noting that if $t - u = b$,

$$\int_u^t \frac{ds}{\sqrt{(t-s)(s-u)}} = \int_0^b \frac{dv}{\sqrt{v(b-v)}} \le 2\int_0^{b/2}\sqrt{\frac{2}{b}}\,\frac{dv}{\sqrt{v}} = 4$$

so that

$$H(t) \le 4K^4C^2 \int_0^t H(s)\,ds.$$

Iterating this, we see it is

$$\le \frac{(4K^4C^2)^{n+1}}{n!} \int_0^t H(s)(t-s)^n ds$$

which tends to zero. Thus $H = 0$, so with probability one, $V^1 = V^2$ a.e.

To prove existence, we take a hint from (3.1) and define

$$(3.12)\qquad \begin{cases} V^0(x,t) = \int_0^L V_0(y)G_t(x,y)dy \\ V^{n+1}(x,t) = V^0(x,t) + \int_0^t\int_0^L f(V^n(y,s))G_{t-s}(x,y)W(dyds) \end{cases}$$

Let $p \ge 2$ and suppose that $\{V_0(y),\ 0\le y\le L\}$ is L^p bounded. We will show that V^n converges in L^p to a solution V. Define

$$F_n(x,t) = E\{|V^{n+1}(x,t) - V^n(x,t)|^p\}$$

and

$$H_n(t) = \sup_x F_n(x,t).$$

From (3.12)

$$F_n(x,t) = E\{|\int_0^t\int_0^L (f(V^n(y,s)) - f(V^{n-1}(y,s)))G_{t-s}(x,y)W(dyds)|^p\}.$$

We are trying to find the pth moment of a stochastic integral. We can bound this in terms of the associated increasing process by Burkholder's inequality.

$$\le C_p E\{(\int_0^t\int_0^L |f(V^n(y,s)) - f(V^{n+1}(y,s))|^2 G_{t-s}^2(x,y)dyds)^{p/2}\}$$

$$\le C_p K\, E\{|\int_0^t\int_0^L (V^n(y,s) - V^{n-1}(y,s))^2 G_{t-s}^2(x,y)dyds|^{p/2}\}$$

where we have used (3.5a), the Lipschitz condition on f. We can bound this using Hölders inequality. To see how to choose the exponents, note from (3.7) that if $0 < r < 3$,

$$(3.13)\qquad \int_0^L G_t^r(x,y)dy \le Ce^{-tr}t^{-r/2}\int_{-\infty}^{\infty} e^{-\frac{ry^2}{2t}}\,dy \le C'e^{-tr}\,t^{\frac{1-r}{2}},$$

which is integrable in t over the interval $(0,\infty)$. Thus we must keep the exponent of G under 3. Set $q = \frac{p}{p-2}$ and choose $0 \le \varepsilon \le 1$ to be strictly between $1 - \frac{3}{p}$ and $\frac{3}{2} - \frac{3}{p}$ ($\varepsilon = 0$ if $p = 2$ and $\varepsilon = 1$ if $p \ge 6$). Then

$$F_n(x,t) \le C\Big(\int_0^t\int_0^L G_s^{2\varepsilon q}(x,y)\Big)^{p/2q} \int_0^t\int_0^L E\{|V^n(y,s)-V^{n-1}(y,s)|^p\}G_{t-s}^{(1-\varepsilon)p}(x,y)\,dyds\,.$$

In this case $2\varepsilon q < 3$, so the first factor is bounded; by (3.13) the expression is

$$\le C\int_0^t H_{n-1}(s)\,(t-s)^a ds,$$

where $a = \frac{1}{2}(1+\varepsilon p-p) > -1$, and C is a constant.

Thus

$$H_n(t) \le C\int_0^t H_{n-1}(s)(t-s)^a ds, \quad t \ge 0 \tag{3.14}$$

for some $a > -1$ and $C > 0$. Notice that if H_{n-1} is bounded on an interval $[0,T]$, so is H_n.

$$H_o(t) \le \sup_x C_p E\{|\int_0^t f(V^o(x,s))^2 G_{t-s}(x,y)dyds|^{p/2}\}.$$

But $V^o(x,s)$ is L^p-bounded since $V_0(y)$ is, hence so is $f(V^o(x,s))$ by (3.5a). An argument similar to the above shows $H_0(t)$ is bounded on $[0,T]$.

Thus the H_n are all finite. We must show they tend to zero quickly. This follows from:

<u>LEMMA 3.3</u>. Let $\{h_n(t),\ n=0,1,\dots\}$ be a sequence of positive functions such that h_0 is bounded on $[0,T]$ and, for some $a > 1$ and constant C_1,

$$h_n(t) \le C_1\int_0^t h_{n-1}(s)(t-s)^a ds, \quad n = 1,2,\dots\,.$$

Then there is a constant C and an integer $k > 1$ such that for each $n \ge 1$ and $t \in [0,T]$,

$$h_{n+mk}(t) \le C^m\int_0^t h_n(s)\,\frac{(t-s)}{(m-1)!}ds, \quad m = 1,2,\dots\,. \tag{3.15}$$

Let us accept the lemma for the moment. It applies to the H_n, and implies that for each n, $\sum_{m=0}^{\infty} (H_{n+mk}(t))^{1/p}$ converges uniformly on compacts, and therefore so does $\sum_{n=0}^{\infty} (H_n(t))^{1/p}$. Thus $V^n(x,t)$ converges in L^p, and the convergence is uniform in $[0,L] \times [0,T]$ for any $T > 0$. In particular, V^n converges in L^2. Let $V(x,t) = \lim V^n(x,t)$.

It remains to show that V satisfies (3.9). (Note that it is easy to show that V satisfies (3.11) - this follows from (3.12). However, we would still have to show that (3.11) implies (3.9), so we may as well show (3.9) directly.)

Consider

$$(3.16) \qquad \int_0^L (V^n(x,t)-V_0(x))\phi(x)dx - \int_0^t\int_0^L V^n(x,s)[\phi''(x)-\phi(x)]dx\,ds - \int_0^t\int_0^L f(V^{n-1}(y,s))\phi(y)W(dyds).$$

By (3.12) this is

$$= \int_0^L\int_0^t\int_0^L f(V^{n-1}(y,s))G_{t-s}(x,y)W(dyds)\phi(x)dx$$

$$+ \int_0^L \Big(\int_0^L G_t(x,y)V_0(y)dy - V_0(x)\Big)\phi(x)dx$$

$$- \int_0^t\int_0^L \Big[\int_0^L G_s(x,y)V_0(y)dy + \int_0^u\int_0^L f(V^{n-1}(y,s))G_{u-s}(x,y)W(dyds)\Big](\phi''(x)-\phi(x))dxdu$$

$$- \int_0^t\int_0^L f(V^{n-1}(y,s))\phi(y)W(dyds).$$

Integrate first over x and collect terms:

$$= \int_0^t\int_0^L f(V^{n-1}(y,s))\Big[G_{t-s}(\phi,y) - \int_s^t G_{u-s}(\phi''-\phi,y)ds - \phi(y)\Big]W(dyds)$$

$$- \int_0^L \Big[G_t(\phi,y) - \phi(y) - \int_0^t G_u(\phi''-\phi,y)du\Big]V_0(y)dy.$$

But this equals zero since both terms in square brackets vanish by (3.8). Thus (3.16) vanishes for each n. We claim it vanishes in the limit too.

Let $n \to \infty$ in (3.16). $V^n(x,s) \to V(x,s)$ in L^2, uniformly in $[0,L]\times[0,T]$ for each $T > 0$, and, thanks to the Lipschitz conditions, $f(V^{n-1}(y,s))$ also converges uniformly in L^2 to $f(V(y,s))$.

It follows that the first two integrals in (3.16) converge as $n \to \infty$. So does the stochastic integral, for

$$E\Big\{\Big(\int_0^t\int_0^L (f(V(y,s)) - f(V^{n-1}(y,s))\phi(y)W(dyds)\Big)^2\Big\}$$

$$\leq K \int_0^t\int_0^L E\{(V(y,s) - V^{n-1}(y,s))^2\}\,\phi(y)dyds$$

which tends to zero. It follows that (3.16) still vanishes if we replace V^n and V^{n-1} by V. This gives us (3.9).

Q.E.D.

We must now prove the lemma.

PROOF (of Lemma 3.3). If $a \geq 0$ take $k = 1$ and $C = C_1$. If $-1 < a < 0$,

$$h_n(t) \leq C_1^2 \int_0^t h_{n-2}(u)\left(\int_u^t (t-s)^a (s-u)^a ds\right)du.$$

If $a = -1 + \varepsilon$, the inner integral is bounded above by

$$2\left(\frac{2}{t-u}\right)^{1-\varepsilon} \int_0^{1/2(t-u)} \frac{dv}{v^{1-\varepsilon}} \leq \frac{4}{\varepsilon}(t-u)^{2\varepsilon-1}$$

so

$$h_n(t) \leq C_1 \int_0^t h_{n-1}(s) \frac{ds}{(t-s)^{1-\varepsilon}} \leq \frac{4}{\varepsilon} C_1^2 \int_0^t h_{n-2}(s) \frac{ds}{(t-s)^{1-2\varepsilon}} .$$

If $2\varepsilon \geq 1$ we stop and take $k = 2$ and $C = \frac{4}{\varepsilon} C_1^2$. Otherwise we continue

$$\leq \frac{16}{\varepsilon^2} C_1^4 \int_0^t h_{n-4}(s) \frac{ds}{(t-s)^{1-4\varepsilon}}$$

until we get (t-s) to a positive power. When this happens, we have

$$h_n(t) \leq C \int_0^t h_{n-k}(s)ds.$$

But now (3.15) follows from this by induction. Q.E.D.

In many cases the initial value $V_0(x)$ is deterministic, in which case $V(x,t)$ will be bounded in L^p for all p. We can then show that V is actually a continuous process, and, even better, estimate its modulus of continuity.

COROLLARY 3.4. Suppose that $V_0(x)$ is L^p-bounded for all $p > 0$. Then for a.e. ω, $(x,t) \to V(x,t)$ is a Hölder continuous function with exponent $\frac{1}{4} - \varepsilon$, for any $\varepsilon > 0$.

PROOF. A glance at the series expansion of G_t shows that it can be written

$$G_t(x,y) = g_t(x,y) + H_t(x,y)$$

where

$$g_t(x,y) = (4\pi t)^{-1/2} e^{-\frac{(y-x)^2}{4t} - t},$$

$H_t(x,y)$ is a smooth function of (t,x,y) on $(0,L) \times (0,L) \times (-\infty, \infty)$, and H vanishes if $t \leq 0$. By (3.11)

$$V(x,t) = \int_0^L V_0(y)G_t(x,y)dy + \int_0^t\int_0^L f(V(y,s))H_{t-s}(x,y)W(dyds)$$
$$+ \int_0^t\int_0^L f(V(y,s))g_{t-s}(x,y)W(dyds).$$

The first term on the right hand side is easily seen to be a smooth function of (x,t) on $(0,L) \times (0,\infty)$. The second term is basically a convolution of W with a smooth function H. It can also be shown to be smooth; we leave the details to the reader.

Denote the third term by U(x,t). We will show that U is Hölder continuous by estimating the moments of its increments and using Corollary 1.4. Now

$$E\{|U(x+h,\ t+k) - U(x,t)|^n\}^{1/n} \le E\{|U(x+h,\ t+k) - U(x,\ t+k)|^n\}^{1/n}$$
$$+ E\{|U(x,t+k) - U(x,t)|^n\}^{1/n}.$$

We will estimate the two terms separately. The basic idea is to use Burkholder's inequality to bound the moments of each of the stochastic integrals. Replacing t+k by t, we see that

$$E\{|U(x+h,t) - U(x,t)|^n\} \le C\, E\{|\int_0^t\int_0^L f^2(V(y,s))(g_{t-s}(x+h,y) - g_{t-s}(x,y))^2 dyds|^{\frac{n}{2}}\}.$$

Apply Hölders inequality with $p = n/2$, $q = \frac{n}{n-2}$:

$$\le C_n\, E\{\int_0^t\int_0^L f(V(y,s))^2 dyds\}\ [\int_0^t\int_0^L |g_s(x+h,y)-g_s(x,y)|^{2q} dyds]^{\frac{n}{2q}}.$$

The expectation is finite for any n by (3.5a) and Theorem 3.2. Letting C be a constant whose value many change from line to line we have

$$\le C\ [\int_0^t\int_{-\infty}^{\infty} \frac{e^{-2qs}}{(4\pi s)^q}\ |e^{\frac{(y+h)^2}{4s}} - e^{-\frac{y^2}{4s}}|^{2q} dy\ ds]^{\frac{n}{2q}}.$$

If we let $y = hz$ and $s = h^2v$, we can see this is

$$\le C[h^{3-2q}\int_0^{\infty}\int_0^{\infty} v^{-q}|e^{-\frac{(z+1)^2}{4v}} - e^{-\frac{z^2}{4v}}|^{2q} dv]^{\frac{n}{2q}}.$$

The integral converges if $q < 3/2$, i.e. if $n > 6$, so that this is

$$= C\, h^{\frac{n}{2}-1}.$$

The first term of (3.15) is thus bounded by $C\, h^{\frac{1}{2}-\frac{1}{n}}$.

Similarly,

$$E\{|U(x,t+k) - U(x,t)|^n\}^{1/n} \le C_n\, E\Big\{\Big|\int_0^t\int_0^L f(V(y,s))^2|g_{t+k}(x,y) - g_t(x,y)|^2 dyds\Big|^{\frac{n}{2}}\Big\}^{\frac{1}{n}}$$

$$+ C_n E\Big\{\Big|\int_t^{t+k}\int_0^L f(V(y,s))^2|g_{t+k}(x,y)|^2 dyds\Big|^{\frac{n}{2}}\Big\}^{\frac{1}{n}}.$$

The first expectation on the right is bounded by

$$C\, E\Big\{\int_0^t\int_0^L f(V(y,s))^n dy\, ds\Big\}^{1/n}\Big[\int_0^t\int_{-\infty}^{\infty} \Big|(s+k)^{-1/2} e^{-\frac{y^2}{4(s+k)}} - s^{-1/2} e^{-\frac{y^2}{4s}}\Big|^{2q}\Big]^{\frac{1}{2q}}$$

The expectation above is finite. If we set $s = ku$, $y = \sqrt{k}\, z$, this becomes

$$\le C\Big[k^{3/2-q}\int_0^{\infty}\int_{-\infty}^{\infty}\Big|\frac{e^{-\frac{-z^2}{4(u+1)}}}{\sqrt{u+1}} - \frac{e^{-\frac{z^2}{4u}}}{\sqrt{u}}\Big|^{2q} dzdu\Big]^{1/2q}.$$

The integral converges if $q < 3/2$, i.e. if $n > 6$, so the expansion is

$$= C\, k^{\frac{1}{4} - \frac{2}{n}}.$$

Finally, the second expectation on the right is bounded by

$$C_n E\Big\{\int_t^{t+k}\int_0^L f^n(V(y,s))dyds\Big\}^{1/n}\Big[\int_0^k\int_0^L g_s(x,y)^2 dyds\Big]^{1/2q}$$

We have seen that $E\{f^n(V(y,s))\}$ is bounded so this is

$$\le C\, k^{1/n}\Big[\int_0^k\int_{-\infty}^{\infty} s^{-q} e^{-\frac{qy^2}{s}} dy\, ds\Big]^{1/2q}.$$

We can do the integral explicitly to get

$$= C\, k^{\frac{1}{4} - \frac{2}{n}}.$$

Putting these bounds together in (3.16) we see that

$$E\{|U(x+h,\, t+k) - U(x,t)|^n\}^{1/n} \le C\Big[h^{\frac{1}{4} - \frac{1}{n}} + 2k^{\frac{1}{4} - \frac{2}{n}}\Big]$$

$$\le C\Big(\sqrt{h^2 + k^2}\Big)^{\frac{1}{4} - \frac{4}{n}}.$$

We can choose n as large as we please, so that the result follows from Kolmogorov's Theorem (Corollary 1.4).

Q.E.D.

The uniqueness theorem gives us the Markov property of the solution, exactly as it does in the classical case. We omit the proof.

THEOREM 3.5. The process $\{V(\cdot,t),\ t\geq 0\}$, considered as a process taking values in $C[0,L]$, is a diffusion process.

Consider the more general equation

$$(3.5b)\qquad \begin{cases} \dfrac{\partial V}{\partial t} = \dfrac{\partial^2 V}{\partial x^2} + g(V,t) + f(V,t)\dot{W}, & t > 0,\ 0 < x < L; \\ \dfrac{\partial V}{\partial x}(0,t) = \dfrac{\partial V}{\partial x}(L,t) = 0, & t > 0; \\ V(x,0) = V_0(x), & 0 < x < L. \end{cases}$$

Exercise 3.3. Find the weak form of (3.5b).

Exercise 3.4. Show that if both f and g satisfy the Lipschitz condition (3.5a) then Theorem 3.2 holds. In particular, (3.5b) has a unique weak solution.

(Hint. The Green's function is as before except that there is no factor e^{-t}, and the Picard iteration formula (3.12) becomes

$$V^{n+1}(x,t) = V^0(x,t) + \int_0^t\int_0^L G_{t-s}(x,y)[g(V^n(y,s))dyds + f(V^n(y,s))W(dyds)].$$

The proof of Theorem 3.2 then needs only a little modification. For instance, for uniqueness, let

$$F(x,t) = 2\int_0^t\int_0^L G^2_{t-s}(x,y)[(f(V_2(y,s)) - f(V_1(y,s)))^2 + Lt(g(V_2(y,s) - g(V_1(y,s)))^2]dyds,$$

show that $E\{|V_2(x,t) - V_1(x,t)|^2\} \leq F(x,t)$,

and conclude that $H(t) \leq K^2C \displaystyle\int_0^t H(s)\,\frac{ds}{\sqrt{t-s}}$.

In order to prove existence, define F_n and H_n as in the proof and note that, once (3.14) is established, the rest of the proof follows nearly word by word. In order to prove (3.14), first show that

$$F_n(x,t) \leq 2^pK\,E\{|\int_0^t\int_0^L |V^n(y,s) - V^{n-1}(y,s)|G_{t-s}(x,y)dyds|^p\}$$
$$+ 2^pC_pK\,E\{|\int_0^t\int_0^L |V^n(y,s) - V^{n-1}(y,s)|^2G_{t-s}(x,y)dyds|^{p/2}\},$$

and then apply Hölder's inequality to each term as in the proof to deduce (3.14).)

<u>Exercise 3.5.</u> Show that Corollary 3.5 also holds for solutions of (3.5b), so that the weak solution of (3.5) is Hölder continuous.

In case f is <u>constant</u>, there is a direct relation between the solutions of (3.5) and (3.5b).

<u>Exercise 3.6.</u> Let $f(x,t) = \sigma$ be constant, and let U and V be solutions of (3.5) and (3.5b) respectively. Show that $V = U + u$, where $u(x,t)$ is the solution of the PDE

$$\begin{cases} \dfrac{\partial u}{\partial t} = \dfrac{\partial^2 u}{\partial x^2} + g(U(x,t) + u(x,t),t); \\ u(0,t) = u(L,t) = 0; \\ u(x,0) = 0. \end{cases}$$

Thus write V explicitly in terms of U.

(The point is that once U is known, one can fix ω and solve this as a classical non-stochastic PDE; the solution u can be written in terms of the Green's function.)

The technique of Picard iteration works for the non-linear wave equation, too. Consider

$$\text{(3.1a)} \qquad \begin{cases} \dfrac{\partial^2 V}{\partial t^2} = \dfrac{\partial^2 V}{\partial x^2} + g(V,t) + f(V,t)\dot{W}, & t > 0,\ x \in \mathbf{R}; \\ V(x,0) = V_0(x), & x \in \mathbf{R}; \\ \dfrac{\partial V}{\partial t}(x,0) = U_0(x), & x \in \mathbf{R}. \end{cases}$$

In this case we let $V^0(x,t)$ be the classical solution of the homogeneous wave equation with initial position $V_0(x)$ and velocity $U_0(x)$ - which we can write explicitly - and define

$$V^{n+1}(x,t) = V^0(x,t) + \int_0^t \int_{\mathbf{R}} C(x,t;y,s)[g(V^{n-1}(y,s))dyds + f(V^{n-1}(y,s))W(dyds)]$$

where $C(x,t;\ y,s) = \begin{cases} 1 & \text{if } s \le t \text{ and } |y-x| \le t-s, \\ 0 & \text{otherwise}. \end{cases}$

Then C is the indicator function of the light cone.

The Picard iteration is in fact easier than it was for the cable equation since C is bounded. We leave it as an exercise for iteration enthusiasts to carry out.

<u>Exercise 3.7.</u> (a) Write (3.1a) in a weak form analogous to (3.2).

(b) Show that if both f and g satisfy the Lipschitz conditions (3.5a), then (3.1a) has a unique weak solution, which has a Hölder continuous version.

THE LINEAR EQUATION

Let us now consider the linear equation ($f \equiv$ constant). This is relatively easy to analyze because the solution is Gaussian and most questions can be answered by computing covariances. (The case $f(x) = ax + b$ is linear too, but the solution, which now involves a product $V\dot{W}$, is no longer Gaussian. This case is often referred to as <u>semi-linear</u>).

The solution can be expanded in eigenfunctions. Assume $L = \pi$, so that (3.5) becomes

(3.17)
$$\begin{cases} V_t = V_{xx} - V + \dot{W}, & 0 < x < \pi,\ t > 0, \\ V_x(0,t) = V_x(\pi,t) = 0, & t > 0; \\ V(x,0) = 0, & 0 < x < \pi. \end{cases}$$

The eigenfunctions and eigenvalues of (3.17) are

$$\phi_0 \equiv 1/\sqrt{\pi}, \quad \phi_k(x) = \sqrt{\frac{2}{\pi}} \cos kx \quad k = 1,2,\ldots$$

$$\lambda_k = k^2 + 1,\ k = 0,1,2,\ldots .$$

The Green's function can be expanded in the ϕ_k:

$$G_t(x,y) = \sum_{k=0}^{\infty} \phi_k(x)\phi_k(y)e^{-\lambda_k t}.$$

For each fixed x, this converges in $L^2[0,\pi]\times[0,T]$ as a function of (y,t).

Thus the unique solution of (3.17) is, by (3.11)

(3.17a)
$$V_t = \int_0^t\int_0^\pi G_{t-s}(x,y)W(dyds)$$
$$= \int_0^t\int_0^\pi \sum_{k=0}^{\infty} \phi_k(x)\phi_k(y)e^{-\lambda_k(t-s)} W(dyds).$$

We can interchange order since the series converges in L^2:

$$= \sum_{k=0}^{\infty} \Big(\int_0^t \int_0^{\pi} \phi_k(y) e^{-\lambda_k(t-s)} W(dyds)\Big)\phi_k(x).$$

<u>Exercise 3.8</u>. Define $B_t^k = \int_0^t \int_0^{\pi} \phi_k(y) W(dyds$

and $A_t^k = \int_0^t \int_0^{\pi} \phi_k(y) e^{-\lambda_k(t-s)} W(dyds).$

(i) Show that $B^1, B^2, \ldots$ are iid standard Brownian motions, and that

$$A_t^k = \int_0^t e^{-\lambda_k(t-s)} dB_s^k.$$

(ii) Show that A^k satisfies $dA_t^k = dB_t^k - \lambda_k A_t^k dt$.

The processes A^k are familiar, for they are Ornstein-Uhlenbeck processes with parameter λ_k - abbreviated $OU(\lambda_k)$ - which are mean zero Gaussian Markov processes. They are independent.

Thus we have

(3.17b) $$V(x,t) = \sum_{k=0}^{\infty} A_t^k \phi_k(x)$$

where A^k is an $OU(\lambda_k)$ process and $A^0, A^1, A^2, \ldots$ are independent. Recall the following well-known facts about Ornstein-Uhlenback processes.

<u>PROPOSITION 3.6</u>. Let $\{A_t, t \geq 0\}$ be an $OU(\lambda)$ process with $A_0 = 0$. A is a mean zero Gaussian process with covariance function

(i) $E\{A_{s+t}A_t\} = \frac{e^{-\lambda s}}{2\lambda}\,[1 - e^{-2\lambda t}]$,

(ii) $E\{(A_{s+t} - A_t)^2\} = \frac{1}{\lambda}(1 - e^{-\lambda s}) - \frac{1}{2\lambda}\, e^{-2\lambda t}(1 - e^{-\lambda s})^2$.

<u>PROPOSITION 3.7</u>. If $0 \leq s \leq t$ and $0 \leq x, y \leq \pi$,

(i) $E\{(V(y,t) - V(x,t))^2\} \leq 4|y-x|$;

(ii) $E\{(V(x,t) - V(y,s))^2\} \leq \frac{4}{\pi}\sqrt{t-s}$.

Before proving this, let us see what we can learn from it. From (i) and (ii),

$$\max\{E\{(V(x,t) - V(y,s))^2\} : |t-s|^2 + |y-x|^2 \leq 2^{-1/2}u\} \leq Cu^{1/4}$$

Thus, let $p(u) = Cu^{1/4}$ in Corollary 1.3 to get

THEOREM 3.8. V has a version which is continuous in the pair (x,t). For T > 0 there is a constant C and random variable Y such that the modulus of continuity $\Delta(\delta)$ of $\{V(x,t),\ t\le T)\}$ satisfies

$$\Delta(\delta) \le Y\delta^{1/4} + C\delta^{1/4}\sqrt{\log 1/\delta},\quad 0 \le \delta \le 1.$$

When we compare Theorem 3.8 with Corollary 3.4, we see that the moduli of continuity are substantially the same in the linear and non-linear cases. The paths are essentially Hölder (1/4).

We will need the following lemma.

LEMMA 3.9. (i) $\displaystyle\sum_{k=1}^{\infty} \frac{(\phi_k(y)-\phi_k(x))^2}{2\lambda_k} \le 4|y-x|$

(ii) $\displaystyle\sum_{k=1}^{\infty} \frac{1-e^{-\lambda_k t}}{2\lambda_k} \le 1 \wedge \sqrt{t}.$

PROOF. $\phi_k(x) = \sqrt{2}\cos kx$ so

$$(\phi_k(y) - \phi_k(x))^2 \le 2(4\wedge k^2(y-x)^2)\ .$$

Since $\lambda_k > k^2$,

$$\sum_{k=1}^{\infty} \frac{(\phi_k(y)-\phi_k(x))^2}{2\lambda_k} \le \int_1^{\infty} \frac{4}{u^2} \wedge (y-x)^2 du$$

$$\le \int_1^{2|y-x|^{-1}} (y-x)^2 du + 4\int_{2|y-x|^{-1}}^{\infty} \frac{du}{u^2}$$

$$\le 4|y-x|.$$

The second series is handled the same way, using $1 - e^{-\lambda_k t} \le 1 \wedge (1+k^2)t$. We leave the details to the reader.

PROOF (of Proposition 3.7). We prove (i), for (ii) is similar. By (3.17b) and Propositon 3.6,

$$E\{(V(y,t) - V(x,t))^2\} = E\{[\sum_{k=0}^{\infty} A_t^k(\phi_k(y) - \phi_k(x))]^2\}$$

$$\le \sum_{k=0}^{\infty} \frac{1}{2\lambda_k}(\phi_k(y) - \phi_k(x))^2$$

$$\le 4|y-x|$$

by the Lemma. Q.E.D.

When we compare Theorem 3.8 with Corollary 3.4, we see that the moduli of continuity are substantially the same in the linear and non-linear cases. The paths are essentially Hölder (1/4).

Note that V is rougher than either Brownian motion or the Brownian sheet. Both of the latter are Hölder continuous of exponent 1/2, while V has exponent 1/4. We can ask if this is due to bad behavior in t, or in x, or in combination. The following exercises eliminate x as a suspect.

Exercise 3.9. Expand Brownian motion B_t in the eigenfunctions ϕ_k to see

$$B_t = \frac{1}{\sqrt{3}}\,\xi_0 + \sum_{k=1}^{\infty} \frac{1}{k}\,\xi_k \phi_k(t), \quad 0 < t < \pi, \tag{3.18}$$

where the ξ_k are iid N(0,1).

Exercise 3.10. Write $V(x,t) = \frac{1}{\sqrt{2}} \sum_{k=0}^{\infty} \frac{\sqrt{2\lambda_k}\; A_k(t)}{\sqrt{\lambda_k}}\, \phi_k(x)$ and compare with (3.18).

Conclude that

$$V(x,t) = \frac{1}{\sqrt{2}}\, B_x + R_x$$

where $\{B_x,\ 0 \leq x < \pi\}$ is a Brownian motion and R is twice-differentiable.

Evidently $x \to V(x,t)$ will have the same local behavior as Brownian motion so that t must be the culprit responsible for the bad behavior of the paths. One striking exhibit of this is the following, according to which $t \to V(x,t)$ has non-trivial quartic (i.e. fourth power) variation, whereas Brownian motion has quadratic variation but zero quartic variation. Define

$$Q_n(t) = (V(0,t) - V(0,[2^n t]2^{-n}))^4 + \sum_{j=1}^{[nt]} (V(0,j2^{-n}) - V(0,(j-1)2^{-n}))^4 .$$

THEOREM 3.10. For a.e. ω, $Q_n(t,\omega)$ converges uniformly on compacts to a limit Qt, where $Q > 0$ is a constant.

THE BARRIER PROBLEM

There is one open problem which deserves mention here, because it is an important question for the neuron, which this equation is meant to describe. That is the problem of finding the distribution of the first hitting time of a given level.

The neuron collects electrical impulses, and when the potential at a certain spot - called the soma and represented here as $x = 0$ - passes a fixed level, called the barrier, the neuron fires and transmits an impulse, the action potential, through the nervous system. The generation of the action potential comes from non-linearities not present in the cable equation, so the cable equation is valid only until the first time τ that $V(0,t)$ exceeds the barrier. However, it can still be used up to τ, and we can ask the question, "what is the distribution of τ?"

We will describe the problem in a bit more detail and show how it is connected with a first-hitting problem for infinite-dimensional diffusions.

Set $\lambda > 0$ and put

$$\tau = \inf\{t > 0 : V(0,t) > \lambda\}.$$

One can show $\tau < \infty$ a.s. and that its moments are finite, and that it even has some exponential moments. Write

$$V(0,t) = \frac{1}{\sqrt{\pi}} A_t^o + \sqrt{\frac{2}{\pi}} \sum_{k=1}^{\infty} A_t^k .$$

Now $V(\cdot,t)$ is a Markov process, but $V(0,t)$ is not, so that the method of studying τ by reducing the problem to a question in differential equations can't be applied directly. However, note that $\{A_t^o, t\geq 0\}$ is a diffusion, and, moreover, if

$$A_\infty(t) = (A_t^0, A_t^1, \ldots)$$

and

$$A_N(t) = (A_t^o, \ldots, A_t^N),$$

then A_N is a diffusion in R^{N+1}, and A_∞ is a diffusion in ℓ_∞. Define

$$\tau_N = \inf\{t\colon A_t^o + \sqrt{2} \sum_{k=1}^{N} A_t^k > \sqrt{\pi\lambda}\}$$

for $N = 0,1,\ldots,\infty$; $(\tau_\infty = \tau)$.

Let H_λ be the half space

$$H_\lambda = \{x \in \ell_\infty : x_0 + \sqrt{2} \sum_1^\infty x_k > \sqrt{\pi\lambda}\}.$$

Then τ is the first hitting time of H_λ by the infinite-dimensional diffusion A. Since the components of A are independent $OU(\lambda_k)$ processes, we can write down its infinitesimal generator and recast the problem in terms of PDE's, at least formally.

Let us see how we would find the expected value of τ, for instance. Suppose F is a smooth function on ℓ_∞ depending on only finitely many coordinates. Then A has the generator G:

$$GF(x) = \sum_{k=0}^{\infty} \left(\frac{1}{2} \frac{\partial^2 F}{\partial x_k^2} - \lambda_k \frac{\partial F}{\partial x_k}\right)(x).$$

Let us proceed purely formally - which means that we will ignore questions about the domain of G and won't look too closely at the boundary values - and set

$$F(x) = E\{\tau \mid A(0) = x\}, \quad x \in \ell_\infty .$$

Then F should satisfy

$$(3.19) \begin{cases} \text{(i)} \quad GF = -1 & \text{in } \ell_\infty - H_\lambda \\ \text{(ii)} \quad F = 0 & \text{on } \partial H_\lambda \\ \text{(iii)} \quad F \text{ is the smallest positive function satisfying (i) and (ii).} \end{cases}$$

Then F(0) is the solution to our problem.

Now (3.19) would hold rigorously for a diffusion in R^N, and in particular, it does hold for each of the $A_N(t)$. But, rigor aside, we can not solve (3.19). We can solve the finite-dimensional analogue. For N = 0 we can solve it in closed form and for small N, we can solve it numerically, but even this becomes harder and harder as N increases. (In this context, N = 1 is a large number, N = 2 is immense, and N = 3 is nearly infinite.)

We have the following:

THEOREM 3.11. $\lim_{N\to\infty} E\{\tau_N\} = E\{\tau\}$.

This might appear to solve the problem, but, in view of the difficulty of finding $E\{\tau_N\}$, one must regard the problem of finding $E\{\tau\}$ as open, and the problem of finding the exact distribution of τ as essentially unattempted.

HIGHER DIMENSIONS

Let us very briefly pose the analogous problem in R^2 and see why the above methods fail.

Consider

$$(3.20)\quad \begin{cases} \frac{\partial V}{\partial t}(x,y;t) = \left(\frac{\partial^2 V}{\partial x^2} + \frac{\partial^2 V}{\partial y^2} - V\right)(x,y;t) + \dot{W}_{xyt} \\ \frac{\partial V}{\partial x}(0,y,t) = \frac{\partial V}{\partial x}(\pi,y,t) = \frac{\partial V}{\partial y}(x,0;t) = \frac{\partial V}{\partial y}(x,\pi;t) = 0 \\ V(x,y,0) = 0 \end{cases}$$

The problem separates, and the eigenfunctions are

$$\phi_{jk}(x,y) = \phi_j(x)\phi_k(y)$$

here the ϕ_j are the eigenfunctions of (3.17), and the eigenvalues are

$$\lambda_{jk} = 1 + j^2 + k^2.$$

roceeding as before, set

$$A_t^{jk} = \int_0^t \int_0^L e^{-\lambda_{jk}(t-s)} \phi_j(x)\phi_k(y)\ W(dxdyds).$$

he A^{jk} are independent $OU(\lambda_{jk})$ processes, as before, and the solution of (3.19) hould be

$$(3.21)\qquad V(x,y,t) = \sum_{j,k=0}^{\infty} A^{jk}(t)\phi_j(x)\phi_k(y).$$

The only problem is that the series on the right hand side does not onverge. Indeed, choose, say, $x = y = 0$ and note that for $t = 1$ and large j and k, hat $\sqrt{\frac{\pi}{2}}\, A^{jk}(t)$ is essentially a $N(0, \frac{1}{\lambda_{jk}})$ random variable by Proposition 3.6. Thus 3.21) converges iff $\sum_{j,k} \frac{1}{\lambda_{jk}} < \infty$. But $\sum_{j,k} \frac{1}{1+j^2+k^2}$ diverges!

One can check that the representation of V as a stochastic integral nalogous to (3.17a) also diverges.

However - and there is a however, or else this course would finish right ere - we can make sense of (3.21) as a Schwartz distribution. Let ψ be a C^∞ unction of compact support in $(0,\pi) \times (0,\pi)$, and write

$$V(\psi,t) = \iint \psi(x,y)\ V(x,y,t)dxdy$$

$$= \sum_{j,k=0}^{\infty} A_{jk}(t)\ \hat{\psi}_{jk}$$

here $\hat{\psi}_{jk} = \int_0^\pi \int_0^\pi \phi_{jk}(x,y)\psi(x,y)dxdy.$

Now the first integral makes no sense, but the sum does, since $\hat{\psi}_{jk}$ tends to ero faster than $j^2 + k^2$ as j and k go to ∞ - a well-known fact of Fourier series - o $\sum \frac{\hat{\psi}_{jk}^2}{1+j^2+k^2} < \infty$. Thus $V(\psi,t)$ makes sense for any test function ψ, and we can use his to define V as a Schwartz distribution rather than as a function.

CHAPTER FOUR

DISTRIBUTION-VALUED PROCESSES

If M_t is a martingale measure and ϕ a test function, put

$$M_t(\phi) = \int_0^L \int_0^t \phi dM.$$

Then M is clearly additive:

$$M_t(a\phi + b\psi) = aM_t(\phi) + bM_t(\psi) \quad \text{a.s.}$$

The exceptional set may depend on a, b, ϕ, and ψ however. We cannot say a priori that $\phi \to M_t(\phi)$ is a continuous linear functional, or even that it is a linear functional. In short, M_t is not yet a distribution. However, it is possible to construct a regular version of M which is. This depends on the fact that spaces of distributions are nuclear spaces.

Let us recall some things about nuclear spaces; we will consider only the simplest setting, which is already sufficient for our purposes.

A norm $\| \ \|$ on a vector space E is <u>Hilbertian</u> if $\|x+y\|^2 + \|x-y\|^2 = 2\|x\|^2 + 2\|y\|^2$, $x, y \in E$. The associated inner product is

$$\langle x,y \rangle = \frac{1}{4} (\|x+y\| - \|x-y\|),$$

so that $(E, \| \ \|)$ is a pre-Hilbert space.

If $\| \ \|_1$ and $\| \ \|_2$ are Hilbertian norms, we say $\| \ \|_1$ is <u>HS weaker than</u> $\| \ \|_2$, and write $\| \ \|_1 \underset{HS}{<} \| \ \|_2$, if

(4.1) $$\sup\Big\{ \sum_k \|e_k\|_1^2 : (e_k) \text{ is } \| \ \|_2\text{-ortho-normal}\Big\} < \infty .$$

(HS stands for Hilbert-Schmidt, for (4.1) is equivalent to the injection map of $(E, \| \ \|_2) \to (E, \| \ \|_1)$ being Hilbert-Schmidt.)

If E is separable relative to $\| \ \|_2$, we can use the Gram-Schmidt procedure to construct a complete ortho-normal basis (f_k) for $(E, \| \ \|_2)$. In this case (4.1) is equivalent to $\sum_k \|f_k\|_1^2 < \infty$.

Let E be a vector space and let $\| \ \|_0 < \| \ \|_1 < \| \ \|_2 < \cdots$ be a sequence of Hilbertian norms on E such that

(i) E is separable with respect to $\| \ \|_n$, all n;

(ii) for each m, there exists $n > m$ such that $\| \ \|_m \underset{HS}{<} \| \ \|_n$.

For each n let $e_{n1}, e_{n2}, \ldots$ be a complete ortho-normal system (CONS) in E,$\| \ \|_n$). Let E'_n be the dual of (E,$\| \ \|_n$) with dual norm $\| \ \|_{-n}$ defined by

$$\|f\|_{-n}^2 = \sum_{k=1}^{\infty} f(e_{nk})^2, \quad f \in E'_n. \tag{4.2}$$

It will be clear shortly why we use -n as an index. Meanwhile, note that $E'_m \subset E'_n$ if $m < n$ for, since $\| \ \|_m < \| \ \|_n$, any linear functional on E which is continuous relative to $\| \ \|_m$ is also continuous relative to the larger norm $\| \ \|_n$. Note also that E'_n is a Hilbert space; denote $H_{-n} = E'_n$. For $n = 0,1,2,\ldots$ let H_n be the completion of E relative to $\| \ \|_n$.

Then H_{-n} is the dual of H_n. (We identify H_0 with its dual H_{-0}, but we do not identify H_n and H_{-n}. In fact we have:

$$\ldots \supset H_{-2} \supset H_{-1} \supset H_0 \supset H_1 \supset H_2 \supset \ldots$$

Let us give E the toplogy determined by the $\| \ \|_n$. A neighborhood basis of is $\{x : \|x\|_n < \varepsilon\}$, $n = 0,1,2,\ldots$, $\varepsilon > 0$.

Let E' be the dual of E. Then $E' = \bigcup_n H_{-n}$. To see this, suppose $f \in E'$. Then there is a neighborhood G of zero such that $|f(y)| < 1$ if $y \in G$. Thus there is member of the basis such that $\{x : \|x\|_n < \varepsilon\} \subset G$. For $\delta > 0$, if $\|x\|_n < \varepsilon\delta$, then $|f(x)| < \delta$. This implies that $\|f\|_{-n} < 1/\varepsilon$, i.e. $f \in H_{-n}$. Conversely, if $\in H_{-n}$, it is a linear functional on E, and it is continuous relative to $\| \ \|_n$, and hence continuous in the topology of E.

Note: The argument above also proves the following more general statement: Let F be linear map of E into a metric space. Then F is continuous iff it is continuous in one of the norms $\| \ \|_n$.

We give E' the strong topology: a set $A \subset E$ is bounded if it is bounded in each norm $\| \ \|_n$, i.e. if $\{\|x\|_n, x \in A\}$ is a bounded set for each n. Define a semi-norm

$$p_A(f) = \sup\{|f(x)| : x \in A\}.$$

The strong topology is generated by the semi-norms $\{p_A : A \subset E \text{ is bounded}\}$. Now E is not in general normable, but its topology is compatible with the metric

$$d(x,y) = \sum_n 2^{-n}(1 + \|y-x\|_n)^{-1}\|y-x\|_n,$$

and we can speak of the completeness of E. If E is complete, then $E = \bigcap_n H_n$.

(Clearly $E \subset \bigcap_n H_n$, and if $x \in \bigcap_n H_n$, then for each n there is $x_n \in E$ such that

$\|x-x_n\|_n < 2^{-n} \Rightarrow \|x-x_n\|_m < 2^{-n}$ $m < n$. Thus $d(x,x_n) < 2^{-n+1}$.)

If E is complete, it is called a <u>nuclear space</u>, and we have

$$E' = \bigcup_n H_{-n} \supset \ldots \supset H_{-1} \supset H_0 \supset H_1 \supset H_2 \supset \ldots \supset \bigcap_n H_n = E$$

where H_n is a Hilbert space relative to the norm $\| \ \|_n$, $-\infty < n < \infty$, E is dense in H_n, H_{-n} is dual to H_n, and for all m there exists $n > m$ such that $\| \ \|_m \underset{HS}{<} \| \ \|_n$.

We may not often use the following explicitly in the sequel, but it is one of the fundamental properties of the spaces H_n.

<u>Exercise 4.1</u>. Suppose $\| \ \|_m \underset{HS}{<} \| \ \|_n$. Then the closed unit ball in H_n is compact in H_m. (Hint: show it is totally bounded.)

REGULARIZATION

Let E be a nuclear space as above. A stochastic process $\{X(x), x \in E\}$ is a <u>random linear functional</u> if, for each $x,y \in E$ and $a,b, \in R$,

$$X(ax + by) = aX(x) + bX(y) \text{ a.s.}$$

<u>THEOREM 4.1</u>. Let X be a random linear functional on E which is continuous in probability in $\| \ \|_m$ for some m. If $\| \ \|_m \underset{HS}{<} \| \ \|_n$, then X has a version which is in H_{-n} a.s. In particular, X has a version with values in E'.

Convergence in probability is metrizable, being compatible with the metric

$$|\!|\!| X(x) |\!|\!| \overset{def}{=} E\{|X(x)| \wedge 1\}.$$

If X is continuous in probability on E, it is continuous in probability in $\| \ \|_m$ for some m by our note. There exists n such that $\| \ \|_m \underset{HS}{<} \| \ \|_n$. Thus we have

<u>COROLLARY 4.2</u>. Let X be a random linear functional which is continuous in probability on E. Then X has a version with values in E'.

PROOF (of Theorem 4.1). Let (e_k) be a CONS in $(E, \| \|_n)$. We will first show that $\sum X(e_k)^2 < \infty$.

For $\varepsilon > 0$ there exists $\delta > 0$ such that $|||X(x)||| < \varepsilon$ whenever $\|x\|_m < \delta$. We claim that

$$\text{Re } E\{e^{iX(x)}\} \geq 1 - 2\varepsilon - 2\varepsilon\delta^{-2}\|x\|_m^2 .$$

Indeed, the left-hand side is greater than $1 - \frac{1}{2} E\{X^2(x)\wedge 4\}$, and if $\|x\|_m \leq \delta$,

$$E\{X^2(x)\wedge 4\} \leq 4E\{|X(x)|\wedge 1\} \leq 4\varepsilon,$$

while if $\|x\|_m > \delta$,

$$E\{X^2(x)\wedge 4\} \leq \|x\|_m^2\delta^{-2}E\{X^2(\delta x/\|x\|_m)\wedge 4\} \leq 4\,\varepsilon\,\delta^{-2}\|x\|_m^2.$$

Let us continue the trickery by letting $Y_1, Y_2, \ldots$ be iid $N(0,\sigma^2)$ r.v. independent of X, and set $x = \sum_{k=1}^{N} Y_k e_k$. Then

$$\text{Re } E\{e^{iX(x)}\} = E\{\text{Re}E\{\exp[i \sum_1^N Y_k X(e_k)]\,|X\}\}.$$

But if X is given, $\sum Y_k X(e_k)$ is conditionally a mean zero Gaussian r.v. with variance $\sigma^2\sum X^2(e_k)$, and the above expectation is its characteristic function:

$$= E\{e^{-\frac{\sigma^2}{2}\{\sum_{k=1}^{N} X^2(e_k)\}}\}.$$

On the other hand, it also equals

$$E\{\text{Re } E\{e^{i\sum Y_k X(e_k)}\,|Y\}\} \geq 1 - 2\varepsilon - 2\delta^{-2}\varepsilon E\{\|x\|_m^2\}$$

$$= 1 - 2\varepsilon - 2\delta^{-2}\varepsilon \sum_{j,k=1}^{N} E\{Y_j Y_k\} \langle e_j, e_k\rangle_m$$

$$= 1 - 2\varepsilon - 2\delta^{-2}\,\varepsilon\,\sigma^2 \sum_{k=1}^{N} \|e_k\|_m^2.$$

Thus

$$E\{e^{-\frac{\sigma^2}{2}\sum_{k=1}^{N} X^2(e_k)}\} \geq 1 - 2\varepsilon - 2\delta^{-2}\varepsilon\sigma^2 \sum_{k=1}^{N} \|e_k\|_m^2.$$

Let $N \to \infty$ and note that the last sum is bounded since $\| \|_m \underset{HS}{<} \| \|_n$. Next let $\sigma^2 \to 0$ to to see that

$$P\{\sum_{k=1}^{\infty} X^2(e_k) < \infty\} \geq 1 - 2\varepsilon.$$

Let $\Omega_1 = \{\omega: \sum_k X^2(e_k,\omega) < \infty\}$. Then $P\{\Omega_1\} = 1$. Define

$$Y(x,\omega) = \begin{cases} \sum_k \langle x,e_k\rangle_n X(e_k) & \text{if } \omega \in \Omega_1. \\ 0 & \text{if } \omega \in \Omega - \Omega_1. \end{cases}$$

The sum is finite by the Schwartz inequality, so Y is well-defined. Moreover, $Y \in H_{-n}$ with norm

$$\|Y\|_{-n} = \sum_k Y^2(e_k) = \sum_k X^2(e_k) < \infty$$

Finally, $P\{Y(x) = X(x)\} = 1$, $x \in E$. Indeed, let $x_N = \sum_{k=1}^{N} \langle x,e_k\rangle_n e_k$. Clearly $X(x_N) = Y(x_N)$ on Ω_1, and $\|x-x_N\|_m \le \|x-x_N\|_n \to 0$. Thus

$$\begin{aligned} Y(x) &= \lim Y(x_N) \\ &= \lim X(x_N) = X(x). \end{aligned}$$

Note: We have followed some notes of Ito in this proof. The tricks are due to Sazanov and Yamazaki.

EXAMPLES

Let us see what the spaces E and H_n are in some special cases.

EXAMPLE 1. Let $G \subset R^d$ be a bounded domain and let $E_0 = \underline{\underline{D}}(G)$ be the set of C^∞ functions of compact support in G. Let $\|\ \|_0$ be the usual L^2-norm on G and set

$$\|\phi\|_n^2 = \|\phi\|_0^2 + \sum_{1\le|\alpha|\le n} \|D^\alpha\phi\|_0^2,$$

where α is a multi-index of length $|\alpha|$, and D^α is the associated partial derivative operator. Let E be the completion of E in the topology induced by the norms $\|\ \|_n$.

In this case $H_0 = L^2(G)$ and H_n is the classical Sobolev space (often denoted $W_0^{n,2}(G)$). By Maurin's theorem, $\|\ \|_m \underset{HS}{<} \|\ \|_n$ if $n > m + d/2$. H_n consists of all L^2-functions whose partials of order n or less are all in L^2. (These are derivatives in the sense of distributions. However, if $n > d/2$ the functions will be continuous; for larger n, they will be differentiable in the usual sense, and $\bigcap_n H_n$ will consist of C^∞ functions.

The spaces H_{-n} - duals to the H_n - consist of derivatives: $f \in H_{-n}$ iff there exist $f_\alpha \in L^2$ such that

$$f = \sum_{|\alpha| \leq n} D^\alpha f_\alpha .$$

EXAMPLE 1a. If we let $G = \mathbb{R}^d$ in Example 1, we can use the Fourier transform to define the H_n in a rather neat way. Let $E = \underline{\underline{S}}(\mathbb{R}^d)$. If $u \in E$, define the Fourier transform $\hat{u}$ of u by

$$\hat{u}(\xi) = \int_{\mathbb{R}^d} e^{-2\pi i x \cdot \xi} u(x)\,dx.$$

If u is a tempered distribution, i.e. if $u \in \underline{\underline{S}}'(\mathbb{R}^d)$, we can define $\hat{u}$ - as a distribution - by $\hat{u}(\phi) = u(\hat{\phi})$, $\phi \in E$.

Define a norm on E by

$$\|u\|_t = \int_{\mathbb{R}^d} (1 + |\xi|^2)^t \, |\hat{u}(\xi)|^2 d\xi \tag{4.2}$$

and let H_t be the completion of E in the norm $\| \ \|_t$.

If u is a distribution whose Fourier transform $\hat{u}$ is a function, then $u \in H_t$ iff (4.2) is finite. The space $H_0 = L^2$ by Plancharel's theorem. For $t > 0$ the elements of H_n are functions. For $t < 0$ they are in general distributions. It can be shown that if t is an integer, say $t = n$, the norms $\| \ \|_n$ defined here and in Example 1 are equivalent, and the spaces H_n in the two examples are identical. Note that (4.2) makes sense for all real t, positive or negative, integer or not, and $\| \ \|_s \underset{HS}{<} \| \ \|_t$ if $t > s + d/2$.

EXAMPLE 2. Let $E = \underline{\underline{S}}(\mathbb{R}^d)$, the Schwartz space of rapidly decreasing functions. Let

$$g_k(x) = (-1)^k e^{x^2} \frac{d^k}{dx^k} e^{-x^2}$$

and set

$$h_k(x) = (\pi^{1/2} 2^k k!)^{-1/2} g_k(x) e^{-x^2/2}.$$

Then $g_0, g_1, \ldots$ are the Hermite polynomials, and $h_0, h_1, \ldots$ are the Hermite functions. The latter are a CONS in $L^2(\mathbb{R}^d)$.

Let $q = (q_1, \ldots, q_d)$ where the q_i are non-negative integers, and for $x = (x_1, \ldots, x_d) \in \mathbb{R}^d$, set

$$h_q(x) = h_{q_1}(x_1) \cdots h_{q_d}(x_d).$$

Then $h_q \in \underline{\underline{S}}(\mathbf{R}^d)$, and they form a CONS in $L^2(\mathbf{R}^d)$. If $\phi \in \underline{\underline{S}}(\mathbf{R}^d)$, let $\hat{\phi}_q = \langle\phi, h_q\rangle$ and write

$$\phi = \sum_q \hat{\phi}_q h_q$$

Define

$$\|\phi\|_n^2 = \sum_q (2|q|+d)^n \hat{\phi}_q^2$$

where $|q|^2 = q_1^2 + \cdots + q_d^2$. One can show $\|\phi\|_n < \infty$ if $\phi \in \underline{\underline{S}}(\mathbf{R}^d)$. Let H_n be the completion of E in $\| \ \|_n$. Note that this makes sense for negative n - in fact for all real n - and

$$\|\phi\|_{-n}^2 = \sum_q (2|q| + d)^{-n} \hat{\phi}_q^2 \quad .$$

H_{-n} is dual to H_n under the inner product

$$\langle\psi, \phi\rangle = \sum_q \hat{\psi}_q \hat{\phi}_q \quad .$$

The Hilbert-Schmidt ordering is easily verified in this example, since the functions $e_q = (2|q|+d)^{-n/2} h_q$ are a CONS under $\| \ \|_n$, and if $m < n$

$$\sum_q \|e_q\|_m = \sum_q (2|q|+d)^{(n-m)}$$

which is finite if $n > m + d/2$. Thus $\| \ \|_m \underset{HS}{<} \| \ \|_n$ if $n > m + \frac{d}{2}$.

<u>EXAMPLE 3</u>. Let us look at an example which is specifically linked to a differential operator.

Let M be a smooth compact d-dimensional differentiable manifold with a smooth (possibly empty) boundary. Let dx be the element of area, and let L be a self-adjoint uniformly strongly elliptic second order differential operator with smooth coefficients, and smooth homogeneous boundary conditions.

$-L$ has a CONS of smooth eigenfunctions $\{\phi_n\}$ with eigenvalues $\{\lambda_n\}$. The eigenvalues satisfy $\sum_j (1+\lambda_j)^{-p} < \infty$ if $p > \frac{d}{2}$.

Let E_0 be the set of f of the form $f(x) = \sum_{j=1}^{N} \hat{f}_j \phi_j(x)$, where the $\hat{f}_j$ are constants. For each integer n, positive or negative, define

$$\|f\|_n = \sum_j (1 + \lambda_j)^n \hat{f}_j^2 .$$

Note that $\| \ \|_m \underset{HS}{<} \| \ \|_n$ if $n > m + d/2$. Indeed, set $e_j = (1 + \lambda_j)^{-n/2} \phi_j$. The e_j form a CONS relative to $\| \|_n$, and

$$\sum_j \|e_j\|_m^2 = \sum_j (1 + \lambda_j)^{m-n} < \infty.$$

Let H_n be the completion of $(E_0, \| \|_n)$. If $f \in H_n$, we can represent f by the formal series

$$f = \sum_j \hat{f}_j \phi_h,$$

where $\sum (1 + \lambda_j)^n \hat{f}_j^2 = \|f\|_n < \infty$. Then H_n and H_{-n} are dual under the inner product $\langle f,g\rangle = \sum_j \hat{f}_j \hat{g}_j$.

Finally, let E be the completion of E_0 in the topology determined by the $\| \|_n$. The ϕ_j are smooth, so that the elements of H_n will be differentiable for large n. Since $E = \bigcap_n H_n$, E will consist of C^∞ functions. Note that if $f \in H_n$, $Lf \in H_{n-2}$ since $Lf = \sum_j \lambda_j \hat{f}_j \phi_j$.

The similarity of Examples 2 and 3 is more than superficial. In fact Example 2 corresponds to the operator $L = -\Delta + |x|^2$.

EXAMPLE 4. At the start of the chapter we raised the question of whether or not a martingale measure could be regarded as a distribution. Let us consider this in the setting of, say, Example 1. Let G be a bounded open set in $\mathbf{R}^d$ with a smooth boundary, let $\underline{\underline{D}}(G)$ be the space of test functions on G, and let M be a worthy martingale measure on G with dominating measure K and let $\mu_t(A\times B) = E\{K(A\times B\times [0,t])\}$. Assume μ_t is finite.

If $\phi \to 0$ in $\underline{\underline{D}}(G)$, $\sup|\phi(x)| \to 0$, hence $E\{M_t^2(\phi)\} = \int\phi(x)\phi(y)\mu_t(dxdy) \to 0$. (Careful! This is not trivial; we have used Sobolev's Theorem.) It follows that $\phi \to M_t(\phi)$ is continuous in probability on $\underline{\underline{D}}(G)'$. By Corollary 4.2, M_t has a version with values in $\underline{\underline{D}}(G)$.

M actually lives in a Sobolev space of negative index. To see why, note that if $n > d/2$, H_n embeds in $C_b(G)$ by the Sobolev embedding theorem and $\| \|_n \underset{HS}{<} \| \|_{2n}$ by Maurin's theorem. By Theorem 4.1, M_t has a version with values in H_{-2n}. In particular, $M_t \in H_{-d-2}$, and if d is odd, we have $M_t \in H_{-d-1}$. (A more delicate analysis here would show that, locally at least, $M_t \in H_{-n}$ for any $n > d/2$.)

<u>Exercise 4.2</u>. Show that under the usual hypotheses (i.e. right continuous filtration, etc.) that the process M_t, considered as a process with values in $\underline{\underline{D}}(G)$, has a right continuous version. Show that it is also right continuous in the appropriate Sobolev space.

Even pathology has its degrees. The martingale measure M_t will certainly not be a differentiable or even continuous function, but it is not infinitely bad. According to the above, it is at worst a derivative of order d + 2 of an L^2 function or, using the embedding theorem again, a derivative of order $\frac{3}{2}$ d + 3 of a continuous function. Thus a distribution in H_{-n} is "more differentiable" than a distribution in H_{-n-1}, and the statement that M does indeed take values in a certain H_{-n} can be regarded as a regularity property of M.

In the future we will discuss most processes as having values in $\underline{\underline{D}}(G)'$ or another relevant nuclear space, and put off the task of deciding which H_{-n} is appropriate until we discuss the regularity of the process. As a practical matter, it is usually easier to do it this way; for it is often much simpler to verify that a process is distribution-valued than to verify it lives in a given Sobolov space...and as an even more practical matter, we shall usually leave even that to the reader.

CHAPTER FIVE

PARABOLIC EQUATIONS IN $\mathbf{R}^d$

Let $\{M_t, \underline{\underline{F}}_t, t \geq 0\}$ be a worthy martingale measure on $\mathbf{R}^d$ with covariation measure $Q(dx\,dy\,ds) = d\langle M(dx), M(dy)\rangle_s$ and dominating measure K. Let $\mu(\Lambda) = E\{K(\Lambda)\}$. Assume that for some $p > 0$ and all $T > 0$

$$\int_{\mathbf{R}^d \times [0,T]} \frac{1}{(1+|x|^p)(1+|y|^p)}\, \mu(dx\,dy\,ds) < \infty.$$

Then $M_t(\phi) = \int_{\mathbf{R}^d \times [0,T]} \phi(x) M(dx\,ds)$ exists for each $\phi \in \underline{\underline{S}}(\mathbf{R}^d)$.

Let L be a uniformly elliptic self-adjoint second order differential operator with bounded smooth coefficients. Let T be a differential operator on $\mathbf{R}^d$ of finite order with bounded smooth coefficients. (Note that T and L operate on x, not on t). Consider the SPDE

$$(5.1)\qquad \begin{cases} \dfrac{\partial V}{\partial t} = LV + T\dot{M} \\ V(x,0) = 0 \end{cases}$$

We will clearly need to let V and M have distribution values, if only to make sense of the term $T\dot{M}$. We will suppose they have values in the Schwartz space $\underline{\underline{S}}'(\mathbf{R}^d)$.

We want to cover two situations: the first is the case in which (5.1) holds in $\mathbf{R}^d$. Although there are no boundary conditions as such, the fact that $V_t \in \underline{\underline{S}}'(\mathbf{R}^d)$ implies a boundedness condition at infinity.

The second is the case in which D is a bounded domain in $\mathbf{R}^d$, and homogeneous boundary conditions are imposed on ∂D.

(There is a third situation which is covered - formally at least - by (5.1), and that is the case where T is an integral operator rather than a differential operator. Suppose, for instance, that $Tf(x) = g(x) \int f(y)h(y)dy$ for suitable functions g and h. In that case, $TM_t(x) = g(x)M_t(h)$. Now $M_t(h)$ is a real-valued martingale, so that (5.1) can be rewritten

$$\begin{cases} dV_t = LV\,dt + g\,dM_t(h) \\ V(x,0) = 0 \end{cases}$$

This differs from (5.1) in that the driving term is a one-parameter martingale rather than a martingale measure. Its solutions have a radically different behavior from those of (5.1) and it deserves to be treated separately.)

Suppose (5.1) holds on $\mathbf{R}^d$. Integrate it against $\phi \varepsilon \underline{\underline{S}}(\mathbf{R}^d)$, and then integrate by parts. Let T^* be the formal adjoint of T. The weak form of (5.1) is then

$$V_t(\phi) = \int_0^t V_s(L\phi)ds + \int_0^t \int_{\mathbf{R}^d} T^*\phi(x)M(dxds), \quad \phi \in \underline{\underline{S}}(\mathbf{R}^d). \tag{5.2}$$

Notice that when we integrate by parts, (5.2) follows easily for ϕ of compact support, but in order to pass to rapidly decreasing ϕ, we must use the fact that V and $T\dot{M}$ do not grow too quickly at infinity.

In case D is a bounded region with a smooth boundary, let B be the operator $B = d(x)D_N + e(x)$, where D_N is the normal derivative on ∂D, and d and e are in $C^\infty(\partial D)$. Consider the initial-boundary-value problem

$$\begin{cases} \dfrac{\partial V}{\partial t} = LV + T\dot{M} & \text{on } D \times [0,\infty); \\ BV = 0 & \text{on } \partial D \times [0,\infty); \\ V(x,0) = 0 & \text{on } D. \end{cases} \tag{5.3}$$

Let $C^\infty(D)$ and $C_0^\infty(D)$ be respectively the set of smooth functions on D and the set of smooth functions with compact support in D. Let $C^\infty(\bar{D})$ be the set of functions in $C^\infty(D)$ whose derivatives all extend to continuous functions on $\bar{D}$. Finally, let

$$\underline{\underline{S}}_B = \{\phi \in C^\infty(\bar{D}): B\phi = 0 \text{ on } \partial D\}.$$

The weak form of (5.3) is

$$V_t(\phi) = \int_0^t V_s(L\phi)ds + \int_0^t \int_D T^*\phi(x)M(dxds), \quad \phi \in \underline{\underline{S}}_B. \tag{5.4}$$

This needs a word of explanation. To derive (5.4) from (5.3), multiply by ϕ and integrate formally over $D \times [0,t]$ - i.e. treat $T\dot{M}$ as if it were a differentiable function - and then use a form of Green's theorem to throw the derivatives over on ϕ. This works on the first integral if both V and ϕ satisfy the boundary condition. Unless T is of zeroth order, it may not work for the second, for $\dot{M}$ may not satisfy the boundary conditions. (It does work if ϕ has compact support in D, however.) Nevertheless, the equation we wish to solve is (5.4), not (5.3).

The requirement that (5.4) hold for all ϕ satisfying the boundary conditions is essentially a boundary condition on V.

The above situation, in which we regard the integral, rather than the differential equation as fundamental, is analogous to many situations in which physical reasoning leads one directly to an integral equation, and then mathematics takes over to extract the partial differential equation. See the physicists' derivations of the heat equation, Navier-Stokes equation, and Maxwell's equation, for instance.

As in the one-variable case, it is possible to treat test functions $\phi(x,t)$ of two variables.

Exercise 5.1. Show that if V satisfies (5.4) and if $\phi(x,t)$ is a smooth function such that for each t, $\phi(\cdot,t) \in \underline{\underline{S}}_B$, then

(5.5) $$V_t(\phi(t)) = \int_0^t V_s(L\phi(s) + \frac{\partial\phi}{\partial s}(s))ds + \int_0^t \int_D T^*\phi(x,s)M(dxds).$$

Let $G_t(x,y)$ be the Green's function for the homogeneous differential equation. If $L = \frac{1}{2}\Delta$, $D = \mathbf{R}^d$, then

$$G_t(x,y) = (2\pi t)^{-d/2} e^{-\frac{|y-x|^2}{2t}}.$$

For a general L, $G_t(x,y)$ will still be smooth except at $t = 0$, $x = y$, and its smoothness even extends to the boundary: if $t > 0$, $G_t(x,\cdot) \in C^\infty(\bar{D})$. It is positive, and for $\tau > 0$,

(5.6) $$G_t(x,y) \le Ct^{-d/2} e^{-\frac{|y-x|^2}{\delta t}}, \quad x, y \in D, \ 0 \le t \le \tau,$$

where $C > 0$ and $\delta > 0$. (C may depend on τ). This holds both for $D = \mathbf{R}^d$ and for bounded D. If $D = \mathbf{R}^d$, $G_t(x,\cdot)$ is rapidly decreasing at infinity by (5.6), so it is in $\underline{\underline{S}}(\mathbf{R}^d)$. Moreover, for fixed y, $(x,t) \to G_t(x,y)$ satisfies the homogeneous differential equation plus boundary conditions. Define $G_t(\phi,y) = \int_D G_t(x,y)\phi(x)dx$.

Then if ϕ is smooth, $G_0\phi = \phi$. This can be summarized in the integral equation:

(5.7) $$G_{t-s}(\phi,y) = \phi(y) + \int_s^t G_{u-s}(L\phi,y)du, \ \phi \in \underline{\underline{S}}_B$$

The smoothness of G then implies that if $\phi \in C^\infty(\bar{D})$, then $G_t(\phi,\cdot) \in \underline{\underline{S}}_B$. In case $D = \mathbf{R}^d$, then $\phi \in \underline{\underline{S}}(\mathbf{R}^d)$ implies that $G_t(\phi,\cdot) \in \underline{\underline{S}}(\mathbf{R}^d)$.

THEOREM 5.1. There exists a unique process $\{V_t, t>0\}$ with values in $\underline{\underline{S}}'(\mathbf{R}^d)$ which satisfies (5.4). It is given by

$$V_t(\phi) = \int_0^t \int_{R^d} T^* G_{t-s}(\phi,y) M(dyds). \tag{5.8}$$

The result for a bounded region is similar except for the uniqueness statement.

THEOREM 5.2 There exists a process $\{V_t, t \geq 0\}$ with values in $\underline{\underline{S}}'(\mathbf{R}^d)$ which satisfies (5.5). V can be extended to a stochastic process $\{V_t(\phi), t \geq 0, \phi \in \underline{\underline{S}}_B\}$; this process is unique. It is given by

$$V_t(\phi) = \int_0^t \int_D T^* G_{t-s}(\phi,y) M(dy\ ds),\ \phi \in \underline{\underline{S}}_B. \tag{5.9}$$

PROOF. Let us first show uniqueness, which we do by deriving (5.9).

Choose $\psi(x,s) = G_{t-s}(\phi,x)$, and suppose that U is a solution of (5.4). Consider $U_s(\psi(s))$. Note that $U_0(\psi(0)) = 0$ and $U_t(\psi(t)) = U_t(\phi)$. Now $G_{t-s}(\phi,\cdot) \in \underline{\underline{S}}_B$, so we can apply (5.5) to see that

$$U_t(\phi) = U_t(\psi(t)) = \int_0^t U_s(L\psi(s) + \frac{\partial\psi}{\partial s}(s))ds + \int_0^t \int_D T^* \psi(x,s) M(dx\ ds).$$

But $L\psi + \frac{\partial\psi}{\partial s} = 0$ by (5.7) so this is

$$= \int_0^t \int_D T^* \psi(x,s)\ M(dx\ ds)$$

$$= \int_0^t \int_D T^* G_{t-s}(\phi,x) M(dx\ ds) = V_t(\phi).$$

Existence: Let $\phi \in \underline{\underline{S}}_B$ and plug (5.9) into the right hand side of (5.4):

$$\int_0^t \Big[\int_0^s \int_D T^* G_{s-u}(L\phi,y)\ M(dy\ du)\Big]ds + \int_0^t \int_D T^* \phi(y) M(dy,du)$$

$$= \int_0^t \int_D \Big[\int_u^t T^* G_{s-u}(L\phi,y)ds + T^*\phi(y)\Big] M(dy\ du).$$

Note that $T^* G_{s-u}(L\phi,y)$ and $T^*\phi(y)$ are bounded, so the integrals exist. By (5.7) this is

$$= \int_0^t \int_D T^* G_{t-u}(\phi,y) M(dy\ du)$$

$$= V_t(\phi).$$

by (5.9). This holds for any $\phi \in \underline{\underline{S}}_B$, but (5.9) also makes sense for ϕ which are not in $\underline{\underline{S}}_B$. In particular, it makes sense for $\phi \in \underline{\underline{S}}(R^d)$ and one can show using Corollary 4.2 that V_t has a version which is a random tempered distribution.

This proves Theorem 5.2. The proof of Theorem 5.1 is nearly identical; just replace D by R^d and $\underline{\underline{S}}_B$ by $\underline{\underline{S}}(R^d)$. Q.E.D.

AN EIGENFUNCTION EXPANSION

We can learn a lot from an examination of the of the case $T \equiv 1$. Suppose D is a bounded domain with a smooth boundary. The operator $-L$ (plus boundary conditions) admits a CONS $\{\phi_j\}$ of smooth eigenfunctions with eigenvalues λ_j. These satisfy

(5.10) $$\sum_j (1+\lambda_j)^{-p} < \infty \quad \text{if } p > d/2.$$

(5.11) $$\sup_j \|\phi_j\|_\infty^2 (1+\lambda_j)^{-p} < \infty \quad \text{if } p > d/2.$$

Let us proceed formally for the moment. We can expand the Green's function:

$$G_t(x,y) = \sum_j \phi_j(x)\phi_j(y)e^{-\lambda_j t}.$$

If ψ is a test function

$$G_t(\psi,y) = \sum_j \hat{\psi}_j \phi_j(y)e^{-\lambda_j t}$$

where $\hat{\psi}_j = \int_D \psi(x)\phi_j(x)dx$, so by (5.9)

$$V_t(\psi) = \int_0^t \int_D \sum_j \hat{\psi}_j \phi_j(y) e^{-\lambda_j (t-s)} M(dyds).$$

Let

$$A_j(t) = \int_0^t \int_D \phi_j(y) e^{-\lambda_j (t-s)} M(dyds).$$

Then

$$V_t(\phi) = \sum \hat{\phi}_j A_j(t). \tag{5.12}$$

This will converge for $\phi \in \underline{\underline{S}}_B$, but we will show more. Let us recall the spaces H_n introduced in Ch. 4, Example 3. H_n is isomorphic to the set of formal eigenfunction series

$$f = \sum_{j=1}^{\infty} a_j \phi_j$$

for which

$$\|f\|_n^2 = \sum a_j^2 (1+\lambda_j)^n < \infty.$$

We see from (5.12) that $V_t \sim \sum_j A_j(t)\phi_j$.

<u>PROPOSITION 5.3</u>. Let V be defined by (5.12). If $n > d$, V_t is a right continuous process in H_{-n}; it is continuous if $t \to M_t$ is. Moreover, V is the solution of (5.4) with $T \equiv 1$. If M is a white noise based on Lebesgue measure then V is a continuous process in H_{-n} for any $n > d/2$.

<u>PROOF</u>. We first bound $E\{\sup_{t \le \tau} A_j^2(t)\}$. Let $X_j(t) = \int_0^t \int_D \phi_j(x)\, M(dxds)$ and note that, as in Exercise 3.3,

$$A_j(t) = \int_0^t e^{-\lambda_j(t-s)}\, dX_j(s).$$

$$= X_t - \int_0^t \lambda_j e^{-\lambda_j(t-s)} X_j(s)ds$$

where we have integrated by parts in the stochastic integral. Thus

$$\sup_{t \le \tau} |A_j(t)| \le \sup_{t \le \tau} |X_j(t)| (1+\int_0^t \lambda_j e^{-\lambda_j(t-s)} ds)$$

$$\le 2 \sup_{t \le \tau} |X_j(t)|.$$

Thus

$$E\{\sup_{t \le \tau} A_j^2(t)\} \le 4E\{\sup X_j^2(t)\}$$

$$\le 16E\{X_j^2(\tau)\}$$

by Doob's inequality. This is

$$(5.13) \qquad = 16 \int_{D\times D\times[0,\tau]} \phi_j(x)\phi_j(y)\mu(dx\ dy\ ds)$$

$$\leq 16\ \mu(D\times D\times[0,\tau])\|\phi_j\|_\infty^2$$

$$\leq C(1+\lambda_j)^p$$

or some constant C and $p > d/2$ by (5.11).

Thus

$$E\{\sum_j \sup_{t\leq\tau} A_j^2(t)(1+\lambda_j)^{-n}\} \leq C\sum_j(1+\lambda_j)^{-n+p}.$$

By (5.10) this is finite if $n - p > d/2$ or, remembering that $p > d/2$, if $n > d$.

Then, clearly,

$$\|V_t\|_{-n}^2 = \sum A_j^2(t)(1+\lambda_j)^{-n}$$

is a.s. finite, hence $V_t \in H_{-n}$. Moreover, if $s > 0$,

$$\|V_{t+s}-V_t\|_{-n}^2 = \sum(A_j(t+s)-A_j(t))^2(1+\lambda_j)^{-n}.$$

The summands are right continuous, and they are continuous if M is. The sum is dominated by

$$4\sum_j \sup_{t\leq\tau} A_j^2(t)(1+\lambda_j)^{-n} < \infty$$

for a.e.ω. Now $A_j(s) \to A_j(t)$ as $s \downarrow t$, hence $\|V_s-V_t\|_{-n} \to 0$ as $s \downarrow t$. If M is continuous, so is A_j, and we can let $s \uparrow t$ to see V is also left continuous, hence continuous.

If M is a white noise the integral in (5.13) reduces to $\int \phi_j^2(x)dxds$. Since the ϕ_j are orthonormal this is just τ. This means we can take $p = 0$ and $n > d/2$ in the remainder of the argument.

<u>REMARKS</u>. The conditions on the Sobolev spaces in Proposition 2.3 can be improved. For instance, if M is a white noise based on Lebesgue measure, V will be continuous in H_{-n} for every $n > d/2 - 1$, not just for $n > d/2$. The same proof shows it, once one improves the estimate of $E\{\sup_{t\leq\tau} A_j^2(t)\}$. In this case, A_j is an $OU(\lambda_j)$ process and one can show that this quantity is bounded by a constant times $\lambda_j^{-1}\log\lambda_j$.

Once we know that V_t is actually a solution of (5.4), the uniqueness result implies that V also satisfies (5.9).

Exercise 5.2. Verify that V (defined by (5.12)) satisfies (5.4).

Exercise 5.3. Treat the case $D = R^d$ using the Hermite expansion of Example 2, Ch.4.

The spaces H_n above are analogous to the classical Sobolev spaces, but they don't explicitly involve derivatives. Here is a result which relates the regularity of the solution directly to differentiability.

THEOREM 5.4. Suppose M is a white noise based on Lebesgue measure. Then there exists a real-valued process $U = \{U(x,t): x \in D,\ t \geq 0\}$ which is Hölder continuous with exponent $1/4 - \varepsilon$ for any $\varepsilon > 0$ such that if $D^{d-1} = \dfrac{\partial^{d-1}}{\partial x_2, \ldots, \partial x_d}$, then

$$V_t = D^{d-1} U_t.$$

Note. This is of course a derivative in the weak sense. A distribution Q is the weak αth derivative of a function f if for each test function ϕ,

$$Q(\phi) = (-1)^{|\alpha|} \int f(x) D^{\alpha} \phi(x)\,dx.$$

If we let H_o^n denote the classical Sobolev space of Example 1 Ch.4, this implies that V_t can be regarded as a continuous process in H_o^{-d+1}.

Note. One must be careful in comparing the classical Sobolev spaces H_0^n with related but not identical spaces H_n of Example 3 in Chapter 4. Call the latter H_3^n for the moment. Theorem 5.4 might lead one to guess that V is in H_3^{-d+1}, but in fact, it is a continuous process in H_3^{-n} for any $n > d/2$ by Proposition 5.3. This is a much sharper result if $d \geq 3$.

This gives us an idea of the behavior of the solution of the equation

$$\frac{\partial V}{\partial t} = LV + \dot{M}.$$

Suppose now that T is a differential operator and suppose both T and L have constant coefficients, so $TL = LT$. Apply T to both sides of the SPDE:

$$\frac{\partial}{\partial t}(TV) = LTV + T\dot{M}$$

i.e. $U = TV$ satisfies $\dfrac{\partial U}{\partial t} = LU + T\dot{M}$. Of course, this argument is purely formal, but the following exercise makes it rigorous.

Exercise 5.4. Suppose T and L commute. Let U be the solution of (5.4) for a general T with bounded smooth coefficients and let V be the solution for $T \equiv 1$. Verify that if we restrict U and V to the space $\underline{\underline{D}}(D)$ of C^∞ functions of compact support in D, that $U = TV$.

Exercise 5.5. Let V solve

$$\begin{cases} \dfrac{\partial V}{\partial t} = \dfrac{\partial^2 V}{\partial x^2} - V + \dfrac{\partial}{\partial x}\dot{W}, & 0 < x < \pi,\ t > 0; \\ \dfrac{\partial V}{\partial x}(0,t) = \dfrac{\partial V}{\partial x}(\pi,t) = 0, & t > 0; \\ V(x,0) = 0, & 0 < x < \pi. \end{cases}$$

Describe $V(\cdot,t)$ for fixed t. (Hint: use Exercises 5.4 and 3.5.)

REMARKS. 1. Theorem 5.2 lacks symmetry compared to Theorem 5.1. V_t exists as a process in $\underline{\underline{S}}(R^d)$ but must be extended slightly to get uniqueness, and this extension doesn't take values in $\underline{\underline{S}}'(R^d)$. It would be nicer to have a more symmetric statement, on the order of "There exists a unique process with values in such and such a space such that ...". One can get such a statement, though it requires a litle more Sobolev space theory and a little more analysis to do it. Here is how.

Let $\| \ \|_n$ be the norm of Example 1, Chapter 4. Let H_B^n be the completion of $\underline{\underline{S}}_B$ in this norm. If n is large enough, one can show that V_t is an element of $(H_B^n)' \overset{\text{def}}{=} H_B^{-n}$. Theorem 5.2 can then be stated in the form: there exists a unique process V with values in H_B^{-n} which satisfies (5.4) for all $\phi \in H_B^n$.

2. Suppose that T is the identity and consider (5.4). Extend V to be a distribution on $D \times R_+$ as follows. If $\phi = \phi(x,t)$ is in $C_0^\infty(D \times (0,\infty))$, let

$$V(\phi) = \int_0^\infty V_s(\phi(s))ds \quad \text{and} \quad T\dot{M}(\phi) = \int_0^\infty \int_D T^* \phi(x,s)\, M(dxds).$$

Then Corollary 4.2 implies that for a.e. ω, V and $T\dot{M}$ define distributions on $D \times (0,\infty)$. Now consider (5.5). For large t, the left-hand side vanishes, for ϕ has compact support. The right-hand side then tells us that $V(L\phi + \frac{\partial \phi}{\partial s}) + TM(\phi) = 0$ a.s. In other words, for a.e. ω, the distribution $V(\cdot,\omega)$ is a distribution solution of the (non-stochastic) PDE

$$\frac{\partial \mu}{\partial t} - L\mu = T\dot{M}.$$

Thus Theorem 5.1 follows from known non-stochastic theorems on PDE's. If T is the identity, the same holds for Theorem 5.2. In general, the translation of (5.4) or (5.5) into a PDE will introduce boundary terms. Still, we should keep in mind that the theory of distribution solutions of deterministic PDE's has something to say about SPDE's.

CHAPTER SIX

WEAK CONVERGENCE

Suppose E is a metric space with metric ρ. Let $\underline{\underline{E}}$ be the class of Borel sets on E, and let (P_n) be a sequence of probability measures on $\underline{\underline{E}}$. What do we really mean by "$P_n \to P_o$"? This is a non-mathematical question, of course. It is asking us to make an intuitive idea precise. Since our intuition will depend on the context, it has no unique answer. Still, we might begin with a reasonable first approximation, see how it might be improved, and hope that our intuition agrees with our mathematics at the end.

Suppose we say:

$$\text{"}P_n \to P_o \text{ if } P_n(A) \to P_o(A), \text{ all } A \in \underline{\underline{E}}.\text{"}$$

This looks promising, but it is too strong. Some sequences which should converge, don't. For instance, consider

<u>PROBLEM 1</u>. Let $P_n = \delta_{1/n}$, the unit mass at 1/n, and let $P_o = \delta_o$. Certainly P_n ought to converge to P_o, but it doesn't. Indeed $0 = \lim P_n\{0\} \neq P_o\{0\} = 1$. Similar things happen with sets like $(-\infty,0]$ and $(0,1)$.

<u>CURE</u>. The trouble occurs at the boundary of the sets, so let us smooth them out. Identify a set A with its indicator function I_A. Then $P(A) = \int I_A dP$. We "smooth out the boundary of A" by replacing I_A by a continuous function f which approximates it, and ask that $\int f dP_n \to \int f dP$. We may as well require this for all f, not just those which approximate indicator functions.

This leads us to the following. Let C(E) be the set of bounded real valued continuous functions on E.

<u>DEFINITION</u>. We say P_n <u>converges weakly</u> to P, and write $P_n \Rightarrow P$, if, for all $f \in C(E)$,

$$\int f dP_n \to \int f dP.$$

PROBLEM 2. Our notion of convergence seems unconnected with a topology.

CURE. We prescribe two definitions:

$$\underline{\underline{P}}(E) \overset{\text{def}}{=} \{P: P \text{ is a probability measure on } \underline{\underline{E}}\}.$$

A fundamental system of neighborhoods is given by sets of the form

$$\{P \in \underline{\underline{P}}(E): |\int f_i dP - \int f_i dP_o| < \varepsilon,\ i=1,\dots,n\},\quad f_i \in C(E),\ i = 1,\dots,n.$$

This notion of convergence may not appear to fill our needs - for we shall be discussing convergence of processes, rather than of random variables - but it is in fact exactly what we need. The reason why it is sufficient is itself extremely interesting, and we shall go into it shortly, but let us first establish some facts.

The first, which gives a number of equivalent characterizations of weak convergence, is sometimes called the Portmanteau Theorem.

THEOREM 6.1. The following are equivalent

(i) $P_n \Rightarrow P$;

(ii) $\int f dP_n \to f dP$, all bounded uniformly continous f;

(iii) $\int f dP_n \to \int f dP$, all bounded functions which are continuous, P-a.e.;

(iv) $\limsup P_n(F) \leq P(F)$, all closed F;

(v) $\liminf P_n(G) \geq P(G)$, all open G;

(vi) $\lim P_n(A) = P(A)$, all $A \in \underline{\underline{E}}$ such that $P(\partial A) = 0$.

Let E and F be metric spaces and $h : E \to F$ a measurable map. If P is a probability measure on E, then Ph^{-1} is a probability measure on F, where $Ph^{-1}(A) = P(h^{-1}(A))$.

THEOREM 6.2 If $h : E \to F$ is continuous (or just continuous P-a.e.) and if $P_n \Rightarrow P$ on E, then $P_n h^{-1} \Rightarrow Ph^{-1}$ on F.

Let $P_1, P_2, \dots$ be a sequence in $\underline{\underline{P}}(E)$. When does such a sequence converge? Here is one answer. Say that a set $K \subset \underline{\underline{P}}(E)$ is relatively compact if every sequence in A has a weakly convergent subsequence. (This should be "relatively sequentially compact," but we follow the common usage.)

Then (P_n) converges weakly if

(i) there exists a relatively compact set $K \subset \underline{\underline{P}}(E)$ such that $P_n \in K$ for all n.

(ii) the sequence has at most one limit point in $\underline{\underline{P}}(E)$.

Since (i) guarantees at least one limit point, (i) and (ii) together imply convergence.

If this condition is to be useful - and it is - we will need an effective criterion for relative compactness. This is supplied by Prohorov's Theorem.

DEFINITION. A set $A \subset \underline{\underline{P}}(E)$ is tight if for each $\varepsilon > 0$ there exists a compact set $K \subset E$ such that for each $P \in A$, $P\{K\} > 1 - \varepsilon$.

THEOREM 6.3. If A is tight, it is relatively compact. Conversely, if E is separable and complete, then if A is relatively compact, it is tight.

Let us return to the question of the suitability of our definition of weak convergence.

PROBLEM 3. We are interested in the behavior of processes, not random variables, so this all seems irrelevant.

CURE. We already know the solution to this. We just have to stand back far enough to recognize it. We often define a process canonically on a function space: if Ω is a space of, say, right continuous functions on $[0,\infty)$, then a process $\{X_t: t>0\}$ can be defined on Ω by $X_t(\omega) = \omega(t)$, $\omega \in \Omega$, for ω, being an element of Ω, is itself a function. X is then determined by its distribution P, which is a measure on Ω. But this means that we are regarding the whole process as a single random variable. The random variable simply takes its values in a space of functions.

With this remark, the outline of the theory becomes clear. We must first put a metric on the function space Ω in some convenient way. The above definition will then apply to measures on Ω.

The Skorokhod space $\underline{\underline{D}} = \underline{\underline{D}}([0,1],E)$ is a convenient function space to use. It is the space of all functions $f : [0,1] \to E$ which are right-continuous and have left limits at each $t \in (0,1]$. We will metrize $\underline{\underline{D}}$. The metric is a bit tricky. It is much like a sup-norm, but the presence of jump discontinuities forces a modification.

First, let Λ be the class of strictly increasing, continuous maps of $[0,1]$ onto itself. If $\lambda \in \Lambda$, then $\lambda(0) = 0$ and $\lambda(1) = 1$. Define

$$\|\lambda\| = \sup_{0 \le s < t \le 1} \left|\log \frac{\lambda(t)-\lambda(s)}{t-s}\right|, \quad \lambda \in \Lambda.$$

(We may have $\|\lambda\| = \infty$. We don't worry about that.) If $\|\lambda\|$ is small, λ must be close to the identity.

Next we define a distance on $\underline{\underline{D}}$ by

$$d_o(f,g) = \inf\{\|\lambda\| + \sup_t \rho(f(t),g(\lambda(t))) : \lambda \in \Lambda\}.$$

The functions λ should be considered as time-changes. The reason we need them can be seen by considering $f(t) = I_{[0,1/2+\varepsilon]}(t)$ and $g(t) = I_{[0,1/2]}(t)$. Both have a single jump of size one, and if ε is small, the jumps nearly coincide, and we would like $d(f,g)$ to be small. Note that $\sup_t |f(t)-g(t)| = 1$ however. The time-change allows us to move the jump of g to coincide with that of f. After doing this, we see that $\sup_t |f(t)-g(\lambda(t))|$ vanishes. (For an exercise, let $\lambda(t) = (1+4\varepsilon)t - 4\varepsilon t^2$ and show that $d(f,g) \le 4\varepsilon/1-4\varepsilon$.)

Let us collect a few miscellaneous facts.

<u>THEOREM 6.4</u>. (i) d_o is a metric on $\underline{\underline{D}}$. If E is a complete separable metric space, so is $\underline{\underline{D}}$.

(ii) $d_o(f_n,f) \to 0$ iff there exist $\lambda_n \in \Lambda$ such that $\|\lambda_n(t)-t\|_\infty \to 0$ and

$$\sup_t \rho(f(t), f_n(\lambda_n(t)))_\infty \to 0.$$

(iii) $C([0,1],E)$ is a closed subspace of $\underline{\underline{D}}$.

We have to be able to characterize compact sets in $\underline{\underline{D}}$ if we want to apply Prohorov's Theorem. Remembering the Arzela-Ascoli Theorem, this should have something to do with equicontinuity. Let us introduce a "modulus of continuity" which is tailored for right continuous processes.

EFINITION. For $f \in \underline{\underline{D}}$, let

6.1) $$w(\delta,f) = \inf_{\{t_i\}} \max_i \sup_{t_i \le s < t < t_{i+1}} \rho(f(s),f(t)),$$

here the infimum is over all finite partitions $0 = t_0 < t_1 < \cdots < t_n = 1$ such that $t_i - t_{i-1} \geq \delta$, for all $i \leq n-1$.

This discounts the large jumps of f, for one can place the partition points here. The effect is to find the modulus of continuity between the jumps. If f is ontinuous, however, this reduces to the ordinary modulus of continuity.

The counterpart of the Arzéla-Ascoli theorem is:

HEOREM 6.5. (Arzéla-Ascoli theorem for $\underline{\underline{D}}$). Let E be a complete separable metric pace. A set A has compact closure in $\underline{\underline{D}}$ iff

(i) for each rational $t \in [0,1]$ there is a compact set $K_t \subset E$ such that if $f \in A$, then $f(t) \in K_t$, all $t \in \mathbf{Q} \cap [0,1]$;

(ii) $\lim_{\delta\to 0} \sup_{f\in A} w(f,\delta) = 0$.

ote: If E is locally compact, the compact sets K_t of (i) can be chosen to be ndependent of t.

Let $(P_n) \subset \underline{\underline{P}}(D)$ be weakly convergent. In order to identify its limit, one ften checks the convergence of the finite-dimensional distributions. If $(x_1,\ldots,x_n)$ is a bounded continuous function on $E \times\cdots\times E$, and if $\leq t_0 \leq\cdots\leq t_n \leq 1$, define a function H on $\underline{\underline{D}}$ by

$$H(\omega) = h(\omega(t_1),\ldots,\omega(t_n)), \quad \omega \in \underline{\underline{D}}.$$

e say that the finite dimensional distributions converge if, for each n and each $\leq t_1 \leq\cdots\leq t_n \leq 1$, there exists a measure $\mu_{t_1\ldots t_n}$ on $E \times\cdots\times E$ such that for each uch h and H

$$\int H dP_n \to \int h d\mu_{t_1\ldots t_n}.$$

PROPOSITION 6.6. A sequence $(P_n) \subset \underline{\underline{P}}(D)$ converges weakly iff

(i) (P_n) is tight;

(ii) the finite-dimensional distributions converge.

All of this is convenient to describe in the language of processes. If $X_n = \{X_n(t),\ 0 \le t \le 1\}$ is a sequence of processes, we say that (X_n) <u>converges weakly</u> if the corresponding distributions (P_n) on $\underline{\underline{D}}$ converge weakly. Similarly, (X_n) is tight if the (P_n) are.

To show the weak convergence of (X_n), we must show (Prop. 6.5) that

(a) (X_n) is tight;

(b) the finite-dimensional distributions of the X_n converge weakly.

Of the two, tightness is often the most difficult to show. It is useful to have easily-checkable criteria. The following theorem, due primarily to Aldous, gives two criteria which are useful in case there are martingales present.

ALDOUS' THEOREM

Let E be a complete separable metric space and let $\underline{\underline{D}} = \underline{\underline{D}}([0,1],E)$. Let ρ be a bounded metric on E. Let $(\underline{\underline{F}}_t)$ be the canonical filtration on $\underline{\underline{D}}$, i.e. $\underline{\underline{F}}_t = \sigma\{\omega(s),\ s \le t,\ \omega \in \underline{\underline{D}}\}$. Let $\underline{\underline{T}}$ be the class of finite-valued stopping times T such that $T \le 1$.

If X is a process defined canonically on $\underline{\underline{D}}$, let

$$\mu(\delta,X) = \sup_{T \in \underline{\underline{T}}} E\{\rho(X_{T+\delta},X_T)\}$$

$$\nu(\delta,X) = \sup_{\alpha \le \delta} \mu(\alpha,X).$$

Results such as Prohorov's Theorem can be extended to some non-metrizable spaces. Here is one such extension due to Le Cam.

<u>THEOREM 6.7.</u> Let E be a competely regular topological space such that all compact sets are metrizable. If (P^n) is a sequence of probability measures on E which is tight, then there exists a subsequence (n_k) and a probability measure Q such that $P^{n_k} \Rightarrow Q$.

<u>THEOREM 6.8.</u> Let (X_n) be a sequence of processes with paths in $\underline{\underline{D}}$. Suppose that for each rational $t \in [0,1]$ the family of random variables $\{X_n(t),\ n=1,2,\dots\}$ is tight.

Then either of the following conditions implies that (X_n) is tight in $\underline{\underline{D}}$.

(a) (Aldous). For every sequence (T_n, δ_n) where $T_n \in \underline{\underline{T}}$ and $\delta_n \to 0$, $\delta_n > 0$,

$$\rho(X_n(T_n+\delta_n), X_n(T_n)) \to 0 \text{ in probability.}$$

(b) (Kurtz). There exists $p > 0$ and processes $\{A_n(\delta), 0<\delta<1\}$, $n = 1,2,\ldots$ such that

(i) $E\{\rho(X_n(t+\delta), X_n(t))^p \mid \underline{\underline{F}}_t\} \leq E\{A_n(\delta) \mid \underline{\underline{F}}_t\}$

and

(ii) $\lim_{\delta\to 0} \limsup_{n\to\infty} E\{A_n(\delta)\} = 0.$

We will follow a proof of T. Kurtz. Most of the work is in establishing the following three lemmas.

LEMMA 6.9. (a) is equivalent to

$$\lim_{\delta\to 0} \limsup_{n\to\infty} \nu(\delta, X_n) = 0 . \tag{6.2}$$

PROOF. If (6.2) holds, (a) follows on noticing that $\delta \to \nu(\delta, X_n)$ is increasing for each n, and that $E\{\rho(X_{T_n+\delta_n}, X_{T_n})\} \leq \nu(\delta_n, X_n)$.

Conversely, if (6.2) does not hold, there is an $\varepsilon > 0$ such that for each n_o and $\delta_o > 0$ there exists $n \geq n_o$ and $\delta_n \leq \delta_o$ such that $\nu(\delta_n, X_n) > \varepsilon/2$. Then $E\{\rho(X_n(T_n+\delta_n'), X(T_n))\} > \varepsilon/2$ for some $T_n \in \underline{\underline{T}}$ and $\delta_n' \leq \delta_n$. Since ρ is bounded, this implies (a) does not hold. Q.E.D.

LEMMA 6.10. If $T_1 \leq T_2 \in \underline{\underline{T}}$ and $T_2 - T_1 \leq \delta$, then

(i) $E\{\rho(X_{T_1}, X_{T_2})\} \leq \frac{2}{\delta}\int_0^{2\delta} \mu(u,X)du \leq 4\nu(2\delta, X)$;

(ii) $\nu(\delta, X) \leq \frac{2}{\delta}\int_0^{2\delta} \mu(u,X)du$.

PROOF. Clearly (i)=>(ii) (let $T_2 = T_1 + \alpha$). To prove (i), use the triangle inequality:

$$\rho(X_{T_1}, X_{T_2}) \leq \frac{1}{\delta}\int_0^{\delta} (\rho(X_{T_1}, X_{T_2+v}) + \rho(X_{T_2+v}, X_{T_2}))dv .$$

If $v \leq \delta$, $T_2 + v = T_1 + u$ for some $u \leq 2\delta$:

$$\leq \frac{1}{\delta}\int_0^{2\delta} \rho(X_{T_1}, X_{T_1+u})du + \frac{1}{\delta}\int_0^{\delta} \rho(X_{T_2+v}, X_{T_2})dv$$

Take expectations of both sides to see that

$$E\{\rho(X_{T_1},X_{T_2})\} \le + \frac{1}{\delta}\int_0^{2\delta}\mu(u,X)du + \frac{1}{\delta}\int_0^{\delta}\mu(v,X)dv$$

which implies (i). Q.E.D.

LEMMA 6.11. (a) implies that

$$\lim_{\delta\to 0}\limsup_{n\to\infty} E\{w(\delta,X_n)\} = 0 . \tag{6.3}$$

PROOF. Fix n, $\varepsilon > 0$ and $0 < \delta < 1$. Set $S_o = 0$ and define

$$S_{k+1} = \inf\{t>S_k : \rho(X_t,X_{S_k}) \ge \varepsilon\} \wedge 1$$

If $t \in [S_k,S_{k+1})$, $\rho(X_t,X_{S_k}) < \varepsilon$ so $\rho(X_s,X_t) < 2\varepsilon$ if $s,t \in [S_k,S_{k+1})$. If the (S_k) form a partition of mesh $\ge\delta$, $w(\delta,X) \le 2\varepsilon$.

Fix K and notice that if $S_K = 1$, there is some $J < K$ for which $S_J < 1 = S_{J+1}$, and if $S_k - S_{k-1} > \delta$ for all $k \le J$, $\{S_o,\ldots,S_j,1\}$ forms a partition of mesh $\ge\delta$, hence $w(\delta,X) \le 2\varepsilon$. In any case, $w(\delta,X) \le 1$. Thus, a very rough estimate of $w(\delta,X)$ is

$$w(\delta,X) \le 2\varepsilon + I_{\{S_k-S_{k-1}\le\delta \text{ some } k\le J,\ J\le K;\ S_K=1\}}.$$

Let $T_k = S_{k+1} \wedge (S_k+\delta)$:

$$\le 2\varepsilon + \sum_{k=0}^{K-1}\frac{1}{\varepsilon}\rho(X_{T_k},X_{S_k}) + I_{\{S_K<1\}}$$

where we use the fact that if $S_{k+1} - S_k < \delta$, $T_k = S_{k+1}$ and $\rho(X_{T_k},X_{S_k}) \ge \varepsilon$ so the sum is $\ge \varepsilon$.

Thus

$$\begin{aligned} E\{w(\delta,X_n)\} &\le 2\varepsilon + \frac{1}{\varepsilon}\sum_{k=0}^{K-1} E\{\rho(X_n(T_k),X_n(S_k))\} + P\{S_K<1\} \\ &\le 2\varepsilon + \frac{4K}{\varepsilon}\nu_n(2\delta,X_n) + P\{S_K<1\} \end{aligned}$$

by Lemma 6.10.

The final probability is independent of δ, and it converges to zero as $K \to \infty$. In fact, the convergence is even uniform in n. To see this, fix δ_o and note

$$P\{S_K<1\} = P\{e^{1-S_K} > 1\} \le E\{e^{1-S_K}\}$$

$$\le e\,E\{e^{-\sum_{1}^{n}(S_{k+1}-S_k)}\}$$

$$= e\,E\{\prod_{k=0}^{K-1} e^{-(S_{k+1}-S_k)}\}$$

$$\le e \prod_{k=0}^{K-1} E\{e^{-K(S_{k+1}-S_k)}\}^{1/K}$$

y Hölder's inequality. The integrand is bounded by $e^{-K\delta_o}$ on $\{S_{k+1}-S_k>\delta_o\}$ and by one therwise, so this is

$$\le e \prod_{k=0}^{K-1} (e^{-K\delta_0}+P\{S_{k+1}-S_k<\delta_0\})^{1/K}$$

$$\le e \prod_{k=0}^{K-1} (e^{-K\delta_o}+\frac{4}{\varepsilon}\nu(2\delta_0,X_n))^{1/K}$$

y Lemma 6.10.

Thus

$$E\{w(\delta,X_n)\} \le 2\varepsilon + \frac{4K}{\varepsilon}\nu(2\delta,X_n) + e^{1-K\delta_0} + \frac{4e}{2\varepsilon}\nu(2\delta_0,X_n).$$

et $n \to \infty$ and $\delta \to 0$ in that order and use (a). Then let $K \to \infty$, $\delta_0 \to 0$ and finally $\to 0$ to get (6.3).

ROOF (of Theorem 6.8). We first show that (b) => (6.2), which implies (a) by Lemma .9, and then we show that (a) implies tightness.

If $0 \le \alpha \le \delta$ and $p > 0$

$$E\{\rho^p(X_n(T+\alpha),X_n(T))\} \le E\{E(\rho^p(X_n(T+\alpha),X_n(T))\mid \underline{\underline{F}}_T\}\}$$

$$\le E\{E\{A_n(\alpha)\mid \underline{\underline{F}}_T\}$$

$$= E\{A_n(\alpha)\}$$

ence $\mu(\alpha,X_n) \le E\{A_n(\alpha)\}$. By the lemma

$$\nu(\delta,X_n) \le \frac{2}{\delta}\int_0^{2\delta} E\{A_n(u)\}\,du .$$

Let $n \to \infty$ and $\delta \to 0$ and use Fatou's lemma and (ii) to see that (6.2) holds.

To prove tightness, note that $w(\delta,X_n) \downarrow 0$ as $\delta \downarrow 0$, so Lemma 6.11 implies

$$\limsup_{\delta\to 0}\ \sup_n E\{w(\delta,X_n)\} = 0 .$$

Let (t_j) be an ordering of $\mathbb{Q} \cap [0,1]$ and let $\varepsilon \to 0$. For each j there is a compact $K_j \subset E$ such that

$$P\{X_n(t_j) \in K_j\} > 1 - \varepsilon/2^{j+1}.$$

Choose $\delta_k \downarrow 0$ such that $\sup_n E\{w(\delta_k, X_n)\} \le \frac{\varepsilon}{k2^{k+1}}$. Thus $\sup_n P\{w(\delta_k, X_n) > \frac{1}{k}\} \le \varepsilon/2^{k+1}$. Let $\Lambda \subset \underline{\underline{D}}$ be

$$\Lambda = \{\omega \in \underline{\underline{D}} : \omega(t_k) \in K_k,\ w(\omega,\delta_k) \le \frac{1}{k},\quad k=1,2,\ldots\}.$$

Now $\lim_{\delta \to 0} \sup_{\omega \in \Lambda} w(\delta,\omega) = 0$. Thus Λ has a compact closure in $\underline{\underline{D}}$ by Theorem 6.5. Moreover

$$\begin{aligned} P\{X_n \in \Lambda\} &\ge 1 - \sum_k P\{X_n(t_k) \in K_k\} \\ &\quad - \sum_k P\{w(\delta_k, X_n) > 1/k\} \\ &\ge 1 - \varepsilon/2 - \varepsilon/2 = 1 - \varepsilon, \end{aligned}$$

hence (X_n) is tight.

MITOMA'S THEOREM

The subject of SPDE's involves distributions in a fundamental way. We will need to know about the weak convergence of processes with values in $\underline{\underline{S}}'$. Since $\underline{\underline{S}}'$ is not metrizable, the preceeding theory does not apply directly.

However, weak convergence of distribution-valued processes is almost as simple as that of real-valued processes. According to a theorem of Mitoma, in order to show that a sequence (X^n) of processes tight, one merely needs to verify that for each ϕ, the real-valued processes $(X^n(\phi))$ are tight.

Rather than restrict ourselves to $\underline{\underline{S}}'$, we will use the somewhat more general setting of Chapter Four. Let

$$E' = \bigcup_n H_n \supset \cdots \supset H_{-1} \supset H_0 \supset H_1 \supset \cdots \supset \bigcap_n H_n = E$$

where H_n is a separable Hilbert space with norm $\| \ \|_n$, E is dense in each H_n, $\| \ \|_n \le \| \ \|_{n+1}$ and for each n there is a $p > n$ such that $\| \ \|_n \underset{HS}{<} \| \ \|_p$. E has the topology determined by the norms $\| \ \|_n$, and E' has the strong topology which is determined by the semi norms

$$p_A(f) = \sup\{|f(\phi)|,\ \phi \in A\}$$

where A is a bounded set in E.

Let $\underline{\underline{D}}([0,1],E')$ be the space of E'-valued right continuous functions which ave left limits in E', and let $C([0,1],E')$ be the space of continuous E'-valued unctions. $C([0,1],H_n)$ and $\underline{\underline{D}}([0,1],H_n)$ are the corresponding spaces of H_n-valued unctions.

If $f,g \in \underline{\underline{D}}([0,1],E')$, let

$$d_A(f,g) = \inf\{\|\lambda\| + \sup_t p_A(f(t)-g(\lambda(t))),\ \lambda \in \Lambda\},$$

ind

$$\bar{d}_A(f,g) = \sup_t p_A(f(t)-g(t)).$$

Give $\underline{\underline{D}}([0,1],E')$ (resp. $C([0,1],E')$) the topology determined by the d_A (resp. $\bar{d}_A$) for bounded A E. They both become complete, separable, completely regular spaces. The $\underline{\underline{D}}([0,1],H_n)$ have already been defined, for H_n is a metric space.

We will need two "moduli of continuity". For $\omega \in \underline{\underline{D}}([0,1],E')$, $\phi \in E$, set

$$w(\delta,\omega;\phi) = \inf_{\{t_i\}} \max_i \sup_{t_i \le s<t<t_{i+1}} |\langle\omega(t) - \omega(s),\phi\rangle|$$

where the infimum is over all finite partitions $0 = t_0<t_1<\cdots<t_n = 1$ such that $t_i - t_{i-1} \geq \delta$ for $i = 1,\dots,n-1$.

Similarly, for $\omega \in \underline{\underline{D}}([0,1],H_n)$, let

$$w_n(\delta,\omega) = \inf_{\{t_i\}} \max_i \sup_{t_i \le s<t<t_{i+1}} \|\omega(t)-\omega(s)\|_n .$$

We will define w and w_n on $C([0,1],E')$ as the ordinary moduli of continuity.

There is a metatheorem which says that anything that happens in E' already happens in one of the H_n. The regularity theorem of Chapter Four is one instance of this. Here is another.

THEOREM 6.12. Let A be compact in $\underline{\underline{D}} = \underline{\underline{D}}([0,1],E')$ (resp. $C([0,1],E')$). Then there is an n such that A is compact in $\underline{\underline{D}}([0,1],H_{-n})$ (resp. $C([0,1],H_{-n})$).

PROOF. We will only prove this for $C([0,1],E')$. The proof for $\underline{\underline{D}}$ involves the same ideas, but is considerably more technical.

Let $\phi \in E$ and map $C([0,1],E') \to C([0,1],\mathbb{R})$ by $\omega \to \{\langle\omega(t),\phi\rangle : 0\le t\le 1\}$. It is easy to see that this map is continuous, so that the image of A is compact in $C([0,1],\mathbb{R})$. By the Arzela-Ascoli Theorem (see Theorem 6.5)

(i) $\sup_{\omega \varepsilon A} \sup_{t \varepsilon [0,1]} |\langle \omega(t), \phi \rangle| < \infty$, and

(ii) $\lim_{\delta \to 0} \sup_{\omega \varepsilon A} w(\delta, \omega, \phi) = 0.$

From (i), we see that the linear functionals $F_{\omega t}$ defined by $F_{\omega t}(\phi) = \langle \omega(t), \phi \rangle$ are bounded at each ϕ, and hence equicontinuous by the Banach-Steinhaus theorem. Thus there is a neighborhood V of zero such that $\phi \varepsilon V \Rightarrow |F_{\omega t}(\phi)| < 1$, all $\omega \varepsilon A$, $t \varepsilon [0,1]$. V contains a basis element, say $\{\phi : \|\phi\|_m < \varepsilon\}$.

Thus if $c = 1/\varepsilon$,

$$|\langle \omega(t), \phi \rangle| \leq c\|\phi\|_m, \quad \text{all } \phi \varepsilon E. \tag{6.4}$$

There exists $p > m$ such that, if (e_j) is a CONS relative to $\| \ \|_p$, $\sum_j \|e_j\|_m^2 = \ell < \infty$. Then

$$\sup_{\omega \varepsilon A} \sup_t \|\omega(t)\|_{-p}^2 = \sup_{\omega \varepsilon A} \sup_t \sum_j \langle \omega(t), e_j \rangle^2 \leq c^2 \ell.$$

Set $K = \{\omega \varepsilon H_{-p} : \|\omega\|_{-p} \leq c^2 \ell\}$. There exists $n > p$ such that $\| \ \|_p \underset{HS}{<} \| \ \|_n$, so K will be compact in H_{-n} by Exercise 4.1. Thus

(i') $\omega \varepsilon A \Rightarrow \omega \varepsilon K$ and K is compact in $C([0,1], H_{-n})$.

Moreover, (6.4) implies that

$$\sup_{\omega \varepsilon A} w(\delta, \omega, e_j) \leq 2c\|e_j\|_m,$$

so

$$\lim_{\delta \to 0} \sup_{\omega \varepsilon A} w_{-m}(\delta, \omega) = \lim_{\delta \to 0} \sup_{\substack{0 \leq s < t \leq 1 \\ |t-s| < \delta}} \Big(\sum_j \langle \omega(t) - \omega(s), e_j \rangle^2\Big)^{1/2}.$$

The sum is bounded by $2c^2 \sum_j \|e_j\|_m^2 < \infty$, so that it converges uniformly in s, t and δ, and it is

$$\begin{aligned} &\leq \lim_{\delta \to 0} \Big(\sum_j \sup_{\omega \varepsilon A} w(\delta, \omega, e_j)^2\Big)^{1/2} \\ &= \Big(\sum_j \lim_{\delta \to 0} \sup_{\omega \varepsilon A} w(\delta, \omega, e_j)^2\Big)^{1/2} \\ &= 0. \end{aligned}$$

Since $\| \ \|_{-n} \leq \| \ \|_{-m}$,

(ii') $\lim_{\delta \to 0} \sup_{\omega \varepsilon A} w_{-n}(\delta, \omega) = 0$

Thus A is compact in $C([0,1], H_{-n})$ by the Arzela-Ascoli theorem. Q.E.D.

HEOREM 6.13 (Mitoma). Let $\{X_t^n, 0\leq t\leq 1\}$, $n = 1,2,\dots$ be a sequence of processes whose ample paths are in $\underline{\underline{D}}([0,1],E')$ a.s. Then the sequence (X^n) is tight iff for each ϵ E, the sequence of real-valued processes $\{X_t^n(\phi), t\geq 0\}$ $n = 1,2,\dots$ is tight in $([0,1], R)$.

All the work in the proof comes in establishing the following lemma.

EMMA 6.14. Suppose $(X^n(\phi))$ is tight for each $\phi \in E$. Then for each $\varepsilon > 0$ there xist p and $M > 0$ such that

$$\sup_n P\{\sup_{0\leq t\leq 1} \|X_t^n\|_{-p} > M\} < \varepsilon.$$

ROOF. We will do this in stages.

1) Let $\varepsilon > 0$. We claim that there exist an m and $\delta > 0$ such that

6.5) $$\|\phi\|_m < \delta \Rightarrow \sup_n \|\sup_t |X_t^n(\phi)|\| < \varepsilon$$

here $\|X\| = E\{|X|\wedge 1\}$.

To see this, consider the function

$$F(\phi) = \sup_n \|\sup_t X_t(\phi)\|, \qquad \phi \in E.$$

hen

(i) $F(0) = 0$;

(ii) $F(\phi) \geq 0$ and $F(\phi) = F(-\phi)$;

(iii) $|a| < |b| \Rightarrow F(a\phi) \leq F(b\phi)$;

(iv) F is lower-semi-continuous on E;

(v) $\lim_{n\to\infty} F(\phi/n) = 0$.

ndeed (i)-(iii) are clear. If $\phi_j \to \phi$ in E, $X_t^n(\phi_j) \to X_t^n(\phi)$ in L^o, hence $X_t^n(\phi_j)|\wedge 1 \to |X_t^n(\phi)|\wedge 1$ in probability, and $\liminf_j [\sup_t |X_t^n(\phi_j)|\wedge 1] \geq \sup_t |X_t^n(\phi_j)|\wedge 1$ .s. Thus

$$\begin{aligned} F(\phi) = \sup_n E\{\sup_t |X_t^n(\phi)|\wedge 1\} &\leq \sup_n \liminf_j E\{\sup_t |X_t^n(\phi)|\wedge 1\} \\ &\leq \liminf_j \sup_n E\{\sup_t |X_t^n(\phi)|\wedge 1\} \\ &= \liminf_j F(\phi_j), \end{aligned}$$

roving (iv).

To see (v), note that $(X^n_t(\phi))$ is tight, so, given ϕ and $\varepsilon > 0$ there exists an M such that $P\{\sup_t |X^n_t(\phi)| > M\} < \varepsilon/2$.

Choose k large enough so that $M/k < \varepsilon/2$. Then

$$\begin{aligned} F(\phi/k) &= \sup_n E\{\sup_t |X^n_t(\phi/k)| \wedge 1\} \\ &\leq \sup_n \left[P\{\sup_t |X^n_t(\phi/k)| > M\} + \frac{M}{k}\right] \\ &< \varepsilon. \end{aligned}$$

Let $V = \{\phi: F(\phi) \leq \varepsilon\}$. V is a closed (by (iv)), symmetric (by (ii)), absorbing (by (v)) set. We claim it is a neighborhood of 0. Indeed, $E = \bigcup_n nV$, so by the Baire category theorem, one, hence all, of the nV must have a non-empty interior. In particular, $\frac{1}{2}V$ does. Then $V \subset \frac{1}{2}V - \frac{1}{2}V$ must contain a neighborhood of zero. This in turn must contain an element of the basis, say $\{\phi: \|\phi\|_m < \delta\}$. This proves (6.5).

The next stages of the argument use the same techniques used in proving Theorem 4.1. (We called them "tricks" there. Now we are using them a second time, we call them "techniques".) However, the presence of the supremum over t makes this proof more delicate than the other.

2) We claim that for all n

$$(6.6) \qquad \operatorname{Re} E\{\sup_t (1 - e^{iX^n_t(\phi)})\} \leq 2\varepsilon(1 + \|\phi\|^2_m/\delta^2), \quad \phi \in E.$$

Indeed, $\operatorname{Re}(1 - e^{iX^n_t(\phi)}) = 1 - \cos X^n_t(\phi) \leq \frac{1}{2}(X_t(\phi)^2 \wedge 2)$. If $\|\phi\|_m < \delta$, then

$$\begin{aligned} \operatorname{Re} E\{\sup_t (1 - e^{iX^n_t(\phi)})\} &\leq \frac{1}{2} E\{\sup_t X^n_t(\phi)^2 \wedge 2\} \\ &\leq 2E\{\sup_t |X^n_t(\phi)| \wedge 1|\} \\ &= 2 |\!|\!| \sup_t |X^n_t(\phi)| |\!|\!| \leq 2\varepsilon. \end{aligned}$$

On the other hand, if $\|\phi\|_m \geq \delta$, replace ϕ by $\psi = \delta\phi/\|\phi\|_m$:

$$\begin{aligned} \operatorname{Re} E\{1 - e^{iX_t(\phi)}\} &\leq \frac{1}{2} E\{\frac{\|\phi\|^2_m}{\delta^2} X^n_t(\psi)^2 \wedge 2\} \\ &\leq \frac{1}{2} \frac{\|\phi\|^2}{\delta^2} E\{X^n_t(\psi)^2 \wedge 2\} \\ &\leq 2\varepsilon \|\phi\|^2_m/\delta^2 \end{aligned}$$

ince $\|\phi\|_m \leq \delta$.

3) We claim that for $M > 0$ there is $p \geq m$ and constant K such that for ll n,

$$E\{1 - e^{-1/M \sup_t \|X_t^n\|_{-p}^2}\} \leq 2\varepsilon(1 + K/M\delta^2). \tag{6.7}$$

ndeed, there exists $p > m$ such that $\| \ \|_m \underset{HS}{<} \| \ \|_p$. Let (e_j) be a CONS in E, relative o $\| \ \|_n$. Let $Y_1, Y_2, \ldots$ be iid $N(0, \frac{2}{M})$ random variables, and put $\phi = \sum_1^N Y_j e_j$.

$$\mathrm{Re}\, E\{\sup_t (1 - e^{iX_t^n(\phi)})\} = \mathrm{Re}\, E\{E\{\sup_t (1 - e^{iX_t^n(\phi)}) | Y\}\}$$

$$\leq E\{2\varepsilon(1 + \|\phi\|_m^2/\delta^2)\}.$$

ut $E\{\|\phi\|_m^2\} = E\{\sum_1^N Y_j^2 \|e_j\|_m^2\} \leq \frac{2}{M} \sum_1^\infty \|e_j\|_m^2 \overset{\text{def}}{=} \frac{K}{M} < \infty$. On the other hand,

$$\mathrm{Re}\, E\{\sup_t (1 - e^{iX_t^n(\phi)})\} = \mathrm{Re}\, E\{\sup_t (1 - e^{i \sum_1^N Y_j X_t^n(e_j)})\}$$

$$\geq \mathrm{Re}\, E\{\sup_t E\{(1 - e^{i\sum_1^N Y_j X_t(e_j)}) | X_t\}\}.$$

ut, given X_t, the conditional distribution of $\sum Y_j X_t(e_j)$ is $N(0, \frac{2}{M} \sum_1^N X_j^n(e_j)^2)$. We now the characteristic function of a normal random variable, so we see this is

$$= E\{\sup_t (1 - e^{-\frac{1}{M} \sum_1^N X_t^n(e_j)^2})\}.$$

Let $N \to \infty$. $\sum X_t^n(e_j)^2 = \|X_t^n\|_{-p}$, so we can combine these inequalities to get 6.7).

4) If $\sup_t \|X_t\|_{-p}^2 > M$, $1 - e^{-\frac{1}{M} \sup_t \|X_t\|_{-p}^2} > \frac{e-1}{e}$, so

$$\frac{e-1}{e} P\{\sup_t \|X_t\|_{-p}^2 > M\} \leq E\{1 - e^{-\frac{1}{M} \sup_t \|X_t\|_{-p}^2}\}$$

$$\leq 2\varepsilon(1 + K/M\delta^2).$$

If $M \geq K/\delta^2$, then

$$P\{\sup_t \|X_t\|_{-p}^2 > M\} \leq \frac{4e}{e-1} \varepsilon.$$

Q.E.D.

PROOF (of Theorem 6.13). Fix $\varepsilon > 0$. Choose M and p as in Lemma 6.14. With probability $1 - \varepsilon$, X_t lies in $B = \{x: \|x\|_{-p} \leq M\}$ for all t. There exists $q > p$ such that $\| \|_p \underset{HS}{<} \| \|_q$, hence $\| \|_{-q} \underset{HS}{<} \| \|_{-p}$. Then A is compact in H_{-q}. Let $K \subset \underline{\underline{D}}([0,1],E')$ be the set $\{\omega : \omega(t) \in A, 0 \leq t \leq 1\}$.

Let (e_j) be a CONS relative to $\| \|_q$. Since $(X^n(e_j))$ is tight by hypothesis, there exists a compact set $K_j \subset \underline{\underline{D}}([0,1], R)$ such that $P\{X^n_{\bullet}(e_j) \varepsilon K_j\} \geq 1 - \varepsilon/2^j$ for all n. Let K'_j be the inverse image of K_j in $\underline{\underline{D}}([0,1],E')$ under the map $\omega \to \{\langle \omega(t), e_j \rangle \; ; \; 0 \leq t \leq 1\}$. By the Arzela-Ascoli Theorem,

$$\lim_{\delta \to 0} \sup_{\omega \varepsilon K'_j} w(\delta,\omega;e_j) = 0;$$

moreover

$$P\{X^n_{\bullet} \varepsilon K'_j\} \geq 1 - \varepsilon/2^j.$$

Set $K' = K \cap \bigcap_j K'_j$. Then

$$P\{X^n_{\bullet} \varepsilon K'\} \geq 1 - \varepsilon - \sum \varepsilon/2^j = 1 - 2\varepsilon.$$

Now,

$$\lim_{\delta \to 0} \sup_{\omega \varepsilon K'} w(\delta,\omega,H_{-q}) = \lim_{\delta \to 0} \sup_{\omega \varepsilon K'} \Big(\sum_j \inf_{\{t_i\}} \max_i \sup_{t_i \leq s < t < t_{i+1}} \langle \omega(t) - \omega(s), e_j \rangle^2\Big)^{1/2}$$

$$\leq \lim_{\delta \to 0} \sup_{\omega \varepsilon K'} \Big(\sum_j w(\delta,\omega,e_j)^2\Big)^{1/2}.$$

But if $\omega \varepsilon K'$, $\|\omega(t)\|_{-p} \leq M$ so the sum is dominated by $2M \sum_j \|e_j\|_p^2 < \infty$ (since $\| \|_p \underset{HS}{<} \| \|_q$). Thus we can go to the limit inside the sum. It is

$$\leq \Big(\sum_j \lim_{\delta \to 0} \sup_{\omega \varepsilon K'} w(\delta,\omega;e_j)^2\Big)^{1/2}$$

$$= 0.$$

A is compact in H_{-q}, so Theorem 6.5 tells us that K' is relatively compact in $\underline{\underline{D}}([0,1],H_{-q})$. The inclusion map of H_{-q} into E' is continuous, so that K' is also relatively compact in $\underline{\underline{D}}([0,1],E')$, and hence (X^n) is tight.

Q.E.D.

This brings us to the convergence theorem.

THEOREM 6.15. Let (X^n) be a sequence of processes with paths in $\underline{\underline{D}}([0,1],E')$. Suppose

(i) for each $\phi \in E$, $(X^n(\phi))$ is tight;

(ii) for each $\phi_1,\ldots,\phi_p$ in E and $t_1,\ldots,t_p \in [0,1]$, the distribution of $(X^n_{t_1}(\phi_1),\ldots,X^n_{t_p}(\phi_p))$ converges weakly on $\mathbb{R}^p$.

Then there exists a process X^o with paths in $\underline{D}([0,1],E')$ such that $X^n \Rightarrow X^o$.

PROOF. (X^n) is tight by Theorem 6.13. The space $\underline{D}([0,1],E')$ is completely regular and each compact subset is in some H_{-n} and is therefore metrizable. By Theorem 6.7, some subsequence converges weakly. But (ii) shows that there is only one possible limit. We conclude by the usual argument that the whole sequence converges.

Q.E.D.

Note: The index p of Lemma 6.14 may depend on ε. If it does not, and if $\| \ \|_p \underset{HS}{<} \| \ \|_q$, then (X^n) then will be tight in $\underline{D}([0,1],H_{-q})$, and we get the following.

COROLLARY 6.16. Suppose that the hypotheses of Theorem 6.15 hold. Let $p < q$ and suppose $\| \ \|_p \underset{HS}{<} \| \ \|_q$. Suppose that for $\varepsilon > 0$ and $M > 0$ there exists $\delta > 0$ such that for all n

$$P\{\sup_t |\langle X^n_t,\phi\rangle| > M\} \le \varepsilon \text{ if } \|\phi\|_p \le \delta.$$

Then (X^n) converges weakly in $\underline{D}([0,1],H_{-q})$.

CHAPTER SEVEN

APPLICATIONS OF WEAK CONVERGENCE

Does the weak convergence of a sequence of martingale measures imply the weak convergence of their stochastic integrals? That is, if $M^n \Rightarrow M$, does $f\cdot M^n \Rightarrow f\cdot M$? Moreover, do the convolution integrals - which give the solutions of SPDE's - also converge?

We will give the beginnings of the answers to these questions in this chapter. We will show that the answer to both is yes, *if* one is willing to impose strong hypotheses on the integrands. Luckily these conditions are satisfied in many cases of interest.

We will confine ourselves to measures on $\mathbf{R}^d$ and on sub-domains of $\mathbf{R}^d$, where we have already discussed the theory of distributions. We will view martingale measures as distribution-valued processes, so that weak convergence means convergence in distribution on the Skorokhod space $\underline{\underline{D}} = \underline{\underline{D}}\{[0,1], \underline{\underline{S}}'(\mathbf{R}^d)\}$.

Our martingale measures may have infinite total mass, but we will require that they not blow up too rapidly at infinity.

Fix $p_0 > 0$ and define $h_0(x) = (1 + |x|^{p_0})^{-1}$, $x \in \mathbf{R}^d$. If M is a worthy martingale measure with dominating measure K, define an increasing process k by

$$k(t) = \int_{\mathbf{R}^{2d}\times[0,t]} h_0(x)\, h_0(y)\, K(dx\, dy\, ds), \tag{7.1}$$

and

$$\gamma(\delta) = \sup_{t\le 1} (k(t+\delta) - k(t)) \tag{7.2}$$

We will assume throughout this chapter that $E\{k(1)\} < \infty$. Note that this means that for any $\phi \in \underline{\underline{S}}(\mathbf{R}^d)$, $M_t(\phi)$ is defined for all $t \le 1$, since ϕ tends to zero at infinity faster than h_0. Thus M_t is a tempered distribution (Corollary 4.2).

For a function f on $\mathbf{R}^d$, define

$$\|f\|_\infty = \sup |f(x)|,$$

$$\|f\|_h = \|f\, h_0^{-1}\|_\infty .$$

Note that $\|\phi\|_h < \infty$ for any $\phi \in \underline{\underline{S}}(\mathbf{R}^d)$, and, moreover, if $\|f\|_h < \infty$, then $M_t(f)$ is defined.

LEMMA 7.1. Let $S \le T$ be stopping times for M, and let $f \in \underline{\underline{P}}_M$. Then

(7.3) $$\langle f\cdot M(E)\rangle_T - \langle f\cdot M(E)\rangle_S \le \int_S^T \|f(s)\|_h^2 \, dk(s).$$

Consequently, if $\phi \in \underline{\underline{S}}(\mathbb{R}^d)$,

(7.4) $$\langle f\cdot M(\phi)\rangle_T - \langle f\cdot M(\phi)\rangle_S \le \|\phi\|_h^2 \int_S^T \|f(s)\|_\infty^2 \, dk(s).$$

PROOF. The left-hand side of (7.3) is

$$\int_{\mathbb{R}^{2d}\times(S,T]} f(x,s)f(y,s)K(dx\,dy\,ds) = \int_{\mathbb{R}^{2d}\times(S,T]} f(x,s)h_0^{-1}(x)f(y,s)h_0^{-1}(y)h_0(x)h_0(y)K(dx\,dy\,ds)$$

$$\le \int_{\mathbb{R}^{2d}\times(S,T]} \|f(s)\|_h^2\, h_0(x)h_0(y)\; K(dx\,dy\,ds)$$

$$= \int_S^T \|f(s)\|_h^2 \, dk(s).$$

Then (7.4) follows since $\|f(s)\phi\|_\infty \le \|\phi\|_h \; \|f(s)\|_\infty$. Q.E.D.

LEMMA 7.2. Let T be a predictable stopping time. If k is a.s. continuous at T then for any bounded Borel set $A \subset \mathbb{R}^d$ and $f \in \underline{\underline{P}}_M$, $P\{f\cdot M(A)$ is continuous at $t\} = 1$.

PROOF. The graph $[T]$ of T is predictable, hence so is $f(x,t)\,I_{[T]}(t)I_A(x)$. By (7.3), if f is bounded

$$E\{(f\cdot M_T(A) - f\cdot M_{T-}(A))^2\} \le E\{\int \|f(t)I_{[T]}(t)\|_h^2 \, dk(t)\}$$

$$= E\{\|f(T)\|_h^2(k(T) - k(T-))\}$$

$$= 0.$$

The result follows for all $f \in \underline{\underline{P}}_M$ by approximation. Q.E.D.

Let $M^0, M^1, M^2, \ldots$ be a sequence of worthy martingale measures and let k_n and γ_n, $n = 0,1,2,\ldots$ be the corresponding quantities defined in (7.1) and (7.2).

PROPOSITION 7.3. Suppose that

(7.5) $$\lim_{\delta\to 0} \limsup_{n\to\infty} E\{\gamma_n(\delta)\} = 0.$$

Then the sequence (M^n) is tight on $\underline{\underline{D}}\{[0,1], \underline{\underline{S}}'(\mathbb{R}^d)\}$.

PROOF. By Mitoma's Theorem, we need only show that $(M^n(\phi))$ is tight for each $\phi \in \underline{\underline{S}}'(R^d)$. By (7.4)

$$E\{(M^n_{t+\delta}(\phi) - M^n_t(\phi))^2 | \underline{\underline{F}}_t\} \leq \|\phi\|_h^2 \, E\{k_n(t+\delta) - k_n(t) | \underline{\underline{F}}_t\}$$
$$\leq \|\phi\|_h^2 \, E\{\gamma_n(\delta) | \underline{\underline{F}}_t\}.$$

The result now follows from Theorem 6.8(b). Q.E.D.

COROLLARY 7.4. Suppose (M^n) is a sequence of worthy martingale measures satisfying (7.5). Let $f_n \in \underline{\underline{P}}_{M^n}$ be a sequence of uniformly bounded functions. Then $(f_n \cdot M^n)$ is tight.

PROOF. $K_{f_n \cdot M^n}(dx\, dy\, ds) = |f_n(x,s) f_n(y,s)| \, K_{N^n}(dx\, dy\, ds)$

$$\leq b^2 \, K_{M^n}(dx\, dy\, ds)$$

where b is the uniform bound for the f_n. Then the $f_n \cdot M^n$ satisfy (7.5) and the result follows from Proposition 7.3. Q.E.D.

WEAK CONVERGENCE OF STOCHASTIC INTEGRALS

In order to talk of the convergence of a sequence of stochastic integrals $f \cdot M^n$, we must be able to define the integrand f for each of the M^n. We can do this by defining all of the M^n on the same probability space. The most convenient space for this is the Skorokhod space $\underline{\underline{D}}$. Thus we will define all our martingale measures canonically on $\underline{\underline{D}}$, so that once we define $f(x,t,w)$ on $R^d \times R^+ \times \underline{\underline{D}}$, we can define all the $f \cdot M^n$.

The stochastic integral is not in general a continuous function on $\underline{\underline{D}}$, so that it is not always true that $M^n \Rightarrow M$ implies that $f \cdot M^n \Rightarrow f \cdot M$, even for classical martingales. Two examples, both of real valued martingales, will illustrate some of the pitfalls.

Example 7.1. Define $M^n_t = \begin{cases} 0 & \text{if } t < 1 + 1/n \\ X & \text{if } t \geq 1 + 1/n \end{cases}$ and $M_t = \begin{cases} 0 & \text{if } t < 1 \\ X & \text{if } t \geq 1 \end{cases}$, where $X = \pm 1$

ith probability 1/2 each. Let $f(t) = I_{[0,1]}(t)$. Then $M^n \Rightarrow M$, but $f \cdot M^n \equiv 0$, $= 1,2,\ldots$ while $f \cdot M_t = X\, I_{[1,\infty)}(t)$, so the integrals don't converge to the right imit.

xample 7.2. Let B_t be a standard real-valued Brownian motion from zero and put $_t = M_0 + B_t$ and $M_t^n = M_0^n + B_t$, where M_0 is a random variable uniformly distributed on 0,1], independent of B, and M_0^n is uniformly distributed on $\{1/n, 2/n, \ldots, 1\}$. Define hese canonically on $\underline{\underline{D}}\{[0,1],\mathbb{R}\}$, and let $f(t,\omega) = I_{\{\omega(0) \in \mathbb{Q}\}}$ (i.e. $f \equiv 1$ or $f \equiv 0$ epending on whether the initial value of the martingale is rational or not.) hen $M^n \Rightarrow M$, but $f \cdot M^n = B_t$ for all n while $f \cdot M = 0$ a.s., so once again the integrals on't converge to the right limit.

EMARKS 1. In Example 7.1, the integrand was deterministic, and the trouble came rom the jumps of M. In Example 7.2, the integrand was simply a badly discontinuous unction of ω on $\underline{\underline{D}}$. Although it might seem that the trouble in Example 7.2 comes ecause the distribution of M is orthogonal to the distribution of the M^n, one can odify it slightly to make the distributions of the M^n and M all equivalent, and till get the same result.

2. It is easy to get examples in which the $f \cdot M^n$ do not just converge to he wrong limit but fail to converge entirely. Just replace every second M^n in ither of the examples above by the limit martingale M.

Let M be a martingale measure defined canonically on $\underline{\underline{D}}$, relative to a robability measure P. Let k be defined by (7.1). Let $\mathcal{P}_S(M)$ be the class of unctions f on $\mathbb{R}^d \times \mathbb{R}_+ \times \underline{\underline{D}}$ of the form

$$f(x,t,\omega) = \sum_{n=1}^{N} a_n(\omega) I_{(s_n,t_n]}(t)\phi_n(x), \tag{7.6}$$

here $0 \le s_n < t_n$, $\phi_n \in C^\infty(\overline{\mathbb{R}}^d)$, and a_n is bounded and $\underline{\underline{F}}_{s_n}$-measurable, $n = 1,\ldots,N$, uch that

(i) a_n is continuous P-a.s. on $\underline{\underline{D}}$;

(ii) $t \rightsquigarrow f(x,t,\omega)$ is continuous at each point of discontinuity of k.

If f is given by (7.6) and if $\phi \in \underline{\underline{S}}(\mathbb{R}^d)$ and $\omega \in \underline{\underline{D}}$, define

$$F_t(\omega)(\phi) = \sum_{n=1}^{N} a_n(\omega) I_{(s_n,t_n]}(t)\,(\omega_{t\wedge t_n}(\phi_n\phi) - \omega_{t\wedge s_n}(\phi_n\phi)).$$

Note that $\omega_t(\cdot)$ is a distribution and $\phi_n\phi \in \underline{\underline{S}}(\mathbb{R}^d)$, so $\omega_t(\phi_n\phi)$ is defined. The map $\omega \to \{F_t(\omega),\ t \geq 0\}$ maps $\underline{\underline{D}} \to \underline{\underline{D}}$. It is continuous at ω if ω is a point of continuity of each of the a_n and if $t \to \omega_t$ is itself continuous at each of the s_n and t_n. Thus if $f \in \mathcal{P}_S(M)$, f is P -a.e. continuous in ω. Moreover, if M is canonically defined on $\underline{\underline{D}}$, then

$$F_t(\omega)(\phi) = f\cdot M_t(\phi).$$

By the continuity theorem (Theorem 6.1 (iii)) we have:

<u>PROPOSITION 7.5</u>. Let $M^0, M^1, \ldots$ be a sequence of worthy martingale measures such that $M^n \Rightarrow M^0$. If $f \in \mathcal{P}_S(M^0)$ then $(M^n, f\cdot M^n) \Rightarrow (M^0, f\cdot M^0)$.

This is too restrictive to be of any real use, so we must extend the class of f. We will do this by approximating more general f by f in $\mathcal{P}_S(M^0)$ and using the fact that L^2 convergence implies convergence in distribution.

The class of f we can treat depends on the sequence (M^n), and, in particular, on the sequence (K_n) of dominating measures. The more we are willing to assume about the K_n, the better description we can give of the class of f. We will start with minimal assumptions on the K_n. This will allow us to give a simple treatment which is sufficient to handle the convergence of the solutions of the SPDE's in Chapter 5. We will then give a slightly deeper treatment under stronger hypotheses on the K_n.

<u>DEFINITION</u>. Let $\overline{\mathcal{P}}_S(M)$ be the closure of $\mathcal{P}_S(M)$ in the norm

$$|||\, f \,||| = \sup_{\substack{\omega \in \underline{\underline{D}} \\ 0 \leq t \leq 1}} \|f(t,\omega)\|_\infty.$$

<u>Note</u>: We will suppress variables from our notation when possible. Thus, we write $\|f(t,\omega)\|$ in place of $\|f(\cdot,t,\omega)\|$.

If $M^n \Rightarrow M^0$, we are interested not only in the convergence of $f\cdot M^n$ to $f\cdot M^0$, but also in the joint convergence of $(M^n, f\cdot M^n)$ to $(M^0, f\cdot M^0)$. We may also want to know about the simultaneous convergence of a number of martingale measures and integrals. This means we will want to state a convergence theorem for martingale measures with values in R^m, and integrands which are $r \times m$ matrices. Weak convergence in this context is convergence in the Skorokhod space $\underline{\underline{D}}^m = \underline{\underline{D}}\{[0,1], S'(R^{md})\}$.

If $M = (M^1, \ldots, M^n)$ and if $f = (f_{ij})$ is matrix valued, we will say $f \in \bar{\mathcal{P}}_S(M)$ resp. $f \in \underline{\underline{P}}(M)$) if for each i, j, $f_{ij} \in \bar{\mathcal{P}}_S(M^j)$, (resp. $f_{ij} \in \underline{\underline{P}}(M^j)$).

PROPOSITION 7.6. Let $M^0, M^1, M^2, \ldots$ be a sequence of R^m-valued worthy martingale measures and suppose that each coordinate satisfies (7.5). Let $f^n(x,t,\omega) = (f^n_{ij}(x,t,\omega))$ be $r \times m$ matrices such that $f^n \in \underline{\underline{P}}(M^n)$, $n = 1,2,\ldots$, and $f^0 \in \bar{\mathcal{P}}_S(M^0)$. Suppose that $M^n \Rightarrow M^0$ (on $\underline{\underline{D}}^m$) and that $\| |f^n_{ij} - f^0_{ij}| \| \to 0$. Then $f^n \cdot M^n \Rightarrow f^0 \cdot M^0$ on $\underline{\underline{D}}^m$.

REMARK. We can choose f^n of the form $\binom{I}{F^n}$ where I is the $m \times m$ identity and F^n is diagonal with $F^n_{jj} = f^n_j$. Then $f \cdot M = (M^{n1}, \ldots, M^{nm}, f^n_1 \cdot M^{n1}, \ldots, f^n_m \cdot M^{nm})$ and the proposition gives the joint convergence of the M^{nj} and the $f^n_j \cdot M^{nj}$.

PROOF. Each coordinate of $(f^n \cdot M^n)$ is tight by Corollary 7.4, so $(f^n \cdot M^n)$ is tight. To show it converges, we need only show convergence of the distributions of $(f^n \cdot M^n_{t_1}, \ldots, f^n \cdot M^n_{t_k})$, where $t_1 \leq t_2 \leq \ldots \leq t_k$ are continuity points of M^0. Note that this vector can be realized by taking another, larger matrix f and looking only at $t = 1$. For instance, if $t_1 < t_2 \leq 1$, $(f \cdot M^n_{t_1}, f \cdot M^n_{t_2}) = \tilde{f} \cdot M^n_1$, where $\tilde{f} = (\tilde{f}_1, \tilde{f}_2)$ and

$$\tilde{f}_i(x,t) = \begin{cases} f(x,t), & t \leq t_i \\ 0 & t > t_i, \quad i = 1,2. \end{cases}$$

Since $f \in \bar{\mathcal{P}}_S(M^0)$ and the t_i are continuity points of M^0, $\tilde{f}_i \in \bar{\mathcal{P}}_S(M^0)$.

By Mitoma's theorem, then, it is enough to check the convergence of $f^n \cdot M_t(\phi)$ for a single test function ϕ and for $t = 1$. By Theorem 6.1, it is enough to show that

$$E\{h(f \cdot M^n_1(\phi))\} \to E\{h(f \cdot M^0_1(\phi))\}$$

for any uniformly continuous bounded h on R^r.

Choose $f^{\varepsilon}_{ij} \in \mathcal{P}_S(M^0)$ such that $\| |f^0_{ij} - f^{\varepsilon}_{ij} \| | < \varepsilon 2^{-n}$ for each i, j and let $f^{\varepsilon} = (f^{\varepsilon}_{ij})$. Then

$$
\begin{aligned}
&|E\{h(f^n \cdot M^n_1(\phi))\} - E\{h(f^0 \cdot M^0_1(\phi))\}| \\
&\le E\{|h(f^n \cdot M^n_1(\phi)) - h(f^0 \cdot M^n_1(\phi))|\} \\
&+ E\{|h(f^0 \cdot M^n_1(\phi)) - h(f^{\varepsilon} \cdot M^n_1(\phi))|\} \\
&+ |E\{h(f^{\varepsilon} \cdot M^n_1(\phi))\} - E\{f^{\varepsilon} \cdot M^0_1(\phi)\}| \\
&+ E\{|h(f^{\varepsilon} \cdot M^0_1(\phi)) - h(f^0 \cdot M^0_1(\phi))|\} \\
&= E_1 + E_2 + E_3 + E_4.
\end{aligned}
$$

Since h is uniformly continuous, given $\rho > 0$ there exists $\delta > 0$ such that $|h(y) - h(x)| < \rho$ if $|y - x| < \delta$. Then

$$
\begin{aligned}
E_1 &\le \rho + 2\|h\|_\infty P\{|f^n \cdot M^n(\phi) - f^0 \cdot M^n(\phi)| > \delta\} \\
&\le \rho + 2\|h\|_\infty \delta^{-2} E\{|(f^n - f^0) \cdot M^n(\phi)|^2\} \\
&\le \rho + 2\|h\|_\infty \delta^{-2} \|\phi\|^2_h \sum_{i,j} \| |f^n_{ij} - f^0_{ij} \| |^2 E\{\gamma^n(1)\}
\end{aligned}
$$

by Lemma 7.1. Now it follows from (7.5) that $E\{\gamma_n(1)\}$ is bounded by, say, C, so that $\limsup_{n \to \infty} E_n \le \rho$. Moreover, the same type of calculation, with $f^n - f^0$ replaced by $f^{\varepsilon} - f^0$, gives

$$E_2 \le \rho + 2mr \; \| |h| \| \; \delta^{-2} \; \|\phi\|^2_h \, C \, \varepsilon,$$

and E_4 satisfies the same inequality since the calculation is valid for n = 0. Finally $E_3 \to 0$ by Proposition 7.5. Since ρ and ε are arbitrary, we conclude that $E_1 + \ldots + E_4 \to 0$. Q.E.D.

Let us now consider the SPDE (5.4). Its solution is

$$
\begin{aligned}
V_t(\phi) &= \int_{D\times[0,t]} T^* G_{t-s}(\phi,y) \, M(dy\,ds) \qquad (7.7) \\
&= \int_{D\times[0,t]} T^* \left(\int_s^t G_{u-s}(L\phi,y)du + \phi(y)\right) M(dy\,ds) \\
&= M_t(T^*\phi) + \int_0^t \left[\int_0^u \int_D T^* G_{u-s}(L\phi,y) \, M(dy\,ds)\right] du,
\end{aligned}
$$

where we have used (5.7), the fundamental equation of the Green's function, and then changed the order of integration.

<u>LEMMA 7.7</u>. $E\{\sup_{t\le 1} V_t(\phi)^2\} \le (8\|T^*\phi\|^2_h + 2\sup_{t\le 1}\|T^* G_t(L\phi)\|^2_h) E\{k(1)\}$.

PROOF. $\sup_t V_t(\phi)^2 \leq 2 \sup_t M_t^2(T^*\phi) + 2 \sup_t \left[\int_0^t \left(\int_0^u \int_D T^*G_{u-s}(L\phi)M(dy\ ds)\right)^2 du\right]$

by Schwartz. By Doob's inequality

$$E\{\sup_t V_t(\phi)^2\} \leq 8\ E\{M_1^2(T^*\phi)\} + 2 \int_0^1 E\{[\int_0^u \int_D T^*G_{u-s}(L\phi)M(dy\ ds)]^2\}du,$$

and the conclusion follows by Lemma 7.1.

Q.E.D.

Let $M^0, M^1, M^2, \ldots$ be a sequence of worthy martingale measures and define V^n by (7.7).

PROPOSITION 7.8. If (M^n) satisfies (7.5), then (V^n) is tight. If, in addition, $M^n \Rightarrow M^0$, then $(M^n, V^n) \Rightarrow (M^0, V^0)$.

PROOF. In order to prove (V^n) is tight, it is enough by Mitoma's theorem to show that $(V_t^n(\phi))$ is tight for any $\phi \in \underline{\underline{S}}(\mathbf{R}^d)$. The (M^n) are tight by Proposition 7.3, so we must show that the sequence (U^n) defined by

$$U_t^n = \int_0^t [\int_0^u \int_D T^*\ G_{u-s}(L\phi,y)M(dy\ ds)du$$

is tight. Let

$$S_n = \sup_{u\leq 1} |\int_0^u \int_D T^*\ G_{u-s}(L\phi,y)M(dy\ ds)|.$$

If τ_n is a stopping time for M^n, $\delta_n > 0$, and if $\tau_n + \delta_n \leq 1$, then

$$|U^n_{\tau_n+\delta_n} - U^n_{\tau_n}| \leq \delta_n S_n.$$

By Lemma 7.7, $E\{S_n^2\} \leq C\ E\{k_n(1)\}$. By (7.5), this expectation is bounded in n, so $U^n_{\tau_n+\delta_n} - U^n_{\tau_n} \to 0$ in L^1, hence in probability. Moreover, $E\{(U_t^n)^2\} \leq E\{S_n^2\}$, so that (U_t^n) is tight for fixed t. By Aldous' Theorem, (U^n) is tight.

We need only check the convergence of the finite-dimensional distributions in order to see that $(M^n, V^n) \Rightarrow (M^0, V^0)$. But for each t, $T^*G_{t-s}(\phi,y)$ is deterministic, is in $\underline{\underline{S}}(\mathbf{R}^d)$ as a function of y, and is a continuous function of s for $s \neq t$. Thus if t is not a discontinuity point of M^0, $(s,y) \to T^*G_{t-s}(\phi,y)$ is in $\mathcal{P}(M^0)$. The finite-dimensional convergence now follows from Proposition 7.6.

Q.E.D.

If ϕ is in the Sobolev space $H_p(D)$ of Example 1, Chapter 4, and if $p > d/2$, then there is a constant C such that $\|\phi\|_\infty \leq C\|\phi\|_{H_p}$. Since h_0 is bounded away from zero on the bounded domain D, $\|\phi\|_h \leq C\|\phi\|_{H_p}$ for another constant C. If α is the order of T (and hence of T*) and if $\phi \in H_{p+\alpha+2}$ for all $0 \leq t \leq 1$, then $\|T^*G_t(L\phi)\|_h \leq C\|\phi\|_{H_{p+\alpha+2}}$ for all $0 \leq t \leq 1$ (and in particular for $t = 0$) so that Lemma 7.7 implies that

$$E\{\sup_{t\leq 1} V_t^n(\phi)^2\} \leq C(\|\phi\|_{H_{p+\alpha+2}})E\{k_n(1)\}.$$

Suppose that $q > p + 2 + d/2$. Now $E\{k_n(1)\}$ is bounded if (7.5) holds, and $H_{p+\alpha+2} \overset{<}{HS} H_q$. By Corollary 6.16, $V^n \Rightarrow V$, as processes with values in H_{-q}. Thus

COROLLARY 7.9. If (M^n) is a sequence of worthy martingale measures which satisfies (7.5) and if $M^n \Rightarrow M$, then $V^n \Rightarrow V^0$ in $\underline{\underline{D}}\{[0,1], H_{-(d+\alpha+3)}\}$.

AN APPLICATION

In many applications - the neurophysiological example of Chapter 3, for instance - the driving noise is basically impulsive, of a Poisson type, but the impulses are so small and so closely spaced that, after centering, they look very much like a white noise. The following results show that for some purposes at least, one can approximate the impulsive model by a continuous model driven by a white noise. One might think of this as a diffusion approximation.

Let us return to the setting of Chapter 5. Let D be a bounded domain in $\mathbf{R}^d$ with a smooth boundary, and consider the initial-boundary value problem (5.3) with two changes: we will allow an initial value given by a measure on $\mathbf{R}^d$, and we will replace the martingale measure M by a Poisson point process Π.

Let Π^n be a sequence of time-homogeneous Poisson point processes on D with characteristic measures μ_n. (Recall that this means that Π^n is a random σ-finite signed measure on $D\times[0,\infty)$ which is a sum of point masses. If $A \subset D$ is Borel and $K \subset \mathbf{R}$ is compact with $0 \notin K$, let $N_t^n(A\times K)$ be the counting process: $N_t^n(A\times K)$ is the number of points in $A\times[0,t]$ whose masses are in K. Then $\{N_t^n(A\times K), t \geq 0\}$ is a oisson process with parameter $\mu_n(A\times K)$, and if $A_1\times K_1 \cap A_2\times K_2 = \phi$, then $N^n(A_1\times K_1)$ and $^n(A_2\times K_2)$ are independent.) Let

$$m_n(A) = \int_{D\times\mathbb{R}} r\,\mu_n(dx\ dr)$$

nd

$$\sigma_n(A) = \int_{D\times\mathbb{R}} r^2\mu_n(dx\ dr)$$

e the mean and intensity measures, respectively, of Π^n. For $\delta > 0$, let

$$Q_n(\delta) = \int_D r^{2+\delta}\mu_n(dx\ dy).$$

Let L, T and B be as in (5.3), let ν_n be a finite measure on D and consider he initial-boundary value problem

$$\text{(7.8)}\qquad \begin{cases} \dfrac{\partial V}{\partial t} = LV + T\dot{\Pi}^n \\ BV = 0 \quad \text{on } \partial D \\ V_0 = \nu_n \end{cases}$$

Note that $M^n_t(A) \overset{\text{def}}{=} \Pi^n(A \times [0,t]) - t\, m_n(A)$ is an orthogonal martingale easure. The solution to (7.8) is, by Theorem 5.2,

$$\text{(7.9)}\qquad V^n_t(\phi) = \int_D G_t(\phi,y)\nu_n(dy) + \int_{D\times[0,t]} G_{t-s}(\phi,y)m_n(dy)ds + \int_{D\times[0,t]} T^*G_{t-s}(\phi,y)M^n(dy\ ds).$$

<u>HEOREM 7.10</u>. Suppose that there exist finite (signed) measures ν, m, and σ^2 on D uch that $\nu_n \Rightarrow \nu$, $m_n \Rightarrow m$, and $\sigma_n \Rightarrow \sigma$, in the sense of weak convergence of measures n D. Suppose further that for some $\delta > 0$, $Q_n(\delta) \to 0$. Then there exists a white oise W on $D\times[0,t)$, based on $d\sigma\, dt$, such that $(M^n, V^n) \Rightarrow (W, V)$, where V is defined y

$$\text{(7.10)}\qquad V_t(\phi) = \int G_t(\phi,y)\nu(dy) + \int_{D\times[0,t]} G_{t-s}(\phi,y)m(dy)ds + \int_{D\times[0,t]} T^*G_{t-s}(\phi,y)W(dy\ ds)$$

<u>ROOF</u>. The first two terms of (7.9) are deterministic and an elementary analysis shows that they converge uniformly in t, $0 \le t \le 1$, to the corresponding terms of (7.10). (Indeed, $y \to G_t(\phi,y)$ is continuous so the integrals with respect to ν_n and m_n converge; one gets the requisite uniform convergence by noticing that the same holds for the integrals of $\frac{\partial G_t}{\partial t}(\phi,y)$.)

It remains to show that the third term converges weakly. Note that the sequences $(M_n(D))$ and $(\sigma_n(D))$, being convergent, are bounded by, say, K. Thus if $0 < s < t \leq 1$:

$$\langle M^n(D)\rangle_t - \langle M^n(D)\rangle_s = (t-s)\sigma_n(D) \leq K(t-s).$$

It follows that (M^n) is of class (ρ, K^ρ) for all $\rho > 0$ and that (7.5) holds, for $\gamma(\delta) = K\delta$. Thus (M^n) is tight by Proposition 7.3. We claim that $M^n \Rightarrow W$. Note that once we establish this claim, the theorem follows by Proposition 7.8.

It is enough to show that the finite-dimensional distributions of $M^n(\phi)$ tend to those of $W(\phi)$ for one ϕ, and, since $M^n_t(\phi)$ and $W_t(\phi)$ are processes of stationary independent increments, we need only check the convergence for one value of t, say t = 1. Thus we have reduced the theorem to a special case of the classical central limit theorem.

Since we know the characteristic functions for the M^n explicitly, we can do this by a direct calculation rather than applying, say, Liapounov's central limit theorem.

Write $e^{ix}-1 = ix - \frac{1}{2}x^2 + f(x)$. Then certainly $|f(x)| \leq 2x^2$ and $\lim_{x\to 0} f(x)/x^2 = 0$. Let

$$\begin{aligned}\phi_n(\lambda) &= \log E\{e^{i\lambda M^n_1(\phi)}\}\\ &= \log E\{e^{i\lambda\Pi^n_1(\phi)}\} - i\lambda\, m_n(\phi),\end{aligned}$$

where $\Pi^n_1(\phi) = \int_{D\times[0,1]} \phi(x)\Pi^n(dx\, ds)$. Let us also write $m_n(\phi) = \int \phi(x)m_n(dx)$, $\sigma_n(\phi^2) = \int \phi^2(x)\sigma_n(dx)$. From the properties of Poisson processes the above is

$$\begin{aligned}&= \lambda \int_{D\times[0,1]} (e^{i\,\phi(x)r}-1)\,\mu_n(dx\,dr) - i\lambda\, m_n(\phi)\\ &= i\lambda \int_{D\times[0,1]} \phi(x)\, r\, \mu_n(dx\,dr) - i\,\lambda\, m_n(\phi)\\ &\quad - \frac{1}{2}\lambda^2 \int_{D\times[0,1]} \lambda^2(x)r^2\mu_n(dx\,dr) + \int_{D\times[0,1]} f(\lambda\phi(x)r)\mu_n(dx\,dr)\\ &= -\frac{1}{2}\lambda^2\sigma_n(\phi^2) + \int_{D\times[0,1]} f(\lambda\,\phi(x)r)\mu_n(dx\,dr)\,.\end{aligned}$$

The first term converges to $-\frac{1}{2}\lambda^2\sigma(\phi^2)$ as $n\to\infty$ since $\sigma_n \Rightarrow \sigma$. We claim the second term tends to zero.

Choose $\varepsilon > 0$ and let $\eta > 0$ be such that if $|x| < \eta$, then $|f(x)| \leq \varepsilon x^2$. The second integral is bounded by

$$\varepsilon \int_{\{\lambda r\|\phi\|_\infty<\eta\}} \lambda^2\phi^2(x)r^2\mu_n(dx\ dr) + \int_{\{\lambda r\|\phi\|_\infty\geq\eta\}} \eta^{-\delta}(\lambda\ \phi^2(x)\ |r|)^{2+\delta}\mu_n(dx\ dr)$$

$$\leq \varepsilon\ \lambda^2\sigma_n(\phi^2) + \eta^{-\delta}\lambda^{2+\delta}\ ||\phi\ ||_\infty^{2+\delta}Q_n(\delta).$$

ince $(\sigma_n(\phi^2))$ is bounded and $Q_n(\delta) \to 0$, we conclude that

$$\phi_n(\lambda) \to -\frac{\lambda^2}{2}\ \sigma_n(\phi^2)$$

hich is the log characteristic function of $W_1(\phi)$.

Q.E.D.

EMARKS. Let α be the degree of T. By Corollary 7.9, we see that $V^n \Rightarrow V$ in $\{[0,1],\ H_{-(d+\alpha+3)}\}$. One can doubtless improve the exponent of the Sobolev space by more careful analysis.

AN EXTENSION

Propositions 7.6 and 7.8 are sufficient for many of our needs, but they are asically designed to handle deterministic integrands, and we need to extend them if e are to handle any reasonably large class of random integrands.

We will look at the case in which the functions $k_n(t)$ of (7.1) are all bsolutely continuous. The treatment unfortunately becomes more complicated. Any eader without a morbid interest in Hölder's inequalities should skip this section ntil he needs it.

Let M be a worthy martingale measure on $\mathbf{R}^d$. Set

$$M^*(\phi) = \sup_{t\leq 1}|M_t(\phi) - M_{t-}(\phi)| \quad\text{and}\quad \Delta M^* = \sup_{A\subset \mathbf{R}^d}\Delta M_t^*(I_A h_0).$$

ote that for any $f \in \underline{\underline{P}}_M$ such that $||f(t)||_h \leq 1$ for all t, we have

$$\sup_{,A}|f\cdot M_t(A) - f\cdot M_{t-}(A)| \leq 2\ \Delta M^*.$$

e say M has L^p-dominated jumps if $\Delta M^* \in L^p$.

Let us recall Burkholder's inequalities for the predictable square function $M(\phi)>_t$.

HEOREM 7.11. (Burkholder-Davis-Gundy) Suppose Φ is a continuous increasing inction on $[0,\infty)$ with $\Phi(0) = 0$, such that for some constant c, $\Phi(2x) \leq c\ \Phi(x)$, for all $x \geq 0$. Then

(i) there exists a constant C such that

$$E\{\Phi(\sup_{s\leq t} |M_s(\phi)|)\} \leq CE\{\Phi(\langle M(\phi)\rangle_t^{1/2})\} + CE\{\Phi(\Delta M^*(\phi))\};$$

(ii) if Φ is concave

$$E\{\Phi(\sup_{s\leq t} |M_s(\phi)|^2\} \leq 5\, E\{\Phi(\langle M(\phi)\rangle_t^{1/2})\}.$$

We will usually apply this to $\Phi(x) = |x|^p$ for $p > 0$. If $0 < p < 1$, we are in case (ii).

In what follows, $M^0, M^1, M^2, \ldots$ is a sequence of worthy martingale measures and $k_0, k_1, k_2, \ldots$ are the increasing processes of (7.1). We will assume, as we may, that all the M^n are defined canonically on $\underline{\underline{D}}$ and we will use P^n and E^n for the distribution and expectation relative to M^n. When there is no danger of confusion we will simply write P and E respectively.

<u>DEFINITION</u>. Let $p > 0$, $K > 0$. The sequence (M^n) is of <u>class (p,K)</u> if for all n there exists a random variable X_n on $\underline{\underline{D}}$ such that

(7.11a) $\quad k_n(t) - k_n(s) \leq X_n |t-s| \quad$ if $0 \leq s \leq t \leq 1$, $n = 0,1,2,\ldots$;

(7.11b) $\quad E^n\{X_n^p\} \leq K$, $n = 0,1,2,\ldots$.

If the M^n are m-dimensional, we say that (M^n) is of class (p,K) if for each n there exists an X_n satisfying (ii) such that (i) holds for each coordinate. (In particular, each coordinate is itself of class (p,K).)

<u>REMARK</u>. If (M^n) is of class (p,K), than each M^n is quasi-left continuous, i.e. M^n has no predictable jumps. This is immediate from Lemma 7.2.

Just as in Proposition 7.6, we want to close $\mathcal{P}_S(M^0)$ in some suitable norm. In this case, the norm depends on the sequence $M = (M^n)$.

<u>DEFINITION</u>. $\bar{\bar{\mathcal{P}}}_S(M)$ is the class of $f \in \bigcap_{n=0}^{\infty} \underline{\underline{P}}_{M^n}$ such that for each $\varepsilon > 0$ there exists $f^\varepsilon \in \mathcal{P}_S(M^0)$ such that

(7.12) $$\sup_n E^n\{1 \wedge \int_0^1 \|f(s) - f^\varepsilon(s)\|_\infty^2 ds\} \leq \varepsilon.$$

Note that $Y \to E\{1 \wedge |Y|\}$ gives a distance compatible with convergence in probability, so that the above says that f can be approximated in P^n-probability for each n, and that the approximation is uniform in n. It is this uniformity which is important. Without it, the condition is trivial, as the following exercise shows.

Exercise 7.1. Let $f \in \bigcap_n \underline{\underline{P}}_{M^n}$. Then for any N and $\varepsilon > 0$ there exists $f^{\varepsilon,N} \in \bigcap_n \mathcal{P}_S(M^n)$ such that

$$E^N\{\int_0^1 \|f(s) - f^{\varepsilon,N}(s)\|_\infty^2 ds\} \le \varepsilon.$$

PROPOSITION 7.12. Let M^n, $n = 0,1,2,\ldots$ be a sequence of m-dimensional worthy martingale measures which is of class (p,k) for some $p > 0$, $K > 0$. For each n, let $f^n(x,t,\omega) = (f^n_{ij}(x,t,\omega))$ be an $r\times m$ matrix such that $f^n \in \underline{\underline{P}}_{M^n}$. We suppose that there exists an $\varepsilon > 0$ such that $E^n\{|\int_0^1 \|f^n_{ij}(s)\|_\infty^{2+\varepsilon} ds|^\varepsilon\}$, $n = 0,1,2,\ldots$ is bounded.

(i) Then $(f^n\cdot M^n)$ is tight on $\underline{\underline{D}}^r$;

(ii) if, further, $M^n \Rightarrow M^0$ in $\underline{\underline{D}}^m$, and if

a) $f^0_{ij} \in \bar{\bar{\mathcal{P}}}_S(M)$, all i,j;

b) $\lim_{n\to\infty} E^n\{|\int_0^1 \|f^n_{ij}(s) - f^0_{ij}(s)\|_\infty^2 ds|^\varepsilon\} = 0,$

then $f^n\cdot M^n \Rightarrow f^0\cdot M^0$ on $\underline{\underline{D}}^r$.

PROOF. If the hypotheses hold for a given ε, they also hold for any $\varepsilon' < \varepsilon$, so we may assume that $\varepsilon \le p \wedge 1$. Thus if X_n, $n = 0,1,2,\ldots$ are the random variables of (7.11), then $E^n\{X_n^\varepsilon\} \le K^{\varepsilon/p}$ by Jensen's inequality. Notice also that for $f \in \underline{\underline{P}}_{M^n}$,

$$|f^n\cdot M^n(\phi)|^\varepsilon = |\sum_i (\sum_j f^n_{ij}\cdot M^{nj}(\phi))^2|^{\varepsilon/2} \le \sum_{i,j} |f^n_{ij}\cdot M^{nj}(\phi)|^\varepsilon \tag{7.13}$$

since $\varepsilon \le 1$.

Let T_n be a stopping time for M^n and let $\delta_n > 0$ be such that $T_n + \delta_n \le 1$. By Burkholder's inequality, (7.13), and (7.4), for any test function ϕ

$$E^n\{|f^n\cdot M^n_{T_n+\delta_n}(\phi) - f^n\cdot M^n_{T_n}(\phi)|^\varepsilon\} \le 5\|\phi\|_h^\varepsilon \sum_{i,j} E^n\{|\int_{T_n}^{T_n+\delta_n} \|f^n_{ij}(s)\|_\infty^2 dh_n(s)|^{\frac{\varepsilon}{2}}\}$$

$$\leq 5 \|\phi\|_h^\varepsilon \sum_{i,j} E^n\{ \int_{T_n}^{T_n+\delta_n} \|f_{ij}^n(s)\|^2 ds)^{\frac{\varepsilon}{2}}\}.$$

Apply Hölder's inequality to the integral with $p = 1 + \varepsilon/2$, $q = 1 + 2/\varepsilon$, then apply Schwartz' inequality:

$$\leq 5 \|\phi\|_h^\varepsilon \quad E^n\{X_n^\varepsilon\}^{1/2} \sum_{i,j} E^n\{| \int_0^1 \|f_{ij}^n(s)\|^{2+\varepsilon} ds \quad |^{2\varepsilon/(2+\varepsilon)}\}^{1/2} \delta^{\varepsilon^2/(4+\varepsilon)}.$$

Now $\frac{2\varepsilon}{2+\varepsilon} < \varepsilon$ so the expectations are bounded, and the above is then

$$\leq C\delta_n^{\varepsilon^2/(4+\varepsilon)}$$

for some constant C. Thus, if $\delta_n \to 0$, $f^n \cdot M^n_{T_n+\delta_n}(\phi) - f^n \cdot M^n_{T_n}(\phi) \to 0$ in probability. Take $T_n = 0$ and $\delta_n = t$ to see that for each t the sequence $(f^n \cdot M_t^n(\phi))$ of random variables is bounded in L^ε, hence tight. Aldous' theorem then implies that the sequence $(f^n \cdot M^n(\phi))$ of real-valued processes is tight, and Mitoma's theorem implies that $(f^n \cdot M^n)$ is tight on $\underline{\underline{D}}^r$.

As in Proposition 7.5 we need only show that $(f^n \cdot M_1^n(\phi))$ converges weakly to prove (ii). Let h be bounded and uniformly continuous on $\mathbb{R}^r$. Then if $\gamma > 0$ there exist $f_{ij}^\gamma \in \mathcal{P}_S(M^0)$ such that

(7.14) $$E^n\{|1 \wedge \int_0^1 \|f_{ij}^\gamma(s) - f_{ij}^0(s)\|^2 ds\} < \gamma, \ n = 0,1,2,\ldots .$$

Now

$$|E^n\{h(f^n \cdot M_1^n(\phi))\} - E^0\{h(f^0 \cdot M_1^0(\phi))\}|$$

$$\leq E^n\{|h(f^n \cdot M_1^n(\phi)) - h(f^0 \cdot M_1^n(\phi))|\} + E^n\{|h(f^0 \cdot M_1^n(\phi)) - h(f^\gamma \cdot M_1^n(\phi))|\}$$
$$+ |E^n\{h(f^\gamma \cdot M_1^n(\phi))\} - E^0\{h(f^\gamma \cdot M_1^0(\phi))\}|$$
$$+ E^0\{|h(f^\gamma \cdot M_1^0(\phi)) - h(f^0 \cdot M_1^0(\phi))|\}$$
$$\stackrel{\text{def}}{=} E_1 + E_2 + E_3 + E_4.$$

Fix $\rho > 0$ and let $\eta > 0$ be such that $|h(y) - h(x)| < \rho$ if $|y-x| < \eta$. Then

$$E_1 \leq \rho + 2 \|h\|_\infty P\{|(f^n - f^0) \cdot M_1^n(\phi)| \geq \eta\}$$
$$\leq \rho + 2\|h\|_\infty \eta^{-\varepsilon} E^n\{|(f^n - f^0) \cdot M_1^n(\phi)|^\varepsilon\}.$$

By (7.13) and Burkholder's inequality (Theorem 7.11 (ii)) this is

$$\leq \rho + 10 \|h\|_\infty \eta^{-\varepsilon} \sum_{ij} E^n\{| \int_0^1 \|f_{ij}^n(s) - f_{ij}^0(s)\|_\infty^2 dk_n(s)|^{\varepsilon/2}\}$$

$$\leq \rho + 10 \|h\|_\infty \eta^{-\varepsilon} \sum_{ij} E^n\{X_n^{\varepsilon/2} | \int_0^1 \|f^n(s) - f^0(s)\|_\infty^2 ds|^{\varepsilon/2}\}$$

$$\leq \rho + 10\,\|h\|_\infty \eta^{-\varepsilon} E^n\{X_n^\varepsilon\}^{1/2} \sum_{ij} E^n\{|\int_0^1 \|f^n(s) - f^0(s)\|_\infty^2 ds|^\varepsilon\}^{1/2}$$

where we have used the fact that (M^n) is of class (p,K) and Schwartz' equality. But $E^n\{X_n^\varepsilon\}$ is bounded, and the last expectation tends to zero, so $\limsup_{n\to\infty} E_1 \leq \rho$.

Similarly,

$$E_2 \leq \rho + 2\,\|h\|_\infty \eta^{-\varepsilon}\, E\{1 \wedge |(f^\gamma - f^0)\cdot M_1^n(\phi)|^\varepsilon\}.$$

Now $\Phi(x) = 1 \wedge |x|^\varepsilon$ is concave so by Burkholder's inequality and (7.4)

$$\leq \rho + 10\,\|h\|_\infty \eta^{-\varepsilon} \sum_{ij} E^n\{1 \wedge (\int_0^1 \|f^\gamma(s) - f^0(s)\|_\infty^2\, dk_n(s)^{\varepsilon/2}\}$$

$$\leq \rho + 10\,\|h\|_\infty \eta^{-\varepsilon}\, E^n\{X_n^\varepsilon\}^{1/2} \sum_{ij} E^n\{(1 \wedge \int_0^1 \|f^\gamma(s) - f^0(s)\|ds)^\varepsilon\}^{1/2}.$$

Apply Jensen's inequality to this last expectation, and use (7.14):

$$\leq \rho + 10mr\ \|h\|_\infty K^{\varepsilon/2}\eta^{-\varepsilon}\gamma^{\varepsilon/2}$$

This is valid for $n = 0$ too, so E_4 has the same bound. Finally, $E_3 \to 0$ as $n \to \infty$ by Proposition 7.5. Since ρ and γ can be made as small as we wish, it follows that $E_1 + \ldots + E_4 \to 0$. Thus $f^n\cdot M_1^n(\phi) \Rightarrow f^0\cdot M_1^0(\phi)$ and, by Mitoma's theorem, we are done.

Q.E.D.

We now look at the convolution integrals. In order to prove convergence of distribution-valued processes such as V^n of (7.9), we usually prove that the real-valued processes $(V_n(\phi))$ converge weakly and appeal to Mitoma's theorem. This means that we only need to deal with the convergence of real-valued processes.

In the interest of simplicity - relative simplicity, that is - we will limit ourselves to the case where the integrand does not depend on n. The extension to the case where it does depend on n requires an additional condition on the order of (ii) of Proposition 7.12, but is relatively straightforward. We will be dealing with processes of the form

$$U_t^n = \int_{R^d\times[0,t]} g(x,s,t)M^n(dx\, ds).$$

Let $M^0, M^1,\ldots$ be a family of worthy martingale measures of class (p,K) for some $p > 0$, $K > 0$. Let $g(x,s,t,\omega)$ be a function on $R^d \times \{(s,t): 0 \leq s \leq t\} \times \underline{\underline{D}}$ such that

(i) for each t_0, $g(\cdot,\cdot,t_0,\cdot)I_{\{s\le t_0\}} \in \bigcap_{n=0}^{\infty} \underline{\underline{P}}_{M^n}$;

(ii) there exist functions Y_n on $\underline{\underline{D}}$ and positive constants β, q, and C such that if $0 \le s \le t \le t'$,

a) $\|g(s,t') - g(s,t)\|_h \le Y_n(\omega)\ |t'-t|^\beta$ P^n-a.s., all n;

b) $\|g(s,t')\|_h \le Y_n(\omega)$ P^n-a.s., all n;

c) $E^n\{Y_n^q\} \le C$, all n.

<u>DEFINITION</u>. If g satisfies (i) and (ii) above we say g is of <u>Hölder class</u> (β,q,C) relative to (M^n).

<u>THEOREM 7.13</u>. Let (M^n) be of class (p,K), where $p > 2$ and $K > 0$. Let g be of Hölder class (β,q,C), where $\frac{1}{p} \le \beta \le 1$ and $q > \frac{2p}{\beta p-1}$. Suppose further that the jumps of M^n are L^{2p}-dominated, uniformly in n. Then

(i) $\{U_t^n,\ 0 \le t \le 1\}$ has a version which is right continuous and has left limits;

(ii) there exists a constant A, depending only on p, q, C and K such that

$$E\{\sup_{t\le 1} |U_t^n|^r\} \le A \quad \text{if } 1 \le r \le \frac{2pq}{2p+q};$$

(iii) if $t \to M_t^n$ is continuous, U^n is Hölder continuous. Moreover there exists a random variable Z_n with $E\{Z_n^r\} < \infty$ such that if $r = \frac{2pq}{2p+q}$,

$$|U_{t+s}^n - U_t^n| \le Z_n\, s^{\frac{1}{2}\wedge\beta - \frac{1}{r}}\left(\log\frac{2}{s}\right)^{2/r}, \quad 0 \le s,\ t \le 1;$$

(iv) the family (U^n) is tight on $\underline{\underline{D}}\{[0,t], R\}$.

(v) Suppose further that $M^n \Rightarrow M^0$ on $\underline{\underline{D}}$ and that for a dense set of $t \in [0,t]$, $g(x,s,t)I_{\{s\le t\}} \in \overline{\underline{\underline{\mathcal{P}}}}_S(M)$. Then $(U^n, M^n) \Rightarrow (U^0, M^0)$.

<u>PROOF</u>. By replacing C and K by $\max(C,K)$ if necessary, we may assume $C = K$. By enlarging K further, we may assume that $E\{|\Delta M^{n*}|^{2p}\} \le K$ for all n.

Set $r = 2pq(2p+q)^{-1}$; since $q > 2p(\beta q-1)^{-1}$, $\beta^{-1} \le p < r < 2p$. If X_n and Y_n are the random variables from (7.11) and from the definition of Hölder class (β,q,C) respectively, then

$$E\{X_n^{r/2} Y_n^r\} \le K. \tag{7.15}$$

ndeed the left-hand side is bounded by

$$E\{X_n^p\}^{r/2p}\, E\{Y_n^q\}^{\frac{2p-r}{2q}} \le K^{r/2p}\, K^{\frac{2p-r}{2p}}.$$

Let us extend the domain of definition of g to include values of $s \ge t$ by etting

$$g(x,s,t) = g(x,s,t \vee s),\ 0 \le s \le 1,\ 0 \le t \le 1.$$

hen $t \to g(x,s,t)$ is constant on $[0,s]$, so g remains predictable in s and still atisfies (i) and (ii). Define

$$V_t^n = \int_{E\times[0,1]} g(x,s,t) M^n(dx\, ds).$$

et us first show that V^n has a continuous version for which $\sup_t V_t^n \in L^r$.

Note that $\Delta((g(t') - g(t))\cdot M^n)$ is bounded by

$\sup_s \|g(s,t') - g(s,t)\|_h\, \Delta M^{n*} \le 2Y\, \Delta M^{n*}(t'-t)^\beta$. By (7.4) and Burkholder's nequality (Thm 7.11 (ii))

$$\begin{aligned} E\{|V_{t'}^n - V_t^n|^r\} &= E\{|\int_{E\times[0,1]} (g(x,s,t') - g(x,s,t))M^n(dx\, ds)|^r\} \\ &\le C_r\, E^n\{|\int_0^1 \|g(s,t') - g(s,t)\|_h^2 dk_n(s)|^{r/2} + |2Y\Delta M^{n*}(t'-t)^\beta|^r\} \\ &\le C_r\, E^n\{Y^r(X_n^{r/2} + (2\Delta M^{n*})^r\}(t'-t)^{\beta r}. \end{aligned}$$

Then (7.15) and a similar calculation with $X_n^{1/2}$ replaced by ΔM^{n*} shows this s

$$\le Cr\, K\, (1 + 2^r)(t'-t)^{\beta r}.$$

the same argument

$$E^n\{|V_t^n|^r\} \le C_r K(1 + 2^r).$$

ince $\beta r > 1$, Corollary 1.2 implies that V^n has a continuous version. More exactly, here exists a random variable Z_n and a constant A', which does not depend on n, such hat for $0 < \gamma < \beta - 1/r$

$$\sup_{0\le t\le t'\le 1} |V_{t'}^n - V_t^n| \le Z_n |t'-t|^\gamma$$

nd

$$E\{\sup_{t\le 1}|V_t^n|^r\} \le A',\quad E\{Z_n^r\} \le A'.$$

Now by the general theory the optional projection of V^n will be right ontinuous and have left limits. But

$$(7.16)\qquad E\{V_t^n|\underline{\underline{F}}_t\} = \int_{R\times[0,t]} g(x,s,t)M^n(dx\,ds) = U_t^n \quad P^n\text{-a.s.},$$

so the optional projection of V is a version of U, and U indeed does have a right continuous version.

Note that

$$|U_t^n| \le E\{|V_t^n|\,|\underline{\underline{F}}_t\} \le E\{\sup_s|V_s^n|\ |\underline{\underline{F}}_t\} \overset{\text{def}}{=} S_t \quad .$$

Thus

$$E\{\sup_t |U_t^n|^r\} \le E\{\sup_t S_t^r\} \le \frac{r^2}{r-1}E\{S_1^r\}$$

by Doob's inequality. But this is

$$\le \frac{r^2}{r-1}A' = A.$$

which proves (ii).

Let us skip (iii) for the moment, and prove (iv). Let $\delta_n > 0$ and let T_n be a stopping time for M^n such that $T_n + \delta_n \le 1$.

$$\begin{aligned}E\{|U^n_{T_n+\delta_n} - U^n_{T_n}|\} &= |E\{V^n_{T_n+\delta_n}|\underline{\underline{F}}_{T_n+\delta_n}\} - E\{V^n_{T_n}|\ \underline{\underline{F}}_{T_n}\}|\\ &\le E\{|V^n_{T_n+\delta_n} - V_{T_n}|\ |\underline{\underline{F}}_{T_n+\delta_n}\} + |E\{V^n_{T_n}|\underline{\underline{F}}_{T_n+\delta_n}\} - E\{V^n_{T_n}|\underline{\underline{F}}_{T_n}\}|\\ &= E\{|V^n_{T_n+\delta_n} - V^n_{T_n}|\underline{\underline{F}}_{T_n+\delta_n}\} + |\int_{R\times(T_n+\delta_n]} g(x,s,T_n)M^n(dx\,ds)|\end{aligned}$$

Similarly,

$$E\{|U^n_{T_n+\delta_n} - U^n_{T_n}|^2\}^{1/2} \le E\{|V^n_{T_n+\delta_n} - V^n_{T_n}|^2\}^{1/2} + E\{\int_{T_n}^{T_n+\delta_n} \|g(s,T_n)\|_\infty^2 dk_n(s)\}^{1/2} .$$

We can estimate the increment of V^n by (ii) and use the hypotheses on g and k_n in the second term:

$$\le \delta^\gamma E^n\{Z_n^2\}^{1/2} + C_r E^n\{Y^2X_n\}^{1/2}\delta_n^{1/2} .$$

Since both expectations are bounded, we see that if $\delta_n \to 0$, then $U^n_{T_n+\delta_n} - U^n_{T_n} \to 0$ in L^2, hence in probability. Since the U^n are bounded in L^r by (ii), the family (U^n) is tight by Aldous' theorem. This proves (iv).

If now $M^n \Rightarrow M^0$, then Proposition 7.12 implies that the finite-dimensional distributions of U_t^n converge to those of U_t^0. Since the (U^n) are tight, this implies that $U^n \Rightarrow U$.

It remains to prove (iii). Suppose M^n is continuous, and compute

$$E\{|U^n_{t+s} - U^n_t|^r\}^{1/r} \le E\{\int_{R^d\times[0,t]} (g(x,u,t+s) - g(x,u,t))M^n(dx\,du)|^r\}^{1/r}$$

$$+ E\{|\int_{R^d\times(t,t+s]} g(x,u,t+s)M(dx\,du)|^r\}^{1/r} .$$

By Burkholder's inequality and (7.4) - and this time M is continuous, so $M^* \equiv 0$ - this is

$$\le C_r E\{|\int_0^t \|g(u,t+s) - g(u,t)\|_h^2 dk_n(u)|^{r/2}\}^{1/r}$$

$$+ C_r E\{|\int_t^{t+s} \|g(u,t+s)\|_h^2 dk_n(u)|^{r/2}\}^{1/r} .$$

Since g is of Hölder class (β,q,K) and M^n is of class (p,K):

$$\le C_r\ E\{Y^r X_n^{r/2}\}^{1/r}(s^\beta t^{1/2} + s^{1/2}) .$$

By (7.14) we conclude that there is a constant B such that for $0 \le s \le 1$

$$E\{|U^n_{t+s} - U^n_t|^r\} \le B\ s^{\frac{r}{2}\wedge\beta r} .$$

But $\beta r > 1$ and $r/2 > p/2 > 1$ so that (iii) follows by Corollary 1.2.

Q.E.D.

CHAPTER EIGHT

THE BROWNIAN DENSITY PROCESS

Imagine a large but finite number - say 6×10^{23} - of Brownian particles B^1, $B^2, \ldots, B^N$ diffusing through a region $D \subset R^d$. Consider the density of particles at a point x at time t. This is of course just the number of particles per unit volume. We can approximate it by counting the number in, say, a small cube centered at x and dividing by the volume. But is this a good approximation? Not really. If we let the cube shrink to the point $\{x\}$, the limit will either be zero, if there is no particle at x, or infinity, if there is one. So we can't take a limit. We'll have to stick to finite sizes of cubes. Since there seems to be no reason to prefer one size cube to another, perhaps we should do it for <u>all</u> cubes. Once that is admitted, one might ask why we should restrict ourselves to cubes, and whether other forms of averaging might be equally relevant. For instance, if ϕ is a positive function of compact support such that $\int \phi(x)dx = 1$, one could define the ϕ-average as

$$\eta_t(\phi) = \sum_{i=1}^{N} \phi(B_t^i). \tag{8.1}$$

Finally, we might as well go all the way and compute (8.1) for all test functions ϕ. This will give us a Schwartz distribution. In short we will describe the density of particles by the Schwartz distribution (8.1).

It is usually easier to deal with a continuous process than a discrete process such as this, so we might let $N \to \infty$, re normalize, and see if the process η_t goes to a limit. It does, and this limit is what we call the <u>Brownian density process</u>.

From what we have said above, one might think that we should describe the particle density by a measure, rather than a distribution, for (8.1) defines both. It is only when we take the limit as $N \to \infty$ that the reason for the choice becomes clear. In general, the limit will be a pure distribution, not a measure.

We will add one complication: we will consider branching Brownian motions, rather than just Brownian motions. The presence of branching gives us a more interesting class of limit processes. However, aside from that, we shall operate in the simplest possible setting. It is possible to generalize to branching diffusions, or even branching Hunt processes, but most of the ideas needed in the general case are already present in this elementary situation.

A branching Brownian motion with parameter μ can be described as follows. A particle performs Brownian motion in some region of R^d. In a time interval $(t, t+h)$ the particle has a probability $\mu h + o(h)$ of branching. If it does branch, it either splits into two identical particles, or it dies out, with probability 1/2 each. If it splits, the two daughters begin their life at the branching point. They continue independently as Brownian motions until the time they themselves branch, and so on.

If we start with a single particle and let N_t be the total number of particles at time t, then N_t is an ordinary branching process, and also a martingale. Note that we are in the critical case, $E\{N_t\} \equiv 1$.

We are going to assume that the initial distribution of particles is a Poisson point process on R^d of parameter λ: the number of particles in a set A is a Poisson random variable with parameter $\lambda|A|$, and the numbers in disjoint sets are independent. This means that the initial number of particles is infinite, but that will not bother us; the Poisson initial distribution makes things easier rather than harder.

Let us first give an explicit construction of branching Brownian motion. There are numerous constructions in the literature, most of which are more sophisticated than this, which is done entirely by hand, but it sets up the process in a useful form.

Let $\underline{\underline{A}}$ be the set of all multi indices, i.e. of strings of the form $\alpha = n_1 n_2 \ldots n_k$ where the n_j are non-negative integers. Let $|\alpha|$ be the length of α. We provide $\underline{\underline{A}}$ with the <u>arboreal ordering</u>: $m_1 \ldots m_p \prec n_1 \ldots n_q$ iff $p \leq q$ and $m_1 = n_1, \ldots, m_p = n_p$. If $|\alpha| = p$, then α has exactly p-1 predecessors, which we shall denote respectively by $\alpha-1, \alpha-2, \ldots, \alpha-|\alpha|+1$. That is, if $\alpha = 2341$, then $\alpha-1 = 234$, $\alpha-2 = 23$ and $\alpha-3 = 2$.

Let Π^λ be a Poisson point process on R^d of parameter λ. The probability that any two points of Π^λ lie exactly the same distance from the origin is zero, so that we can order them by magnitude. Thus the initial values can be denoted by

$$\{X^\alpha(0),\ \alpha \in \underline{\underline{A}},\ |\alpha| = 1\}.$$

Define three families

$$\{B_t^\alpha,\ t \geq 0,\ \alpha \in \underline{\underline{A}}\},\quad \{S_t^\alpha,\ \alpha \in \underline{\underline{A}}\} \quad \text{and} \quad \{N^\alpha,\ \alpha \in \underline{\underline{A}}\},$$

where the B^α are independent standard Brownian motions in $\mathbf{R}^d$ with $B_0^\alpha = 0$, all α; the S^α are i.i.d. exponential random variables with parameter μ, which will serve as lifetimes; and the N^α are i.i.d. random variables with $P\{N^\alpha = 0\} = P\{N^\alpha = 2\} = 1/2$. The families (B^α), (S^α), (N^α), and $(X^\alpha(0))$ are independent.

The <u>birth time</u> $\beta(\alpha)$ of X^α is

$$\beta(\alpha) = \begin{cases} \sum_{1}^{|\alpha|-1} S^{\alpha-j} & \text{if } N^{\alpha-j} = 2,\ j = 1,\dots,|\alpha|-1 \\ \infty & \text{otherwise} \end{cases}.$$

The <u>death time</u> $\zeta(\alpha)$ of X^α is

$$\zeta(\alpha) = \beta(\alpha) + S^\alpha.$$

Define $h^\alpha(t) = I_{\{\beta(\alpha) \leq t < \zeta(\alpha)\}}$, which is the indicator function of the lifespan of X^α.

If $\alpha = n_1 n_2 \dots n_p$ (so $\alpha - p+1 = n_1$) then the <u>birthplace</u> $X^\alpha(\beta(\alpha))$ is

$$X^\alpha(\beta(\alpha)) = X^{n_1}(0) + \sum_{i=1}^{|\alpha|-1} (B^{\alpha-i}_{\zeta(\alpha-i)} - B^{\alpha-i}_{\beta(\alpha-i)}).$$

Now let ∂ - the cemetary - be a point disjoint from $\mathbf{R}^d$, and put

$$X^\alpha(t) = \begin{cases} \partial & \text{if} \quad t < \beta(\alpha) \text{ or } t \geq \zeta(\alpha) \\ X^\alpha(\beta(\alpha)) + \int_0^t h^\alpha(s)\,dB_s^\alpha & \text{otherwise.} \end{cases}$$

Note that since $h^\alpha = 1$ between β and ζ, X^α is a Brownian motion on the interval $[\beta(\alpha), \zeta(\alpha))$, and $X^\alpha = \partial$ outside it. We make the convention that any function f on $\mathbf{R}^d$ is extended to $\mathbf{R}^d \cup \{\partial\}$ by $f(\partial) = 0$.

Finally, define

$$\eta_t(\phi) = \sum_{\alpha \in \underline{\underline{A}}} \phi(X_t^\alpha) \tag{8.2}$$

for any ϕ on $\mathbf{R}^d$ for which the sum makes sense.

The branching process η_t^n starting at the single point $X^n(0)$ is constructed from the X^α for $n \prec \alpha$. That is,

$$\eta_t^n(\phi) = \sum_{\substack{\alpha \in \underline{\underline{A}} \\ n \prec \alpha}} \phi(X_t^\alpha); \quad N_t^n = \sum_{\substack{\alpha \in \underline{\underline{A}} \\ n \prec \alpha}} h^\alpha(t). \tag{8.3}$$

Note that N_t^n is the number of particles alive at time t, and it is a classical branching process of the critical case - we leave it to the reader to verify that his does indeed follow from our construction - hence $E\{N_t^n\} = 1$. In particular, $\eta_t^n(\phi)$ is finite for all t.

Furthermore, a symmetry argument shows that, given $h^\alpha(t) = 1$ (so X^α is live at t),

$P\{X_t^\alpha \in A | X^n(0) = x\} = P\{x + B_t \in A\}$, where $n < \alpha$. (This relies on the fact that all irth and death times are independent of the B_t^α.) Thus

$$E\{ \sum_{n<\alpha} \phi(X_t^\alpha) | X^n(0) = x\} = \sum_{n<\alpha} E\{\phi(X_t^\alpha) | X^n(0) = x,\ h^\alpha(t) = 1\}\ P\{h^\alpha(t) = 1 | X^n(0) = x\}$$

$$= E\{\phi(B_t + x)\} \sum_{n<\alpha} P\{h^\alpha(t) = 1\}$$

$$= E\{\phi(B_t + x)\}.$$

Suppose now that $\phi \geq 0$. We can integrate over the initial values, which re Poisson(λ), to see that

$$E\{ \sum_\alpha \phi(X_t^\alpha)\} = \int \lambda\ dx\ E\{\phi(x + B_t)\}$$

$$= \lambda \iint (2\pi t)^{-d/2} e^{-\frac{|y-x|^2}{2t}} \phi(y) dx\ dy .$$

ince Lebesgue measure is invariant this is

$$= \lambda \int \phi(x) dx .$$

hus, for any positive ϕ,

$$E\{\eta_t(\phi)\} = \lambda \int \phi(x) dx. \tag{8.4}$$

his makes it clear that $\eta_t(\phi)$ makes sense for any integrable ϕ.

THE FUNDAMENTAL NOISES

There are two distinct sources of randomness for our branching Brownian otion. The first comes from the diffusion, and the second comes from the branching. e will treat these separately.

Recall that λ is the parameter of the initial distribution and μ is the ranching rate. Define, for a Borel set $A \subset R^d$,

$$(8.5)\quad \begin{cases} W(A \times (0,t]) = \lambda^{-1/2} \sum_{\alpha \in \underline{\underline{A}}} \int_0^t h^\alpha(s) I_{\{X_s^\alpha \in A\}}\, dB_s^\alpha \\ \\ Z(A \times (0,t]) = (\lambda\mu)^{-1/2} \sum_{\alpha \in \underline{\underline{A}}} (N^\alpha - 1) I_{\{\zeta(\alpha) \leq t,\ X^\alpha(\zeta(\alpha)-) \in A\}} \end{cases}$$

(If $\mu = 0$, $Z \equiv 0$).

These take some explanation. In the expression for W, $h^\alpha \equiv 1$ when X^α is alive so $dB^\alpha = dX^\alpha$, and the integral gives us the total amount that X^α diffuses while in A; the sum over α gives the total diffusion of all the particles while they are in A. Thus W keeps track of the diffusion and ignores the branching.

Z does just the opposite. N^α is the size of family foaled by X^α when it branches, at time $\zeta(\alpha)$. $N^\alpha - 1$ counts +1 if X^α splits in two, -1 if it dies without progeny. Thus $Z(A \times (0,t])$ is the number of births minus the number of deaths occurring in A up to time t.

Let us put

$$W_t(A) = W(A \times (0,t]); \qquad Z_t(A) = Z_t(A \times (0,t]).$$

<u>PROPOSITION 8.1</u> W and Z are orthogonal martingale measures. W is continuous and R^d-valued while Z is purely discontinuous and real-valued. Moreover,

(i) $\langle W(A)\rangle_t = \langle Z(A)\rangle_t I = \frac{1}{\lambda}\left(\int_0^t \eta_s(A)ds\right)I;$

(ii) $\langle W_{ij}(A), Z(C)\rangle_t = 0$ for all Borel A and C of finite measure and all $i, j \leq n$;

(iii) both W and Z have the same mean and covariance as white noises (based on Lebesgue measure).

<u>Note</u>. W_t has values in R^d, so $\langle W\rangle_t$ is a $d\times d$ matrix whose ijth component is $\langle W^i, W^j\rangle_t$.

<u>PROOF</u>. We must first show W and Z are square integrable. This done, the rest is easy. First look at the branching diffusion starting from one of the $x^n(0)$. Define

$$W_t^n(A) = \lambda^{-1/2} \sum_{n<\alpha} \int_0^t h^\alpha(s)\, I_{\{X_s^\alpha \in A\}}\, dB_s^\alpha.$$

Since the B^α are independent, hence orthogonal,

$$\langle W^n(A)\rangle_t = \lambda^{-1}\left(\sum_{n<\alpha} \int_0^t h^\alpha(s)\, I_{\{X_s^\alpha \in A\}}\, ds\right)I$$

$$= \lambda^{-1}\left(\int_0^t \eta_s^n(A)ds\right)I .$$

The interchange is justified since the sum is dominated by $\int_0^t N_s^n ds$ which has xpectation t. (I is the d×d identity matrix.)

t follows that $E\{|W_t^n(A)|^2\} = E\{|\langle W^n(A)\rangle_t|\} \le td/\lambda$.

Turning to Z, let

$$M_t^\alpha = (N^\alpha - 1)I_{[\zeta(\alpha),\infty)}(t)I_{\{X^\alpha(\zeta(\alpha)-)\in A\}}.$$

ote that this is adapted to the natural σ-fields ($\underline{\underline{F}}_t$). Since $N^\alpha-1$ has mean zero and s independent of the indicator functions, M^α is a martingale. Noticing that $(\alpha) = \beta(\alpha) + S^\alpha$, where S^α is exponential(μ), it is easy to see that

$$\langle M^\alpha\rangle_t = \int_0^t h^\alpha(s)I_{\{X_s^\alpha \in A\}}\mu ds.$$

The N^α are independent, so the M^α are orthogonal for different α and

$$\langle Z_t^n(A)\rangle = \frac{1}{\lambda\mu}\sum_{n<\alpha}\langle M^\alpha\rangle = \frac{1}{\lambda}\sum_{n<\alpha}\int_0^t h^\alpha(s)\, I_{\{X_s^\alpha \in A\}}ds$$
$$= \frac{1}{\lambda}\int_0^t \eta_s^n(A)ds.$$

The processes η^n, n = 0,1,2,.. are conditionally independent given $\underline{\underline{F}}_0$, so f m ≠ n

$$\langle W^m(A), W^n(A)\rangle_t = \langle Z^m(A), Z^n(A)\rangle_t = 0.$$

hen

(8.6)
$$\langle W(A)\rangle_t = \frac{1}{\lambda}\sum_n \langle W^n(A)\rangle_t$$
$$= \frac{1}{\lambda}\Big(\sum_n \int_0^t \eta_s^n(A)ds\Big)I$$
$$= \frac{1}{\lambda}\Big(\int_0^t \eta_s(A)ds\Big)I.$$

f A has finite Lebesgue measure, this is finite and even integrable by (8.4). The ame reasoning applies to $Z_t(A)$.

Now that we know $W_t(A)$ and $Z_t(A)$ are square-integrable, we can read off heir properties directly from the definition. If $A \cap C = \phi$ and if they have finite easure,

$$\langle W(A), W(C)\rangle_t = \sum_{\alpha,\gamma\in\underline{\underline{A}}}\int_0^t h^\alpha(s)h^\gamma(s)\, I_{\{X_s^\alpha \in A,\, X_s^\gamma \in C\}}\, d\langle B^\alpha, B^\gamma\rangle_s.$$

ow $\langle B^\alpha, B^\gamma\rangle_t = 0$ unless α = γ, and if α = γ, the indicator function vanishes, so $W(A), W(C)\rangle_t = 0$. Similarly, $\langle Z(A), Z(C)\rangle_t = 0$. Furthermore both $A \to W_t(A)$ and $\to Z_t(A)$ are L^2-valued measures. This is clear for Z from (8.5) and almost clear for W, so we leave the verification to the reader. This proves (i).

If A and C are two sets of finite Lebesgue measure, $W_t(A)$ is a continuous martingale while $Z_t(A)$ is purely discontinuous, so the two are orthogonal, proving (ii).

To see (iii), recall W and Z have mean zero and note that by (i)

$$E\{W_s(A)W_t(C)^T\} = E\{\langle W(A),\ W(C)\rangle_{s\wedge t}\}I$$

$$= (\frac{1}{\lambda}\int_0^{s\ t} E\{\eta_s(A\cap C)\}ds)I$$

$$= (s\ t)|A\cap C|I$$

by (8.4). Exactly the same calculation holds for Z, except that I = 1 since Z is real-valued. This is the covariance of white noise.

Q.E.D.

<u>Corollary 8.2</u> If $\phi(x)$ is deterministic

$$E\{(\int_0^t \int_{R^d} \phi(x)W(dxds))^2\} = E\{\int_0^t \int_{R^d} (\phi(x)Z(dxds))^2\} = t\int_{R^d} \phi^2(x)dx\ .$$

It is clear how to integrate with respect to Z. Here is a fundamental identity for integrals with respect to W.

<u>PROPOSITION 8.3</u> Let $\phi(x) \in L^2(R^d)$. Then

$$\sum_{\alpha\in\underline{\underline{A}}} \int_0^t \phi(X_s^\alpha)h^\alpha(s)dB_s^\alpha = \sqrt{\lambda}\int_0^t \int_{R^d} \phi(x)\ W(dxds)\ . \tag{8.7}$$

<u>PROOF.</u> If $\phi(x) = I_A(x)$, this is true by the definition of W, for both sides equal $W_t(A)$. It follows that (8.7) holds for finite sums of such processes, hence for all $\phi \in L^2(R^d)$ by the usual argument. Q.E.D.

<u>AN INTEGRAL EQUATION FOR η</u>

Let ϕ be a real-valued $C^{(2)}$ function on R^d. By convention, $\phi(X_t^\alpha) = 0$ unless $t \in [\beta(\alpha),\ \zeta(\alpha))$; for t in this interval we can apply Ito's formula:

$$\phi(X_t^\alpha) = \phi(X_\beta^\alpha) + \int_0^t h^\alpha(s)\ \nabla\phi(X_s^\alpha)\cdot dB_s^\alpha + \frac{1}{2}\int_0^t h^\alpha(s)\ \Delta\phi(X_s^\alpha)ds.$$

If $t > \zeta(\alpha)$, we must subtract $\phi(X^\alpha_{\zeta-})$ from this.) Thus for any test function ϕ:

$$\eta_t(\phi) = \eta_0(\phi) - \sum_{|\alpha|=1} \phi(X^\alpha_{\zeta-}) I_{\{\zeta(\alpha)\le t\}}$$
$$+ \sum_{|\alpha|\ge 2} \left(\phi(X^\alpha_\beta) I_{\{\beta(\alpha)\le t\}} - \phi(X^\alpha_{\zeta-}) I_{\{\zeta(\alpha)\le t\}}\right)$$
$$+ \sum_\alpha \int_0^t h^\alpha(s) \nabla\phi(X^\alpha_s)\cdot dB^\alpha_s$$
$$+ \sum_\alpha \frac{1}{2} \int_0^t h^\alpha(s)\, \Delta\phi(X^\alpha_s) ds\ .$$

e can identify all these terms. $\eta_0(\phi)$ is of course the initial value. The next two ums give all the births and deaths, and combine into $\sqrt{\lambda\mu} \int_0^t \int_{\mathbf{R}^d} \phi(x) Z(dxds)$.

roposition 8.3 applies to the stochastic integral, which equals

$\int_0^t \int_{\mathbf{R}^d} \nabla\phi(x)\cdot W(dxds)$. The final integral is $\frac{1}{2} \int_0^t \eta_s(\Delta\phi) ds$. Thus we have proved

ROPOSITION 8.4. Let $\phi \in \underline{\underline{S}}(\mathbf{R}^d)$. Then

3.8) $$\eta_t(\phi) = \eta_0(\phi) + \frac{1}{2} \int_0^t \eta_s(\Delta\phi) ds + \sqrt{\lambda\mu} \int_0^t \int_{\mathbf{R}^d} \phi(x) Z(dxds)$$
$$+ \sqrt{\lambda} \int_0^t \int_{\mathbf{R}^d} \nabla\phi(x)\cdot W(dxds)\ .$$

If we check equations (5.3) and (5.4), we see this translates into the SPDE

8.9) $$\begin{cases} \dfrac{\partial\eta}{\partial t} = \dfrac{1}{2}\Delta\eta + \sqrt{\lambda\mu}\, \dot{Z} + \sqrt{\lambda}\, \nabla\cdot\dot{W} \\ \eta_0 = \Pi^\lambda \end{cases}$$

f $G_t(x,y) = (2\pi t)^{-d/2} e^{-\frac{|y-x|^2}{2t}}$, Theorem 5.1 tells us the solution of (8.8) and 8.9) is

8.10) $$\eta_t(\phi) = \int_{\mathbf{R}^d} G_t(\phi,y)\Pi^\lambda(dy) + \sqrt{\lambda\mu} \int_0^t \int_{\mathbf{R}^d} G_{t-s}(\phi,y) Z(dyds)$$
$$+ \sqrt{\lambda} \int_0^t \int_{\mathbf{R}^d} \nabla G_{t-s}(\phi,y)\cdot W(dyds).$$

ote that in the last integral the ith component of $\nabla G_t(\phi,y)$ is $G_t(\frac{\partial\phi}{\partial x}, y)$ so we will rite $\nabla G_t(\phi,y) = G_t(\nabla\phi,y)$ below.

WEAK CONVERGENCE IN THE GAUSSIAN CASE

If $\phi \in L^1(R^d)$, let $\langle\phi\rangle = \int_{R^d} \phi(x)dx$. From (8.4)

$$E\{\eta_t(\phi)\} = \lambda\langle\phi\rangle, \quad t \geq 0.$$

We want to estimate the higher moments - at least the second moments - of η.

Let $\eta_t^*(\phi) = \sup_{s\leq t} |\eta_s(\phi)|$.

PROPOSITION 8.5. There exist continuous functions $C_1(\phi,t)$ and $C_2(\phi,t)$ such that

$$E\{\eta_t^*(\phi)^2\} \leq \lambda^2 C_1(\phi,t) + \lambda(\mu + 1)C_2(\phi,t) \tag{8.11}$$

PROOF. Note that the three terms on the right-hand side of (8.10) are orthogonal, so, remembering that Z and N have the same covariance as white noise, (Cor. 8.2) we have

$$\begin{aligned}
E\{\eta_t(\phi)^2\} &= E\{(\int_{R^d} G_t(\phi,y)\Pi^\lambda(dy))^2\} \\
&\quad + \lambda\mu \int_0^t \int_{R^d} G^2_{t-s}(\phi,y)dyds \\
&\quad + \lambda \int_0^t \int_{R^d} |G_{t-s}(\nabla\phi,y)|^2 dyds \\
&= \lambda(\langle G_t^2(\phi,\cdot)\rangle + \lambda\langle G_t(\phi,\cdot)\rangle^2) + \lambda\mu \int_0^t \langle G^2_{t-s}(\phi,\cdot)\rangle ds \\
&\quad + \lambda \int_0^t \langle|G_{t-s}(\nabla\phi,\cdot)|^2\rangle ds .
\end{aligned}$$

Now $\langle G_t^2(\phi,\cdot)\rangle$ is bounded uniformly in t, so the above is bounded by

$$E\{\eta_t(\phi)^2\} \leq \lambda^2 c_1(\phi) + t\lambda(\mu + 1)c_2(\phi) \tag{8.12}$$

for some c_1 and c_2. From (8.8)

$$\begin{aligned}
E\{\eta_t^*(\phi)^2\} &\leq 9\, E\{\eta_0(\phi)^2\} + \frac{9}{4} E\{(\int_0^t \eta_s(|\phi|)ds)^2\} \\
&\quad + 9\, E\{\sup_{s\leq t} (\sqrt{\lambda\mu} \int_0^s \int_{R^d} \phi dZ + \sqrt{\lambda} \int_0^s \int_{R^d} \nabla\phi \cdot dW)^2\} .
\end{aligned}$$

Now

$$\begin{aligned}
E\{\eta_0(\phi)^2\} &= E\{(\int \phi(x)\, \Pi^\lambda(dx))^2\} \\
&= \lambda \langle\phi^2\rangle + \lambda^2\langle\phi\rangle^2.
\end{aligned}$$

Apply Schwartz' inequality and (8.12) to the second term:

$$E\{(\int_0^t \eta_s(\phi)ds)^2\} \leq t \int_0^t E\{\eta_s^2(\phi)\}ds$$

$$\leq \lambda^2 t^2 c_1(\phi) + \lambda(\mu + 1)\frac{t^3}{2} c_2(\phi).$$

pply Doob's inequality to see the third term is bounded by

$$36\, E\{(\sqrt{\lambda\mu} \int_0^t \int_{R^d} \phi\, dZ + \sqrt{\lambda} \int_0^t \int_{R^d} \nabla\phi \cdot dW)^2\} .$$

he two stochastic integrals are orthogonal, so by Cor.8.2:

$$= 36\ (t\ \lambda\mu\langle\phi^2\rangle + \lambda t\ \langle|\nabla\phi|^2\rangle).$$

ut these three estimates together to get (8.11).

xercise 8.1. Show η_t^* is bounded in L^p for all $p > 0$. (Hint: By Prop. 8.1 and urkholder's inequality, η_t L^p => W and Z are in L^{2p}. Use induction on $p = 2^n$ to ee η_t L^p for all p, then use (8.8) and Doob's L^p inequality as above.)

HEOREM 8.6. Let (μ_n, λ_n) be sequence of parameter values and let (W^n, Z^n) be the orresponding processes. Let $V^n(dx) = \lambda_n^{-1/2}(\Pi^{\lambda_n}(dx) - \lambda_n dx)$ be the normalized nitial measure. If the sequence $((\mu_n+1)/\lambda_n)$ is bounded, then (V^n, W^n, Z^n) is tight n $D\{[0,1], S'(R^{d^2+2d})\}$.

ROOF. We regard V^n as a constant process: $V_t^n \equiv V^n$, in order to define it on $\{[0,1], S'(R^{d^2+2d})\}$. It is enough to prove the three are individually tight.

By Mitoma's theorem it is enough to show that $(V^n(\phi))$, $(W^n(\phi))$, and $(Z^n(\phi))$ re each tight for each $\phi \in S(R^d)$.

In the case of V^n, which is constant in t, it is enough to notice that $\{V^n(\phi)^2\} = \langle\phi^2\rangle$ is uniformly bounded in n.

We will use Kurtz' criterion (Theorem 6.8b) for the other two. Set

$$A_n(\delta) = (\delta d/\lambda_n)\eta_1^{n*}(\phi),\ 0 \leq \delta \leq 1 .$$

hen $\sup_{t\leq 1} (|\langle W^n(\phi)\rangle_{t+\delta} - \langle W^n(\phi)\rangle_t|) \leq A_n(\delta)$ and

$$\sup_{t\leq 1} (\langle Z^n(\phi)\rangle_{t+\delta} - \langle Z^n(\phi)\rangle_t) \leq A_n(\delta)$$

y Proposition 8.1. By Jensen's inequality and (8.11),

$$\lim_{\delta\to 0}\limsup_{n\to\infty} E\{A_n(\delta)\} \le \lim_{\delta\to 0}\limsup_{n\to\infty} \frac{\delta d}{\lambda_n} E\{\eta_1^{n*}(\phi)^2\}^{1/2}$$
$$\le \lim_{\delta\to 0} \delta d(c_1 + \frac{\mu_n+1}{\lambda_n} c_2)$$
$$= 0$$

since $(\mu_n + 1)/\lambda_n$ is bounded. Furthermore

$$E\{|W_t^n(\phi)|^2\} = d\, E\{Z_t^n(\phi)^2\} = td < \phi^2>$$

for all n so that for each t, $(W_t^n(\phi))$ and $(Z_t^n(\phi))$ are tight on R^d and R respectively. By Theorem 6.8 the processes $(W^n(\phi))$ and $(Z^n(\phi))$ are each tight.

Q.E.D.

THEOREM 8.7. If $\lambda_n \to \infty$, $\mu_n\lambda_n \to \infty$, and $\mu_n/\lambda_n \to 0$, then $(V^n, Z^n, W^n) \Rightarrow (V^0, Z^0, W^0)$, where V^0, Z^0 and W^0 are white noises based on Lebesgue measure on R^d, $R^d \times R_+$, and $R^d \times R_+$ respectively; V^0 and Z^0 are real-valued and W^0 has values in R^d. If $\lambda_n \to \infty$ and $\mu_n/\lambda_n \to 0$, $(V^n, W^n) \Rightarrow (V^0, W^0)$.

PROOF. Suppose λ_n is an integer. Modifications for non-integral λ are trivial. To show weak convergence, we merely need to show convergence of the finite-dimensional distributions and invoke Theorem 6.15.

The initial distribution is Poisson (λ_n) and can thus be written as a sum of λ_n independent Poisson (1) point processes.

Let $\hat\eta^1, \hat\eta^2, \dots$ be a sequence of iid copies with $\lambda = 1$, $\mu = \mu_n$. (We have changed notation: these are not the $\hat\eta^n$ used in constructing the branching Brownian motion.) Then the branching Brownian motion corresponding to λ_n, μ_n has the same distribution as $\hat\eta^1 + \hat\eta^2 + \dots + \hat\eta^{\lambda_n}$. Define $\hat V^1, \hat V^2, \dots, \hat W^1, \hat W^2, \dots$ and $\hat Z^1, \hat Z^2, \dots$ in the obvious way. Then

$$V^n = \hat V^1 + \dots + \hat V^{\lambda_n}, \quad W^n = \frac{\hat W^1 + \dots + \hat W^{\lambda_n}}{\sqrt{\lambda_n}}, \quad Z^n = \frac{\hat Z^1 + \dots + \hat Z^{\lambda_n}}{\sqrt{\lambda_n}}.$$

We have written everything as sums of independent random variables. To finish the proof, we will call on the classical Lindeberg theorem.

Let $\phi_1, \phi_1', \phi_1'', \dots, \phi_p, \phi_p', \phi_p'' \in \underline{\underline{S}}'(R)$, $t_1 \le t_2 \le \dots \le t_p$. We must show weak convergence of the vector

$$U^n \overset{\text{def}}{=} (V^n(\phi_1),\ldots,V^n(\phi_p),\ W^n_{t_1}(\phi'_1),\ldots,W^n_{t_p}(\phi'_p),\ Z^n_{t_1}(\phi''_1),\ldots,Z^n_{t_p}(\phi''_p))$$

This can be written as a sum of iid vectors, and the mean and covariance of he vectors are <u>independent of n</u> (Prop. 8.1).

It is enough to check the Lindeberg condition for each coordinate. The istribution of V^n does not depend on μ_n, so we leave this to the reader.

Fix i and look at $W^n_{t_i}(\phi'_i) = \lambda^{-1/2} \sum_{k=1}^{\lambda_n} \hat{W}^k_{t_i}(\phi'_i)$. Now $(\hat{W}^k_t(\phi'_i))$ is a R^d-valued continuous martingale, so by Burkholder's inequality

$$E\{|\hat{W}^k_{t_i}(\phi'_i)|^4\} \le c_4\, E\{|\langle\hat{W}^k(\phi_i)\rangle|^2\}$$

$$\le td\, c_4 \int_0^t E\{\hat{\eta}^k_s(\phi_i^2)^2\}ds \quad .$$

ow $t \le 1$ so by Proposition 8.5 with $\lambda = 1$, there is a C independent of k and λ_n such at this is

$$\le C(\mu_n + 1).$$

or $\varepsilon > 0$,

$$E\{|\hat{W}^k_{t_i}(\phi_i)|^4;\ |\hat{W}^k_{t_i}(\phi)|^2 \ge \lambda_n\varepsilon\} \le E\{|\hat{W}^k_{t_i}(\phi_i)|^4\}^{1/2} P\{|\hat{W}^k_{t_i}|^4 \ge \lambda_n^2\varepsilon^2\}^{1/2}$$

y Schwartz. Use Chebyshev with the above bound:

$$\le [C(1 + \mu_n)]^{1/2}[C(1 + \mu_n)/\lambda_n^2\varepsilon^2]^{1/2}$$

$$\le C(1 + \mu_n)/\lambda_n\varepsilon.$$

hus

$$\sum_{k=1}^{\lambda_n} E\{|\lambda_n^{-1/2}\hat{W}^k_{t_i}(\phi_i)|^2;\ |\lambda_n^{-1/2}\hat{W}^k_{t_i}(\phi_i)|^2 > \varepsilon\}$$

$$= E\{|\hat{W}^1_{t_i}(\phi_i)|^2;\ |\hat{W}^1_{t_i}(\phi_i)|^2 > \lambda_n\varepsilon\}$$

$$\le C_3(1 + \mu_n)/\lambda_n\varepsilon \to 0.$$

hus the Lindeberg condition holds for each of the $W^n_{t_i}(\phi'_i)$. The same argument holds or the $Z^n_{t_i}(\phi''_i)$. In this case, while $(\hat{Z}^n_t(\phi''_i))$ is not a continuous martingale, its umps are uniformly bounded by $(\lambda_n\mu_n)^{-1/2}$, which goes to zero, and we can apply urkholder's inequality in the form of Theorem 7.11(i). Thus the finite-dimensional istributions converge by Lindeberg's theorem, implying weak convergence.

The only place we used the hypothesis that $\lambda_n \mu_n \to \infty$ was in this last statement, so that if we only have $\lambda_n \to \infty$, $\mu_n/\lambda_n \to 0$, we still have $(V^n, W^n) \Rightarrow (V^0, W^0)$.

Q.E.D.

We have done the hard work and have arrived where we wanted to be, out of the woods and in the cherry orchard. We can now reach out and pick our results from the nearby boughs.

Define, for n = 0, 1, ...

$$R_t^n(\phi) = \int_0^t \int_{R^d} G_{t-s}(\nabla\phi, y) \cdot W^n(dy\, ds)$$

$$U_t^n(\phi) = \int_0^t \int_{R^d} G_{t-s}(\phi, y) Z^n(dy\, ds).$$

Recall from Proposition 7.8 that convergence of the martingale measures implies convergence of the integrals. It thus follows immediately from Theorem 8.7 that

COROLLARY 8.8. Suppose $\lambda_n \to \infty$ and $\mu_n/\lambda_n \to 0$. Then

(i) $(V^n, W^n, R^n) \Rightarrow (V^0, W^0, R^0)$;

(ii) if, in addition, $\lambda_n \mu_n \to \infty$,

$$(V^n, W^n, Z^n, R^n, U^n) \Rightarrow (V^0, W^0, Z^0, R^0, U^0).$$

Rewrite (8.10) as

$$\frac{\eta_t(\phi) - \lambda\langle\phi\rangle}{\sqrt{\lambda}} = V(G_t(\phi, \cdot)) + \sqrt{\mu}\, U_t(\phi) + R_t(\phi). \tag{8.13}$$

In view of Corollary 8.6 we can read off all the weak limits for which $\frac{\mu}{\lambda} \to 0$.

THEOREM 8.9 (i) If $\lambda_n \to \infty$ and $\mu_n \to 0$, then $\frac{\eta_t^n(\phi) - \lambda_n\langle\phi\rangle}{\sqrt{\lambda_n}}$ converges in $\underline{D}\{[0,1], \underline{S}'(R^d)\}$ to a solution of the SPDE

$$\begin{cases} \frac{\partial\eta}{\partial t} = \frac{1}{2}\Delta\eta + \nabla\cdot\dot{W} \\ \eta_0 = V^0 \end{cases}.$$

ii) If $\lambda_n \to \infty$, $\mu_n \to \infty$ and $\mu_n/\lambda_n \to 0$, then

$\frac{\eta^n_t(\phi) - \lambda_n\langle\phi\rangle}{\sqrt{\lambda_n\mu_n}}$ converges in $\underline{\underline{D}}\{[0,1], \underline{\underline{S}}'(R^d)\}$ to a solution of the SPDE

$$\begin{cases} \frac{\partial\eta}{\partial t} = \frac{1}{2}\Delta\eta + \dot{Z} \\ \eta_0 = 0 \,. \end{cases}$$

iii) If $\lambda_n \to \infty$, $\lambda_n\mu_n \to \infty$ and $\mu_n \to c^2 \geq 0$, then $\frac{\eta^n_t(\phi) - \lambda_n\langle\phi\rangle}{\sqrt{\lambda_n}}$ converges in

$\{[0,1], \underline{\underline{S}}'(R^d)\}$ to a solution of the SPDE

$$\begin{cases} \frac{\partial\eta}{\partial t} = \frac{1}{2}\Delta\eta + c\dot{Z} + \nabla\cdot\dot{W} \\ \eta_0 = V_0. \end{cases}$$

Theorem 8.9 covers the interesting limits in which $\lambda \to \infty$ and $\mu/\lambda \to 0$. hese are all Gaussian. The remaining limits are in general non Gaussian. those in hich μ and λ both tend to finite limits are trivial enough to pass over here, which eaves us two cases

(iv) $\lambda \to \infty$ and $\mu/\lambda \to c^2 > 0$;

(v) $\mu \to \infty$ and $\mu/\lambda \to \infty$.

The limits in case (v) turn out to be zero, as we will show below. Thus e only non-trivial, non-Gaussian limit is case (iv), which leads to measure-valued ocesses.

A MEASURE DIFFUSION

EOREM 8.10 Suppose $\lambda_n \to \infty$ and $\mu_n/\lambda_n \to c^2 > 0$. Then $\frac{1}{\lambda_n}\eta^n_t$ converges weakly in $[0,1], \underline{\underline{S}}'(R^d)\}$ to a process $\{\eta_t, t \in [0,1]\}$ which is continuous and has easure-values.

There are a number of proofs of this theorem in the literature (see the otes), but all those we know of use specific properties of branching processes which don't want to develop here, so we refer the reader to the references for the roof, and limit ourselves to some formal remarks.

We can get some idea of the behavior of the limiting process by rewriting

(8.13) in the form

$$\frac{1}{\lambda}\,\eta_t(\phi) = \langle\phi\rangle + cU_t(\phi) + \frac{1}{\sqrt{\lambda}}\,(V(G_t(\phi,\cdot)) + \frac{1}{\lambda}\,R_t(\phi)). \tag{8.14}$$

If (λ_n,μ_n) is any sequence satisfying (iv), $\{(V^n,W^n,Z^n,R^n,U^n,\frac{1}{\lambda_n}\,\eta^n)\}$ is tight by Theorem 8.6 and Proposition 7.8, hence we may choose a subsequence along which it converges weakly to a limit (V, W, Z, R, U, η). From (8.14)

$$\begin{aligned}\eta_t(\phi) &= \langle\phi\rangle + cU_t(\phi) \\ &= \langle\phi\rangle + c\int_0^t \int_{R^d} G_{t-s}(\phi,y)Z(dy,ds).\end{aligned} \tag{8.15}$$

In SPDE form this is

$$\begin{cases}\dfrac{\partial\eta}{\partial t} = c\Delta\eta + \dot{Z} \\ \eta_0(dx) = dx\end{cases} \tag{8.16}$$

We can see several things from this. For one thing, η^n is positive, hence so is η. Consequently, η_t, being a positive distribution, is a measure. It must be non-Gaussian - Gaussian processes aren't positive - so Z itself must be non-Gaussian. In particular, it is not a white noise.

Now η_0 is Lebesgue measure, but if $d > 1$, Dawson and Hochberg have shown that η_t is purely singular with respect to Lebesgue measure for $t > 0$. If $d = 1$, Roelly-Coppoletta has shown that η_t is absolutely continuous.

To get some idea of what the orthogonal martingale measure Z is like, note from Proposition 8.1 that

$$\langle Z^n(A)\rangle = \int_0^t \frac{1}{\lambda_n}\,\eta_s^n(A)ds,$$

which suggests that in the limit

$$\langle Z(A)\rangle_t = \int_0^t \eta_s(A)ds,$$

or, in terms of the measure ν of Corollary 2.8,

$$\nu(dx,ds) = \eta_s(dx)ds. \tag{8.17}$$

This indicates why the SPDE (8.16) is not very useful for studying η: the statistics of Z are simply too closely connected with those of η, for Z vanishes wherever η does, and η vanishes on large sets - in fact on a set of full Lebesgue measure if $d \geq 2$. In fact, it seems easier to study η, which is a continuous state branching process, than Z, so (8.16) effectively expresses η in terms of a process hich is even less understood. This contrasts with cases (i)-(iii), in which Z and W ere white noises, processes which we understand rather well.

Nevertheless, there is a heuristic transformation of (8.16) into an SPDE nvolving a white noise which is worthwhile giving. This has been used by Dawson to ive some intuitive understanding of η, but which has never, to our knowledge, been ade rigorous. And which, we hasten to add, will certainly not be made rigorous ere.

Let W be a real-valued white noise on $R^d \times R_+$. Then (8.17) indicates hat Z has the same mean and covariance as Z', where

$$Z'_t(A) = \int_0^t \int_{R^d} \sqrt{\eta_s(y)}\, W(dy\, ds).$$

(If $d = 1$, $\eta_s(dy) = \eta_s(y)dy$, so $\sqrt{\eta_s(y)}$ makes sense. If $d > 1$, η_s is a ingular measure, so it is hard to see what $\sqrt{\eta_s}$ means, but let's not worry about t.)

In derivative form, $\dot{Z}' = \sqrt{\eta_s}\dot{W}$, which makes it tempting to rewrite the SPDE 8.16) as

8.18) $$\frac{\partial \eta}{\partial t} = c\Delta\eta + \sqrt{\eta}\,\dot{W}.$$

It is not clear that this equation has any meaning if $d \geq 2$, and even if $= 1$, it is not clear what its connection is with the process η which is the weak imit of the infinite particle system, so it remains one of the curiosities of the ubject.

THE CASE $\mu/\lambda \to \infty$

EMARKS. One of the features of Theorem 8.9 is that it allows us to see which of the hree sources - initial measure, diffusion, or branching - drives the limit process. n case (i), the branching is negligeable and the noise comes from the initial istribution and the diffusion. In case (ii) the initial distribution washes out ompletely, the diffusion becomes deterministic and only contributes to the drift erm $\frac{1}{2}\Delta\,\eta$, while the noise comes entirely from the branching. In case (iii), all hree effects contribute to the noise term. In case (iv), the measure-valued diffusion, we see from (8.16) that the initial distribution and diffusion both become deterministic, while the randomness comes entirely from the branching.

In case (v), which we will analyze now, it turns out that all the sources wash out. Notice that Theorem 8.6 doesn't apply when $\mu/\lambda \to \infty$, and in fact we can't affirm that the family is tight. Nevertheless, η tends to zero in a rather strong way. In fact the unnormalized process tends to zero.

THEOREM 8.11. Let $\lambda \to \infty$ and $\mu/\lambda \to \infty$. Then for any compact set $K \subset R^d$ and $\varepsilon > 0$

(i) $P_{\lambda,\mu}\{\eta_t(K) = 0, \text{ all } t \in [\varepsilon, 1/\varepsilon]\} \to 1$

and, if $d = 1$,

(ii) $P_{\lambda,\mu}\{\eta_t(K) = 0, \text{ all } t \geq \varepsilon\} \to 1.$

Before proving this we need to look at first-hitting times for branching Brownian motions. This discussion is complicated by the profusion of particles: many of them may hit a given set. To which belongs the honor of first entry?

The type of first hitting time we have in mind uses the implicit partial ordering of the branching process - its paths form a tree, after all - and those familiar with two parameter martingales might be interested to compare these with stopping lines.

Suppose that $\{X^\alpha, \alpha \in \underline{\underline{A}}\}$ is the family of processes we constructed at the beginning of the chapter, and let $A \subset R^d$ be a Borel set. For each α, let $\tau_A^\alpha = \inf\{t > 0: X_t^\alpha \in A\}$, and define T_A^α by

$$T_A^\alpha = \begin{cases} \tau_A^\alpha & \text{if } \tau_A^\beta = \infty \text{ for all } \beta \prec \alpha,\ \beta \neq \alpha; \\ \infty & \text{otherwise} \end{cases} .$$

The time T_A^α is our analogue of a first hitting time. Notice that T_A^α may be finite for many different α, but if $\alpha \prec \beta$, T_A^α and T_A^β can't both be finite. Consider, for example, the first entrance T_E of the British citizenery to an earldom. If an individual - call him α - is created the first Earl of Emsworth, some of his descendants may inherit the title, but his elevation is the vital one, so only T_E^α is finite. On the other hand, a first cousin - call him β - may be created the first Earl of Ickenham; then T_E^β will also be finite.

In general, if $\alpha \neq \beta$ and if T^α_A and T^β_A are both finite, then the descendants f X^α and of X^β form disjoint families. (Why?) By the strong Markov property and the dependence of the different particles, the post-T^α_A and post-T^β_A processes are nditionally independent given $X^\alpha(T^\alpha_A)$ and $X^\beta(T^\beta_A)$.

Let P^x_μ be the distribution of the branching Brownian motion with branching te μ which starts with a single particle X^1 at x. Under P^x_0, then, X^1_t is an dinary (non-branching) Brownian motion.

The following result is a fancy version of (8.4). While it is true for the me reason (symmetry), it is more complicated and rates a detailed proof.

ROPOSITION 8.12. Let $\phi(x,t)$ be a bounded Borel function $R^n \times \overline{R}_+$, with $\phi(x,\infty) = 0$, R^d. For any Borel set $A \subset R^d$

$$E^x_\mu\{\sum_\alpha \phi(X^\alpha(T^\alpha_A), T^\alpha_A)\} = E^x_0\{\phi(X^1(\tau^1_A),\tau^1_A)\} .$$

OOF. By standard capacity arguments it is enough to prove this for the case where is compact and ϕ has compact support in $R^d \times [0,\infty)$. We will drop the subscript d write T^α and τ^α instead of T^α_A and τ^α_A.

Define $u(x,t) = E^x_0\{\phi(X^1_{\tau^1}, t + \tau^1)\}$. Note that $\{u(X^1_{t\wedge\tau^1}, t \wedge \tau^1), t \geq 0\}$ s a martingale, so that we can conclude that $u \in C^{(2)}$ on the open set $A^c \times R_+$ and $u + \frac{1}{2}\Delta u = 0$. Thus by Ito's formula

$$Y_t \overset{\text{def}}{=} \sum_\alpha u(X^\alpha_{t\wedge T^\alpha}, t \wedge T^\alpha)$$

$$= u(x,0) + \sum_\alpha \int_0^t h^\alpha(s) I_{\{s<T^\alpha\}} \nabla u(X^\alpha_s,s) \cdot dB^\alpha_s$$

$$+ \sum_\alpha u(X^\alpha_{\zeta(\alpha)-}, \zeta(\alpha)) I_{\{\zeta(\alpha)\leq t\ T^\alpha\}} (N^\alpha - 1),$$

ere h^α, B^α, $\zeta(\alpha)$, and N^α are the quantitites used to define the branching process. te that $Y_t = \eta_t(u(\cdot,t))$ as long as $t < \inf_\alpha T^\alpha$. Since u is bounded, Y has all ments. We claim it is a martingale.

Certainly the stochastic integrals are martingales, hence so is their sum. see that the second sum is also a martingale, put

$$v_t^\alpha = u(X^\alpha(\zeta(\alpha)-), \zeta(\alpha))(N^\alpha - 1)I_{\{\zeta(\alpha) \le t \wedge T^\alpha\}}.$$

Then $v_t^\alpha = 0$ if $t < \zeta(\alpha)$, it is constant on $[\zeta(\alpha), \infty)$, and it is identically zero on the set $\{T^\alpha < \zeta(\alpha)\}$. Thus if $s < t$, $E\{v_t^\alpha | \underline{\underline{F}}_t\}$ vanishes on $\{t < \zeta(\alpha)\}$, and equals $v_t^\alpha = v_s^\alpha$ on $\{s \ge \zeta(\alpha), T^\alpha \ge \zeta(\alpha)\}$. One can use the fact that $N^\alpha - 1$ has mean zero and is independent of (X^α, ζ^α) to see that the expectation also vanishes on the set $\{s < \zeta(\alpha), t \ge \zeta(\alpha), T^\alpha \ge \zeta(\alpha)\}$. Thus in all cases $E\{v_t^\alpha | \underline{\underline{F}}_s\} = v_s^\alpha$, proving the claim.

If x is a regular point of A, i.e. if $P_0^x\{\tau^1 = 0\} = 1$, then $u(x,t) = \phi(x,t)$ for all $t \ge 0$. A Brownian motion hitting A must do so at a regular point, so $u(X_{T^\alpha}^\alpha, T^\alpha) = \phi(X_{T^\alpha}^\alpha, T^\alpha)$. (This even holds if $T^\alpha = \infty$, since both sides vanish then.)
Thus

$$u(x,0) = E_\mu^x\{\lim Y_t\} = E_\mu^x\{\sum_\alpha \phi(X_{T^\alpha}^\alpha, T^\alpha)\}.$$

Q.E.D.

REMARKS. This implies that the hitting probabilities of the branching Brownian motion are dominated by those of Brownian motion - just take $\phi \equiv 1$ and note that the left hand side of (8.14) dominates $E_\mu^x\{\sup_\alpha \phi(X_{T^\alpha}, T^\alpha)\} = P_\mu^x\{T^\alpha < \infty, \text{ some } \alpha\}$. It also implies that the left hand side of (8.14) is independent of μ.

We need several results before we can prove Theorem 8.11. Let us first treat the case $d = 1$. Let D be the unit interval in R^1 and put

$$H_\mu(x) = P_\mu^x\{T_D^\alpha < \infty, \text{ some } \alpha\}.$$

PROPOSITION 8.14. $H_\mu(x) = \frac{6}{\mu}(x - 1 - \sqrt{6/\mu})^{-2}$ if $x \ge 1$.

PROOF. This will follow once we show that H_μ is the unique solution of

$$(8.19)\qquad \begin{cases} u'' = \mu u^2 & \text{on } (1,\infty) \\ u(1) = 1 \\ 0 \le u \le 1 & \text{on } (1,\infty), \end{cases}$$

since it is easily verified that the given expression satisfies (8.19).

Let $T = \inf_\alpha T^\alpha$. If $x > 1$, Proposition 8.12 implies

(8.20) $$P^x_\mu\{T < h\} = P^x_0\{T < h\} = o(h) \quad \text{as } h \to 0.$$

Let ζ be the first branching time of the process. Then

$$H_\mu(x) = P^x_\mu\{T \le \zeta \wedge h\} + P^x_\mu\{\zeta > h,\ \zeta \wedge h < T < \infty\}$$
$$+ P^x_\mu\{\zeta \le h,\ \zeta \wedge h < T < \infty\}.$$

The first probability is o(h) by (8.20). Apply the strong Markov property at $\zeta \wedge h$ to the latter two. If $\zeta > h$, there is still only one particle, X^1, alive, so $T = T^1$ and the probability equals $E^x_\mu\{\zeta > h,\ H_\mu(X^1_{\zeta\wedge h})\} + o(h)$, where the o(h) comes from ignoring the possibility that $T < \zeta \wedge h$. X^1 either dies or splits into two independent particles, X^{11} and X^{12}, at ζ, so if $\zeta \le h$,

$I_{\{T<\infty\}} = I_{\{T^{11}<\infty\}} + I_{\{T^{12}<\infty\}} - I_{\{T_1^{11}<\infty, T^{12}<\infty\}}$. Since T^{11} and T^{12} are independent given $\mathcal{F}_\zeta$, we see the second term is

$$\frac{1}{2} E^x_\mu\{\zeta \le h,\ 2H_\mu(X^1_{\zeta\wedge h-})\} - \frac{1}{2} E^x_\mu\{\zeta \le h;\ H^2_\mu(X^1_{\zeta\wedge h-})\} + o(h),$$

where we have used the fact that $X^{11}_\zeta = X^{12}_\zeta = X^1_{\zeta-}$.

Add these terms together to see that

$$E^x_\mu\{H_\mu(X^1_{\zeta\wedge h-})\} - H_\mu(x) = \frac{1}{2} E^x_\mu\{\zeta \le h;\ H^2_\mu(X^1_{\zeta\wedge h-})\} + o(h).$$

X^1 is Brownian motion up to ζ, which is exponential (μ), so we can calculate this. Divide by $E^x_\mu\{\zeta \wedge h\} > (1 - e^{-\mu h})/\mu$, let $h \to 0$, and use Dynkin's formula. The left had side tends to $H''_\mu(x)/2$ while, since $H^2_\mu(X^1_{\zeta\wedge h-}) \to H^2_\mu(x)$, the right hand side tends to $\mu\, H^2_\mu(x)/2$. Thus H_μ satisfies (8.19).

To see that the solution of (8.19) is unique, suppose that u_1 and u_2 are both solutions and that $u_1'(1) < u_2'(1)$. Let $v = u_2 - u_1$. Then $v'' = \mu(u_1 + u_2)v$, which is strictly positive on $\{x\colon v(x) > 0\}$. Thus v' is increasing on $\{v > 0\}$, while $v(1) = 0$ and $v'(1) > 0$. This implies that v' is increasing on $[1,\infty)$, hence $v(x) \to \infty$ as $x \to \infty$, contradicting the fact that $0 \le u_1, u_2 \le 1$. Thus $u_1'(1) = u_2'(1)$, $u_1(1) = u_2(1)$, and the usual uniqueness result for the initial value problem implies that $u_1 \equiv u_2$. Q.E.D.

Moving to the d-dimensional case, let D_r be the ball of radius r centered at 0, let $T = \inf_\alpha T^\alpha_{D_1}$, and put

$$f_\mu(x,t) = P^x_\mu\{T \le t\}.$$

Let $Q(x,r,ds) = P_0^x\{\tau^1_{D_r} \in ds\}$ be the distribution of τ_{D_r} for ordinary Brownian motion.

<u>LEMMA 8.15</u>. Let $r > 1$ and let $x, y \in \mathbb{R}^d$ be such that $|x| > r$ and $|y| = r$. Then

$$f_\mu(x,t) \leq \int_0^t Q(x,r,ds) f_\mu(y,t-s).$$

<u>PROOF</u>. In order for a particle to reach D_1 from x either it or one of its ancestors must first reach D_r. Now

$$f_\mu(x,t) = 1 - P_\mu^x\{T^\alpha_{D_1} > t, \text{ all } \alpha\}$$

and

$$P_\mu^x\{T^\alpha_{D_1} > t \text{ all } \alpha\} = P_\mu^x\{\text{for all } \alpha\text{: } T^\beta_{D_1} > t \text{ for all } \beta \succ \alpha \text{ if } T^\alpha_{D_r} \leq t\}.$$

Let us apply the strong Markov property at $T^\alpha_{D_r}$. Since $|y| = r$ and $f_\mu(y,t)$ is symmetric in y, the conditional probability that the particle - or some descendant - reaches D_1 before t given that $T^\alpha_{D_r} < t$ is $f_\mu(y,t-T^\alpha_{D_r})$. Since the different post-T^α processes are independent, the above probability equals

$$E_\mu^x\{\prod_\alpha (1 - f_\mu(y, t-T^\alpha_{D_r}))\},$$

where $f_\mu(y, t-T^\alpha_{D_r}) = 0$ if $T^\alpha_{D_r} > t$. Thus

$$\begin{aligned} f_\mu(x,t) &= 1 - E_\mu^x\{\prod_\alpha (1 - f_\mu(y, t-T^\alpha_{D_r}))\} \\ &\leq 1 - E_\mu^x\{1 - \sum_\alpha f_\mu(y, t-T^\alpha_{D_r})\} \\ &= E_\mu^x\{\sum_\alpha f_\mu(y, t-T^\alpha_{D_r})\} \\ &= E_0^x\{f_\mu(y, \tau_{D_r})\} \end{aligned}$$

by Proposition 8.12. Q.E.D.

This brings us to Theorem 8.11.

<u>PROOF</u> of Theorem 8.11. Suppose without loss of generality that K is the unit ball D_1. Write $\eta^\lambda_t = \vec{\eta}^\lambda_t + \bar{\bar{\eta}}^\lambda_t$ where $\vec{\eta}^\lambda_t$ comes from those initial particles inside D_2, and

$\bar{\bar{\eta}}_t^\lambda = \eta_t^\lambda - \bar{\eta}_t^\lambda$.

Now if we start a single particle from x, $\eta_t^\lambda(1)$ is an ordinary branching process hence (see e.g. Harris [28]).

$$P^x\{\eta_t(1) > 0\} \sim \frac{c}{\mu t} \quad \text{as } \mu \to \infty \quad \text{for } t > 0.$$

Thus

$$P_\mu\{\bar{\eta}_t^\lambda = 0 \quad \text{all } t > \varepsilon\} = P_\mu\{\bar{\eta}_t^\lambda(1) = 0\}$$
$$= \int_{D_2} \frac{c}{\mu\varepsilon}\lambda dx \sim \frac{c}{\varepsilon}\frac{\lambda}{\mu} \to 0.$$

Next

$$P_\mu\{\bar{\bar{\eta}}_t^\lambda(D_1) > 0 \quad \text{some } t < 1/\varepsilon\}$$
$$= \int_{R^d - D_2} P_\mu^x\{T_{D_1}^\alpha < \frac{1}{\varepsilon}\}\lambda \, dx$$
$$\leq \int_{R^d - D_2} \int_0^t Q(x,\tfrac{3}{2},ds) f_\mu(y,t-s)\lambda ds$$

by Lemma 8.15.

Now in order for a particle to hit D_1, its first coordinate must hit $[-1,1]$, so that $f_\mu(y, t-s) \leq H_\mu(|y|)$. Thus this is

$$\leq \lambda \, H_\mu(\tfrac{3}{2}) \int_{R^d - D_2} Q(x, \tfrac{3}{2}, [0,t])dx .$$

This integral is finite - indeed, it is bounded by

$P_0^x\{\sup_{0 \leq s \leq t} |X_s^1 - x| \leq |x| - \frac{3}{2}\} \leq C \, e^{-\frac{(|x|-3/2)^2}{2t}}$, so by Proposition 8.14, this is

$$= C\,\frac{\lambda}{\mu}\,(\tfrac{1}{2} + \sqrt{6/\mu})^{-2} \to 0.$$

Putting these together gives (i). In case d = 1,

$$P\{\bar{\bar{\eta}}_t^\lambda(D_1) > 0 \text{ some } t \geq 0\} = 2 \int_1^\infty \frac{6}{\mu}\,(x - 1 + \sqrt{\frac{6}{\mu}}\,)^{-2}\lambda \, dx$$
$$\leq 12\,\frac{\lambda}{\mu} \to 0$$

giving (ii).

Q.E.D.

THE SQUARE OF THE BROWNIAN DENSITY PROCESS

Now that we have seen how the Brownian density process can be squeezed out of an infinite particle system, we can't resist the temptation to look at its square. We renormalize during its construction, so that it is actually the square by analogy rather than by algebra, but it is at least closely related.

We return to the system of discrete particles to set up the process and then take a weak limit at the end. We will show that the limiting process satisfies a stochastic partial differential equation whose solution can be written in terms of multiple Wiener integrals. In particular, it is non-Gaussian. The end result is in Theorem 8.18.

Let $\{X^\alpha, \alpha \in \mathbf{N}\}$ be a family of i.i.d. standard (i.e. non-branching) Brownian motions in $\mathbf{R}^d$, with initial distribution given by a Poisson point process Π^λ of parameter λ. Set, as before,

$$\eta_t(\phi) = \sum_\alpha \phi(X_t^\alpha).$$

Now $\eta_t^2(\phi) = \sum_{\alpha,\beta} \phi(X_t^\alpha)\phi(X_t^\beta)$. We will first symmetrize this, then throw away the terms with $\alpha = \beta$, to get a new process, Q_t.

Let $\{\xi^\alpha, \alpha \in \mathbf{N}\}$ be a sequence of i.i.d. random variables, independent of the X^α, such that $P\{\xi^\alpha = 1\} = P\{\xi^\alpha = -1\} = 1/2$. Define

$$\tilde{\eta}_t(\phi) = \sum_{\alpha \in \mathbf{N}} \xi^\alpha \, \phi(X_t^\alpha) \tag{8.21}$$

and, for any function ψ on $\mathbf{R}^d \times \mathbf{R}^d$, set

$$Q_t(\psi) = \lambda^{-1} \sum_{\alpha \neq \beta} \xi^\alpha \xi^\beta \psi(X_t^\alpha, X_t^\beta). \tag{8.22}$$

This is the process of interest. We define it on $\mathbf{R}^{2d}$ rather than $\mathbf{R}^d$,; to see its connection with $\tilde{\eta}^2$, set $\psi(x,y) = \phi(x)\phi(y)$. Then $Q_t(\psi) = \lambda^{-1}(\tilde{\eta}_t^2(\phi) - \tilde{\eta}_t(\phi^2))$.

<u>Notation:</u> Let $D \subset \mathbf{R}^{2d}$ be the set $\{(x,y): x \in \mathbf{R}^d, y \in \mathbf{R}^d, x = y\}$. If μ is a measure on $\mathbf{R}^d$, define a measure $\mu \otimes \mu$ on $\mathbf{R}^{2d}$ by $\mu \otimes \mu(A) = \mu \times \mu(A-D)$, where $A \subset \mathbf{R}^{2d}$, and $\mu \times \mu$ is the product measure on $\mathbf{R}^{2d}$. If we let $\tilde{\Pi}^\lambda = \tilde{\eta}_0$, which is the symmetrized version of Π^λ, then $Q_0 = \tilde{\Pi}^\lambda \otimes \tilde{\Pi}^\lambda$ and $Q_t = \tilde{\eta}_t \otimes \tilde{\eta}_t$.

If we try to write Q in differential form, we would expect that $dQ = \tilde{\eta}_t \times$

t $+ d\tilde{\eta}_t \otimes \tilde{\eta}_t$. This is roughly what happens. To see exactly what happens, tho, we ust analyze the system from scratch.

Define $W_t(A) = \lambda^{-1/2} \sum_\alpha \int_0^t I_A(X_s^\alpha)dX_s^\alpha$ as before, and let its symmetrized ersion be

$$\tilde{W}_t(A) = \lambda^{-1/2} \sum_\alpha \xi^\alpha \int_0^t I_A(X_s^\alpha)dX_s^\alpha.$$

ince there is no branching, (8.10) becomes

$$\eta_t(\phi) = \int_{R^d} G_t(\phi,y)\eta_0(dy) + \lambda^{1/2} \int_0^t \int_{R^d} \nabla G_{t-s}(\phi,y)\cdot W(dy\, ds).$$

ts symmetrized version is

$$\tilde{\eta}_t(\phi) = \int_{R^d} G_t(\phi,y)\, \tilde{\eta}_0(dy) + \lambda^{1/2} \int_0^t \int_{R^d} \nabla G_{t-s}(\phi,y)\cdot\tilde{W}(dyds). \tag{3.23}$$

Let us use ∇_1 and ∇_2 to indicate the gradients $\nabla_1\phi(x,y) = \nabla_x\phi(x,y)$ and $_2\phi(x,y) = \nabla_y\phi(x,y)$. Similarly we define the Laplacians $\Delta_1 = \Delta_x$ and $\Delta_2 = \Delta_y$, so that ne Laplacian on R^{2d} is $\Delta = \Delta_1 + \Delta_2$.

If $\phi = C^2(R^{2d})$ has compact support, then by Ito's formula

$$Q_t(\phi) = Q_0(\phi) + \lambda^{-1} \sum_{\alpha\neq\beta} \xi^\alpha\xi^\beta \Big[\int_0^t \nabla_1\phi(X_s^\alpha,X_s^\beta)\cdot dX_s^\alpha + \int_0^t \nabla_2\phi(X_s^\alpha,X_s^\beta)\cdot dX_s^\beta$$
$$+ (2\lambda)^{-1} \sum_{\alpha\neq\beta} \xi^\alpha\xi^\beta \int_0^t \Delta\phi(X_s^\alpha,X_s^\beta)ds. \tag{.24}$$

Each of these sums can be identified in terms of η, Q and the martingale easures W and $\tilde{W}$. The last term, for instance, is just $(2\lambda)^{-1} \int_0^t Q_s(\Delta\phi)ds$, while

$$\sum_{\alpha\neq\beta} \xi^\alpha\xi^\beta \int_0^t \nabla_1\phi(X_s^\alpha,X_s^\beta)\cdot dX_s^\alpha$$
$$= \sum_\beta \xi^\beta \int_0^t \xi^\alpha\nabla_1\phi(X_s^\alpha,X_s^\beta)\cdot dX_s^\alpha - \sum_\alpha \int_0^t \nabla_1\phi(X_s^\alpha,X_s^\alpha)\cdot dX_s^\alpha$$
$$= \lambda^{1/2} \sum_\beta \xi^\beta \int_0^t \int_{R^d} \nabla_1\phi(x,X_s^\beta)\cdot d\tilde{W}_{sx} - \lambda^{1/2} \sum_\alpha \int_0^t \int_{R^d} \nabla_1\phi(x,x)\cdot dW_{sx}$$
$$= \lambda^{1/2} \int_0^t \int_{R^d} \tilde{\eta}_s(\nabla_1\phi(x,\cdot))\cdot d\tilde{W}_{sx} - \lambda^{1/2} \int_0^t \int_{R^d} \nabla_1\phi(x,x)\cdot dW_{sx}.$$

et $\chi(x) = (\nabla_1\phi)(x,x) + (\nabla_2\phi)(x,x)$. Then (8.24) becomes

$$Q_t(\phi) = Q_0(\phi) + \frac{1}{2}\int_0^t Q_s(\Delta\phi)ds + \lambda^{-1/2} \int_{R^d\times[0,t]} \tilde{\eta}_s(\nabla_1\phi(x,\cdot))d\tilde{W}_{sx}$$

$$+ \lambda^{-1/2} \int_{R^d \times [0,t]} \tilde{\eta}_s(\nabla_2 \phi(\cdot,y)) \cdot d\tilde{W}_{sy} - \lambda^{-1/2} \int_{R^d \times [0,t]} \chi(x) \cdot dW_{sx}.$$

Let us write $Q = \tilde{Q} + R$ where

$$\tilde{Q}_t(\phi) = Q_0(\phi) + \frac{1}{2} \int_0^t \tilde{Q}_s(\Delta\phi)ds + \lambda^{-1/2} \int_{R^d \times [0,t]} \tilde{\eta}_s(\nabla_1 \phi(x,\cdot) \cdot d\tilde{W}_{sx} \tag{8.25}$$

$$+ \lambda^{-1/2} \int_{R^d \times [0,t]} \tilde{\eta}_s(\nabla_2 \phi(\cdot,y)) \cdot d\tilde{W}_{sy};$$

$$R_t(\phi) = \frac{1}{2} \int_0^t R_s(\Delta\phi)ds + \lambda^{-1/2} \int_{R^d \times [0,t]} \chi(x) \cdot dW_{sx}. \tag{8.26}$$

These are integral forms of SPDE's which we can solve by Theorem 5.1. To get them in the form (5.4), define a pair of martingale measures on R^{2d} by

$$M_t^1(\phi) = \int_{R^d \times [0,t]} \Big(\int_{R^d} \phi(x,y)\tilde{\eta}_s(dx)\Big) d\tilde{W}_{sy}$$

and

$$M_t^2(\phi) = \int_{R^d \times [0,t]} \Big(\int_{R^d} \phi(x,y)\tilde{\eta}_s(dy)\Big) d\tilde{W}_{sx}.$$

Note that these are worthy martingale measures. The covariance measure for M^1, for instance, is $\tilde{\eta}_s(dx)\tilde{\eta}_s(dx')\delta_y(y')dydy'dsI$, hence its dominating measure is

$$K^1(dx\ dy\ dx'dy'ds) = \eta_s(dx)\eta_s(dx')\delta_y(y')dydy'ds,$$

which is clearly positive definite.

Notice also that the M^i are neither orthogonal nor of nuclear covariance. M^1 and M^2 have values in R^d (since $\tilde{W}$ does) and, if we abuse notation by writing $\int \nabla_i \phi(x,y) \cdot M_t^i(dx\ dy) = M_t^i(\nabla_i \phi)$, (8.25) becomes

$$\tilde{Q}_t(\phi) = Q_0(\phi) + \frac{1}{2} \int_0^t \tilde{Q}_s(\Delta\phi)ds + \lambda^{-1/2} M^1(\nabla_1\phi) + \lambda^{-1/2} M^2(\nabla_2\phi). \tag{8.27}$$

By Theorem 5.1

$$\tilde{Q}_t(\phi) = Q_0(G_t\phi) + \int_{R^d \times [0,t]} \nabla_1 G_{t-s}(\phi)(x,y,s) \cdot M^2(dx\ dy\ ds)$$

$$+ \int_{R^d \times [0,t]} \nabla_2 G_{t-s}(\phi)(x,y,s) \cdot M^1(dx\ dy\ ds).$$

If we write $M^1(\phi)$ as $\int \eta_s(\phi(\cdot,y)) d\tilde{W}^1_{sy}$ this becomes

$$\tilde{Q}_t(\phi) = Q_0(G_t\phi) + \int_{R^d \times [0,t]} \tilde{\eta}_s(\nabla_1 G_{t-s}\phi(x,\cdot)) \cdot d\tilde{W}_{sx} \tag{8.28}$$

$$+ \int_{R^d \times [0,t]} \tilde{\eta}_s(\nabla_2 G_{t-s}\phi(\cdot,y)) \cdot d\tilde{W}_{sy}.$$

Next, define $N_t(\phi) = \int_0^t \phi(x,x)d\tilde{W}_{sx}$. N is an orthogonal R^d-valued artingale measure on R^{2d}, and (8.26) becomes

$$R_t(\phi) = \frac{1}{2} \int_0^t R_s(\Delta\phi)ds + \lambda^{-1/2} N_t(\nabla_1\phi+\nabla_2\phi).$$

Let $P_t(x,y) = (2\pi t)^{-d/2} e^{-\frac{|y-x|^2}{2t}}$ and let $G_t(x,y;x',y') = P_t(x,y)P_t(x',y')$. hen G is the Green's function on R^{2d} for this problem, so that, by Theorem 5.1 it s

$$= \lambda^{-1/2} \int_{R^d\times[0,t]} (\nabla_1 G_{t-s}\phi(x,y) + \nabla_2 G_{t-s}\phi(x,y)\cdot N(dx\,dy\,ds)$$

r

$$\text{(8.29)} \qquad R_t(\phi) = \lambda^{-1} \int_{R^d\times[0,t]} [\nabla_1 G_{t-s}(\phi)(y,y) + \nabla_2(G_{t-s}(\phi)(y,y)]\cdot W(dyds)$$

If we let $\lambda \to \infty$, we will see that $\tilde{Q}$ and R have weak limits, and in fact => 0. Most of the work has already been done. Let us make the dependence on λ xplicit, writing W^λ, η^λ, Q^λ etc. Note that W^λ and $\tilde{W}^\lambda$ have the same distribution, nd thus both have the same means and covariances as a white noise, independent of λ. hey are orthogonal, hence W^λ and $\tilde{W}^\lambda$ must converge weakly to independent white noises y Theorem 8.7. We know about the moments of η from (8.4), (8.12), Proposition 8.5 nd Exercise 8.1. In the case of $\tilde{\eta}$,

$$\text{(8.30)} \qquad E\{\tilde{\eta}_t(\phi)\} = 0, \qquad\qquad E\{\tilde{\eta}_t^2(\phi)\} = \lambda\langle\phi^2\rangle,$$

he latter following since the left hand side is

$$E\{\sum_{\alpha,\beta} \xi^\alpha\xi^\beta\phi(X_t^\alpha)\phi(X_t^\beta)\} = E\{\sum_\alpha \phi^2(X_t^\alpha)\}.$$

nce can show as in Exercise 8.1 that $\lambda^{-1/2}\tilde{\eta}^\lambda$ is L^p bounded, independent of λ, for ll $p < \infty$.

In order to establish that Q converges weakly, we need to show that R => 0 nd that the various terms of (8.28) converge weakly. Let us dispatch the easy parts f the convergence argument first.

<u>ROPOSITION 8.16</u>. The processes $\lambda^{-1/2}\tilde{\Pi}^\lambda$, $\lambda^{-1}\tilde{\Pi}^\lambda\otimes\tilde{\Pi}^\lambda$, $\tilde{W}^\lambda$, $\lambda^{-1}\tilde{\eta}^\lambda$, R^λ and Q^λ are tight n the appropropriate space $\underline{\underline{D}}\{([0,1],\underline{\underline{S}}'(R_p)\}$. Moreover, if V_0 and W_0 are independent

white noises on R^d and $R^d \times R_+$ respectively, with values in R and R^d respectively, then, as $\lambda \to \infty$

$$\left(\lambda^{-1/2}\tilde{\Pi}^\lambda,\ \lambda^{-1}\tilde{\Pi}^\lambda \otimes \tilde{\Pi}^\lambda,\ \tilde{W}^\lambda,\ \lambda^{-1/2}\eta^\lambda,\ R^\lambda\right) \Rightarrow (V^0,\ V^0 \otimes V^0,\ W^0,\ \eta, 0),$$

where η is a solution of the SPDE

$$\begin{cases} \dfrac{\partial \eta}{\partial t} = \dfrac{1}{2}\Delta\eta + \nabla\cdot W^0 \\ \eta_0 = V^0 . \end{cases}$$

<u>Note</u>. $V^0 \otimes V^0$ is a multiple Wiener integral. We can define $V^0 \otimes V^0(A) = V^0(A_1)V^0(A_2)$ if $A = A_1 \times A_2$ and $A_1 \cap A_2 = \phi$, and extend it to all Borel A by the usual approximation arguments.

<u>PROOF</u>. As remarked above $\tilde{W}^\lambda \Rightarrow W^0$, and we have proved $\lambda^{-1/2}\Pi^\lambda$ converges weakly. We leave it as an exercise to show that $\lambda^{-1/2}\tilde{\Pi}^\lambda \Rightarrow V^0$. If ϕ and ψ are test functions of disjoint support in R^d,

$$\left(\lambda^{-1/2}\Pi^\lambda(\phi),\ \lambda^{-1/2}\Pi^\lambda(\psi)\right) \Rightarrow (V^0(\phi),\ V^0(\psi)).$$

Multiplication is continuous on R^2, so this implies that

$$\lambda^{-1}\,\tilde{\Pi}^\lambda \otimes \tilde{\Pi}^\lambda(\phi,\psi) = \lambda^{-1}\tilde{\Pi}^\lambda(\phi)\tilde{\Pi}^\lambda(\psi) \Rightarrow V^0(\phi)V^0(\psi) = V^0 \otimes V^0(\phi\psi).$$

This holds for finite sums of such functions, hence for all $\psi(x,y) \in \underline{\underline{S}}(R^{2d})$ by approximation.

In view of (8.23), the convergence of $\lambda^{-1/2}\,\tilde{\eta}^\lambda$ to η follow from Proposition 7.8, and the limit is identified as in Theorem 8.9(i).

The tightness of R^λ follows from (8.29) and Proposition 7.8, and, from (8.29)

$$E\left\{R_t^\lambda(\psi)^2\right\} \le \frac{C}{\lambda} \to 0,$$

hence $R^\lambda \Rightarrow 0$.

We leave it to the reader to use Proposition 7.6 to show joint convergence of these processes. Q.E.D.

This leaves us the question of the weak convergence of Q or, equivalently, of $\tilde{Q}$. In view of (8.28), it is enough to show the convergence of the stochastic integrals there.

Let $U_t^\lambda = \lambda^{-1/2} \int_{R^d \times [0,t]} \tilde{\eta}_s^\lambda(\nabla_1 G_{t-s}\phi(x,\cdot))\cdot d\tilde{W}_{xs}^\lambda .$

f $\psi(x,y) = \phi(x)\chi(y)$, then

$$U_t^\lambda = \int_{R^d\times[0,t]} \lambda^{-1/2}\,\tilde\eta_s^\lambda(\chi)\,\nabla_1 G_{t-s}\phi(x)\cdot d\tilde W_{xs}^\lambda.$$

Let (λ_n) be a sequence tending to infinity.

ROPOSITION 8.17. $\left(\tilde W^{\lambda_n},\ \lambda^{-1/2}\,\tilde\eta^{\lambda_n},\ U^{\lambda_n}\right) \Rightarrow \left(W^0,\ \eta,\ U^0\right)$

here

$$U_t^0(\psi) = \int_{R^d\times[0,t]} \eta_s(\nabla_1 G_{t-s}\psi(x,\cdot))\cdot dW_{xs}^0,\ \psi \in \underline{\underline{S}}(R^{2d}).$$

ROOF. We already know from Proposition 8.16 that $\left(\tilde W^{\lambda_n},\ \lambda^{-1/2}\eta^{\lambda_n}\right) \Rightarrow \left(W^0,\ \eta\right)$. It emains to treat U. The idea of the proof is the same: the martingale measure onverges, hence so does the stochastic integral. However, we can't use Propositions .6 and 7.8, for the integrand is not in $\bar{\mathcal{P}}_s$. We will use 7.12 and 7.13 instead.

Define (V^0, W^0) and $(\tilde V^{\lambda_n}, \tilde W^{\lambda_n})$ canonically on $\underline{\underline{D}} = \underline{\underline{D}}([0,1], \underline{\underline{S}}'(R^{2d}))$, and lenote their probability distributions by P^0 and P respectively. By (8.23) we can lefine $\tilde\eta_t^{\lambda_n}$ on $\underline{\underline{D}}$ simultaneously for n = 0,1,2,... as a continuous process. That is, he stochastic integrals in (8.23) are consistent up to sets which are of P^n-measure ero for all $n \geq 0$. Thus we can also define

$$g(x,s,t) = \int_{R^d\times[0,t]} \eta_s(\phi)\nabla P_{t-s}\chi(x)\cdot d\tilde W_{xs}^{\lambda_n}$$

or each s,t and x, independent of n. You can do better.

xercise 8.2. Show $g(\cdot,\cdot,t_0)I_{\{s\leq t_0\}} \in \bar{\bar{\mathcal{P}}}_S(W)$, where W is the sequence $W^0, W^{\lambda_1}, W^{\lambda_2}, \ldots$,

int. It is not quite enough to approximate $\eta_s(\phi)$ by the step function $\eta_{\frac{[ms]}{m}}(\phi)$ here [t] is the greatest integer in t, since it is not clear that this will be a ontinuous function of ω on $\underline{\underline{D}}$. Go back one step further and approximate $\eta_{\frac{[ms]}{m}}(\phi)$

itself by the integral of a deterministic step function with respect to dW_{xs}. This will be continuous in ω.

Now apply Theorem 7.13 with $p = q = 4$ and $\beta = 1$. Note that g is of Hölder class (1,4,C) for some C. Indeed (i) of the definition was verified in the above exercise; if we take $Y_n(\omega) = \sup_{t\leq 1}[\,\|\nabla P_t\chi\|_\infty + \|P_t\chi\|_\infty]$ then parts (a) and (b) of (ii) hold, while (c) follows from the uniform L^p-boundedness of $\lambda^{-1/2}\eta^\lambda$.

The family (W^{λ_n}) is of class (4,K) (see Chapter 7) for some K, and β, p, and q satisfy the necessary inequalities, so that Theorem 7.13 (v) implies that $U_t^n(\psi) \Rightarrow U_t(\psi)$ for ψ of the form $\chi(x)\phi(y)$. It follows that there is also convergence for finite sums of such functions. Since any $\psi \in \underline{S}(\mathbb{R}^{2d})$ can be approximated uniformly and in L^p by such sums, it is easy to see that $U^n(\psi) \Rightarrow U(\psi)$ for all $\psi \in \underline{S}(\mathbb{R}^{2d})$. We leave the details to the reader. It now follows from Mitoma's theorem that $U^{\lambda_n} \Rightarrow U^0$ on $\underline{D}\{[0,1],\ \underline{S}'(\mathbb{R}^{2d})\}$.

To show that the triple converges jointly, let $f^n(x,t,\omega)$ be the 3×2 matrix

$$\begin{pmatrix} \phi_1, & P_{t-s}\phi_2, & \eta_s(\phi_3)\nabla P_{t-s}\chi \\ 0 & P_t\phi_2 & 0 \end{pmatrix}^T$$

and let $M^n = (\tilde{W}^{\lambda_n}, V^{\lambda_n})^T$, in which case $(f^n\cdot M^n)_t = \left(\tilde{W}_t^{\lambda_n}(\phi_1),\ \eta_t^{\lambda_n}(\phi_2),\ U_t^{\lambda_n}(\phi_3\chi)\right)^T$, and $f^n\cdot M^n \Rightarrow f^0\cdot M^0$ by Proposition 7.12, which implies joint convergence. Q.E.D

<u>REMARK</u>. We took $p = q = 4$ in the above proof, but as W and η are L^p bounded for all p, we could let p and $q \to \infty$ with $p = q$. In that case, Theorem 7.13(iii) tells us that U is L^p bounded for all p, and Hölder continuous with exponent $\frac{1}{2} - \varepsilon$ for any $\varepsilon > 0$.

The second integral in (8.28) also converges, so, combining Propostions 8.16 and 8.17, we have

<u>THEOREM 8.18</u>. The process Q^λ converges weakly in $\underline{D}\{[0,1],\ \underline{S}'(\mathbb{R}^{2d})\}$ to a solution of the SPDE

$$(8.31)\qquad \begin{cases} \dfrac{\partial Q}{\partial t}(x,y) = \dfrac{1}{2}\Delta Q(x,y) + \eta(x)\nabla_2\cdot W^0_{ys} + \eta(y)\nabla_1\cdot W^0_{xs} \\ Q_0 = V^0 \otimes V^0 \end{cases}$$

PROOF. There is very little to prove. (8.31) is just the differential form of (8.25) (with $\tilde{W}$ replaced by W^0 and $\tilde{\eta}$ by η) and (8.28) is the unique solution of (8.25). By Propositions 8.16 and 8.17, all the stochastic integrals converge to the right limits, so equation (8.28) is valid for the limiting process, hence so is (8.25).

Q.E.D.

If we plug (8.23) - which remains valid if we replace $\tilde{\eta}_0$ by V^0 and $\tilde{W}$ by W^0 - into (8.25) we get Q in all its glory. For simplicity's sake we will only write it for ψ of the form $\psi(x,y) = \chi(x)\ \phi(y)$.

$$Q_t(\psi) = \int_{R^{2d}-D} P_{t-s}\chi(x)P_{t-s}\phi(y)V^0(dx)V^0(dy)$$

$$+ \int_{R^d\times[0,t]} \Big[\int_{R^d} P_s\chi(x)\ V^0(dx) + \int_{R^d\times[0,t]} \nabla P_{s-u}\chi(x)\cdot W^0(dxdu)\Big]\nabla\phi(y)\cdot W^0(dyds)$$

$$+ \int_{R^d\times[0.t]} \Big[\int_{R^d} P_s\phi(x)V^0(dx) + \int_{R^d\times[0.t]} \nabla P_{s-u}\phi(x)\cdot W^0(dxdu)\Big]\nabla\chi(y)\cdot W^0(dyds).$$

The first integral is a classical multiple Wiener integral. The next two could also be called multiple Wiener integrals, as they are iterated stochastic integrals with respect to white noise.

CHAPTER NINE

We have spent most of our time on parabolic equations; non-parabolic equations have made only token appearances, such as at the beginning of these notes when we took a brief glance at the wave equation, which is hyperbolic. It is fitting to end with a brief glance at a token elliptic, Laplace's equation.

We will give one existence and uniqueness theorem for bounded regions, and then see how such equations arise as the limits of parabolic equations. In particular, we will look at the limits of the Brownian density process as $t \to \infty$.

Let D be a bounded domain in $\mathbf{R}^d$ with a smooth boundary. Consider

$$\text{(9.1)} \qquad \begin{cases} \Delta u = f & \text{in } D \\ u = 0 & \text{on } \partial D . \end{cases}$$

If f is bounded and continuous, the solution to (9.1) is

$$\text{(9.2)} \qquad u(y) = \int K(x,y)\, f(x)dx = K(f,y),$$

where K is the Green's function for (9.1). Notice that in particular, $\Delta K(f,y) = f(y)$.

Let M be an L^2-valued measure on $\mathbf{R}^d$ (not a martingale measure, for there is no t in the problem!) Set $Q(A,B) = E\{M(A)M(B)\}$ and suppose that there exists a positive definite measure $\tilde{Q}$ on $\mathbf{R}^d \times \mathbf{R}^d$ such that $|Q(A,B)| \leq \tilde{Q}(A\times B)$ for all Borel A, $B \subset \mathbf{R}^d$. This assures us of a good integration theory. We also assume for convenience that $M(\partial D) = 0$.

Let T be a kth order differential operator on $\mathbf{R}^d$ with smooth coefficients ($0 \leq k < \infty$) and consider the SPDE

$$\text{(9.3)} \qquad \begin{cases} \Delta U = T\dot{M} & \text{in } D \\ U = 0 & \text{in } \partial D . \end{cases}$$

Let us get the weak form of (9.3). Multiply by a test function ϕ and integrate over $\mathbf{R}^d$, pretending $\dot{M}$ is smooth. Suppose $\phi = 0$ on ∂D. We can then do two integrations by parts to get

$$\text{(9.4)} \qquad \int U(x)\Delta\phi(x)dx = \int_D T\, \dot{M}(x)\phi(x)dx.$$

Let T* be the formal adjoint of T. If T is a zeroth or first order perator, or if ϕ has compact support in D, we can integrate by parts on the ight to get

(9.5) $$U(\Delta\phi) = \int_D T^*\phi(x)\, M(dx).$$

Let H_n be the Sobolev space introduced in Example 1, Chapter 4. We say U is an $\underline{H_n \text{ solution}}$ of (9.3) if for a.e.ω, $U(\cdot,\omega)$ takes values in H_{-n} and (9.5) holds for all $\phi \in H_n$. We say U is a weak solution if $U \in \underline{\underline{S}}'(R^d)$ a.e. nd if (9.5) holds for all $\phi \in \underline{\underline{S}}_0$, where

$$\underline{\underline{S}}_0 = \{\phi \in \underline{\underline{S}}(R^d): \phi = 0 \text{ on } \partial D\}.$$

ROPOSITION 9.1. Let k = order of T. If $n > d + k$, then (9.3) has a unique H_n-solution, defined by

(9.6) $$U(\phi) = \int T^*K(\phi,y)M(dy).$$

his also defines a weak solution of (9.3).

PROOF. Uniqueness. By the general theory of PDE, $\phi \in H_n \Rightarrow K(\phi,\cdot) \in H_{n+2} \subset H_n$. If U is any H_n-solution, apply (9.5) to $\psi(y) = K(\phi,y)$:

$$U(\phi) = U(\Delta\psi) = \int_D T^*\psi(y)M(dy) = \int_D T^*K(\phi,y)M(dy).$$

Existence. Define U by (9.6). If $\phi \in C^\infty(R^d)$,

$$E\{|U(\phi)|^2\} = E\{[\int_D T^*K(\phi,y)M(dy)]^2\}$$

$$= \int_D\int_D T^*K(\phi,x)T^*K(\phi,y)Q(dx\,dy)$$

$$\leq C_1 \|T^*K(\phi,\cdot)\|^2_{L^\infty}$$

where $C_1 = \tilde{Q}\{D \times D\}$. By the Sobolev embedding theorem, if $q > d/2$ this is

$$\leq C_2 \|T^*K(\phi,\cdot)\|^2_q .$$

T is a differential operator of order k, hence it is bounded from $H_{q+k} \to H_q$, while K maps $H_{q+k-2} \to H_{q+k}$ boundedly. Thus the above is

$$\leq C_4 \|\phi\|^2_{q+k-2}$$

It follows that U is continuous in probability on H_{q+k-2} and, by Theorem 4.1, it is a random linear functional on H_p for any $p > q + k - 2 + d/2$. Fix a $p > d + k - 2$ and let $n = p + 2$. Then $U \in H_{-n}$. (It is much easier to see that $U \in \underline{\underline{S}}'(R^d)$. Just note that $T^*K(\phi,\cdot)$ is bounded if $\phi \in \underline{\underline{S}}(R^d)$ and apply Corollary 4.2).

If $\phi \in \underline{\underline{S}}_0$,

$$U(\Delta\phi) = \int_D T^*K(\Delta\phi,y)M(dy)$$

$$= \int_D T^*\phi(y)M(dy).$$

On the other hand, $C_0^\infty \subset \underline{\underline{S}}_0$, and C_0^∞ is dense in all the H_t. U is continuous on H_p so the map $\phi \to \Delta\phi \to U(\Delta\phi)$ of $H_n \to H_p \to R$ is continuous, while on the right-hand side of (9.5)

$$E\{|\int_{R^d} T^*\phi dM|^2\} \leq C\|T^*\phi\|^2_{L^\infty} \leq C\|\phi\|^2_{k+q},$$

which tells us the right-hand side is continuous in probability on H_{k+q}, hence, by Theorem 4.1, it is a linear functional on $H_{d+k} = H_n$. Thus (9.5) holds for $\phi \in H_n$. Q.E.D.

LIMITS OF THE BROWNIAN DENSITY PROCESS

The Brownian density process η_t satisfies the equation

$$\frac{\partial\eta}{\partial t} = \frac{1}{2}\Delta\eta + a\,\nabla\cdot\dot{W} + b\dot{Z} \tag{9.7}$$

where W is a d-dimensional white noise and Z is an independent one-dimensional white noise, both on $R^d \times R_+$, and the coefficients a and b are constants. (They depend on the limiting behavior of μ and λ.)

Let us ask if the process has a weak limit as $t \to \infty$. It is not too hard to see that the process blows up in dimensions d = 1 and 2, so suppose $d \geq 3$. The Green's function G_t for the heat equation on R^d is related to the Green's function K for Laplace's equation by

$$K(x,y) = -\int_0^\infty G_t(x,y)dt \tag{9.8}$$

and K itself is given by

$$K(x,y) = \frac{C_d}{|y-x|^{d-2}},$$

where C_d is a constant. The solution of (9.7) is

$$\eta_t(\phi) = \eta_0 G_t(\phi) + a \int_0^t \int_{R^d} \nabla G_{t-s}(\phi,y)\cdot W(dy\,ds) + b \int_0^t \int_{R^d} G_{t-s}(\phi,y) Z(dyds)$$

$$\underset{=}{\text{def}}\ \ \eta_0 G_t(\phi) + a\, R_t(\phi) + b\, U_t(\phi).$$

R_t and U_t are mean-zero Gaussian processes. The covariance of R_t is

$$E\{R_t(\phi)R_t(\psi)\} = \int_0^t \int_{R^d} (\nabla_x G_{t-s})(\phi,y)\cdot(\nabla_x G_{t-s})(\psi,y)dy\, ds\ .$$

ow $\nabla_x G = -\nabla_y G$; if we then integrate by parts

$$= - \int_0^t \int_{R^d} \Delta_y G_{t-s}(\phi,y) G_{t-s}(\psi,y) dy\, ds$$

$$= - \int_0^t \int_{R^d} G_{t-s}(\Delta\phi,y) G_{t-s}(y,\psi) dy\, ds$$

$$= - \int_0^t G_{2t-2s}(\Delta\phi,\psi) ds$$

$$= - \frac{1}{2} \int_{R^d} \psi(y) \int_0^{2t} G_u(\Delta\phi,y) ds\, dy$$

$$= - \frac{1}{2} \int_{R^d} \psi(y)[-\phi(x) + G_{2t}(x,\phi)]dy$$

y (5.7). Since $d \geq 3$, $G_t \to 0$ as $t \to \infty$ so

9.9) $$E\{R_t(\phi)R_t(\psi)\} \to \frac{1}{2} \langle\phi,\psi\rangle.$$

The calculation for U is easier since we don't need to integrate by arts:

$$E\{U_t(\phi)U_t(\psi)\} = \int_0^t \int_{R^d} G_{t-s}(\phi,y) G_{t-s}(\psi,y) dy\, ds$$

$$= \frac{1}{2} \int_0^{2t} G_{2t-u}(\phi,\psi) du \to - \frac{1}{2} K(\phi,\psi)$$

s $t \to \infty$. Taking this and (9.9) into account, we see:

ROPOSITION 9.3. Suppose $d \geq 3$. As $t \to \infty$, $\sqrt{2}\, R_t$ converges weakly to a white oise and $\sqrt{2}\, U_t$ converges to a random Gaussian tempered distribution with ovariance function

9.10) $$E\{U(\phi)U(\psi)\} = -K(\phi,\psi).$$

n particular, η_t converges weakly as $t \to \infty$. The convergence is weak onvergence of $\underline{\underline{S}}'(R^d)$-valued random variables in all cases.

Exercise 9.1. Fill in the details of the convergence argument.

DEFINITION. The mean zero Gaussian process $\{U(\phi): \phi \in \underline{\underline{S}}(\mathbf{R}^d)\}$ with covariance (9.10) is called the Euclidean free field.

CONNECTION WITH SPDE's

We can get an integral representation of the free field U from Proposition 9.3, for the weak limit of $\sqrt{2}\,U_t$ has the same distribution as

$$\int_0^\infty \int_{\mathbf{R}^d} G_s(\phi,y)Z(dy\,ds).$$

This is not enlightening; we would prefer a representation independent of time. This is not hard to find. Let $\dot{W}$ be a d-dimensional white noise on $\mathbf{R}^d$ (not on $\mathbf{R}^d \times \mathbf{R}_+$ as before) and, for $\phi \in \underline{\underline{S}}(\mathbf{R}^d)$, define

$$U(\phi) = \int_{\mathbf{R}^d} \nabla K(\phi,y)\cdot W(dy). \tag{9.11}$$

If $\phi, \psi \in \underline{\underline{S}}(\mathbf{R}^d)$,

$$\begin{aligned} E\{U(\phi)U(\psi)\} &= \int_{\mathbf{R}^d} \nabla K(\phi,y)\cdot\nabla K(\psi,y)dy \\ &= -\int_{\mathbf{R}^d} K(\phi,y)\Delta K(\psi,y)dy \\ &= -\int_{\mathbf{R}^d} K(\phi,y)\psi(y)dy \\ &= -K(\phi,\psi). \end{aligned}$$

(This shows a posteriori that $U(\phi)$ is defined!) Thus, as $U(\phi)$ is a mean zero Gaussian process, it is a free field.

PROPOSITION 9.4. U satisfies the SPDE

$$\Delta U = \nabla\cdot\dot{W} \tag{9.12}$$

PROOF. $U(\Delta\phi) = \int_{\mathbf{R}} \nabla K(\Delta\phi,y)\cdot W(dy)$

$$= \int_{\mathbf{R}} \nabla\phi(y)\cdot W(dy)$$

ince for $\phi \in S(R^d)$, $K(\Delta\phi,y) = \phi(y)$. But this is the weak form of (9.12).

Q.E.D.

xercise 9.1. Convince yourself that for a.e.ω, (9.12) is an equation in istributions.

SMOOTHNESS

Since we are working on R^d, we can use the Fourier transform. Let H_t e the Sobolev space defined in Example 1a, Chapter 4. If u is any istribution, we say $u \in H_t^{loc}$ if for any $\phi \in C_0^\infty$, $\phi u \in H_t$.

ROPOSITION 9.5. Let $\varepsilon > 0$. Then with probability one, $\dot{W} \in H^{loc}_{-d/2-\varepsilon}$ and $\in H^{loc}_{1-d/2-\varepsilon}$, where U is the free field.

ROOF. The Fourier transform of ϕW is a function:

$$\widehat{W}(\xi) = \int_R e^{-2\pi i\xi\cdot x}\phi(x)W(dx)$$

nd

$$E\{|(1+|\xi|^2)^{t/2}\widehat{\phi W}(\xi)|^2\} = (1+|\xi|^2)^t \iint \phi(x)\bar{\phi}(y)e^{2\pi i(y-x)\cdot\xi}dydx,$$

o

$$E\{\|\phi W\|_t^2\} = \int(1+|\xi|^2)^t|\hat{\phi}(\xi)|^2d\xi$$
$$\leq C\int(1+|\xi|^2)^t d\xi$$

hich is finite if $2t < -d$, in which case $\|\phi W\|_t$ is evidently finite a.s.

Now $\nabla\cdot W \in H^{loc}_{-d/2-1-\varepsilon}$ so, since U satisfies $\Delta U = \nabla\cdot\dot{W}$, the elliptic egularity theorem of PDE's tells us $U \in H^{loc}_{1-\varepsilon-d/2}$. Q.E.D.

THE MARKOV PROPERTY OF THE FREE FIELD

We discussed Lévy's Markov and sharp Markov properties in Chapter One, n connection with the Brownian sheet. They make sense for general distribution-valued processes, but one must first define the σ-fields $\underline{\underline{G}}_D$ and $\underline{\underline{G}}^*_D$. This involves extending the distribution.

Since U takes values on a Sobolev space, the Sobolev embedding theorem tells us that it has a trace on certain lower-dimensional manifolds. But since we want to talk about its values on rather irregular sets, we will use a more direct method.

If μ is a measure on R^d, let us define $U(\mu)$ by (9.11). This certainly works if μ is of the form $\mu(dx) = \phi(x)dx$, and it will continue to work if μ is suffciently nice. By the calculation following (9.11), "sufficiently nice" means that

$$\|\mu\|_K^2 = -\int\int \mu(dx)K(x,y)\mu(dy) < \infty. \tag{9.13}$$

Let $\underline{\underline{E}}_+$ be the class of measures on R^d which satisfy (9.13), and let $\underline{\underline{E}} = \underline{\underline{E}}_+ - \underline{\underline{E}}_+$.

If $B \subset R^d$ is Borel, one version of the restriction of U to B would be

$$\{U(\mu)\colon \mu \in \underline{\underline{E}},\ \ U(A) > 0, \text{ all } A \subset R^d - B\}.$$

Of course, this requires that there be measures in $\underline{\underline{E}}$ which sit on B. This will always be true if B has positive capacity, for if B is bounded and has positive capacity, its equilibrium measure has a bounded potential and is thus in E. (For the potential of μ is $K(\mu,\cdot)$ and $\|\mu\|_K^2 = \int K(\mu,y)\mu(dy)$.

Thus, define

$$\underline{\underline{G}}_B = \sigma\{U(\mu)\colon \mu \in \underline{\underline{E}},\ \mu(A) = 0, \text{ all } A \subset R^d - B\}$$

$$\underline{\underline{G}}^*_B = \bigcap_{\substack{A \supset B \\ A \text{ open}}} \underline{\underline{G}}_A .$$

<u>PROPOSITION 9.6</u>. The free field U satisfies Lévy's sharp Markov property relative to bounded open sets in R^d.

<u>PROOF</u>. This follows easily from the balayage property of K: if $D \subset R^d$ is an open set and if μ is supported by D^c, there exists a measure ν on ∂D such $-K(\nu,y) \le -K(\mu,y)$ for all y, and $K(\nu,y) = K(\mu,y)$ for all $y \in D$, and all but a set of capacity zero in ∂D. We call ν the <u>balayage of μ on ∂D</u>.

Suppose $\mu \in \underline{\underline{E}}$ and supp $\mu \subset D^c$. If ν is the balayage of μ on ∂D, we laim that

(9.14) $$E\{U(\mu)|\underline{\underline{G}}_{\bar{D}}\} = U(\nu).$$

his will do it since, as $U(\nu)$ is $\underline{\underline{G}}_{\partial D}$-measurable, the left-hand side of (9.14) must be $E\{U(\mu)|\underline{\underline{G}}_{\partial D}\}$.

Note that $\nu \in \underline{\underline{E}}$ (for μ is and $-K(\nu,\cdot) \leq -K(\mu,\cdot)$) so if $\lambda \in \underline{\underline{E}}$, upp$(\lambda) \subset \bar{D}$,

$$\begin{aligned} E\{(U(\mu) - U(\nu))U(\lambda)\} &= \int [K(\mu,y) - K(\nu,y)]\lambda(dy) \\ &= 0 \end{aligned}$$

ince $K(\mu,x) = K(\nu,x)$ on $\bar{D}$, except possibly for a set of capacity zero, and , being of finite energy, does not charge sets of capacity zero. Thus the ntegrand vanishes λ-a.e. But we are dealing with Gaussian processes, so his implies (9.14). Q.E.D.

NOTES

We omitted most references from the body of the text - a consequence of putting off the bibliography till last - and we will try to remedy that here. Our references will be rather sketchy - you may put that down to a lack of scholarship - and we list the sources from which we personally have learned things, which may not be the sources in which they originally appeared. We apologize in advance to the many whose work we have slighted in this way.

CHAPTER ONE

The Brownian sheet was introduced by Kitagawa in [37], though it is usually credited to others, perhaps because he failed to prove the underlying measure was countably additive. This omission looks less serious now than it did then.

The Garsia-Rodemich-Rumsey Theorem (Theorem 1.1) was proved for one-parameter processes in [23], and was proved in general in the brief and elegant article [22], which is the source of this proof. This commonly gives the right order of magnitude for the modulus of continuity of a process, but doesn't necessarily give the best constant, as, for example, in Proposition 1.4. The exact modulus of continuity there, as well as many other interesting sample-path properties of the Brownian sheet, may be found in Orey and Pruitt [49].

Kolmogorov's Theorem is usually stated more simply than in Corollary 1.2. In particular, the extra log terms there are a bit of an affectation. We just were curious to see how far one can go with non-Gaussian processes. Our version is only valid for real-valued processes, but the theorem holds for metric-space valued processes. See for example [44, p.519].

The Markov property of the Brownian sheet was proved by L. Pitt [52]. The splitting field is identified in [59]; the proof there is due to S. Orey (private communication.)

The propagation of singularities in the Brownian sheet is studied in detail in [56]. Orey and Taylor showed the existence of singular points of the Brownian path and determined their Hausdorff dimension in [50]. Proposition 1.7 is due to G. Zimmerman [63], with a quite different proof.

The connection of the vibrating string and the Brownian sheet is due to E. Cabaña [8], who worked it out in the case of a finite string, which is harder than the infinite string we treat. He also discusses the energy of the string.

CHAPTER TWO

In terms of the mathematical techniques involved, one can split up much of the study of SPDE's into two parts: that in which the underlying noise has nuclear covariance, and that in which it is a white noise. The former leads naturally to Hilbert space methods; these don't suffice to handle white noise, which leads to some fairly exotic functional analysis. This chapter is an attempt to combine the two in a (nearly) real variable setting. The integral constructed here may be technically new, but all the important cases can also be handled by previous integrals.

(We should explain that we did not have time or space in these notes to cover SPDE's driven by martingale measures with nuclear covariance, so that we never take advantage of the integral's full generality).

Integration with respect to orthogonal martingale measures, which include white noise, goes back at least to Gihman and Skorohod [25]. (They assumed as part of their definition that the measures are worthy, but this assumption is unnecessary; c.f. Corollary 2.9.)

Integrals with respect to martingale measures having nuclear covariance have been well-studied, though not in those terms. An excellent account can be found in Métivier and Pellaumeil [46]. They handle the case of "cylindrical processes", (which include white noise) separately.

The measure ν of Corollary 2.8 is a Doléans measure at heart, although we haven't put it in the usual form. True Doléans measures for such processes have been constructed by Huang, [31].

Proposition 2.10 is due to J. Watkins [61]. Bakry's example can be found in [2].

CHAPTER THREE

The linear wave and cable equations driven by white and colored noise have been treated numerous times. Dawson [13] gives an account of these and similar equations.

The existence and uniqueness of the solution of (3.5) were established by Dawson [14]. The L^p-boundednes and Hölder continuity of the paths are new. See [57] for a detailed account of the sample path behavior in the linear case and for more on the barrier problem.

The wave equation has been treated in the literature of two-parameter processes, going back to R. Cairoli's 1972 article [9]. The setting there is special because of the nature of the domain: on these domains, only the initial position need be specified, not the velocity.

As indicated in Exercises 3.4 and 3.5, one can extend Theorem 3.2 and Corollary 3.4, with virtually the same proof, to the equation

$$\frac{\partial V}{\partial t} = \frac{\partial^2 V}{\partial x^2} + g(V,t) + f(V,t)\dot{W},$$

where both f and g satisfy Lipschitz conditions. Such equations can model physical systems in which g is potential term. Faris and Jona-Lasinio [19] have used similar equations to model the "tunnelling" of a system from one stable state to another.

We chose reflecting boundary conditions in (3.5) and (3.5b) for convenience. They can be replaced by general linear homogeneous boundary conditions; the important point is that the Green's funciton satisfies (3.6) and (3.7), which hold in general [27].

CHAPTER FOUR

We follow some unpublished lecture notes of the Ito here. See also [24] nd [34].

CHAPTER FIVE

The techniques used to solve (5.1) also work when L is a higher order lliptic operator. In fact the Green's function for higher order operators has a ower order pole, so that the solutions are better behaved than in the second-order ase.

We suspect that Theorem 5.1 goes back to the mists of antiquity. Ito tudies a special case in [33]. Theorem 5.4 and other results on the sample paths of he solution can be found in [58]. See Da Prato [12] for another point of view on hese and similar theorems.

CHAPTER SIX

The basic reference on weak convergence remains Billingsley's book [5]. ldous' theorem is in [1], and Kurtz' criterion is in [42]. We follow Kurtz' reatment here. Mitoma's theorem is proved in [47], but the article is not elf-contained. Fouque [21] has generalized this to a larger class of spaces of istributions, which includes the familiar spaces $D(\Omega)$. His proof is close to that f Mitoma.

CHAPTER SEVEN

It may not be obvious from the exposition - in fact we took care to hide it but the first part of the chapter is designed to handle deterministic integrands. he accounts for its relatively elementary character.

Theorems general enough to handle the random integrands met in practice seem to be delicate; we were surprised to find out how little is known, even in the classical case. Our work in the section "an extension" is just a first attempt in that direction.

Peter Kotelenez showed us the proof of Proposition 7.8. Theorem 7.10 is due to Kallianpur and Wolpert [36]. An earlier, clumsier version can be found in [57]. The Burkholder-Davis-Gundy theorem is summarized in its most highly developed form in [7].

CHAPTER EIGHT

This chapter completes a cycle of results on weak limits of Poisson systems of branching Brownian motion due to a number of authors. "Completes" is perhaps too strong a word, for these point in many directions and we have only followed one: to find all possible weak limits of a certain class of infinite particle systems, and to connect them with SPDE's.

These systems were investigated by Martin-Löf [45] who considered non-branching particles ($\mu = 0$ in our terminology) and by Holley and Stroock [29], who considered branching Brownian motions in R^d with parameters $\lambda = \mu = 1$; their results look superficially different since, instead of letting μ and λ tend to infinity, they rescale the process in both space and time by replacing x by x/α and t by $\alpha^2 t$. Because of the Brownian scaling, this has the same effect as replacing λ by α^d and μ by α^2, and leaving x and t unscaled. The critical parameter is then $\mu/\lambda = \alpha^{2-d}$, so their results depend on the dimension d of the space. If $d \geq 3$, they find a Gaussian limit (case (ii) of Theorem 8.9), if d = 2 they have the measure-valued diffusion (case (iv)) and if d = 1, the process tends to zero (Theorem 8.11). The case $\mu = 0$, investigated by Martin-Löf and, with some differences, by Ito [33], [34], also leads to a Gaussian limit (Theorem 8.9 (i)).

Gorostitza [26] treated the case where μ is fixed and $\lambda \to \infty$ (Theorem 8.9(iii) if $\mu > 0$). He also gets a decomposition of the noise into two parts, but it is different from ours; he has pointed out [26, Correction] that the two parts are not in fact independent.

The non-Gaussian case (case (iv)) is extremely interesting and has been nvestigated by numerous authors. S. Watanabe [60] proved the convergence of the ystem to a measure-valued diffusion. Different proofs have been given by Dawson 13], Kurtz [42], and Roelly-Copoletta [53]. Dawson and Hochberg [15] have looked at he Hausdorff dimension of the support of the measure and showed it is singular with espect to Lebesgue measure if $d \geq 2$. It is absolutely continuous if $d = 1$ Roelly-Copoletta [53]). A related equation which can be written suggestively as

$$\frac{\partial \eta}{\partial t} = \frac{1}{2} \Delta\eta + \eta(1-\eta)\dot{W}$$

as been studied by Fleming and Viot [20].

The case in which $\mu/\lambda \to \infty$ comes up in Holley and Stroock's paper if $d = 1$. he results presented here, which are stronger, are joint work with E. Perkins and J. atkins, and appear here with their permission. The noise W of Proposition 8.1 is ue to E. Perkins who used it to translate Ito's work into the setting of SPDE's elative to martingale measures (private communication.)

A more general and more sophisticated construction of branching diffusions an be found in Ikeda, Nagasawa, and Watanabe [32]. Holley and Stroock also give a onstruction.

The square process Q is connected with U statistics. Dynkin and Mandelbaum 17] showed that certain central limit theorems involving U statistics lead to ultiple Wiener integrals, and we wish to thank Dynkin for suggesting that our ethods might handle the case when the particles were diffusing in time. In fact heorem 8.18 might be viewed as a central limit theorem for certain U-statistics volving in time.

We should say a word about generalizations here. We have treated only the implest settings for the sake of clarity, but there is surprisingly little hange if we move to more complex systems. We can replace the Brownian particles by ranching diffusions, or even branching Hunt processes, for instance, without hanging the character of the limiting process. (Roelly-Copoletta [Thesis, U. of aris, 1984]). One can treat more general branching schemes. If the family size N has a finite variance, Gorostitza [26] has shown that one gets limiting equations of the form $\frac{\partial\eta}{\partial t} = \frac{1}{2}\Delta\eta + \alpha\eta + \beta Z + \gamma\nabla\cdot W$, where $\alpha = 0$ if $E\{N - 1\} = 0$, so that the only new effect is to add a growth term, $\alpha\eta$.

If $E\{N^2\} = \infty$, however, things do change. For example, η can tend to zero in certain cases when μ/λ has a finite limit. This needs further study.

CHAPTER NINE

The term "random field" is a portmanteau word. At one time or another, it has been used to cover almost any process having more than one parameter - and some one-parameter processes, too. It seems to be used particularly for elliptic systems, though why it should be used more often for elliptic than parabolic or hyperbolic systems is something of a mystery. (As is the term itself, for that matter). At any rate, this chapter is about random fields.

We have used some heavy technical machinery here. Frankly, we were under deadline pressure and didn't have time to work out an easier approach. For Sobolev spaces, see Adams [64]; for the PDE theorems, see Folland [67] and Hormander [30]. The classical potential theory and the energy of measures can be found in Doob [66].

The exponent n of the Sobolev space in Proposition 7.1 can doubtless be improved. If $\dot{M}$ is a white noise, one can bypass the Sobolev embedding in the proof and get $n > k + d/2$ rather than $n > k + d$.

The free field was introduced by Nelson in [48]. He proved the sharp Markov property, and used it to construct the quantum field which describes non-interacting particles. He also showed that it can be modified to describe certain interacting systems.

Rozanov's book [54] is a good reference for Lévy's Markov property. See Evstigneev for a strong Markov property, and Kusuoka [43] for results which also apply to parabolic systems in which, contrary to the claim in [57], one commonly finds that Lévy's Markov property holds but the sharp Markov property does not.

CHAPTER TEN

There is no Chapter Ten in these notes. For some reason that hasn't stopped us from having notes on Chapter Ten. We will use this space to collect some remarks which didn't fit in elsewhere. Since the chapter under discussion doesn't exist, no one can accuse us of digressing.

We did not have a chance to discuss equations relative to martingale measures with a nuclear covariance. These can arise when the underlying noise is smoother than a white noise or, as often happens, it is white noise which one has approximated by a smoothed out version. If one thinks of a white noise, as we did in the introduction, as coming from storm-driven grains of sand bombarding a guitar string, one might think of nuclear covariance noise as coming from a storm of ping-pong balls. The solutions of such systems tend to be better-behaved, and in particular, they often give function solutions rather than distributions. This makes it possible to treat non-linear equations, something rather awkward to do otherwise (how does one take a non-linear function of a distribution?) Mathematically, these equations are usually treated in a Hilbert-space setting. See for instance Curtain and Falb [11], Da Prato [12], and Ichikawa [68].

There have been a variety of approaches devised to cope with SPDE's driven by white noise and related processes. See Kuo [41] and Dawson [13] for a treatment based on the theory of abstract Wiener spaces. The latter paper reviews the subject of SPDE's up to 1975 and has extensive references. Balakrishnan [3] and Kallianpur and Karandikar [35] have used cylindrical Brownian motions and finitely additive measures. See also Métivier and Pellaumail [46], which gives an account of the integration theory of cylindrical processes. Gihman and Skorohod [25] introduced orthogonal martingale measures. See also Watkins [61]. Ustunel [55] has studied nuclear space valued semi-martingales with applications to SPDE's and stochastic flows. The martingale problem method can be adapted to SPDE's as well as ordinary SDE's. It has had succes in handling non-linear equations intractable to other methods. See Dawson [65] and Fleming and Viot [20], and Holley and Stroock [29] for the linear case.

Another type of equation which has generated considerable research is the SPDE driven by a single one-parameter Brownian motion. (One could get such an equation from (5.1) by letting T be an integral operator rather than a differential operator.) An example of this is the Zakai equation which arises in filtering theory. See Pardoux [51] and Krylov and Rosovski [39].

Let us finish by mentioning a few more subjects which might interest the reader: fluid flow and the stochastic Navier-Stokes equation (e.g. Bensoussan and Temam [4]); measure-valued diffusions and their application to population growth (Dawson [65], Fleming and Viot [20]); reaction diffusion equations in chemistry (Kotelenz [38]) and quantum fields (Wolpert [70] and Dynkin [16]).

REFERENCES

[1] Aldous, D., Stopping times and tightness, Ann. Prob. 6 (1978), 335-340.

[2] Bakry, D., Semi martingales à deux indices, Sem. de Prob. XV, Lecture Notes in Math 850, 671-672.

[3] Balakrishnan, A. V., Stochastic bilinear partial differential equations, in Variable Structure Systems, Lecture Notes in Economics and Mathematical Systems 3, Springer Verlag, 1975.

[4] Bensoussan, A. and Temam, R., Equations stochastiques du type Navier-Stokes, J. Fcl. Anal. 13 (1973), 195-222.

[5] Billingsley, P., Convergence of Probability Measures, Wiley, New York, 1968.

[6] Brennan, M. D., Planar semimartingales, J. Mult. Anal. 9 (1979), 465-486.

[7] Burkholder, D. L., Distribution function inequalities for martingales, Ann. Prob. 1 (1973), 19-42.

[8] Cabaña, E., On barrier problems for the vibrating string, ZW 22 (1972), 13-24.

[9] Cairoli, R., Sur une equation differentielle stochastique, C.R. 274 (1972), 1738-1742.

[10] Cairoli, R. and Walsh, J. B., Stochastic integrals in the plane, Acta Math 134 (1975), 111-183.

[11] Curtain, R. F. and Falb, P. L., Stochastic differential equations in Hilbert spaces, J. Diff. Eq. 10 (1971), 434-448.

[12] Da Prato, G., Regularity results of a convolution stochastic integral and applications to parabolic stochastic equations in a Hilbert space (Preprint).

[13] Dawson, D., Stochastic evolution equations and related measure processes, J. Mult. Anal. 5 (1975), 1-52.

[14] Dawson, D., Stochastic evolution equations, Math Biosciences 15, 287-316.

[15] Dawson, D. and Hochberg, K. J., The carrying dimension of a stochastic measure diffusion. Ann. Prob. 7 (1979).

[16] Dynkin, E. B., Gaussian and non-Gaussian random fields associated with Markov processes, J. Fcl. Anal. 55 (1984), 344-376.

[17] Dynkin, E. B. and Mandelbaum A., Symmetric statistics, Poisson point processes, and multiple Wiener integrals, Ann. Math. Stat 11 (1983), 739-745.

[18] Evstigneev, I. V., Markov times for random fields, Theor. Prob. Appl. 22 (1978), 563-569.

[19] Faris, W. G., and Jona-Lasinio, G., Large fluctuations for a nonlinear heat equation with white noise, J. Phys. A: Math, Gen. 15 (1982), 3025-3055.

[20] Fleming, W. and Viot, M., Some measure-valued Markov processes in population genetics theory, Indiana Univ. Journal 28 (1979), 817-843.

[21] Fouque, J-P., La convergence en loi pour les processus à valeurs dans un éspace nucleaire, Ann. IHP 20 (1984), 225-245.

[22] Garsia, A., Continuity properties of Gaussian processes with multidimensional time parameter, Proc. 6th Berkeley Symposium, V.II, 369-374.

[23] Garsia, A., Rodemich, G., and Rumsey, H. Jr., A real variable lemma and the continuity of paths of some Gaussian processes, Indiana U. Math. J. 20 (1970), 565-578.

[24] Gelfand, I. M. and Vilenkin, N. Ya., Generalized Functions, V.4, Academic Press, New York-London 1964.

[25] Gihman, I. I., and Skorohod, A. V., The Theory of Stochastic Processes, III, Springer-Verlag Berlin (1979).

[26] Gorostitza, L., High density limit theorems for infinite systems of unscaled branching Brownian motions, Ann. Prob. 11 (1983), 374-392; Correction, Ann. Prob. 12 (1984), 926-927.

[27] Greiner, P., An asymptotic expansion for the heat equation, Arch. Ratl. Mech. Anal. 41 (1971), 163-218.

[28] Harris, T. E., The Theory of Branching Processes, Prentice-Hall, Englewood Cliffs, N.J., 1963.

[29] Holley, R. and Stoock, D., Generalized Ornstein-Uhlenbeck processes and infinite particle branching Brownian motions, Publ. RIMS Kyoto Univ. 14 (1978), 741-788.

[30] Hormander, L. Linear Partial Differential Operators, Springer Verlag, Berlin, Heidelberg, New York, 1963.

31] Huang, Zhiyuan, Stochastic integrals on general topological measurable spaces, Z.W. 66 (1984), 25-40.

32] Ikeda, N., Nagasawa, M., and Watanabe, S., Branching Markov processes, I, II and III, J. Math. Kyoto. Univ. 8 (1968) 233-278, 365-410; 9 (1969), 95-110.

33] Ito, K., Stochastic anaysis in infinite dimensions; in Stochastic Analysis (A. Friedman and M. Pinsky, eds.) Academic Press, New York, 1980, 187-197.

34] Ito, K., Distribution-valued processes arising from independent Brownian motions. Math Z. 182 (1983) 17-33.

35] Kallianpur, G. and Karandikar, R. L., A finitely additive white noise approach to non-linear filtering, J. Appl. Math and Opt. (to appear).

36] Kallianpur, G. and Wolpert, R., Infinite dimensional stochastic differential equation models for spatially distributed neurons.

37] Kitagawa, T., Analysis of variance applied to function spaces, Mem. Fac. Sci. Kyusu Univ. 6 (1951), 41-53.

38] Kotelenz, P., A stopped Doob inequality for stochastic convolution integrals and stochastic evolution equations; (preprint).

39] Krylov, N. V. and Rosovski, B. L. Stochastic evolution equations, J. Soviet Math. (1981).

40] Kuo, H. H., Gaussian measures in Banach spaces. Lecture notes in Math. 463, Springer Verlag 1975.

41] Kuo, H. H., Differential and stochastic equations in abstract Wiener space. J. Fcl. Anal. 12 (1973), 246-256.

42] Kurtz, T., Approximation of Population Processes CBMS-NSF Reg. Conf. Ser. in Appl. Math. 36, 1981.

43] Kusuoka, S., Markov fields and local operators, J. Fac. Sci., U of Tokyo 26, (1979), 199-212.

44] Loève, Probability Theory, Van Nostrand, Princeton, 1963.

45] Martin-Löf, A., Limit theorems for motion of a Poisson system of independent Markovian particles with high density, Z.W. 34 (1976), 205-223.

46] Métivier, M. and Pellaumail, J., Stochastic Integration. Academic Press, New York, 1980.

[47] Mitoma, I., Tightness of Probabilities on C([0,1], S') and D([0,1], S'), Ann. Prob. 11 (1983), 989-999.

[48] Nelson, E., The free Markoff field, J. Fcl. Anal 12 (1973), 211-227.

[49] Orey, S. and Pruitt, W., Sample functions of the N-parameter Wiener process, Ann. Prob. 1 (1973), 138-163.

[50] Orey, S. and Taylor, S. J., How often on the Brownian path does the law of iterated logarithm fail? Proc. Lond. Math. Soc. 28 (1974), 174-192.

[51] Pardoux, E., Stochastic PDE and filtering of diffusion processes, Stochastics 3 (1979) 127-167.

[52] Pitt, L. D., A Markov property for Gaussian processes with a multidimensional parameter. Arch. Rat. Mech. Anal 43 (1971), 367-391.

[53] Roelley-Coppoletta, S., Un critère de convergence pour les lois de processus à valeurs mesures. (1984) (preprint).

[54] Rozanov, Yu. A., Markov Random Fields, Springer-Verlag, New York, Heidelberg, Berlin, 1982.

[55] Ustunel, S., Stochastic integration on nuclear spaces and its applications. Ann. Inst. H. Poincaré 28 (1982), 165-200.

[56] Walsh, J. B., Propagation of singularities in the Brownian sheet, Ann. Prob. 10 (1982), 279-288.

[57] Walsh, J. B., A stochastic model of neural response, Adv. Appl. Prob. 13 (1981), 231-281.

[58] Walsh, J. B., Regularity properties of a stochastic partial differential equation, Seminar on Stochastic Processes, 1983, Birkhauser, Boston, (1984), 257-290.

[59] Walsh, J. B., Martingales with a multidimensional parameter and stochastic integrals in the plane. To appear, Springer Lecture Notes in Math.

[60] Watanabe, S., A limit theorem of branching processes and continuous state branching processes, J. Math. Kuoto. Univ. 8 (1968), 141-167.

[61] Watkins, J., A stochastic integral representation for random evolutions. To appear, Ann. Prob.

[62] Zakai, M., On the optimal filtering of diffusion processes, Z.W. 11 (1969), 230-243.

[63] Zimmerman, G. J., Some sample function properties of the two parameter Gaussian process, Ann. Math. Stat. 43 (1972), 1235-1246.

[64] Adams, R. A., Sobolev Spaces, Academic Press, New York, (1975).

[65] Dawson, D. A., Geostochastic calculus, Can. J. Stat. 6 (1978), 143-168.

[66] Doob, J. L., Classical Potential Theory and its Probabilistic Counterpart, Springer-Verlag, New York, Berlin, Heidelberg, Tokyo, 1984.

[67] Folland, G. B., Introduction to Partial Differential Equations, Mathematical Notes 17, Princeton Univ. Press, 1976.

[68] Ichikawa, A., Linear stochastic evolution equations in Hilbert spaces, J. Diff. Eq. 28 (1978), 266-283.

[69] Kotelenz, P., Thesis, U. of Bremen, 1982.

[70] Wolpert, R. L., Local time and a particle picture for Euclidean field theory, J. Fcl. Anal. 30 (1978), 341-357.

INDEX

l. 1034: J. Musielak, Orlicz Spaces and Modular Spaces. V, 2 pages. 1983.

l. 1035: The Mathematics and Physics of Disordered Media. oceedings, 1983. Edited by B.D. Hughes and B.W. Ninham. , 432 pages. 1983.

l. 1036: Combinatorial Mathematics X. Proceedings, 1982. Edited L.R.A. Casse. XI, 419 pages. 1983.

l. 1037: Non-linear Partial Differential Operators and Quantization ocedures. Proceedings, 1981. Edited by S.I. Andersson and -D. Doebner. VII, 334 pages. 1983.

l. 1038: F. Borceux, G. Van den Bossche, Algebra in a Localic pos with Applications to Ring Theory. IX, 240 pages. 1983.

l. 1039: Analytic Functions, Błażejewko 1982. Proceedings. Edited J. Ławrynowicz. X, 494 pages. 1983

l. 1040: A. Good, Local Analysis of Selberg's Trace Formula. III, 8 pages. 1983.

l. 1041: Lie Group Representations II. Proceedings 1982–1983. ited by R. Herb, S. Kudla, R. Lipsman and J. Rosenberg. IX, 0 pages. 1984.

l. 1042: A. Gut, K. D. Schmidt, Amarts and Set Function Processes. 258 pages. 1983.

l. 1043: Linear and Complex Analysis Problem Book. Edited by Havin, S.V. Hruščëv and N.K. Nikol'skii. XVIII, 721 pages. 1984.

. 1044: E. Gekeler, Discretization Methods for Stable Initial Value oblems. VIII, 201 pages. 1984.

l. 1045: Differential Geometry. Proceedings, 1982. Edited by M. Naveira. VIII, 194 pages. 1984.

l. 1046: Algebraic K-Theory, Number Theory, Geometry and Analy-. Proceedings, 1982. Edited by A. Bak. IX, 464 pages. 1984.

l. 1047: Fluid Dynamics. Seminar, 1982. Edited by H. Beirão da iga. VII, 193 pages. 1984.

l. 1048: Kinetic Theories and the Boltzmann Equation. Seminar, 81. Edited by C. Cercignani. VII, 248 pages. 1984.

l. 1049: B. Iochum, Cônes autopolaires et algèbres de Jordan. 247 pages. 1984.

l. 1050: A. Prestel, P. Roquette, Formally p-adic Fields. V, 167 pages. 84.

l. 1051: Algebraic Topology, Aarhus 1982. Proceedings. Edited by Madsen and B. Oliver. X, 665 pages. 1984.

l. 1052: Number Theory, New York 1982. Seminar. Edited by D.V. udnovsky, G.V. Chudnovsky, H. Cohn and M.B. Nathanson. V, 309 ges. 1984.

l. 1053: P. Hilton, Nilpotente Gruppen und nilpotente Räume. 221 pages. 1984.

l. 1054: V. Thomée, Galerkin Finite Element Methods for Parabolic oblems. VII, 237 pages. 1984.

l. 1055: Quantum Probability and Applications to the Quantum eory of Irreversible Processes. Proceedings, 1982. Edited by Accardi, A. Frigerio and V. Gorini. VI, 411 pages. 1984.

l. 1056: Algebraic Geometry. Bucharest 1982. Proceedings, 1982. ited by L. Bădescu and D. Popescu. VII, 380 pages. 1984.

. 1057: Bifurcation Theory and Applications. Seminar, 1983. Edited L. Salvadori. VII, 233 pages. 1984.

l. 1058: B. Aulbach, Continuous and Discrete Dynamics near anifolds of Equilibria. IX, 142 pages. 1984.

l. 1059: Séminaire de Probabilités XVIII, 1982/83. Proceedings. ité par J. Azéma et M. Yor. IV, 518 pages. 1984.

l. 1060: Topology. Proceedings, 1982. Edited by L. D. Faddeev and A. Mal'cev. VI, 389 pages. 1984.

l. 1061: Séminaire de Théorie du Potentiel. Paris, No. 7. Pro-edings. Directeurs: M. Brelot, G. Choquet et J. Deny. Rédacteurs: Hirsch et G. Mokobodzki. IV, 281 pages. 1984.

Vol. 1062: J. Jost, Harmonic Maps Between Surfaces. X, 133 pages. 1984.

Vol. 1063: Orienting Polymers. Proceedings, 1983. Edited by J.L. Ericksen. VII, 166 pages. 1984.

Vol. 1064: Probability Measures on Groups VII. Proceedings, 1983. Edited by H. Heyer. X, 588 pages. 1984.

Vol. 1065: A. Cuyt, Padé Approximants for Operators: Theory and Applications. IX, 138 pages. 1984.

Vol. 1066: Numerical Analysis. Proceedings, 1983. Edited by D.F. Griffiths. XI, 275 pages. 1984.

Vol. 1067: Yasuo Okuyama, Absolute Summability of Fourier Series and Orthogonal Series. VI, 118 pages. 1984.

Vol. 1068: Number Theory, Noordwijkerhout 1983. Proceedings. Edited by H. Jager. V, 296 pages. 1984.

Vol. 1069: M. Kreck, Bordism of Diffeomorphisms and Related Topics. III, 144 pages. 1984.

Vol. 1070: Interpolation Spaces and Allied Topics in Analysis. Proceedings, 1983. Edited by M. Cwikel and J. Peetre. III, 239 pages. 1984.

Vol. 1071: Padé Approximation and its Applications, Bad Honnef 1983. Prodeedings. Edited by H. Werner and H.J. Bünger. VI, 264 pages. 1984.

Vol. 1072: F. Rothe, Global Solutions of Reaction-Diffusion Systems. V, 216 pages. 1984.

Vol. 1073: Graph Theory, Singapore 1983. Proceedings. Edited by K. M. Koh and H. P. Yap. XIII, 335 pages. 1984.

Vol. 1074: E. W. Stredulinsky, Weighted Inequalities and Degenerate Elliptic Partial Differential Equations. III, 143 pages. 1984.

Vol. 1075: H. Majima, Asymptotic Analysis for Integrable Connections with Irregular Singular Points. IX, 159 pages. 1984.

Vol. 1076: Infinite-Dimensional Systems. Proceedings, 1983. Edited by F. Kappel and W. Schappacher. VII, 278 pages. 1984.

Vol. 1077: Lie Group Representations III. Proceedings, 1982–1983. Edited by R. Herb, R. Johnson, R. Lipsman, J. Rosenberg. XI, 454 pages. 1984.

Vol. 1078: A.J.E.M. Janssen, P. van der Steen, Integration Theory. V, 224 pages. 1984.

Vol. 1079: W. Ruppert. Compact Semitopological Semigroups: An Intrinsic Theory. V, 260 pages. 1984

Vol. 1080: Probability Theory on Vector Spaces III. Proceedings, 1983. Edited by D. Szynal and A. Weron. V, 373 pages. 1984.

Vol. 1081: D. Benson, Modular Representation Theory: New Trends and Methods. XI, 231 pages. 1984.

Vol. 1082: C.-G. Schmidt, Arithmetik Abelscher Varietäten mit komplexer Multiplikation. X, 96 Seiten. 1984.

Vol. 1083: D. Bump, Automorphic Forms on GL (3,IR). XI, 184 pages. 1984.

Vol. 1084: D. Kletzing, Structure and Representations of Q-Groups. VI, 290 pages. 1984.

Vol. 1085: G.K. Immink, Asymptotics of Analytic Difference Equations. V, 134 pages. 1984.

Vol. 1086: Sensitivity of Functionals with Applications to Engineering Sciences. Proceedings, 1983. Edited by V. Komkov. V, 130 pages. 1984

Vol. 1087: W. Narkiewicz, Uniform Distribution of Sequences of Integers in Residue Classes. VIII, 125 pages. 1984.

Vol. 1088: A.V. Kakosyan, L.B. Klebanov, J.A. Melamed, Characterization of Distributions by the Method of Intensively Monotone Operators. X, 175 pages. 1984.

Vol. 1089: Measure Theory, Oberwolfach 1983. Proceedings. Edited by D. Kölzow and D. Maharam-Stone. XIII, 327 pages. 1984.

Vol. 1090: Differential Geometry of Submanifolds. Proceedings, 1984. Edited by K. Kenmotsu. VI, 132 pages. 1984.

Vol. 1091: Multifunctions and Integrands. Proceedings, 1983. Edited by G. Salinetti. V, 234 pages. 1984.

Vol. 1092: Complete Intersections. Seminar, 1983. Edited by S. Greco and R. Strano. VII, 299 pages. 1984.

Vol. 1093: A. Prestel, Lectures on Formally Real Fields. XI, 125 pages. 1984.

Vol. 1094: Analyse Complexe. Proceedings, 1983. Edité par E. Amar, R. Gay et Nguyen Thanh Van. IX, 184 pages. 1984.

Vol. 1095: Stochastic Analysis and Applications. Proceedings, 1983. Edited by A. Truman and D. Williams. V, 199 pages. 1984.

Vol. 1096: Théorie du Potentiel. Proceedings, 1983. Edité par G. Mokobodzki et D. Pinchon. IX, 601 pages. 1984.

Vol. 1097: R.M. Dudley, H. Kunita, F. Ledrappier, École d'Éte de Probabilités de Saint-Flour XII – 1982. Edité par P.L. Hennequin. X, 396 pages. 1984.

Vol. 1098: Groups – Korea 1983. Proceedings. Edited by A.C. Kim and B.H. Neumann. VII, 183 pages. 1984.

Vol. 1099: C.M. Ringel, Tame Algebras and Integral Quadratic Forms. XIII, 376 pages. 1984.

Vol. 1100: V. Ivrii, Precise Spectral Asymptotics for Elliptic Operators Acting in Fiberings over Manifolds with Boundary. V, 237 pages. 1984.

Vol. 1101: V. Cossart, J. Giraud, U. Orbanz, Resolution of Surface Singularities. Seminar. VII, 132 pages. 1984.

Vol. 1102: A. Verona, Stratified Mappings – Structure and Triangulability. IX, 160 pages. 1984.

Vol. 1103: Models and Sets. Proceedings, Logic Colloquium, 1983, Part I. Edited by G.H. Müller and M.M. Richter. VIII, 484 pages. 1984.

Vol. 1104: Computation and Proof Theory. Proceedings, Logic Colloquium, 1983, Part II. Edited by M.M. Richter, E. Börger, W. Oberschelp, B. Schinzel and W. Thomas. VIII, 475 pages. 1984.

Vol. 1105: Rational Approximation and Interpolation. Proceedings, 1983. Edited by P.R. Graves-Morris, E.B. Saff and R.S. Varga. XII, 528 pages. 1984.

Vol. 1106: C.T. Chong, Techniques of Admissible Recursion Theory. IX, 214 pages. 1984.

Vol. 1107: Nonlinear Analysis and Optimization. Proceedings, 1982. Edited by C. Vinti. V, 224 pages. 1984.

Vol. 1108: Global Analysis – Studies and Applications I. Edited by Yu.G. Borisovich and Yu.E. Gliklikh. V, 301 pages. 1984.

Vol. 1109: Stochastic Aspects of Classical and Quantum Systems. Proceedings, 1983. Edited by S. Albeverio, P. Combe and M. Sirugue-Collin. IX, 227 pages. 1985.

Vol. 1110: R. Jajte, Strong Limit Theorems in Non-Commutative Probability. VI, 152 pages. 1985.

Vol. 1111: Arbeitstagung Bonn 1984. Proceedings. Edited by F. Hirzebruch, J. Schwermer and S. Suter. V, 481 pages. 1985.

Vol. 1112: Products of Conjugacy Classes in Groups. Edited by Z. Arad and M. Herzog. V, 244 pages. 1985.

Vol. 1113: P. Antosik, C. Swartz, Matrix Methods in Analysis. IV, 114 pages. 1985.

Vol. 1114: Zahlentheoretische Analysis. Seminar. Herausgegeben von E. Hlawka. V, 157 Seiten. 1985.

Vol. 1115: J. Moulin Ollagnier, Ergodic Theory and Statistical Mechanics. VI, 147 pages. 1985.

Vol. 1116: S. Stolz, Hochzusammenhängende Mannigfaltigkeiten und ihre Ränder. XXIII, 134 Seiten. 1985.

Vol. 1117: D.J. Aldous, J.A. Ibragimov, J. Jacod, Ecole d'Été d Probabilités de Saint-Flour XIII – 1983. Édité par P.L. Hennequi IX, 409 pages. 1985.

Vol. 1118: Grossissements de filtrations: exemples et application Seminaire, 1982/83. Edité par Th. Jeulin et M. Yor. V, 315 pages. 198

Vol. 1119: Recent Mathematical Methods in Dynamic Programmin Proceedings, 1984. Edited by I. Capuzzo Dolcetta, W.H. Flemin and T. Zolezzi. VI, 202 pages. 1985.

Vol. 1120: K. Jarosz, Perturbations of Banach Algebras. V, 118 page 1985.

Vol. 1121: Singularities and Constructive Methods for Their Treatmer Proceedings, 1983. Edited by P. Grisvard, W. Wendland and J. Whiteman. IX, 346 pages. 1985.

Vol. 1122: Number Theory. Proceedings, 1984. Edited by K. Allad VII, 217 pages. 1985.

Vol. 1123: Séminaire de Probabilités XIX 1983/84. Proceedings. Edi par J. Azéma et M. Yor. IV, 504 pages. 1985.

Vol. 1124: Algebraic Geometry, Sitges (Barcelona) 1983. Procee ings. Edited by E. Casas-Alvero, G.E. Welters and S. Xamb Descamps. XI, 416 pages. 1985.

Vol. 1125: Dynamical Systems and Bifurcations. Proceedings, 198 Edited by B.L.J. Braaksma, H.W. Broer and F. Takens. V, 129 page 1985.

Vol. 1126: Algebraic and Geometric Topology. Proceedings, 198 Edited by A. Ranicki, N. Levitt and F. Quinn. V, 523 pages. 1985.

Vol. 1127: Numerical Methods in Fluid Dynamics. Seminar. Edited F. Brezzi, VII, 333 pages. 1985.

Vol. 1128: J. Elschner, Singular Ordinary Differential Operators an Pseudodifferential Equations. 200 pages. 1985.

Vol. 1129: Numerical Analysis, Lancaster 1984. Proceedings. Edite by P.R. Turner. XIV, 179 pages. 1985.

Vol. 1130: Methods in Mathematical Logic. Proceedings, 1983. Edite by C.A. Di Prisco. VII, 407 pages. 1985.

Vol. 1131: K. Sundaresan, S. Swaminathan, Geometry and Nonline Analysis in Banach Spaces. III, 116 pages. 1985.

Vol. 1132: Operator Algebras and their Connections with Topolo and Ergodic Theory. Proceedings, 1983. Edited by H. Araki, C. Moore, Ş. Strătilă and C. Voiculescu. VI, 594 pages. 1985.

Vol. 1133: K.C. Kiwiel, Methods of Descent for Nondifferentiable Op mization. VI, 362 pages. 1985.

Vol. 1134: G.P. Galdi, S. Rionero, Weighted Energy Methods in Flu Dynamics and Elasticity. VII, 126 pages. 1985.

Vol. 1135: Number Theory, New York 1983–84. Seminar. Edited D.V. Chudnovsky, G.V. Chudnovsky, H. Cohn and M.B. Nathanso V, 283 pages. 1985.

Vol. 1136: Quantum Probability and Applications II. Proceeding 1984. Edited by L. Accardi and W. von Waldenfels. VI, 534 page 1985.

Vol. 1137: Xiao G., Surfaces fibrées en courbes de genre deux. IX, 1 pages. 1985.

Vol. 1138: A. Ocneanu, Actions of Discrete Amenable Groups on v Neumann Algebras. V, 115 pages. 1985.

Vol. 1139: Differential Geometric Methods in Mathematical Physic Proceedings, 1983. Edited by H.D. Doebner and J.D. Hennig. VI, 3 pages. 1985.

Vol. 1140: S. Donkin, Rational Representations of Algebraic Grour VII, 254 pages. 1985.

Vol. 1141: Recursion Theory Week. Proceedings, 1984. Edited H.-D. Ebbinghaus, G.H. Müller and G.E. Sacks. IX, 418 pages. 198

Vol. 1142: Orders and their Applications. Proceedings, 1984. Edit by I. Reiner and K.W. Roggenkamp. X, 306 pages. 1985.

Vol. 1143: A. Krieg, Modular Forms on Half-Spaces of Quaternio XIII, 203 pages. 1985.

Vol. 1144: Knot Theory and Manifolds. Proceedings, 1983. Edited D. Rolfsen. V, 163 pages. 1985.